STATISTICAL ECOLOGY

About ECOSTAT

The software ECOSTAT was developed to perform computations for the statistical methods not readily available in standard statistical packages. It may be downloaded free from the University of Nebraska website (http://www.ianr.unl.edu/ianr/biometry/faculty/linda/lyoung.html). The zip file should be placed in a temporary folder. Then the individual files should be extracted (unzipped) from this zip file. The temporary folder(s) used for the zip file and the extracted files should NOT be called ECOSTAT. Once the files have been extracted, ECOSTAT must be installed on the machine before it can be run. To do so, run SETUP.EXE. The downloaded zip file and the files that were extracted from the zip file may then be deleted from the temporary folder(s).

ECOSTAT was developed for IBM-compatibles and a Windows95 environment. At least 16 megabytes of RAM are needed to run the full program, although some portions require less memory. The program is written in Microsoft's Visual Basic.

ECOSTAT has seven parts. *Probability Distributions* has the set of programs that complement the first three chapters of the book. *Sequential Sampling* contains the programs for both the fixed and sequential methods in Chapters 4-6. *Spatial Statistics* is the portion relating to Chapters 7 and 8. *Capture-Recapture* is the program corresponding to Chapters 9 and 10. *Transect Sampling* may be used for the material covered in Chapter 11. The temperature data of Chapter 12 can be analyzed using *Temperature Models*. The portion of the program relating to Chapter 13 is in *Life Stage*. An eighth program, DISCRETE (Gates 1989), has been included with ECOSTAT for use in Chapter 2.

ECOSTAT is able to help the user identify some problems in running the program. However, at the present time, it is not designed to detect errors in the input file. These include the errors that arise because the wrong format was used for the data. Care should be taken at this point.

Should you encounter an error, please contact us at the following e-mail address: biom025@unlvm.unl.edu.

A Population Perspective

STATISTICAL ECOLOGY

Linda J. Young
Biometry Department
University of Nebraska—Lincoln

Jerry H. Young
Entomology Department (Emeritus)
Oklahoma State University

KLUWER ACADEMIC PUBLISHERS
Boston/Dordrecht/London

Library of Congress Cataloging-in-Publication Data

Young, Linda J.
    Statistical ecology : a population perspective / Linda J. Young, Jerry H. Young
       p. cm.
    Includes bibliographical references and index.
    ISBN 0-412-04711-X (alk. paper)
    1. Ecology--Statistical methods. 2. Population biology--Statistical methods. I. Young, Jerry H. II. Title.
QH541.15.S72Y68   1998
577'.01'5195--dc21   97-25603
                                         CIP

British Library Cataloguing in Publication Data available

To John, Shamar, and RaQwin

Table of Contents

Preface

This book is a collection of formulae, techniques, and methods developed for use in field ecology. We try to illustrate treatment of ecological data, from sampling through modeling, for single-species populations. The material has been chosen because of its frequent application by researchers and workers in pest management, forestry, wildlife, plant protection and environmental studies. These methods are not always the strongest ones statistically because ease of application is a primary consideration. At times, the statistical properties of well-accepted procedures are unknown. By giving an awareness of the statistical foundation for existing methods, we hope that biologists will become more aware of the strengths, and possible weaknesses, of the procedures and that statisticians will more fully appreciate the needs of the field ecologist.

The book is designed as a reference or entry level text for biologists or statisticians that are developing a better understanding of statistical ecology. It is assumed that readers have an understanding of basic ecological principles and a statistical foundation in estimation, hypothesis testing, and regression. The ECOS-TAT software that accompanies the text will hopefully permit the focus to change from the computations to the concepts underlying the methods.

Most of the equations are presented in calculator or computer programming form. The notations are minimal and are taken, for the most part, from biology. An effort has been made to relate the notation used to that of standard statistics, mathematics and physics. Extensive examples for most methods have been included.

These materials have been used in statistical ecology courses taught at Oklahoma State University and the University of Nebraska—Lincoln. Numerous students from these courses have aided in the development of this manual. Several graduate students from the Departments of Statistics and Entomology at Oklahoma State and the Biometry Department at the University of Nebraska—

Lincoln have researched key areas in this book. We have attempted to note their contributions where appropriate.

The treatment of Bose–Einstein Statistics in Chapter 3 is a rather sharp departure from the traditional view of population dynamics and should be read to give the reader an insight into the philosophy of the authors.

The chapters are arranged in a natural progression of identifying the applicable distributions, developing sampling programs, and modeling populations. An effort has been made to make each chapter complete, allowing workers to readily focus on particular interests.

Acknowledgments

Statistical ecology is an exciting area of study from both the statistical and ecological perspectives. The enormous variation encountered in nature enhances its beauty and challenges those who wish to quantify its characteristics. The discipline has benefitted from the intense efforts of pioneers in this field. These individuals have made great inroads into our understanding and have been generous in suggesting avenues they believe will lead to further progress. We have made an effort to identify at least some of these leaders within each chapter.

This text grew out of a statistical ecology class developed at Oklahoma State University and later taught at the University of Nebraska–Lincoln. Our graduate students and students taking the class have added greatly to this work. Bob Hill and Bill Ruth helped census the insects on 11 quarter-acres of cotton. This work shaped our views of insect distribution. Michelle Strabala investigated the performance of normal-based methods applied to discrete distributions. Be-ny Wu conducted a study of goodness-of-fit tests for discrete distributions. The models of insect movement in ECOSTAT are based on an initial program written by Alan Stark. Katherine Seebeck developed software for studying the properties of the sequential probability ratio test. Lim Siew with the input of Madhuri Mulekar produced the foundational software for the 2-SPRT. Our views of deciding among three hypotheses were shaped by Mark Payton's work.

In developing this book, we benefitted from the unselfish help and support from many colleagues. From the beginning, Bill Drew provided strong support and encouragement. Nitis Mukhopadhyay introduced us to sequential analysis. Igo Kotlarski provided insight and references leading to the models in Chapter 3. Colleagues and students in the Departments of Statistics and Entomology at Oklahoma State University and the Department of Biometry at the University of Nebraska—Lincoln shaped our views through frequent discussions and questions. They also provided data for a number of the examples and exercises. Charles Gates provided us with the DISCRETE program that we have included with

ECOSTAT. Noel Cressie was always encouraging and pointed out the Power Series family of test statistics. Carol Gotway read and gave detailed suggestions for strengthening Chapters 7 and 8. Svata Louda provided ideas for improving the presentation of Chapter 8 that were useful in other chapters as well. Ken Burnham gave insightful comments on Chapters 9, 10, and 11. Leon Higley not only read and commented on Chapters 12 and 13 but made significant changes in Chapter 12 that greatly improved it. Chris Heinzle helped us make the transition to Visual Basic from our DOS-based programs. Others offered encouragement and help.

Bea Shube provided great insight and useful suggestions in the early stages of development. We still review materials she provided. Henry Flesh, Kendall Harris, Lisa LaMagna, and Deslie B. Lawrence were instrumental in this book reaching completion.

Patsy Lang helped collect materials from the library, developed figures for some later chapters, and provided general support. Linda Pavlish, Daryll Travnicek and Leona Barratt developed a number of the figures in the first six chapters.

Friends and neighbors have given us invaluable help and support, especially during the last stages of writing. They have driven for us in the car pools, taken our children to numerous events, and treated our children as their own. At the same time, they have encouraged us to persevere. We are truly blessed.

John, Shamar, and RaQwin helped us keep a perspective on the truly important aspects of life. They were understanding and helped keep our household functioning during this process.

1

Probability Distributions

Introduction

Statistical ecology involves a quantitative approach to collecting, analyzing, and interpreting data. Central to this approach is the assumption that a random sample can be taken from the ecosystem of interest. This random sample may be used to draw inference about the shape of the population distribution or about characteristics of that distribution. Unfortunately, the population distribution is not the same under all circumstances. Various probability distributions have been found useful in describing these population distributions. These distributions, their properties, and common applications are the focus of this chapter.

Most analysis methods taught in beginning statistics courses are based on the normal distribution, primarily for two reasons. First, numerous types of measurements do result in normally or near-normally distributed data. The yield of crops, lengths, or weights of organisms, and so forth, do tend to be normally distributed. Second, by the Central Limit Theorem, we know that as long as the distribution from which we are sampling has a finite mean and variance, the distribution of the sample mean approaches the normal distribution as the sample size increases. It is often stated that a sample size of 30 to 60 is sufficient to invoke the *Central Limit Theorem*. However, if the distribution is markedly skewed, particularly if a few values are far from the mean, sample sizes in excess of 100 may be needed before the normal approximation can be used with any assurance (*see* Snedecor and Cochran, 1989). Most population data and other ecological measurements do have markedly skewed distributions, with a few values that are far from the mean. In our experience, unless the population mean exceeds 10, large sample sizes are required for the distribution of sample means arising from population counts to be approximated with the normal distribution. For example, when sampling for the cotton fleahopper (a small myrid pest of cotton), samples in excess of 2,000 are needed when the mean is near the economic threshold. Sample sizes of this magnitude are not feasible under most applied situations, and many ecological studies do not allocate the resources to take sample sizes this large. The options are to use the additional resources needed to take larger samples or to use alternate plans. One such plan is to use methods based on the population distribution that is being sampled. Sampling methods that do this are covered in Chapters 4, 5, and 6.

Population density is well defined in both ecology and statistics. Unfortunately, the definitions differ. In ecology, population density is the average number of organisms per unit area. In statistics, it refers to the probability density function associated with the population distribution. Here we use population density in the ecological sense. *Population density function* is used to capture the statistical meaning of the term.

Population density estimates are required for most population ecology studies. These estimates are based on counting some portion of the population being studied and using these counts to draw inference to the population. These estimates, and all studies that use counts, produce data that are discrete. *Discrete,* as used here, means that the data values take some definite number (an integer). This is contrasted with continuous data that may take (theoretically) any value. Measurements such as weight, height, and length are continuous. The only limitation is the accuracy of the measuring device. Measuring tools are limited by the smallest unit that can be measured but, for most considerations, the resulting measurements can be considered continuous and assumed to take any value. If organisms, or their effects, are counted, the data are from discrete distributions. Examples of discrete and continuous distributions are the subject of this chapter.

Discrete distributions are discussed first because an understanding of how to handle data derived from them is important to interpreting population dynamics. Admittedly, some investigators try to avoid the distribution problem by using various indices but, even if these are used, a general knowledge of distributions is needed. The treatment here is simple. (For a more complete coverage of the distributions found in population and ecological studies consult Patil et al., 1984; Rao, 1970; and Johnson et al., 1992. Texts on probability and statistical inference may also be consulted.)

Discrete Distributions

The general area of applied ecology, especially in such areas as pest management systems, where information is used to determine when and how to allocate large amounts of resources, needs more carefully conducted investigations. Population dynamics in areas like agroecosystems will not be understood until the interaction of the crop, the pest, and entomophagus organisms are understood, and they will not be understood unless they are defined. Unfortunately, it is true that we do not have good techniques that allow a researcher to enter a field and come up with an accurate estimate of populations of some of the more important pests. A field of commercial crops is generally a simple ecosystem. The problem is more acute when defining the population density of organisms in a complicated system, like a lake, or a river. However, we can begin by learning more about the distributions of counts we expect to encounter when sampling.

The term *distribution* usually means something quite different to an ecologist than it does to a statistician. To ecologists, distribution is generally taken to mean the spatial arrangement of the organisms within the ecosystem. We refer to this as *spatial pattern,* or simply *pattern.* Statisticians tend to associate distribution with the proportion of sampling units that have 0 organisms, 1 organism, 2 organisms, 3 organisms, and so forth, without regard to the surface arrangement of these counts. We call this the *probability distribution, or simply distribution.* In this chapter, our focus is on the probability distribution of counts of organisms in ecosystems. Chapters 7 and 8 consider measures of spatial correlation and spatial pattern. A relationship sometimes, but not necessarily, occurs between spatial pattern and probability distribution, but the present treatment is strictly with numerical distributions.

The most common distributions found in populations of organisms are the binomial, Poisson, negative binomial, and geometric. These distributions are related, and an understanding of how they relate is essential for ecological studies of populations of organisms. Much of the controversy surrounding the distributions of organisms is in the definition of these distributions. The subject of randomness, aggregation, and clumping, as well as how these concepts relate to distributions, are discussed in more detail in Chapter 3, but a brief explanation is warranted here.

The traditional views of biologists have been that a Poisson distribution is a random distribution and that this is the expected distribution if a population of organisms is allowed to distribute naturally over a field or other habitats. Random, as used here, means that every microhabitat has an equal opportunity of being occupied by any organism. For example, if insects invade a field and each insect has an equal opportunity to occupy any plant, the probability distribution of the number of insects on a randomly chosen plant follows a Poisson, or random, distribution.

A Poisson distribution has only one parameter, lambda (λ).

$$\lambda = \mu = \sigma^2 \tag{1.1}$$

where

$$\mu = E(X) \tag{1.2}$$

and

$$\sigma^2 = E[(X - \mu)^2] \tag{1.3}$$

Here μ is the population mean and σ^2 is the population variance.

The symbol $E(X)$ represents the expectation of the random variable X; that is,

$$E(X) = \sum_x xP(x)$$

if X is discrete and

$$E(X) = \int_x xf(x)dx$$

if X is continuous. When taking a random sample of size n from this population, we obtain estimates of the parameters:

$$\hat{\lambda} = \bar{X} \approx s^2 \tag{1.4}$$

where the sample mean, an estimate of μ, is

$$\bar{X} = \frac{\sum_{i=1}^{n} X_i}{n} \tag{1.5}$$

and the sample variance, an estimate of σ^2, is

$$s^2 = \frac{\sum_{i=1}^{n} X_i^2 - \frac{\left(\sum_{i=1}^{n} X_i\right)^2}{n}}{n - 1}. \tag{1.6}$$

Little and Hills (1978), Steel and Torrie (1980), and Snedecor and Cochran (1989) give accounts of the basic statistical information as used in this manual and should be consulted on questions of applied statistics.

Although $\bar{X}$ and s^2 are rarely exactly the same, they should be close if the numerical distribution of the sampled population is truly Poisson. How far apart should the estimates of mean and variance be before one begins to doubt the assumption that the mean and variance are equal? To answer this question, consider the test statistic

$$\chi_R^2 = \frac{(n - 1)s^2}{\bar{x}}. \tag{1.7}$$

If the null hypothesis that the mean and variance are equal is true, this test statistic has an approximate χ^2 distribution with $(n - 1)$ degrees of freedom. Note that we want to reject the null hypothesis if the value of the test statistic is too small or too large. Thus, a test at the 5% significance level rejects the null hypothesis if the test statistic is below the 0.025 quantile or above the 0.975 quantile of the χ^2 distribution with $(n - 1)$ degrees of freedom. A normal approximation works well if the degrees of freedom of the test statistic exceeds 30. The expression

$$\sqrt{2\chi_R^2} - \sqrt{2(n - 1) - 1} \qquad (1.8)$$

has an approximate standard normal distribution if the hypothesis that the mean and variance are equal is true. The null hypothesis is rejected at the 5% significance level if this function of the test statistic is less than the 0.025 quantile or above the 0.975 quantile of the standard normal distribution.

It should be noted that this test is useful as a measure of the strength of the evidence against the assumption of equality of mean and variance. If the null hypothesis is rejected, we can conclude that the mean and variance are not equal, with a known probability of error. However, this test does not prove that the mean and variance are equal. Suppose through experience or theoretical development, one believes that the variance should exceed the mean. Further assume that the estimate of the variance exceeds that of the mean, but there was not enough evidence to reject the hypothesis of equality of mean and variance. In fact, if the estimated mean is slightly greater than the estimated variance and the mean is small, one should not be too quick to conclude from one sample that the variance is not greater than the mean. [For a sample of size 50, as many as one in five samples from the negative binomial with a mean of one and a k of 1 may have the sample mean greater than the variance (*see* Willson et al., 1984).] This test based on a single sample is not sufficient for us to conclude that the mean and variance are equal. Samples taken at different times and locations that consistently indicate that the variance is not significantly different from the mean are needed before we can have much confidence in using this hypothesis as a basis for inference.

Example 1.1

A random sample of foxtails, a common weed in soybeans, was taken from a soybean field. Twenty-five quadrats 1 meter (m) long and 0.3 m wide resulted in an estimated mean of 1.88 and an estimated variance of 4.73. To test the hypothesis that the population mean equals the population variance, we construct the test statistic

$$\chi_R^2 = \frac{(25 - 1)4.73}{1.88} = 60.4$$

If the population mean and variance are equal, 60.4 should be an observed value from the χ^2 distribution with $(25 - 1) = 24$ degrees of freedom (df). The accompanying software ECOSTAT can be used to perform these calculations and find the *p*-value. From the *Probability Distributions* menu, choose *Variance–Mean*. Then enter the sample size, sample mean, and sample variance. After pressing the *Variance–Mean Test* button, the test statistic, degrees of freedom, and *p*-value are displayed. ECOSTAT always computes the *p*-value for a two-sided test. If a one-sided test is desired and the relationship of the sample mean and variance is consistent with the alternative hypothesis, the displayed *p*-value should be divided by 2. For this example, $p = 0.0001$. Therefore, we reject the null hypothesis and conclude that in fact the variance exceeds the mean in this population of foxtails.

Example 1.2

European Corn Borer (ECB) is a common pest of corn. The number of ECB egg masses was recorded on each of 250 plants in a corn field. The estimated mean based on this sample was 0.166, and the estimated variance was 0.182. To test the hypothesis that the population mean equals the population variance, we first construct the test statistic

$$\chi_R^2 = \frac{(250 - 1)0.182}{0.166} = 273$$

Although this statistic has an approximate χ^2 distribution with 249 degrees of freedom, most χ^2 tables do not extend to that many degrees of freedom. Thus, we compute

$$\sqrt{2(273)} - \sqrt{2(250 - 1) - 1} = 1.07$$

If the population mean and variance are equal, the observed test statistic of 1.07 is an observation from an approximate standard normal distribution. The probability of observing a value of 1.07 or more or of -1.07 or less can be found using a standard normal table. Alternatively, ECOSTAT could be used in the same manner as in the first example. The *p*-value is found to be 0.28. Hence, we do not reject the hypothesis that the population mean and variance are equal.

If the variance is greater than the mean, the population is said to be aggregated, clumped, or overdispersed. If the variance is less than the mean, the population is said to be underdispersed. These terms have often caused many to equate probability distribution and spatial pattern. However, as we discuss in Chapter 7, this does not have to be the case, and often is not. An aggregated probability distribution may have a random spatial pattern. Similarly, a random probability distribution may have an aggregated spatial pattern. Care should be taken to avoid the pitfall of equating the two. The relationship between the mean and variance is one of the primary ways of discerning among distributions:

Binomial: $\quad\quad\quad\quad\quad\quad\quad\quad\mu > \sigma^2$

Poisson: $\quad\quad\quad\quad\quad\quad\quad\quad\mu = \sigma^2$

Negative binomial, geometric: $\quad\mu < \sigma^2$

Geometric: $\quad\quad\quad\quad\quad\quad\quad\quad\mu = \sigma^2 - \mu^2$

Negative Binomial Distribution

We begin with the negative binomial distribution because of its wide application in population dynamics, especially in agroecosystems studies. Taylor (1984) summarizes much of the integrated pest management (IPM) literature and gives evidence that it is the most common distribution found in insect control studies. Although other distributions, such as the Neyman type A, have variances that exceed the mean, the negative binomial is usually the first distribution considered when this property is observed. The negative binomial distribution is often called a clumped or aggregated distribution (Southwood, 1978).

The negative binomial distribution can be derived in many ways (*see* Boswell and Patil, 1970). The parameterization method used here is that of Anscombe (1949) and is the one most frequently encountered in the ecological literature. The negative binomial has two parameters, k and μ. μ is the mean and k is often referred to as an aggregation, or clumping, parameter. High values of k are less clumped; conversely, lower k values have greater clumping. As stated above, this does not necessarily imply spatial aggregation. k does express a type of mean (μ) to variance (σ^2) relationship, which is

$$k = \frac{\mu^2}{\sigma^2 - \mu} \tag{1.9}$$

Several methods can be used to estimate k. The maximum likelihood estimator (MLE) is the value that maximizes the probability of the observed data (*see* Hogg and Craig, 1995; Bain and Engelhardt, 1992, for more details). The MLE of μ is the sample mean. The MLE of k is the root of the following equation in $\hat{k}$:

$$n \ln\left(1 + \frac{\bar{X}}{\hat{k}}\right) = \sum_{j=1}^{\infty} m_j\left(\frac{1}{\hat{k}} + \frac{1}{\hat{k} + 1} + \cdots + \frac{1}{\hat{k} + j - 1}\right) \tag{1.10}$$

where m_j is the number of times a j occurs in a sample of size n. The method-of-moments estimators (MME) of μ and k are obtained by equating the first two sample moments to the first two population moments. The sample mean is the MME of μ, just as it was for the MLE of μ. The MME of k is simpler to compute than the MLE and more commonly used (*see* Southwood, 1978; Anscombe, 1949; Willson, 1981):

$$\hat{k} = \frac{\bar{X}^2}{s^2 - \bar{X}}. \tag{1.11}$$

The MLE of k has better asymptotic properties and is generally considered superior to the MME, although this may not always be the case (*see* Willson et al., 1984). Its limited use is probably due to computational considerations. However, ECOSTAT computes both the MLE and the MME of k so the choice should be made on the properties of the estimators. Willson and Young (1983) have shown that a good field estimator of k is a multistage routine which is described in Chapter 4.

As k approaches infinity, the negative binomial approaches the Poisson distribution. In fact, if k exceeds 25, the Poisson distribution will usually provide a good approximation to the negative binomial. When the zeroes of the negative binomial distribution are truncated (dropped and their probability distributed proportionately among the positive integers) and k approaches 0, the limiting distribution is the logarithmic (*see* Johnson et al., 1992).

Expected Frequencies

The first step in determining whether a negative binomial distribution adequately describes the data is to compare the observed frequencies of the sample values to the expected frequencies of a negative binomial distribution with the same values of the parameters k and μ; that is, compare the number of samples with 0, 1, 2, etc., with the expected number of 0, 1, 2, etc., of a negative binomial distribution. To accomplish this, the expected frequencies of the negative binomial must be generated (for a discussion of goodness-of-fit tests, *see* Chapter 2). The following method is widely used in biology and can be programmed on a handheld calculator or personal computer.

We begin by computing the proportions (or probabilities) of 0's, 1's, 2's, ..., of a negative binomial distribution. In general, the probability of obtaining $X = x$ organisms in an observation is

$$P(X = x) = \binom{k + x - 1}{k - 1}\left(\frac{k}{\mu + k}\right)^k\left(\frac{\mu}{\mu + k}\right)^x, \qquad x = 0, 1, 2, \ldots . \tag{1.12}$$

In the above,

$$\binom{k + x - 1}{k - 1} = \frac{(k + x - 1)!}{(k - 1)!x!}$$

where factorials $(x!)$ are positive integers multiplied consecutively to the level indicated. Thus

$$5! = 1 \times 2 \times 3 \times 4 \times 5 = 120$$

and

$$3! = 1 \times 2 \times 3 = 6.$$

The probabilities given in equation (1.12) can be computed recursively. To do this, the probability of observing 0 or no organisms in a randomly selected sampling unit is

$$P(0) = \left(\frac{k}{\mu + k}\right)^k. \tag{1.13}$$

The general equation for all values greater than 0 is

$$P(x) = \left(\frac{k + x - 1}{x}\right)\left(\frac{\mu}{\mu + k}\right)P(x - 1), \qquad x = 1, 2, \ldots. \tag{1.14}$$

The expected frequencies are found by multiplying the probabilities by the sample size. Thus,

$$\text{Expected frequency of } 0 = nP(0)$$

$$\text{Expected frequency of } 1 = nP(1)$$

and, in general,

$$\text{Expected frequency of } j = nP(j).$$

Example 1.3

Suppose we want to calculate the expected frequencies of a negative binomial distribution with parameters $k = 2$ and $\mu = 1$. The probability of 0 is computed first and found to be

$$P(0) = \left(\frac{k}{\mu + k}\right)^k = \left(\frac{2}{1 + 2}\right)^2 = 0.4444.$$

Then the probabilities of 1, 2, 3, etc., are computed recursively:

$$P(1) = \left(\frac{k + x - 1}{x}\right)\left(\frac{\mu}{\mu + k}\right)P(0) = \left(\frac{2 + 1 - 1}{1}\right)\left(\frac{1}{1 + 2}\right)(0.4444) = 0.2963$$

$$P(2) = \left(\frac{k + x - 1}{x}\right)\left(\frac{\mu}{\mu + k}\right)P(1) = \left(\frac{2 + 2 - 1}{2}\right)\left(\frac{1}{1 + 2}\right)(0.2963) = 0.1481$$

$$P(3) = \left(\frac{k + x - 1}{x}\right)\left(\frac{\mu}{\mu + k}\right)P(2) = \left(\frac{2 + 3 - 1}{3}\right)\left(\frac{1}{1 + 2}\right)(0.1481) = 0.0658$$

$$P(4) = \left(\frac{k + x - 1}{x}\right)\left(\frac{\mu}{\mu + k}\right)P(3) = \left(\frac{2 + 4 - 1}{4}\right)\left(\frac{1}{1 + 2}\right)(0.0658) = 0.0274$$

$$P(5) = \left(\frac{k + x - 1}{x}\right)\left(\frac{\mu}{\mu + k}\right)P(4) = \left(\frac{2 + 5 - 1}{5}\right)\left(\frac{1}{1 + 2}\right)(0.0274) = 0.0110$$

$$P(6) = \left(\frac{k + x - 1}{x}\right)\left(\frac{\mu}{\mu + k}\right)P(5) = \left(\frac{2 + 6 - 1}{6}\right)\left(\frac{1}{1 + 2}\right)(0.0110) = 0.0043$$

$$P(7) = \left(\frac{k + x - 1}{x}\right)\left(\frac{\mu}{\mu + k}\right)P(6) = \left(\frac{2 + 7 - 1}{7}\right)\left(\frac{1}{1 + 2}\right)(0.0043) = 0.0016$$

$$P(8) = \left(\frac{k + x - 1}{x}\right)\left(\frac{\mu}{\mu + k}\right)P(7) = \left(\frac{2 + 8 - 1}{8}\right)\left(\frac{1}{1 + 2}\right)(0.0016) = 0.0006$$

$$P(9) = \left(\frac{k + x - 1}{x}\right)\left(\frac{\mu}{\mu + k}\right)P(8) = \left(\frac{2 + 9 - 1}{9}\right)\left(\frac{1}{1 + 2}\right)(0.0006) = 0.0002$$

ECOSTAT may be used to compute and graph these probabilities. First, choose *Discrete Distributions* and then *Mass Functions* from the *Probability Distributions* menu. Then select *Negative Binomial* from the menu. After setting the parameters by moving the scroll bar, press the *Probabilities* button. The probabilities are displayed with more than four decimal places so that you have the option of deciding how many significant digits are needed for your purpose. The graph can be displayed by pressing the *Graph* button.

Calculating probabilities to four decimals is generally sufficient. What is the significance of these probabilities? The probabilities are expressed in proportions and must sum to 1. In other words, if you have 10,000 observations, you expect 4,444 of the observations to be 0 or no organisms, 2,963 to be 1 organism, etc. The expected number of samples with $X = x$ organisms is calculated by multi-

Table 1.1. *Expected Frequencies for the Negative Binomial Distribution:* k = 2,
μ = 1

X	Sample size*	P(x)	Expected frequency
0	10,000	0.4444	4,444
1	10,000	0.2963	2,963
2	10,000	0.1481	1,481
3	10,000	0.0658	658
4	10,000	0.0274	274
5	10,000	0.0110	110
6	10,000	0.0043	43
7	10,000	0.0016	16
8	10,000	0.0006	6
9	10,000	0.0002	2
>9	10,000	0.0003	3

*Large sample sizes are used here to show the extent of the probabilities.

plying the probability by the sample size, as shown in Table 1.1, for $x = 0, 1, 2, 3, 4, 5, 6, 7, 8$, and 9.

Figure 1.1 shows the effect of increasing values of the mean on the probability of observing counts of 0, 1, 2, etc., in a sampling program when k is fixed at 2.

The variance of the negative binomial distribution is a quadratic function of the mean:

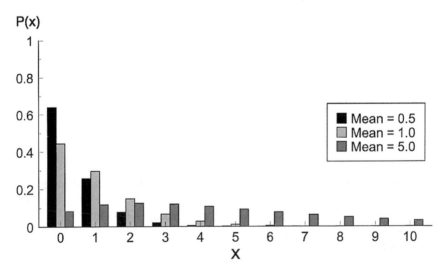

Figure 1.1. The effect of increasing the mean of a negative binomial distribution when $k = 2$.

$$\sigma^2 = \mu + \frac{\mu^2}{k}. \tag{1.15}$$

If k is known, an intuitive estimate of the variance is

$$\hat{\sigma}^2 = \bar{X} + \frac{\bar{X}^2}{k}. \tag{1.16}$$

However, this estimator is biased; that is, it tends to overestimate the mean. An unbiased estimator is

$$V_n = \frac{nk}{nk + 1} \bar{X} + \frac{n}{nk + 1} \bar{X}^2. \tag{1.17}$$

If k is unknown, the variance may be estimated using

$$\hat{\sigma}^2 = \bar{X} + \frac{\bar{X}^2}{\hat{k}}. \tag{1.18}$$

Note that if k is estimated using the MME of k, this estimate of the variance is the same as that in equation (1.6). However, if the MLE of k is used, this estimate differs from that in equation (1.6).

The negative binomial distribution has several other parameterizations. One that will be useful when we develop sequential sampling procedures involves the parameter P:

$$P = \frac{\mu}{k}. \tag{1.19}$$

P may be estimated by

$$\hat{P} = \frac{\bar{X}}{\hat{k}}. \tag{1.20}$$

As before, the estimate differs depending on whether k is estimated using maximum likelihood or method of moments. We can express the negative binomial distribution in terms of the parameters P (in equation (1.19)) and $Q = P + 1$. The probability of observing the value x on a randomly selected unit can then be expressed as

$$P(x) = \binom{k + x - 1}{k - 1} Q^{-k-x} P^x, \qquad x = 0, 1, 2, \ldots. \tag{1.21}$$

These probabilities may also be computed recursively. Begin by calculating the probability of 0 as

$$P(0) = \frac{1}{Q^k}. \tag{1.22}$$

The remaining probabilities can then be determined using the usual relationship, as shown in equation (1.14) and expressed here in terms of P and Q:

$$P(x) = \left(\frac{k + x - 1}{x}\right)\left(\frac{P}{Q}\right)P(x - 1), \qquad x = 1, 2, 3, \ldots. \tag{1.23}$$

Both of the preceding parameterizations are commonly found in the biological literature. A form of the distribution is more frequently found in elementary probability courses is the following:

$$P(x) = \binom{k + x - 1}{k - 1} p^k q^x, \qquad x = 0, 1, 2, \ldots. \tag{1.24}$$

where $q = 1 - p$. This parameterization arises when independent Bernoulli trials are performed, each with the probability p of success. Let X be the number of failures prior to the kth success. Then X has the negative binomial distribution given above. If we let

$$p = \frac{k}{\mu + k}$$

in equation (1.12), we obtain the parameterization in (1.24).

Additional information on the usages of the negative binomial distribution can be obtained from Fisher (1941), Anscombe (1949, 1950), Beall (1942), Bliss and Owens (1958), Gurland (1959), Patil (1960), Iwao (1970a,b), Patil and Stiteler (1974), Hill et al. (1975), Willson (1981), Patil et al. (1984), Willson et al. (1985, 1987), Young and Young (1987), and Johnson et al. (1992).

Geometric Distribution

The geometric distribution is a special case of the negative binomial distribution, where $k = 1$. This distribution has wide applications in populations studies as shown by Willson et al. (1987), Young and Wilson (1987), and Young and Young (1989). The geometric is the limiting case of Bose–Einstein statistics as used in statistical mechanics and is discussed in more detail in Chapter 3. The specific relationship of the mean to the variance gives the geometric distribution some special properties that simplify calculations of probabilities and expected frequencies.

The geometric distribution has only one parameter, μ. The variance is

$$\sigma^2 = \mu + \mu^2. \qquad (1.25)$$

The geometric distribution has no central tendency. The number of samples or cells that have 0 or no organisms is always largest regardless of the size of the mean. The distribution always has a similar shape. As the mean increases, it flattens, and the probability of observing a 0 decreases. As the mean surpasses 100, the distribution becomes so flat that the probability of observing any particular value is very close to the probability of observing any other value. This is the most probable distribution of freely moving organisms in a uniform habitat (Willson et al., 1987). As can be seen in Figure 1.2, this allows a maximum diversity in the numbers of organisms in the sampling units. This aspect is covered in more detail in Chapter 3.

Expected Frequencies

The probability of samples with 0, 1, 2, . . . organisms can be calculated with the same equations as for the negative binomial. However, the fact that $k = 1$ allows the probability of observing the value $X = x$ to be expressed as

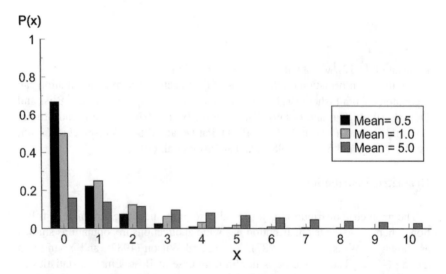

Figure 1.2. The effect of increasing the mean of the geometric distribution.

$$P(X = x) = \frac{\mu^x}{(\mu + 1)^{x+1}}, \qquad x = 0, 1, 2, \ldots. \qquad (1.26)$$

The recursive formula for computing these probabilities simplifies to

$$P(0) = \frac{1}{\mu + 1} \qquad (1.27)$$

and

$$P(x) = \left(\frac{\mu}{\mu + 1}\right) P(x - 1), \qquad x = 1, 2, 3, \ldots. \qquad (1.28)$$

A second method of calculating the probabilities uses parameters $P = \mu$ and $Q = 1 + P$, as in equation (1.21), for the negative binomial. Then the probability of observing the value x on a randomly selected sampling unit is

$$P(x) = Q^{-1}\left(\frac{P}{Q}\right)^x, \qquad x = 0, 1, 2, \ldots. \qquad (1.29)$$

This method has the advantage of being simple to use. It can be defined as a function in a computer program and gives quick results for any mean.

A third parameterization encountered most frequently in quality control and elementary probability is

$$P(x) = pq^x, \qquad x = 0, 1, 2, \ldots. \qquad (1.30)$$

This arises when performing independent Bernoulli trials each with the probability p of success. If X is the number of failures prior to the first success, X has the geometric distribution with the above parameterization.

Examples of the geometric distribution are common in nature, especially in insect populations. They have been recorded for the cotton fleahoppers, greenbugs, boll weevils, convergent lady beetles, *Collops* spp., and bollworms (Young and Willson, 1987).

Example 1.4

Suppose we want to compute the probabilities for a geometric distribution with $\mu = 1$ (remember k is always 1). Then we have $P = \mu = 1$ and $Q = P + 1 = 1 + 1 = 2$. The probabilities of 0, 1, 2, ... are then computed.

$$P(0) = Q^{-1}\left(\frac{P}{Q}\right)^x = 2^{-1}\left(\frac{1}{2}\right)^0 = 0.5$$

$$P(1) = Q^{-1}\left(\frac{P}{Q}\right)^x = 2^{-1}\left(\frac{1}{2}\right)^1 = 0.25$$

$$P(2) = Q^{-1}\left(\frac{P}{Q}\right)^x = 2^{-1}\left(\frac{1}{2}\right)^2 = 0.125$$

$$P(3) = Q^{-1}\left(\frac{P}{Q}\right)^x = 2^{-1}\left(\frac{1}{2}\right)^3 = 0.0625$$

$$P(4) = Q^{-1}\left(\frac{P}{Q}\right)^x = 2^{-1}\left(\frac{1}{2}\right)^4 = 0.0313$$

$$P(5) = Q^{-1}\left(\frac{P}{Q}\right)^x = 2^{-1}\left(\frac{1}{2}\right)^5 = 0.0156$$

$$P(6) = Q^{-1}\left(\frac{P}{Q}\right)^x = 2^{-1}\left(\frac{1}{2}\right)^6 = 0.0078$$

$$P(7) = Q^{-1}\left(\frac{P}{Q}\right)^x = 2^{-1}\left(\frac{1}{2}\right)^7 = 0.0039$$

$$P(8) = Q^{-1}\left(\frac{P}{Q}\right)^x = 2^{-1}\left(\frac{1}{2}\right)^8 = 0.0020$$

$$P(9) = Q^{-1}\left(\frac{P}{Q}\right)^x = 2^{-1}\left(\frac{1}{2}\right)^9 = 0.0010$$

$$P(10) = Q^{-1}\left(\frac{P}{Q}\right)^x = 2^{-1}\left(\frac{1}{2}\right)^{10} = 0.0005$$

$$P(11) = Q^{-1}\left(\frac{P}{Q}\right)^x = 2^{-1}\left(\frac{1}{2}\right)^{11} = 0.0002$$

$$P(12) = Q^{-1}\left(\frac{P}{Q}\right)^x = 2^{-1}\left(\frac{1}{2}\right)^{12} = 0.0001$$

As with the negative binomial distribution, these probabilities may be computed using ECOSTAT. For the geometric, this may be done either by choosing the geometric distribution or by choosing the negative binomial and setting $k = 1$.

Goodness-of-fit tests are used to determine whether the expected frequencies provide an adequate model for the observed data. For a discussion of these tests, *see* Chapter 2.

Note that the expected frequencies are in a decreasing geometric progression (except for round-off error); that is, the number of observations with 1 organism is one-half the number with 0 organisms and the number of observations with 2 is one-half the number with 1, etc. This is characteristic of this distribution and is the origin of the name geometric. As the mean increases, the common ratio of the progressions increases. These common ratios are always $\mu/(\mu + 1)$.

$$\text{For} \quad \mu = 1: \qquad \frac{\mu}{\mu + 1} = \frac{1}{1 + 1} = 0.5.$$

$$\text{For} \quad \mu = 2: \qquad \frac{\mu}{\mu + 1} = \frac{2}{2 + 1} = 0.667.$$

A mean of 1 has a 0.5 geometric progression, and a mean of 2 gives a 0.667 geometric progression.

An important feature of the geometric and the negative binomial distributions is that they are additive when there is no spatial correlation. For example, consider a geometric distribution with $\mu = 1$ (remember that $k = 1$) when the sampling unit is a single microhabitat for the organism being sampled. Further assume that the number of organisms on one microhabitat is independent of the number on other microhabitats. If the sampling unit changes to 2 microhabitats, the result is a negative binomial distribution with $\mu = 2$ and $k = 2$; if 10 microhabitats comprise a sampling unit, $\mu = 10$ and $k = 10$. This may help explain why we so often get the negative binomial distribution as a best fit for data from populations. The sampling unit may not express the discrete microhabitat of the organism, or the organism may not have a well-delimited habitat.

Binomial Distribution

The binomial distribution is commonly described as the probability of X successes in n independent Bernoulli trials. A Bernoulli trial is a single test with two possible outcomes. Is the paper black or white; is the baby boy or girl? In biology, we often determine if a fruit is infested or not infested. Is the grain of wheat damaged or not damaged, or is a female fertile or not fertile; pregnant or not pregnant? If we select a single unit from the population and observe the characteristic of interest, then we have a Bernoulli trial. However, usually, our interest is not in the infestation of a single fruit, the damage of a single wheat grain, or the fertility of a single female, but in the proportion of fruit, wheat, or females exhibiting the characteristic of interest. If n units are randomly selected from the population and the number X exhibiting that characteristic is recorded, the distribution of X is binomial.

Field populations with binomial distributions are easy to sample with precision because the variance is lower than that of a negative binomial, geometric or Poisson distribution.

Expected Frequencies

First consider the classical parameterization of the binomial distribution. Suppose n independent Bernoulli trials are performed each with the probability p of success. Then

$$P(x) = \binom{n}{x} p^x q^{n-x}, \qquad x = 0, 1, 2, \ldots, n \tag{1.31}$$

where $q = 1 - p$. The mean is np and the variance is npq. This parameterization of the binomial seems particularly appropriate when the purpose of sampling is to estimate a proportion, such as the proportion of infested sampling units, the proportion of damaged fruit, etc. The probabilities may be computed easily using a recursive relationship. The probability of 0 occurrences in n observations is

$$P(0) = q^n. \tag{1.32}$$

The probabilities of observing $X = x$ occurrences (or successes) in n trials is then given by

$$P(x) = \left(\frac{n - x + 1}{x} \right) \left(\frac{p}{q} \right) P(x - 1), \qquad x = 1, 2, \ldots . \tag{1.33}$$

Figure 1.3 displays graphs of the binomial distribution with $n = 15$ and $p = 0.1$, 0.5, and 0.9. Notice that for $p = 0.5$, the distribution is symmetric.

Example 1.5

Suppose that 5% of the apples in an orchard have sustained some type of damage. Compute the probability of observing 0, 1, 2, ..., 10 damaged apples when 10 apples are randomly selected from the orchard for inspection.

First, we should note that $p = 0.05$, $q = 1 - 0.05 = 0.95$, and $n = 10$. Then the probability of finding no damaged apples in the 10 is

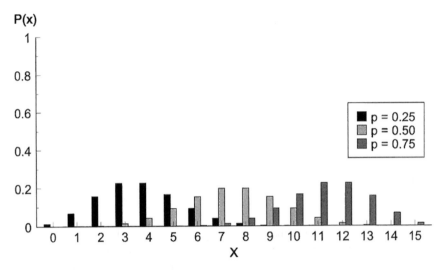

Figure 1.3. The effect of changing p of the binomial distribution when $n = 15$.

$$P(0) = q^n = (1 - 0.05)^{10} = 0.5987.$$

The probability of finding x of the 10 apples to be damaged is

$$P(1) = \left(\frac{n - x + 1}{x}\right)\left(\frac{p}{q}\right)P(0) = \left(\frac{10 - 1 + 1}{1}\right)\left(\frac{0.05}{0.95}\right)(0.5987) = 0.3151$$

$$P(2) = \left(\frac{n - x + 1}{x}\right)\left(\frac{p}{q}\right)P(1) = \left(\frac{10 - 2 + 1}{2}\right)\left(\frac{0.05}{0.95}\right)(0.3151) = 0.0746$$

$$P(3) = \left(\frac{n - x + 1}{x}\right)\left(\frac{p}{q}\right)P(2) = \left(\frac{10 - 3 + 1}{3}\right)\left(\frac{0.05}{0.95}\right)(0.0746) = 0.0105$$

$$P(4) = \left(\frac{n - x + 1}{x}\right)\left(\frac{p}{q}\right)P(3) = \left(\frac{10 - 4 + 1}{4}\right)\left(\frac{0.05}{0.95}\right)(0.0105) = 0.0010$$

$$P(5) = \left(\frac{n - x + 1}{x}\right)\left(\frac{p}{q}\right)P(4) = \left(\frac{10 - 5 + 1}{5}\right)\left(\frac{0.05}{0.95}\right)(0.0010) = 0.0001$$

$$P(6) = \left(\frac{n - x + 1}{x}\right)\left(\frac{p}{q}\right)P(5) = \left(\frac{10 - 6 + 1}{6}\right)\left(\frac{0.05}{0.95}\right)(0.0001) = 0.0000$$

$$P(7) = \left(\frac{n - x + 1}{x}\right)\left(\frac{p}{q}\right)P(6) = \left(\frac{10 - 7 + 1}{7}\right)\left(\frac{0.05}{0.95}\right)(0.0000) = 0.0000$$

$$P(8) = \left(\frac{n - x + 1}{x}\right)\left(\frac{p}{q}\right)P(7) = \left(\frac{10 - 8 + 1}{8}\right)\left(\frac{0.05}{0.95}\right)(0.0000) = 0.0000$$

$$P(9) = \left(\frac{n - x + 1}{x}\right)\left(\frac{p}{q}\right)P(8) = \left(\frac{10 - 9 + 1}{9}\right)\left(\frac{0.05}{0.95}\right)(0.0000) = 0.0000$$

$$P(10) = \left(\frac{n - x + 1}{x}\right)\left(\frac{p}{q}\right)P(9) = \left(\frac{10 - 10 + 1}{10}\right)\left(\frac{0.05}{0.95}\right)(0.0000) = 0.0000$$

Note that because the probability of 6 is 0 to 4 decimal places, the probability of any value greater than 6 will also be 0 to at least 4 decimal places. Thus, we could have stopped at 6 and simply recorded 0 for the remaining values.

In the above example, the parameters n and p have a definite and understandable meaning; however, these parameters lose meaning when the goal of sampling is to estimate population density. In this case, we have found that a simple method of calculating the probabilities of binomial data from populations is to use the same method as that used for the negative binomial distribution. This may be done because if the k in the negative binomial becomes negative, the distribution becomes binomial. That k is negative for the binomial may be seen by noting from equation (1.9) that

$$k = \frac{\mu^2}{\sigma^2 - \mu}$$

and recalling that $\sigma^2 < \mu$ for the binomial distribution.

Example 1.6

Suppose we wish to calculate the probabilities of a binomial distribution with $k = -6$ and $\mu = 2$. Using the same method that we used for the negative binomial, we have

$$P(0) = \left(\frac{k}{\mu + k}\right)^k = \left(\frac{-6}{2 + (-6)}\right)^{-6} = 0.0878$$

$$P(1) = \left(\frac{k + x - 1}{x}\right)\left(\frac{\mu}{\mu + k}\right)P(x - 1) = \left(\frac{-6 + 1 - 1}{1}\right)\left(\frac{2}{2 - 6}\right)(0.0878) = 0.2634$$

$$P(2) = \left(\frac{k + x - 1}{x}\right)\left(\frac{\mu}{\mu + k}\right)P(x - 1) = \left(\frac{-6 + 2 - 1}{2}\right)\left(\frac{2}{2 - 6}\right)(0.2634) = 0.3293$$

$$P(3) = \left(\frac{k + x - 1}{x}\right)\left(\frac{\mu}{\mu + k}\right)P(x - 1) = \left(\frac{-6 + 3 - 1}{3}\right)\left(\frac{2}{2 - 6}\right)(0.3293) = 0.2195$$

$$P(4) = \left(\frac{k + x - 1}{x}\right)\left(\frac{\mu}{\mu + k}\right)P(x - 1) = \left(\frac{-6 + 4 - 1}{4}\right)\left(\frac{2}{2 - 6}\right)(0.2195) = 0.0823$$

$$P(5) = \left(\frac{k + x - 1}{x}\right)\left(\frac{\mu}{\mu + k}\right)P(x - 1) = \left(\frac{-6 + 5 - 1}{5}\right)\left(\frac{2}{2 - 6}\right)(0.0823) = 0.0165$$

$$P(6) = \left(\frac{k + x - 1}{x}\right)\left(\frac{\mu}{\mu + k}\right)P(x - 1) = \left(\frac{-6 + 6 - 1}{6}\right)\left(\frac{2}{2 - 6}\right)(0.0165) = 0.0014$$

ECOSTAT may be used to compute these probabilities as described in Example 1.3, with the change that the *Binomial* is the selected distribution.

The goodness-of-fit tests discussed in Chapter 2 should be used to verify that the expected frequencies fit the observed values. The probabilities are expressed in proportions as with the negative binomial and geometric distributions. For a sample size of 10,000, Table 1.2 lists the expected frequencies for 0, 1, 2, 3, 4, 5, and 6. Table 1.2 illustrates the central tendency in binomial expected frequencies. Thus smaller sample sizes are needed to estimate the density of populations described by the binomial distribution with the same precision when compared to populations fit by the negative binomial distribution. Also, methods based on the assumption of normality perform better. Few population distributions found in nature are fit well by the binomial distribution. This distribution is used exten-

Table 1.2. Binomial Distribution Expected Frequencies of Population Data

X	Sample size	P(x)	Expected frequency
0	10,000	0.0878	878
1	10,000	0.2634	2,634
2	10,000	0.3293	3,293
3	10,000	0.2195	2,195
4	10,000	0.0823	823
5	10,000	0.0165	165
6	10,000	0.0014	14

sively in presence–absence sampling. For more information on the binomial distribution, *see* Feller (1945), Somerville (1957), Patil et al. (1984), and Johnson et al. (1992).

Poisson Distribution

The Poisson distribution is a limiting distribution for both the binomial and negative binomial distributions. If k approaches infinity for the negative binomial distribution or negative infinity for the binomial distribution, the variance approaches the mean, and the limiting distribution is Poisson. Some have noted that if, instead of k, the parameter $\alpha = 1/k$ was used, then we would have the binomial if $\alpha < 0$, the Poisson if $\alpha = 0$, and the negative binomial if $\alpha > 0$, reflecting a movement from underdispersed, to random, to overdispersed, respectively. Further, α is easier to estimate with precision. These features may eventually lead to the adoption of this parameterization but, at present, the form involving k is the standard in biology.

The Poisson distribution has only one parameter, lambda, and as shown in (1.1):

$$\sigma^2 = \mu = \lambda.$$

Biologists and biometricians generally call the Poisson distribution a random distribution. What is meant by this statement is that each organism's behavior is completely independent of that of every other organism. As an example, suppose we are sampling a field with 34,345,212 plants using a m^2 quadrat as the sampling unit. If every point in the field is equally likely to have a plant on it, we expect the Poisson to describe the distribution of the number of plants in the population of quadrat counts. As another example, suppose insects are moving into a field. If each plant is equally likely to be selected by each insect with complete indifference to any feature of the plant, position of the plant, or other insects, we expect the Poisson distribution. This does agree with our intuition about randomness, but it seems highly improbable that organisms would be that unresponsive to the environment. As has been so aptly pointed out by Kac (1983), no truly random process has ever been described.

The Poisson is not often encountered in sampling organisms, but Kogan and Herzog (1980) report it for several soybean pests. Often, the Poisson is reported to fit when population densities are low, in which case it is difficult to discern among numerous distributions. In many cases, it is an artifact of sampling procedures. As pointed out in the geometric distribution, the parameters are additive. If the population distribution is binomial, negative binomial, or Poisson when the micro-habitat is the sampling unit, the sampling distribution will be of the same form but with parameters that depend on the number of microhabitats in a sample unit. For example, suppose six feet of linear row is the sampling unit for a crop-infesting pest. Generally, numerous microhabitats are within each 6-foot sampling unit. The additive effects of the negative binomial could produce a Poisson, or something close to a Poisson. In addition, goodness-of-fit tests are notoriously weak in discerning among discrete distributions. Often data may be adequately fit by more than one distribution, especially if few observations were taken.

Expected Frequencies

As stated above, the Poisson distribution has one parameter, lambda (λ). The probability of observing the value $X = x$ for a randomly selected sampling unit is

$$P(x) = \frac{e^{-\lambda}\lambda^x}{x!}, \qquad x = 0, 1, 2, \ldots. \tag{1.34}$$

Then

$$P(0) = e^{-\lambda} \tag{1.35}$$

and

$$P(x) = \frac{\lambda}{x} P(x - 1), \qquad x = 1, 2, 3, \ldots. \tag{1.36}$$

Example 1.7

Calculate the probabilities of a Poisson distribution with

$$\lambda = \mu = \sigma^2 = 2$$

$$P(0) = e^{-\lambda} = (2.71828)^{-2} = 0.1353$$

$$P(1) = \frac{\lambda}{x} P(0) = \frac{2}{1}(0.1353) = 0.2707$$

$$P(2) = \frac{\lambda}{x} P(1) = \frac{2}{2}(0.2707) = 0.2707$$

$$P(3) = \frac{\lambda}{x} P(2) = \frac{2}{3}(0.2707) = 0.1804$$

$$P(4) = \frac{\lambda}{x} P(3) = \frac{2}{4}(0.1804) = 0.0902$$

$$P(5) = \frac{\lambda}{x} P(4) = \frac{2}{5} (0.0902) = 0.0361$$

$$P(6) = \frac{\lambda}{x} P(5) = \frac{2}{6} (0.0361) = 0.0120$$

$$P(7) = \frac{\lambda}{x} P(6) = \frac{2}{7} (0.0120) = 0.0034$$

$$P(8) = \frac{\lambda}{x} P(7) = \frac{2}{8} (0.0034) = 0.0009$$

As described in Example 1.3, ECOSTAT may also be used to compute probabilities for the Poisson distribution.

Suppose we have a quadrat that is 1 m long on each side. Let X be the number of pigweeds within a quadrat. The average number of pigweeds in a square meter of the 10,000-square meter sampling area is 2. How many of the quadrats are expected to have 0, 1, 2, . . . , pigweeds. The results for 0 to 8 are in Table 1.3.

Methods for the use of these frequencies with a goodness-of-fit test are explained in Chapter 2. When the mean is small, the probabilities for the Poisson distribution are similar to those for the geometric and negative binomial distributions (*see* Figure 1.4) As the mean rises, the distribution develops a central tendency and begins to look more normal. When the mean is small, many observations must be taken to reach an acceptable level of accuracy. As the mean rises, fewer observations are necessary.

Confidence Intervals

Suppose the goal of sampling is to estimate the population density μ. Then the best estimator is the sample mean $\bar{X}$. The sample mean is best in the sense that it is unbiased, and of all unbiased estimators, it has the smallest variance. If we must choose a single value as a guess at the population mean, the sample mean is best. However, it is very unlikely that the sample mean in fact equals the population mean. This leads us to ask whether we can determine a set of values that are likely to include the population mean. The answer is yes. To do so, we determine an interval that has a known probability of covering the true population mean; that is, we set a confidence interval on the mean. Thus we have changed from a point estimate (the sample mean) to an interval estimate (the confidence interval). In this section, we discuss methods of setting confidence intervals.

Traditionally, $(1 - \alpha)100\%$ confidence intervals have been based on the normal distribution, taking the form:

Table 1.3   Poisson Distribution Probabilities and Expected Frequencies

X	Sample size	P(x)	Expected frequency
0	10,000	0.1353	1,353
1	10,000	0.2707	2,707
2	10,000	0.2707	2,707
3	10,000	0.1804	1,804
4	10,000	0.0902	902
5	10,000	0.0361	361
6	10,000	0.0120	120
7	10,000	0.0034	34
8	10,000	0.0009	9

$$\bar{X} \pm t_{n-1,\alpha/2} \frac{s}{\sqrt{n}}$$

where $t_{n-1,\alpha/2}$ is the $(1 - \alpha/2)$ quantile of the Student's distribution with $(n - 1)$ degrees of freedom, s is the sample standard deviation, and n is the sample size. Confidence intervals based on the normal distribution are approximate for non-normal distributions, including the discrete ones we have discussed. If the distribution is symmetric or nearly symmetric, the nominal and actual confidence

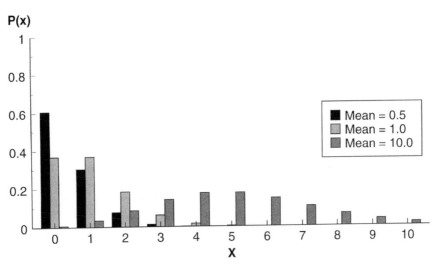

Figure 1.4.   The effect of changing λ of the Poisson distribution.

levels tend to be close. For the highly skewed distributions often encountered in ecological studies, the actual confidence level may be substantially below the nominal level. With today's computing power, it is no longer necessary to use the normal approximation. If the form of the probability distribution is known, exact confidence intervals based on that distribution can be set.

A confidence interval on the mean of a negative binomial or geometric distribution can be based on the fact that the sum of independent observations from these distributions is negative binomial and on the relationship between the distribution function of the negative binomial distribution and the incomplete beta function (*see*, for example, Bain and Engelhardt, 1992; Kennedy and Gentle, 1980). Similarly, the sum of independent observations from a binomial distribution is binomial. This and the relationship of the distribution function to the incomplete beta function forms the foundation to set confidence intervals for the mean or proportion of the binomial distribution (*see*, for example, Kennedy and Gentle, 1980; Bain and Engelhardt, 1992). Finally, the sum of independent observations from a Poisson distribution has a Poisson distribution. This fact permits us to establish exact confidence intervals for the Poisson mean.

It should be noted here that the discrete nature of the distribution may not permit a confidence interval of, say, exactly 95%. We may have to choose between a slightly lower level or a slight higher level. Here the conservative approach of setting confidence intervals with at least the stated level of confidence is taken.

Continuous Distributions

The emphasis throughout much of this text is on discrete distributions because they are essential when studying population dynamics. However, continuous variables are observed on numerous occasions in ecological work. The age, height, length, and weight of organisms represent continuous variables. The time an organism spends in a particular life stage, or the distance it has moved since it was last observed, are also continuous variables. Although the normal distribution is central, other continuous distributions are extremely useful. In addition to the normal, we consider the lognormal, exponential, gamma, and Weibull distributions. Other continuous distributions are useful in certain applications. For a more complete listing, *see* Rao (1970), Patil et al. (1984), and Johnson et al. (1992).

Continuous random variables differ from discrete random variables in that they may conceivably assume any value in an interval. For example, for a particular organism, the time spent in a life stage may be any value from 0 to 4,824 hours. In reality, the measurement of this time becomes discrete due to the finite scale of the measuring device. The organism may be observed hourly, in which case all deaths or transitions to the next stage occurring in the preceding hour are recorded on the hour; the organism entering the next life stage at 12.76654922201 hours is recorded as having made the transition at 13 hours. However, as long as the measurements are not unduly coarse, the continuous distributions are appropriate.

Distribution functions are particularly useful when working with continuous random variables. The distribution function, F, evaluated at x is the probability that the random variable X is less than or equal to x: $F(x) = Pr(X \leq x)$. The probability that the random variable X assumes a value between x_1 and x_2 is then

$$P(x_1 \leq X \leq x_2) = P(X \leq x_2) - P(X < x_1)$$
$$= F(x_2) - F(x_1) \tag{1.37}$$

Distribution functions are defined for both discrete and continuous random variables. For discrete random variables, the distribution function is a step function with the points of discontinuity (the steps) occurring at the points of positive probability. If the distribution function is continuous, the random variable is continuous.

As you know, discrete random variables have distinct values of X that have positive probabilities of occurring. The sum of these probabilities is 1. With continuous random variables, the probability of any given value x is 0. Further, probabilities can be determined from the distribution function as in equation (1.37) or by finding the corresponding area under the *probability density function* $f(x)$. The probability density function (*pdf*) may be viewed as the population relative frequency curve. It may be found by taking the derivative of the distribution function. Given the *pdf* of a random variable, the distribution function can be determined by integration.

Normal Distribution

Abraham de Moivre first published the normal distribution in 1733 as an approximation for the distribution of a sum of binomial random variables. The extent to which the normal distribution has dominated both statistical practice and theory is amazing. Several reasons exist for this (Snedecor and Cochran, 1989). First, several random variables, such as heights, yields, and weights, are normally distributed. Further, a simple transformation may induce approximate normality when a random variable is not normally distributed. Third, theory relating to the normal distribution is relatively easy to develop, and normal-based methods often are robust to moderate departures from normality. Finally, according to the central limit theorem, as long as a population has a finite mean and variance, the distribution of the sample mean tends toward normality as the sample size increases under random sampling. However, as mentioned earlier, the highly skewed distributions often encountered in ecological work require extremely large sample sizes before the distribution of the sample mean is approximately normal. The increasing accessibility of computers has spurred the development of statistical methods for non-normal distributions, and alternative methods are becoming increasingly popular. However, the normal-based methods continue to be the most widely used.

A random variable X follows the normal distribution if it has the probability density function (pdf)

$$f(x) = \frac{1}{\sigma\sqrt{2\pi}} e^{-[(x-\mu)/\sigma]^2/2}, \qquad -\infty < x < \infty. \tag{1.38}$$

The mean is μ and the variance is σ^2. Figure 1.5 graphically depicts two normal distributions, each with a variance of 1. The effect of changing the mean from 0 to 1 is shown. We can see that a change in the mean results in a shift of the distribution, but to no change in its shape. This has resulted in the parameter μ being called the location parameter. The effect of changing the variance while holding the mean fixed is shown in Figure 1.6. When the variance is increased from 1 to 2, the curve is lower at the mean and has a broader spread. Thus σ^2 is called the scale parameter.

The symbols

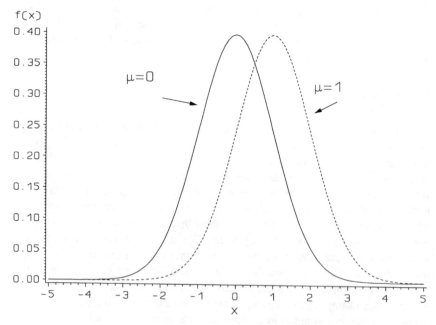

Figure 1.5. The effect of changing μ on the normal distribution.

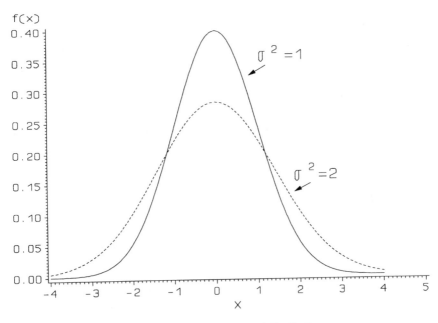

Figure 1.6. The effect of changing σ^2 on the normal distribution.

$$X \sim N(\mu, \sigma^2) \qquad (1.39)$$

indicate that the random variable X has the normal distribution, with mean μ and variance σ^2.

No closed form exists for the distribution function of the normal random variable. This means that, in order to compute the distribution function at a given value, one must use either numerical techniques or tables that have been developed using these techniques. Fortunately, any normal random variable can be standardized to a mean of 0 and a variance of 1 by computing

$$z = \frac{x - \mu}{\sigma}. \qquad (1.40)$$

This permits the tabulation of all probabilities associated with any normal random variable in a single table. Further, the value of a standard normal random variable z can be transformed to a normal random variable X with any given mean μ and variance σ^2 using

$$X = \mu + \sigma z.$$

Example 1.9

The African estrildid finch *Pyrenestes ostrinus* exhibits a polymorphism in bill size that is unrelated to sex (Smith, 1987). Assuming that the lower bill width of the large-billed morphs is normally distributed with a mean of 15.9 mm and a variance of 0.0044 mm², find the proportion of the population that has a lower bill width in excess of 15.92 mm.

We begin by writing the probability of interest in standard notation. Then, in order to use standard normal tables, the probability is rewritten in terms of a standard normal variate:

$$P(X > 15.92) = P\left(z > \frac{(15.92 - 15.9)}{\sqrt{0.0044}}\right) = P(z > 0.30).$$

The value of the standard normal distribution function at 0.30 is 0.62 (see Figure 1.7); that is,

$$F(0.30) = Pr(z \le 0.30) = 0.62$$

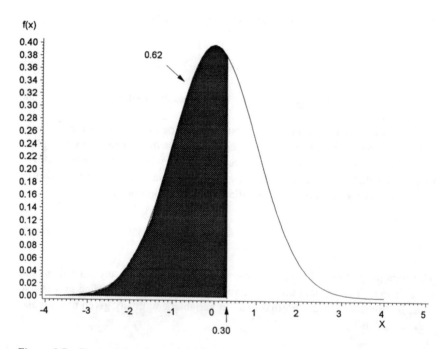

Figure 1.7. Determining the probability that X is less than 0.30 when X is normally distributed.

This tells us that about 62% of the population has a lower bill width of less than 15.92 mm. Therefore, $1 - 0.62 = 0.38$, that is, 38% of the population of large-billed morphs has a lower bill width in excess of 15.92 mm.

An African finch with a lower bill width of less than 14 mm is classified as a small morph, and one with a lower bill width in excess of 14 mm is considered a large morph. One concern may be the possible misclassification of morph if there is an overlap in the two populations. For the large morph, what is the lower bill width, for which only 1 finch in 1,000 of the morph would have a smaller lower bill width? Because the tabulated values are in terms of the standard normal, we first begin by finding the value of the standard normal that has a probability of 0.001 (1 in a 1,000) of obtaining a smaller value when sampling from a standard normal. Then this value is transformed to the normal distribution with the mean and variance in our study. The value of the standard normal variate of interest is that value of z that satisfies

$$F(z) = 0.001$$

The corresponding value of z is -3.08. This should next be transformed into a variate from a distribution with a mean of 15.9 and a variance of 0.0044:

$$x = \mu + \sigma z = 15.9 + \sqrt{0.0044}(-3.08) = 15.7.$$

Only 1 in a 1,000 randomly sampled large-morph African finches have a lower bill width of less than 15.7 mm. Therefore, the probability of classifying a large morph as a small morph (lower bill width of less than 14 mm) is negligible.

ECOSTAT can be used to do the computations for the normal and the other continuous distributions discussed in this chapter. Begin by selecting *Continuous Distributions* and then *Probabilities* on the *Probability Distributions* menu. For this example, we select *Normal* from the menu. For the first part, we want to find the probability so we choose *Find p* on the submenu. After entering the mean and variance of the distribution and the value of X, the probability of a smaller value can be displayed by pressing *Compute*. For the second part, *Find x* is chosen. The mean and variance of the distribution and the probability p of a smaller value of X is entered. Pressing *Compute* results in the value of X for which $F(x) = p$.

Interval estimates of the mean are usually based on the normal distribution. Suppose we take a random sample of size n from a population with mean μ and variance σ^2. Then, by invoking the central limit theorem, we have

$$z = \frac{\bar{X} - \mu}{\sigma} \tag{1.49}$$

has a standard normal distribution if the population distribution is normal and an approximate standard normal if the population is non-normal. Therefore,

$$P\left(\bar{X} - z_{\alpha/2} \frac{\sigma}{\sqrt{n}} < \mu < \bar{X} + z_{\alpha/2} \frac{\sigma}{\sqrt{n}}\right) \approx 1 - \alpha \qquad (1.50)$$

where $(1 - \alpha)$ is the specified confidence level and $z_{\alpha/2}$ is the $(1 - \alpha/2)$ quantile of the standard normal distribution. Equality holds in equation (1.50) if the population distribution is normal.

The difficulty with applying the above result is that the variance is usually unknown, and the Central Limit Theorem is not directly applicable. These cases use an estimate of the variance. If the population distribution is normal

$$\frac{\bar{X} - \mu}{s/\sqrt{n}} \sim t_{n-1}. \qquad (1.51)$$

This relationship may be used to develop a confidence interval because

$$P\left(\bar{X} - t_{n-1,\alpha/2} \frac{s}{\sqrt{n}} < \mu < \bar{X} + t_{n-1,\alpha/2} \frac{s}{\sqrt{n}}\right) \approx 1 - \alpha \qquad (1.52)$$

where $t_{n-1,\alpha/2}$ is the $(1 - \alpha/2)$ quantile value of a t-distribution with $(n - 1)$ degrees of freedom. Equality holds in equation (1.52) if the population distribution is normal. If it is not normal, the approximation is generally not as good as that based on the Central Limit Theorem. This results because the distribution of s^2 changes, and the independence of the estimated mean and variance is violated.

Example 1.10

Interspecific competition and resource use were studied for two temperate marine reef fishes: black surfperch (*Embiotoca jacksoni*) and striped surfperch (*E. lateralis*) (Holbrook and Schmitt, 1989). Because it was anticipated that competition could depend on food abundance, estimates of the prey density and size were desired for warm-water and cold-water seasons and for shallow and deep areas. Data from 36 quadrats, each with area of 0.1 m², were collected for each season–habitat combination. The average number of prey in warm-water shallow areas was 5461, and the standard deviation of the observed counts was 9330. Setting a 95% confidence interval on the mean prey density/0.1 m², the confidence interval has the form

$$\bar{X} \pm t_{35,025} \frac{s}{\sqrt{n}}$$

Substituting the sample and t-values, we obtain

$$5{,}461 \pm 2.030 \, \frac{9330}{\sqrt{36}} = 5{,}461 \pm 3{,}157$$

Therefore, the estimated mean prey density/0.1 m² in warm-water shallow areas in the region sampled is 5,461, and we are 95% confident that this estimate is within 3,157 prey of the true mean prey density/0.1 m²; that is, we are 95% confident that the mean prey density/0.1 m² is between 2,304 and 8,618.

Lognormal Distribution

The distribution of many biological measurements is not symmetric as in the normal distribution. Often the distribution is skewed to the right. For example, consider the distribution of lengths of a species of fish in a riverine system. Numerous fish perish shortly after birth, and thus the extremely small fish are not very common. However, once a critical size has been attained, an increasing portion of the population may be found with this size; then, as a result of death, and perhaps harvesting, few large fish remain in the population (*see* Figure 1.8). Numerous continuous distributions exhibit these general tendencies. One commonly used in ecological studies is the lognormal distribution.

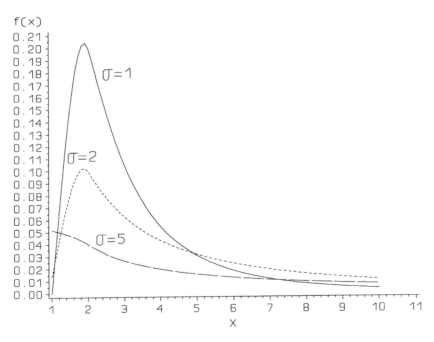

Figure 1.8. The effect of changing σ on the lognormal distribution.

The random variable X is said to be lognormal if there is a number θ such that $Y = \log(X - \theta)$ is normally distributed. Often θ is taken to be zero. The log-normal has been used to model the distribution of critical dose (minimal dose causing reaction) for a number of drug applications (*see* Gaddum, 1933; Bliss, 1934). Preston (1948) proposed using the lognormal distribution in his model of species abundance.

Exponential Distribution

Historically, the exponential was the first widely used lifetime distribution. The exponential and Poisson distributions have a unique relationship, in that the time (or space) between the occurrence of *random events* (those events occurring according to a Poisson process) has an exponential distribution. For example, suppose that births in a population occur randomly, according to a Poisson process. The number of births in a given period of time has a Poisson distribution, and the time between births is exponentially distributed. As another example, assume that grass seeds are randomly dispersed in a region; that is, each grass seed is as likely to fall on any one point as on any other point in the region, and this placement is independent of that for every other seed. A point P in the region is selected at random, and the grass seed nearest to P is found. (The obvious problem of locating the seed will not be considered.) A circle with center at the selected point P is formed, with the radius being the distance from P to the nearest seed. The circle is then centered at P, and the nearest seed lies on the circle. The area of the circle has an exponential distribution. Therefore, the exponential distribution may represent either the time or space between events in a Poisson process.

The exponential distribution has one parameter, λ. The probability density function has the form

$$f(x) = \lambda e^{-\lambda x}, \qquad x > 0 \tag{1.53}$$

(*see* Figure 1.9). The mean of the distribution is $\mu = 1/\lambda$, and the variance is $\sigma^2 = 1/\lambda^2$. The distribution function has the closed form

$$
\begin{aligned}
F(x) &= 0, & x < 0 \\
&= 1 - e^{-\lambda x}, & x \geq 0
\end{aligned}
\tag{1.54}
$$

The exponential distribution is unique among continuous distributions, in that it has the *memoryless property;* that is, the probability that at least x_2 units of time remain until the next event occurs given that x_1 units of time have already elapsed is the same as the original probability that x_2 units of time will occur between events. From the standpoint of scientific validation and modeling, the memoryless property provides some advantages. Suppose we believe that the time

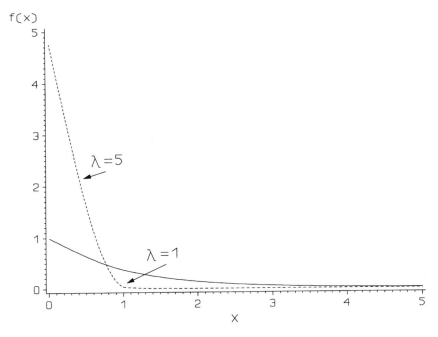

Figure 1.9. The effect of changing λ on the exponential distribution.

an insect inhabits a plant is exponentially distributed. Then if we find an insect already on the plant, the distribution of time until it moves is, at least theoretically, the same as when it initially inhabited the plant, and records can be made from that point in time.

The memoryless property is also the reason that the use of the exponential distribution is increasingly questioned, especially in lifetime studies. For example, assume the lifetime of a pocket gopher is modeled using the exponential distribution. Then the probability that it will live another year, when it is 5 years old, is the same as the probability it would live a year from birth. The biological implications of such an assumption have led many researchers to consider other continuous distributions, especially the gamma and Weibull, for these applications. For a discussion of the applications of the exponential distribution, *see* Chapter 8.

Gamma Distribution

Suppose the time (or space) between random events has the exponential distribution. Then the time (or space) until the occurrence of the rth event occurs has the gamma distribution. The gamma has also been useful in modeling lifetime data in cases where the exponential distribution has not described the data well.

The gamma distribution has also been useful in describing biological phenomena that have positively skewed continuous distributions. The form of the probability density function is

$$f(x) = \frac{\lambda^r}{\Gamma(r)} x^{r-1} e^{-\lambda x}, \qquad x > 0. \tag{1.55}$$

The mean of the distribution is $\mu = r/\lambda$, and the variance is r/λ^2. Note that the exponential is a special case of the gamma distribution with $r = 1$. The effect of the parameters on the shape of the distribution is illustrated in Figure 1.10. The fact that a random variable X has a gamma distribution with parameters r and λ is denoted by

$$X \sim Ga(r, \lambda).$$

The gamma distribution function does not have a closed form. Tables have been constructed in terms of parameters r and λ (Beyer, 1968). We can also use the χ^2 distribution, with its more readily available tables, to obtain any value of the gamma distribution function. Our ability to use the χ^2 distribution is based on

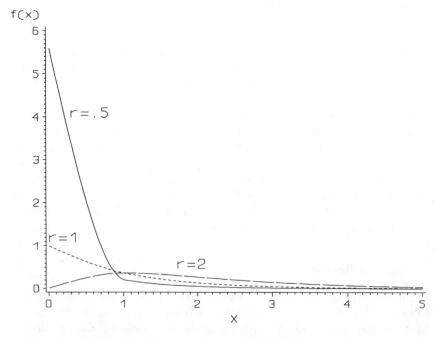

Figure 1.10.   The effect of changing r on the gamma distribution.

two facts. First, the χ^2 distribution with k degrees of freedom is a special case of the gamma distribution with $r = k/2$ and $\lambda = 1/2$. Further, if $X \sim Ga(r, \lambda)$, $2\lambda X \sim Ga(r, 0.5)$; that is, $2\lambda X$ has the χ^2 distribution with $2r$ degrees of freedom. Although r must be a multiple of 0.5 for the degrees of freedom of the χ^2 to be integer, allowing ready use of existing tables, the χ^2 distribution need not have integer degrees of freedom to be well defined. Most computer packages and several calculators permit computation of chi-square probabilities with fractional degrees of freedom. Most broad-based computer packages today have functions that can be used to evaluate probabilities for the gamma distribution. ECOSTAT can be used to evaluate gamma probabilities.

Example 1.11

Consider a random variable X that has a gamma distribution with $r = 5$ and $\lambda = 3$. Find the probability that X is less than 2 using the χ^2 distribution. We first want to transform the probability statement from one involving the gamma distribution to one involving the chi-squared distribution:

$$P(X < 2) = P[2\lambda X < 2(3)(2)]$$
$$= P(\chi^2_{10} < 12).$$

From ECOSTAT, we can now find the probability is 0.7149.

Weibull Distribution

The physicist W. Weibull first proposed a distribution for numerous applications, including fatigue and breaking strength of materials. Known as the Weibull distribution, this approach has been used increasingly in biology to model survival, physiological development, and other biological phenomena. The probability density function has the form

$$f(x) = \beta\lambda^\beta x^{\beta-1}e^{-(\lambda x)^\beta}, \qquad x > 0 \qquad (1.56)$$

where $\lambda > 0$, $\beta > 0$. The mean of the Weibull distribution is $\Gamma(1 + 1/\beta)/\lambda$, and the variance is $[\Gamma(1 + 2/\beta) - \Gamma(1 + 1/\beta)]/\lambda^2$. Note that the special case of $\beta = 1$ is the exponential distribution. As with the exponential distribution, an advantage of the Weibull distribution is the distribution function can be written in closed form as

$$F(x) = 1 - e^{-(\lambda x)^\beta}, \qquad x > 0. \qquad (1.57)$$

The Weibull is often used because of its flexibility (*see* Figure 1.11), and seldom due to theoretical justification.

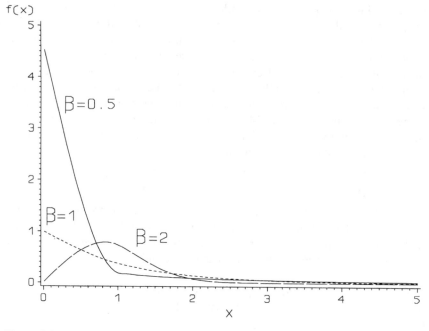

Figure 1.11. The effect of changing β on the Weibull distribution.

Summary

Probability distributions often serve as good models of real-world phenomena. Because many of the ecological data we encounter are count data, this chapter has focused on those distributions. If the mean and variance are equal, the Poisson distribution is the distribution that we first try to use. In most ecological data, the variance exceeds the mean. The negative binomial, geometric, and logarithmic distributions are the primary distributions used in these circumstances. The binomial distribution is most useful in presence–absence sampling. Although numerous other discrete distributions have been applied in ecology, the extent of their use has been more limited. Johnson et al. (1992) serves as a good reference for alternative distributions if one has data that are not described well by those presented here.

Lengths, heights, weights, time, and other variables are continuous. The normal distribution is the most widely-used distribution, not only because many populations are normally distributed but also because the Central Limit Theorem assures us that, if we are sampling from *any* population with a finite mean and variance, the distribution of the sample mean approaches the normal as the sample size approaches infinity. However, other continuous distributions are extremely useful. Data are often described well by the lognormal distribution. The exponential distribution is particularly useful in the study of spatial point patterns and lifetime analysis. The gamma distribution has wide application in lifetime analysis. Temperature and other models are fit well by the Weiibull distribution.

Exercises

The ECOSTAT software probability section can be used for these exercises.

1. Antelope bitterbrush is an important food plant for domestic and wild mammal browsers in the western United States. Wall (1994) conducted a study to determine the fates of antelope bitterbrush seeds in an effort to better understand the population biology of the plant. The seed fate with the highest probability of seedling establishment at the Nevada study site was harvest, dispersal and scatter hoarding by chipmunks, followed by subsequent neglect of the caches. Of the 55 caches formed directly from seeds put out under a source shrub in 1989, the observed mean number of seeds per cache was 4.9 with an estimated standard deviation of 3.4.

    A. Test the hypothesis that the distribution of the number of seeds per cache has equal mean and variance.

    B. Use the normal approximation to determine the observed significance level for the chi-squared statistic computed in A and compare the result.

2. A New Hampshire study was designed to determine the effect of a dominant predatory fish (brook charr, *Salvelinus fontinalis* Mitchill) on mayfly drift rates and benthic densities (Forrester, 1994). The study was conducted in small patches of run habitat within a temperate stream. Measurement of mayfly drift rates and benthic densities were made at 20 unmanipulated locations. The sampling unit was a square-meter quadrat. The estimated mean number of *Baetis* spp. per square quadrat was 253, with a standard error of 20. Test the following hypothesis; for the distribution of the number of *Baetis* spp. per square quadrat, the mean and variance are equal.

3. For your area of interest, find an example of count data from the literature. Give a short summary of the study and a complete citation of the work. Test the hypothesis that the mean and variance are equal for the distribution of counts.

4. Calculate the probabilities of the following negative binomial distributions.

    A. $k = 0.5$; $\mu = 1$

    B. $k = 1$; $\mu = 0.5$

    C. $k = 5$; $\mu = 1$

    D. $k = 1$; $\mu = 5$

    E. $k = 10$; $\mu = 12$

    F. $k = 1.2$; $\mu = 1.5$

5. Compute the expected frequencies of the sets in exercise 4 for samples of the following sizes.

    A. $n = 100{,}000$

    B. $n = 12{,}345$

    C. $n = 56{,}123$

    D. $n = 230$

    E. $n = 30$

6. Plot the probabilities for each set in exercise 4. Use these plots to describe the change in the shape of the negative binomial distribution as μ and k change.

7. Plot the probability of 0 of a negative binomial distribution as k goes from 0.1 to 10 if μ is constant at the following values.

    A. $\mu = 0.5$

    B. $\mu = 1$

    C. $\mu = 5$

    D. $\mu = 10$

Use these plots to describe the effect of k on the probability of 1. Does this effect depend on the value of μ?

8. Plot the probability of 1 of a negative binomial distribution as μ goes from 0.1 to 10 if k is constant at the following values.

    A. $k = 1$

    B. $k = 2$

    C. $k = 5$

    D. $k = 10$

Use these plots to describe the effect of μ on the probability of 1. Does that effect depend on k?

9. Calculate the probabilities of Poisson distributions for the following values of λ.

    A. $\lambda = 0.5$

    B. $\lambda = 0.75$

    C. $\lambda = 1$

    D. $\lambda = 1.4$

    E. $\lambda = 2$

F. $\lambda = 3.3$

G. $\lambda = 4.2$

H. $\lambda = 5$

I. $\lambda = 10$

10. Plot the distributions in exercise 9. How does λ affect the shape of the distribution? Describe the effect of λ on the probability of 0.

11. Develop a graph of the probabilities of a Poisson distribution with $\lambda = 1$ and a negative binomial distribution with $k = 25$ and $\mu = 1$. What conclusions can be drawn from this graph?

12. Produce a graph with a Poisson distribution with $\lambda = 1$ and a binomial distribution with $k = -25$ and $\mu = 1$. What conclusions can be drawn from this graph?

13. Calculate the probabilities of a negative binomial distribution with $k = 5$ and $\mu = 0.5, 1, 2,$ and 5. How do they compare with the probabilities in exercise 9?

2

Goodness-of-Fit Tests

Introduction

Goodness-of-fit tests are used to validate the use of a particular distribution to describe data arising from sampling or experimentation. Numerous goodness-of-fit tests have been developed. The power divergence family of test statistics includes Pearson's chi-squared test, the likelihood ratio test, and the Freeman–Tukey chi-squared test. We consider these special cases as well as the family of test statistics and the Nass test. These may also be applied to test the fit of continuous distributions by forming discrete intervals on the continuous scale of measurement. However, the Kolmogorov–Smirnov test is generally preferred when the distribution is continuous and is described in this chapter. Like all data handling systems, they are applied best when the problems and limitations of their use are understood. We begin by focusing on the application of goodness-of-fit tests to discrete distributions.

Little and Hills (1978) pointed out that many statistical methods such as analysis of variance (ANOVA) and t-tests are designed primarily for the analysis of measurements such as weight, yield, or height. Measurements are used in ecological studies, but those involved with population density are, instead, counts. Examples of measurements are the length of tarsal segments on insects, pounds of cotton lint per plot, time of pupal development in *Heliothis zea* (Boddie), and feed consumption by cattle. Count data do not fit the conditions for most analyses based on the normal distribution. For example, consider the t-test. One of the assumptions made when forming the test statistic is that the estimates of mean and variance are independent; that is, as the estimated mean increases or decreases, the estimate of variance is not affected, and vice versa. This is true only for the normal distribution. Because the t-test is robust to mild departures in the assumptions, it is still useful, especially when the normal distribution serves as a good approximation to the population distribution. The normal distribution may provide a good approximation to count data with high densities and low variation as in the binomial distribution with large n and p close to 0.5. The normal approximation of Poisson distributions with means above 5 is sufficient for most practical applications. The normal distribution does not provide a good approximation for the geometric distribution or for the negative binomial distribution when the parameter k is below 1. Unfortunately, most count data collected from populations of organisms are from these distributions.

The problem of non-normality can sometimes be dealt with by transformations. Each observation in the data set is transformed by taking the log, the square root, or some other function of the observation. Then the analysis is performed on these transformed values. In many cases, the focus of these transformations is to produce data with mean and variance approximately independent. Strabala (1984) performed an extensive Monte Carlo study of the size and power of the analysis of variance of a randomized complete block design based on raw count data, transformed count data, and the ranks of the count data. She found that for the analysis of the transformed data, the significance level tended to be substantially below the stated level, and the power was lower than when the analyses were based on the count data or the ranks of those data. Strabala (1984) did show that the use of the rank transformation resulted in more power than an analysis of the raw data when the number of observations per experimental unit is small or when the distribution is highly skewed as in the negative binomial or geometric distributions. Therefore, because skewed distributions are common in ecological studies, perhaps an alternative to the use of normal-based methods should be sought. One approach is to determine the form of the population distribution being sampled and then to use statistical methods (e.g., generalized linear models) for that distribution. This will be the goal of much of our work. Goodness-of-fit tests play a vital role in the first step of determining the form of the population distribution.

What is the value of various goodness-of-fit tests and how are they used? As the name implies, they can be used to test how well some predetermined model fits the data. Most biologists are familiar with genetic tests that use goodness-of-fit tests. A simple example is the factor for horns or poll (no horns) in cattle. A cow homozygous for horns is crossed with a bull homozygous for poll. All the offspring will be heterozygous and have one gene for horned and one for poll. All will be poll because it is dominant and suppresses the horn condition when present. If a heterozygous bull is crossed with a heterozygous cow the offspring can be either horned or poll, in a ratio of 3 poll to 1 horn. A goodness-of-fit test is used to confirm or validate such a model. In a similar fashion, discrete distributions can be used to determine whether the observed proportions of 0's, 1's, 2's, ..., agree with those expected when the population distribution is negative binomial, geometric, Poisson, or binomial.

For goodness-of-fit tests, the data are considered to be a random sample from a population. The null hypothesis is that the population distribution takes a specific form. The null hypothesis may completely specify the population distribution. For example, "The population distribution is Poisson with a mean of 2" or "The population distribution is binomial with $n = 15$ and $p = 0.4$." More often the general form of the distribution is specified, but the parameters are not, such as "The population distribution is Poisson" or "The population distribution is binomial." The alternative hypothesis is that the population distribution is different from that specified in the null hypothesis. The probability of a type II error, deciding in favor of the null hypothesis when it is not true, depends on the true population distribution. When the null hypothesis is false, the population distribution and, consequently, the type II error rate, are unknown. Therefore, we could not state with great assurance that the data are a random sample from the null population distribution. We can conclude that the model adequately describes the data. Using goodness-of-fit tests will illustrate tendencies of data sets if they are present. However, broad-reaching biological conclusions should not be based on a goodness-of-fit test for a single data set. The strength of the biological conclusions grows if the same distribution is found to consistently describe data collected through time and space.

All goodness-of-fit tests have limitations and should be used with caution. A biometrician should be consulted before applying any of these tests. The small-sample properties of Pearson's chi-squared test, the likelihood ratio test and the Freeman–Tukey test were investigated through Monte Carlo simulation by Larntz (1978). Pearson's chi-squared statistic was found to be the most desirable based on a criterion of the closeness of the small-sample distribution to the asymptotic chi-squared approximation. Both the likelihood ratio and Freeman–Tukey tests resulted in too many rejections under the null distribution; that is, the probability of declaring that the sample was not drawn from the hypothesized distribution, when in fact it was, is greater than the stated level. Wu (1985) conducted a Monte Carlo study to evaluate the significance level and power of the Pearson chi-squared, likelihood ratio, Freeman–Tukey chi-squared, Nass's, and Hinz–Gurland goodness-of-fit tests as they are applied to the negative binomial and Neyman type A distributions. No test was best overall. The Pearson chi-squared and Nass tests behaved in a similar manner and consistently resulted in the nominal significance level of the test being close to the exact one. As in Larntz's earlier study, Wu found the likelihood ratio and Freeman–Tukey tests resulted in too many type I errors; that is, the probability of rejecting the hypothesized distribution, when in fact it was the population distribution, was larger than stated. It was suggested that parallel analyses using several tests be compared on data sets.

Pearson's Chi-Squared Test

Pearson's chi-squared test was first presented by Pearson (1900) and is often referred to as the chi-squared test. It is the oldest, the best known, and probably the most widely applied goodness-of-fit test. Pearson's chi-squared test is used to determine whether the hypothesized probability distribution adequately describes, or fits, the sample data. The statistic used in this test has an approximate chi-squared distribution if the data were collected from a population with the hypothesized probability distribution. The chi-squared approximation assumes a *large* sample.

In exploratory data analysis, the fit of several competing distributions may be compared for the same data set. The results may not be conclusive. For example, two or more distributions may fit within error; that is, there is not enough evidence to conclude that these distributions do not fit the data. One distribution may appear to fit better than the others. However, we do not know whether that same distribution would give the best fit for another sample from the same population. Be cautions about assigning population tendencies based on inconclusive chi-squared tests from limited data. The fact that you can not reject a distribution does not necessarily mean that it is the population distribution. The strongest results are obtained when the same distribution consistently fits data from a population that has been sampled over time and/or space.

Pearson's chi-squared test measures the agreement in the observed data with what is expected if a sample of the same size had been drawn from the hypothesized population distribution. For example, suppose we have hypothesized that the distribution is Poisson with a mean of one. A sample is drawn. Are the number of 0's in the sample the same as we expect in a sample from a Poisson with a mean of one? Is there agreement in what we observed and expected for the 1's, the 2's, the 3's, etc.? We do not expect the observed and expected frequencies to be exactly the same, but they should be close. Pearson's chi-squared test provides a measure of how close these observed and expected frequencies are to each other. First, the data are grouped into r classes. For the discrete distributions, the classes are based on the observed values that are possible when sampling from the hypothesized distribution; that is, for the Poisson with a mean of one, possible classes are the 0 class consisting of the 0's, the 1 class consisting of the 1's, the 2 class consisting of the 2's, etc. However, for the chi-squared approximation to be sufficiently accurate, the expected frequency within each of these classes should not be too small. A general rule of thumb is that no more than 20% of the classes should have an expected frequency less than 5.0, and none should have an expected frequency less than 1.0. When some classes have expectations smaller than these guidelines suggest, adjacent classes are grouped together to provide a single class with a larger expectation. For example, the 3's and 4's may form a single class. All values of 5 or more may constitute another class.

The general equation for Pearson's chi-squared goodness-of-fit test is

$$\chi_P^2 = \sum_{t=1}^{r} \frac{(O_i - E_i)^2}{E_i}, \tag{2.1}$$

where O_i is the observed frequency in class i, E_i is the expected frequency in class i, and r represents the number of classes after grouping (if this is needed). If the hypothesized distribution is, in fact, the population distribution, then this test statistic has an approximate χ^2 distribution. The degrees of freedom associated with the distribution is ($r - 1 -$ the number of estimated parameters). If we hypothesized that the population distribution is Poisson with a mean of one, the degrees of freedom are ($r - 1$) because the parameter λ is specified and does not need to be estimated. However, if we hypothesize that the population distribution is Poisson without restricting the mean to a specific value (a more common hypothesis), we need to estimate the mean before calculating the expected frequencies. In this case, we have ($r - 2$) degrees of freedom for the test statistic, having lost an additional degree of freedom in the estimation of the mean. Notice that if the hypothesized distribution provides a perfect fit of the data, $O_i = E_i$ for all i, the value of the test statistic is 0. As the differences in the expected and observed frequencies increase, the value of the test statistic also increases. Thus, if the test statistic gets too large, we reject the hypothesis that the sample was drawn from the specified population distribution.

Example 2.1

A farmer counted the number of greenbugs on each of 100 seedling oats. He recorded the number of seedlings with 0 greenbugs, the number with 1, the number with 2, etc. The data are given in Table 2.1. Determine which discrete distribution fits best. To provide insight into which distribution might be the most appropriate, we first estimate the mean and the variance. The estimate of the mean is

$$\bar{x} = \frac{\Sigma x}{N} = \frac{109}{100} = 1.09.$$

The estimated variance is

Table 2.1   Number of Greenbugs Observed on Seedling Oats

No. of greenbugs	Observed frequency
0	43
1	29
2	16
3	6
4	3
5	1
6	1
7	1

$$s^2 = \frac{\Sigma x^2 - \left(\dfrac{(\Sigma x)^2}{n}\right)}{n-1} = \frac{305 - \left(\dfrac{109^2}{100}\right)}{99} = 1.88.$$

Because the estimated variance is larger than the estimated mean, we suspect that the data might be from a population with a negative binomial distribution. Therefore, we test the hypothesis that the population distribution is negative binomial. The sample mean is the estimate of the parameter μ. Because k is unknown, it must also be estimated. The method-of-moments estimator of k is

$$\hat{k} = \frac{\bar{x}^2}{s^2 - \bar{x}} = \frac{1.09^2}{1.88 - 1.09} = 1.50$$

However, the chi-squared approximation is based on maximum likelihood estimates. Solving equation (1.10) for $\hat{k}$, we find the MLE of k to be 1.62. This may be accomplished using ECOSTAT. From the *Probability Distributions* menu, select *Discrete Distributions* and then *Estimation*. To identify the ASCII data file, press *Enter Data*. The first line of the data file has n, the number of observations. Each of the next n lines has an observed value. The data for Example 2.1 are in gbug.dat. After selecting the distribution, the parameter estimates are displayed. Using the estimated values for the parameter values, the next step is to estimate the probabilities as shown in equations (1.22) and (1.27) of Chapter 1:

$$\hat{P}(0) = \left(\frac{\hat{k}}{\hat{k} + \bar{X}}\right)^k = \left(\frac{1.62}{1.62 + 1.09}\right)^{1.62} = 0.4345$$

$$\hat{P}(1) = \left(\frac{\hat{k} + x - 1}{x}\right)\left(\frac{\bar{x}}{\bar{x} + \hat{k}}\right)\hat{P}(x-1) = \left(\frac{1.62 + 1 - 1}{1}\right)\left(\frac{1.09}{1.09 + 1.62}\right)(0.4345) = 0.2831$$

$$\hat{P}(2) = \left(\frac{\hat{k} + x - 1}{x}\right)\left(\frac{\bar{x}}{\bar{x} + \hat{k}}\right)\hat{P}(x-1) = \left(\frac{1.62 + 2 - 1}{2}\right)\left(\frac{1.09}{1.09 + 1.62}\right)(0.2831) = 0.1492$$

$$\hat{P}(3) = \left(\frac{\hat{k} + x - 1}{x}\right)\left(\frac{\bar{x}}{\bar{x} + \hat{k}}\right)\hat{P}(x-1) = \left(\frac{1.62 + 3 - 1}{3}\right)\left(\frac{1.09}{1.09 + 1.62}\right)(0.1492) = 0.0724$$

$$\hat{P}(4) = \left(\frac{\hat{k} + x - 1}{x}\right)\left(\frac{\bar{x}}{\bar{x} + \hat{k}}\right)\hat{P}(x-1) = \left(\frac{1.62 + 4 - 1}{4}\right)\left(\frac{1.09}{1.09 + 1.62}\right)(0.0724) = 0.0336$$

$$\hat{P}(5) = \left(\frac{\hat{k} + x - 1}{x}\right)\left(\frac{\bar{x}}{\bar{x} + \hat{k}}\right)\hat{P}(x-1) = \left(\frac{1.62 + 5 - 1}{5}\right)\left(\frac{1.09}{1.09 + 1.62}\right)(0.0336) = 0.0152$$

$$\hat{P}(6) = \left(\frac{\hat{k} + x - 1}{x}\right)\left(\frac{\bar{x}}{\bar{x} + \hat{k}}\right)\hat{P}(x-1) = \left(\frac{1.62 + 6 - 1}{6}\right)\left(\frac{1.09}{1.09 + 1.62}\right)(0.0152) = 0.0067$$

$$\hat{P}(X \geq 7) = 1 - [\hat{P}(0) + \hat{P}(1) + \hat{P}(2) + \hat{P}(3) + \hat{P}(4) + \hat{P}(5) + \hat{P}(6)]$$
$$= 1 - [0.4345 + 0.2831 + 0.1492 + 0.0724 + 0.0336 + 0.0152 + 0.0067]$$
$$= 1 - 0.9947 = 0.0053$$

Notice that instead of computing the probability of 7, we estimated the probability of getting a value of 7 or larger. In this way, we have accounted for the probability of all possible values from this negative binomial distribution, whether they were observed in the sample or not.

These estimated probabilities may be used to compute the expected frequencies in the same manner that was used in Chapter 1; that is, the expected frequency of 0 is the product of the sample size and the probability of 0, the expected frequency of 1 is the product of the sample size and the probability of 1, etc. Using a subscript to denote the class, we obtain

$$E_0 = 100 \times 0.4345 = 43.45$$

$$E_1 = 100 \times 0.2831 = 28.31$$

$$E_2 = 100 \times 0.1492 = 14.92$$

$$E_3 = 100 \times 0.0724 = 7.24$$

$$E_4 = 100 \times 0.0336 = 3.36$$

$$E_5 = 100 \times 0.0152 = 1.52$$

$$E_6 = 100 \times 0.0067 = 0.67$$

$$E_{\geq 7} = 100 \times 0.0053 = 0.53.$$

The expected frequencies are less than 5 for values of 4, 5, 6, and ≥ 7. To improve the χ^2 approximation, these classes will be combined into a single class of values greater than 3; that is,

$$E_{\geq 4} = E_4 + E_5 + E_6 + E_{\geq 7} = 3.36 + 1.52 + 0.67 + 0.53 = 6.08.$$

Recall Pearson's chi-squared test statistic has the form

$$\chi_P^2 = \sum_{i=0}^{4} \frac{(O_i - E_i)^2}{E_i}.$$

Each term in the summation may be evaluated to determine the contribution of each class to the test statistic:

$$\chi_P^2(0) = \frac{(O_0 - E_0)^2}{E_0} = \frac{(43 - 43.45)^2}{43.45} = 0.0047$$

$$\chi_P^2(1) = \frac{(O_1 - E_1)^2}{E_1} = \frac{(29 - 28.31)^2}{28.31} = 0.0168$$

$$\chi_P^2(2) = \frac{(O_2 - E_2)^2}{E_2} = \frac{(16 - 14.92)^2}{14.92} = 0.0782$$

$$\chi_P^2(3) = \frac{(O_3 - E_3)^2}{E_3} = \frac{(6 - 7.24)^2}{7.24} = 0.2124$$

$$\chi_P^2(\geq 4) = \frac{(O_{\geq 4} - E_{\geq 4})^2}{E_{\geq 4}} = \frac{(6 - 6.08)^2}{6.08} = 0.0011.$$

Pearson's chi-squared test statistic is then

$$\chi_P^2 = \chi_P^2(0) + \chi_P^2(1) + \chi_P^2(2) + \chi_P^2(3) + \chi_P^2(\geq 4)$$
$$= 0.0047 + 0.0168 + 0.0782 + 0.2124 + 0.0011 = 0.3132.$$

What does a chi-squared statistic of 0.3132 mean? The smaller the value of the test statistic, the better the distributions fits. A perfect fit yields a chi-squared statistic of 0. Poor fits have larger chi-squared statistics. To determine whether 0.3132 is large enough to indicate a poor fit, a chi-squared table or computer software must be used. In either case, the degrees of freedom associated with the test statistic must be known. Recall, the number of degrees of freedom is

$$\text{df} = \text{number of classes} - 1 - \text{number of estimated parameters.} \quad (2.2)$$

We estimated two parameters: μ and k. The number of classes is 5 (0, 1, 2, 3, ≥ 4). Thus the degrees of freedom are

$$\text{df} = 5 - 1 - 2 = 2.$$

Using ECOSTAT, we find that the probability that a χ^2 random variable with 2 degrees of freedom is greater than 0.3132 is 0.8551. Therefore, if the sample was drawn from a negative binomial distribution, the probability of obtaining a test statistic value of 0.3132 or more is approximately 0.86. This indicates that the negative binomial distribution provides a good fit to this data set. How large does the test statistic have to be in order for us to reject the hypothesis that the negative binomial fits? In most cases, we would not reject unless the value of the test statistic exceeds the 0.95 quantile; that is, the probability of a larger value of the test statistic when the population distribution is negative binomial is less than 0.05. In our example, this means that the value of the test statistic would need to exceed 5.99.

DISCRETE is a FORTRAN program developed to fit various discrete distributions (Gates and Ethridge, 1972; Gates et al., 1987; Gates, 1989). We have chosen to include it instead of fully developing ECOSTAT in this area. More discrete distributions than we covered in Chapter 1 can be fit to the data, but that is the choice of the user. We have included most of the information from the user's manual in the Appendix to this chapter. This will provide additional references to the distributions in the program not covered in this text, as well as program details. We generally fit all the available distributions. DISCRETE performs a test of the hypothesis that the mean and variance are equal, using the test statistic given in equation (1.7). For each distribution to be fit to a given data set, DISCRETE estimates the parameters. Based on the parameter estimates, expected frequencies are computed, and Pearson's chi-squared test statistic with its associated p-value computed.

ECOSTAT has some limited support for goodness-of-fit tests. However, ECOSTAT assumes that the expected frequencies have been computed and that classes have been grouped if needed. An ASCII file is needed. The first line of the file has the number of classes after grouping (r) and the number of estimated parameters. The next r lines each have a numeric identification of the class, the observed class frequency, and the expected class frequency separated by spaces and in the order given. From the *Probability Distributions* menu, *Discrete* and then *Fit* are chosen. After entering the data, choose the test to be conducted from the menu.

Pearson's chi-squared approximation is theoretically based on using maximum likelihood estimates of the parameters for the grouped data. Some researchers use the method-of-moments estimator of k instead of the maximum likelihood estimator simply because it is easier to compute. For this example, using the MME would have made very little difference, yielding a test statistic of 0.4147 instead of 0.3132. This is not always the case. Derron (1962) suggested that when the method-of-moments estimators differ from the maximum likelihood estimators, the χ^2 distribution still yields a satisfactory fit to the test statistic. However, Albrecht (1980) noted that in these cases, the asymptotic distribution of the test statistic is not known, but is generally not χ^2. To be conservative, we use the maximum likelihood estimator. However, we computed the estimates of μ and k based on the observed data, not the data after it had been grouped into five classes. To our knowledge, no study exists in the literature that considers the impact of using the MLE of the raw data instead of the grouped data on the asymptotic distribution of the chi-squared test statistic.

Likelihood Ratio Test

Neyman and Pearson introduced the likelihood ratio test, also called the G-test, in 1928. It is sometimes used as an alternative to Pearson's chi-squared test.

The likelihood ratio is the ratio of the likelihood function maximized in the restricted parameter space of the null hypothesis to the maximum of the likelihood function in the unrestricted parameter space. For a goodness-of-fit test, the likelihood function in the restricted parameter space is maximized by computing the likelihood function using the maximum likelihood estimators for the hypothesized distribution. The likelihood function in the unrestricted parameter space is maximized using the observed data. The likelihood ratio is the ratio of these two likelihood functions. By taking (-2) times the natural logarithm of this ratio and simplifying, we obtain a test statistic of the following form:

$$\chi^2_{LR} = 2 \sum_{i=1}^{r} O_i \ln\left(\frac{O_i}{E_i}\right) \tag{2.3}$$

where O_i is the observed frequency in the ith class, E_i is the expected frequency in the ith class, and the summation is over the r classes. The asymptotic distribution of χ^2_{LR} is chi-squared if the data are a random sample from the hypothesized distribution. The degrees of freedom of the test statistic is the same as in the chi-squared test; that is, the degrees of freedom equals $(r - 1 - $ the number of estimated parameters).

Example 2.2

Consider again the sample of 100 seedling oats discussed in Example 2.1. The number of greenbugs on each seedling was observed. We group some of the classes for the same reasons and in the same manner as in Example 2.1. Consequently, the observed and expected frequencies are the same as in that example (Table 2.2). The contribution of each class to the χ^2 test statistic can now be calculated:

$$\chi^2_{LR}(0) = O_0 \ln\left(\frac{O_0}{E_0}\right) = 43 \ln\left(\frac{43}{43.45}\right) = -0.4477$$

$$\chi^2_{LR}(1) = O_1 \ln\left(\frac{O_1}{E_1}\right) = 29 \ln\left(\frac{29}{28.31}\right) = 0.6983$$

$$\chi^2_{LR}(2) = O_2 \ln\left(\frac{O_2}{E_2}\right) = 16 \ln\left(\frac{16}{14.92}\right) = 1.1182$$

$$\chi^2_{LR}(3) = O_3 \ln\left(\frac{O_3}{E_3}\right) = 6 \ln\left(\frac{6}{7.24}\right) = -1.1272$$

$$\chi^2_{LR}(\geq 4) = O_{\geq 4} \ln\left(\frac{O_{\geq 4}}{E_{\geq 4}}\right) = 6 \ln\left(\frac{6}{6.08}\right) = -0.0795.$$

The likelihood statistic is

Table 2.2 *Observed and Expected Frequencies of Greenbugs on Seeding Oats for the Likelihood Ratio Test*

Class (i)	Observed frequency (O_i)	Expected frequency (E_i)
0	43	43.45
1	29	28.31
2	16	14.92
3	6	7.24
≥ 4	6	6.08

$$\chi^2_{LR} = (2)[(-0.4477) + 0.6983 + 1.1182 + (-1.1272) + (-0.0795)]$$
$$= 0.3243.$$

The degrees of freedom associated with the test are

$$df = 5 - 1 - 2 = 2.$$

The distribution of the likelihood ratio test statistic is asymptotically chi-squared under the null hypothesis. The likelihood ratio value of 0.3243 has an associated p-value of 0.8503. (The p-value was found by choosing *Likelihood* from the *Fit* menu as described in Example 2.1.)

Wald (1943) stated that the likelihood ratio test has some advantages due to the asymptotically optimal properties when the error probabilities are bounded away from 0. Hoeffding (1965) noted that, in all simple and some composite hypotheses, the likelihood ratio test was superior to the chi-squared test. However, Larntz (1978) and Wu (1984) concluded that for sample sizes of 100 or less, the likelihood ratio test tends to have a higher type I error rate than the chi-squared test, especially when the mean is large, and that this type I error rate is greater than the stated level. Their conclusions indicated that the likelihood ratio test is less desirable overall than the chi-squared test.

Freeman–Tukey Chi-Squared Test

Freeman and Tukey (1950) reported the use of a goodness-of-fit test based on the statistic

$$\chi^2_{FT} = \sum_{i=1}^{r} (O_i^{0.5} + (O_i + 1)^{0.5} - (4E_i + 1)^{0.5})^2 \qquad (2.4)$$

where O_i is the observed frequency of the ith class, E_i is the expected frequency of the ith class, and r is the number of classes. Under the null hypothesis, the asymptotic distribution of the test statistic is chi-squared with the degrees of freedom being ($r - 1 -$ the number of estimated parameters); that is, the degrees of freedom are the same as those for the two tests discussed previously.

Example 2.3

Consider the same data as in Examples 2.1 and 2.2. A farmer has counted the number of greenbugs on each of 100 seedling oats. The observed and expected frequencies are given in Table 2.3. The contribution of each class to the test statistic may now be computed:

$$\chi^2_{FT}(0) = (43^{0.5} + (43 + 1)^{0.5} - (4(43.45) + 1)^{0.5})^2 = 0.0009$$

$$\chi^2_{FT}(1) = (29^{0.5} + (29 + 1)^{0.5} - (4(28.31) + 1)^{0.5})^2 = 0.0303$$

$$\chi^2_{FT}(2) = (16^{0.5} + (16 + 1)^{0.5} - (4(14.92) + 1)^{0.5}) = 0.1111$$

$$\chi^2_{FT}(3) = (6^{0.5} + (6 + 1)^{0.5} - (4(7.24) + 1)^{0.5})^2 = 0.1431$$

$$\chi^2_{FT}(\geq 4) = (6^{0.5} + (6 + 1)^{0.5} - (4(6.08) + 1)^{0.5})^2 = 0.0040.$$

The Freeman–Tukey chi-squared statistic is

$$\chi^2_{FT} = \chi^2_{FT}(0) + \chi^2_{FT}(1) + \chi^2_{FT}(2) + \chi^2_{FT}(3) + \chi^2_{FT}(\geq 4)$$
$$= 0.0009 + 0.0303 + 0.1111 + 0.1431 + 0.0040 = 0.2897.$$

The degrees of freedom are $5 - 1 - 3 = 2$. Using ECOSTAT as discussed in Example 2.1 and indicating the Freeman–Tukey test, we find a p-value of 0.8651.

For this example, the fit appears even better for this test than it did for Pearson's chi-squared test. Studies indicate that the opposite happens more often. Sylwester (1974) concluded that this test has true significance levels closer to the nominal

Table 2.3   Observed and Expected Frequencies for Greenbugs on Seedling Oats for the Freeman–Tukey Test

Class (i)	Observed frequency (O_i)	Expected frequency (E_i)
0	43	43.45
1	29	28.31
2	16	14.92
3	6	7.24
≥ 4	6	6.08

levels than the regular chi-squared for certain models. However, Larntz (1978) found that under a number of models, the Freeman–Tukey test has a larger significance level than the stated one for small samples, indicating the value of the test statistic tended to be inflated. Wu's (1985) results agreed with those of Larntz. She stated that the Freeman–Tukey test had more type I errors than the chi-squared when applied to the negative binomial distribution. This fact should be kept in mind when applying the Freeman–Tukey test.

Power Divergence Statistic

Pearson's chi-squared test, the likelihood ratio test, and Freeman–Tukey test are all members of the power divergence family of test statistics (Cressie and Read, 1984; Read and Cressie, 1988). The general form of the test statistics for members of this family is

$$I^\lambda = \frac{2}{\lambda(\lambda + 1)} \sum_i O_i \left[\left(\frac{O_i}{E_i} \right)^\lambda - 1 \right].$$ (2.5)

If $\lambda = 1$, the test is equivalent to Pearson's chi-squared test statistic. If $\lambda \to 0$, the test statistic converges to the likelihood ratio test statistic X_{LR}^2. Further, if $\lambda = -0.5$, it is equivalent to the Freeman–Tukey test statistic. Cressie and Read (1984) investigated the properties of this family of statistics. Under regularity conditions, each member of the family has the same asymptotic distribution. (Recall that we have had the same asymptotic distribution for Pearson's chi-squared test, the likelihood ratio test, and the Freeman–Tukey test). Cressie and Read (1984) recommended using $\lambda = 2/3$ because the chi-squared approximation to the sampling distribution under the null hypothesis is superior at this point.

Example 2.4

Consider the data used in the previous three examples again. Notice that the expected values do not change. The contribution of each class to the test statistic is now

$$\chi_{CR}^2(0) = 43 \left[\left(\frac{43}{43.45} \right)^{0.6667} - 1 \right] = -0.2973$$

$$\chi_{CR}^2(1) = 29 \left[\left(\frac{29}{28.31} \right)^{0.6667} - 1 \right] = 0.4668$$

$$\chi_{CR}^2(2) = 16 \left[\left(\frac{16}{14.92} \right)^{0.6667} - 1 \right] = 0.7642$$

$$\chi^2_{CR}(3) = 6\left[\left(\frac{6}{7.24}\right)^{0.6667} - 1\right] = -0.7064$$

$$\chi^2_{CR}(\geq 4) = 6\left[\left(\frac{6}{6.08}\right)^{0.667} - 1\right] = -0.0513.$$

The power divergence test statistic with $\lambda = 0.6667$ is

$$\chi^2_{CR} = 1.8(\chi^2_{CR}(0) + \chi^2_{CR}(1) + \chi^2_{CR}(2) + \chi^2_{CR}(3) + \chi^2_{CR}(\geq 4)) = 0.1408.$$

The 1.8 above is the constant in front of the summation in equation (2.5) evaluated at $\lambda = 0.6667$. Using ECOSTAT to conduct the test, we find the p-value associated with the test statistic is 0.9320, indicating a very good fit of the negative binomial distribution to the data.

Nass Test

Nass (1959) suggested that a correction factor to the chi-squared test be used when expectations are small:

$$\chi^2_{Na} = C\chi^2_{P}. \tag{2.6}$$

The correction factor C for χ^2_{Na} is defined as

$$C = \frac{2E(\chi^2_P)}{V(\chi^2_P)} \tag{2.7}$$

where $E(\chi^2_P)$ and $V(\chi^2_P)$ are, respectively, the mean and variance of χ^2_P. Nass developed approximations for the mean and variance of χ^2_P. He showed that

$$E(\chi^2_P) \approx r - s - 1 \tag{2.8}$$

and

$$V(\chi^2_P) \approx \left(\frac{r - s - 1}{r - 1}\right)\left[2(r - 1) - \frac{(r^2 + 2r - 2)}{n} + \sum_{i=1}^{r} E_i^{-1}\right] \tag{2.9}$$

where n is the sample size, E_i is the expectation of the ith class, r is the number of classes, and s is the number of estimated parameters. The distribution of χ^2_{Na} is asymptotically chi-squared with q degrees of freedom under the null hypothesis, where

$$q = CE(\chi_P^2). \tag{2.10}$$

Example 2.5

Again consider the data discussed in Examples 2.1, 2.2, 2.3, and 2.4. The farmer determined the number of greenbugs on each of 100 seedling oats. The observed and expected frequencies are given in Table 2.4.

Notice that we permitted expected frequencies below 5 without grouping. Although Nass suggests that the expected frequencies may be below 1, we will not. We need to compute the contributions to χ^2 for the classes of 4, 5, and ≥ 6, as these were combined in the earlier examples:

$$\chi_P^2(4) = \frac{(O_4 - E_4)^2}{E_4} = \frac{(3 - 3.36)^2}{3.36} = 0.0386$$

$$\chi_P^2(5) = \frac{(O_5 - E_5)^2}{E_5} = \frac{(1 - 1.52)^2}{1.52} = 0.1779$$

$$\chi_P^2(\geq 6) = \frac{(O_{\geq 6} - E_{\geq 6})^2}{E_{\geq 6}} = \frac{(2 - 1.20)^2}{1.20} = 0.5333.$$

Then, χ_P^2 is

$$\chi_P^2 = \chi_P^2(0) + \chi_P^2(1) + \chi_P^2(2) + \chi_P^2(3) + \chi_P^2(4) + \chi_P^2(5) + \chi_P^2(\geq 6)$$
$$= 0.0047 + 0.0168 + 0.0782 + 0.2124 + 0.0386 + 0.1779 + 0.5333$$
$$= 1.0619.$$

The reason the chi-squared test has the guideline that no more than 20% of the classes should have expected frequencies of less than 5 is clearly demonstrated

Table 2.4 *Observed and Expected Frequencies for Greenbugs on Seedling oats for the Nass Test*

Class (i)	Observed frequency (O_i)	Expected frequency (E_i)
0	43	43.45
1	29	28.31
2	16	14.92
3	6	7.24
4	3	3.36
5	1	1.52
≥ 6	2	1.20

here. When compared to Example 2.1, two more classes were added, resulting in three of the seven classes having expected frequencies of less than 5. Although each of the observed frequencies for classes 4, 5, and ≥ 6 is close to the corresponding expected frequencies, differing by less than 1, the value of the test statistic more than tripled, increasing from 0.3131 in Example 2.1 to 1.0619 here. Because the contribution to the chi-squared for each class involves dividing by the expected frequencies, small expected frequencies accentuate the differences in observed and expected frequencies. Notice that the degrees of freedom for χ_P^2, as computed here, are now the number of classes minus the number of estimated parameters minus 1 or $7 - 2 - 1 = 4$, two more degrees of freedom than in Example 2.1.

The correction factor is designed to correct for the inflation in χ_P^2 while keeping the increased power from the larger number of classes. We now determine the value of the correction factor, C. To calculate C, we must first approximate the mean and variance of χ_P^2. The approximate mean is

$$E(\chi_P^2) \approx r - s - 1 = 7 - 2 - 1 = 4.$$

where $r = 7$ is the number of classes and $s = 2$ is the number of estimated parameters. The approximate variance is

$$V(\chi_P^2) \approx \left(\frac{r - s - 1}{r - 1}\right)\left(2(r - 1) - \frac{(r^2 + 2r - 2)}{n} + \sum_{i=1}^{r} E_i^{-1}\right)$$

$$\approx \left(\frac{7 - 2 - 1}{7 - 1}\right)\left(2(7 - 1) - \frac{(7^2 + 2(7) - 2)}{100}\right.$$

$$\left. + \left(\frac{1}{43.45} + \frac{1}{28.31} + \frac{1}{14.92} + \frac{1}{7.24} + \frac{1}{3.36} + \frac{1}{1.52} + \frac{1}{1.20}\right)\right)$$

$$= 8.9616.$$

Then the correction factor C is

$$C = \frac{2E(\chi_P^2)}{V(\chi_P^2)} = \frac{2(4)}{8.9616} = 0.8927.$$

Therefore, the value of the test statistic is

$$\chi_{Na}^2 = C\chi_P^2 = 0.8927(1.0619) = 0.9479.$$

The degrees of freedom associated with this test statistic are approximately

$$q = CE(\chi_P^2) = (0.8927)(4) = 3.57.$$

As shown in this example, it is not necessary for the degrees of freedom of a chi-squared random variable to be an integer. Fractional degrees of freedom are not available in tables but are accepted by ECOSTAT. In our example, the value of the test statistic, 0.9479, has a p-value of 0.8817.

Wu (1985) found the behavior of the Pearson chi-squared test and of the Nass test to be quite similar. However, in that study, the data were grouped in classes with expected frequencies of five or more. The Nass test is designed to address those cases for which the expected value within one or more cells is small. Therefore, the full power of the Nass test was not explored, and its behavior may not be the same when grouping is not performed.

Kolmogorov–Smirnov Test

The chi-squared, likelihood ratio, Freeman–Tukey, and Nass goodness-of-fit tests could be used to test the fit of continuous distributions to data. The next example illustrates the use of the chi-squared test in this manner.

Example 2.6

One hundred newborn female guppies were placed in an aquarium and exposed to a pathogen. At 10:00 a.m. each day, the dead guppies were removed from the tank and counted. This process continued until all guppies were dead. The results are presented in Table 2.5.

First note that the underlying survival time distribution is continuous. The fact that deaths were only recorded on a daily basis has resulted in the data having

Table 2.5   *Number of Dead Guppies Found Each Day After 100 Newborn Female Guppies Are Placed in an Aquarium*

Day	No. of dead guppies	Day	No. of dead guppies
1	20	14	2
2	14	15	2
3	10	16	1
4	8	17	2
5	6	18	1
6	5	19	1
7	4	20	1
8	5	21	2
9	3	22	1
10	2	23	1
11	3	24	0
12	2	25	1
13	2	26	1

the appearance of being discrete. However, only continuous distributions will be considered as possible models for the survival time distribution. As discussed in Chapter 1, several distributions have been used to model survival time. For this example, we test whether the exponential distribution provides an adequate fit.

The exponential distribution has one parameter, λ. Since it is unknown, λ must be estimated from the data. To do this, we must first compute the sample mean, the average number of days that the guppies survived. By summing the number of days each survived and dividing by the total number of guppies, we find that the guppies survived an average of 6.67 days. The estimated value of λ can now be found using

$$\hat{\lambda} = \frac{n-1}{n\bar{x}} = \frac{99}{100(6.67)} = 0.148.$$

Based on the estimate of λ, the estimated exponential distribution function has the form

$$F_x^*(x) = P(X \le x) = 1 - e^{-0.148x}, \qquad x > 0.$$

The expected frequencies can now be computed. To do this, first recall how the data were recorded. The guppies that died between the time they entered the tank and 10:00 a.m. the next day were recorded as having died on day one. This corresponds to the probability they survive from 0 to 1 day; that is,

$$\begin{aligned} P(0 \le X \le 1) &= F_X^*(1) - F_X^*(0) \\ &= (1 - e^{-0.148(1)}) - (1 - e^{-0.148(0)}) = 0.1376 - 0 = 0.1376. \end{aligned}$$

The expected frequency of the number of dead guppies on the first day is then

$$E_1 = nP(0 \le X \le 1) = 100 \times 0.1376 = 13.76.$$

As another example of the computations, consider the number of guppies expected to die on the 10th day. The probability of the ith guppy being observed dead on day 10 is estimated to be

$$\begin{aligned} P(9 \le X \le 10) &= F_X^*(10) - F_X^*(9) \\ &= (1 - e^{-0.148(10)}) - (1 - e^{-0.148(9)}) = 0.7724 - 0.7361 \\ &= 0.0363. \end{aligned}$$

The expected frequency of the number of dead guppies on the tenth day is then

$$E_{10} = nP(9 \le X \le 10) = 100 \times 0.0363 = 3.63.$$

These computations result in the observed and expected frequencies shown in Table 2.6.

The last class (≥ 26) includes the expected frequency associated with dying on or after the 26th day in the aquarium; that is, the probability that a guppy will survive more than 25 days is

$$P(X > 25) = 1 - F_X^*(25) = 1 - (1 - e^{-0.148(25)}) = 0.0247.$$

The expected frequency is then $100(0.0247) = 2.47$.

Some grouping needs to be done but, as is usually the case with continuous distributions, the choice of groups is not evident. It has been shown that it is better to have about the same expected frequency within each class after grouping than it is to have about the same number of days within each class. Therefore, after grouping, we have the values shown in Table 2.7.

In Table 2.7, O_i represents the observed frequency in the ith class, E_i is the expected frequency in the ith class, and $\chi_P^2(i)$ is the contribution of the ith class to the chi-squared test statistic. As an example of the computations made for this last quantity, we calculate the contribution to the chi-squared test statistic made by the data in the class of day 7:

$$\chi_P^2(7) = \frac{(O_7 - E_7)^2}{E_7} = \frac{(4 - 5.66)^2}{5.66} = 0.487.$$

Then, the total value of the test statistic is the sum of these individual contributions to the chi-squared statistic; that is,

Table 2.6 *Observed and Expected Frequencies for Dead Guppies, Based on the Exponential Distribution*

Day	Observed	Expected	Day	Observed	Expected
1	20	13.76	14	2	2.01
2	14	11.86	15	2	1.73
3	10	10.23	16	1	1.49
4	8	8.82	17	2	1.29
5	6	7.61	18	1	1.11
6	5	6.56	19	1	0.96
7	4	5.66	20	1	0.83
8	5	4.88	21	2	0.71
9	3	4.21	22	1	0.61
10	2	3.63	23	1	0.53
11	3	3.13	24	0	0.46
12	2	2.70	25	1	0.39
13	2	2.33	≥ 26	1	2.47

Table 2.7 Observed and Expected Frequencies of Dead Guppies After Grouping

Day (i)	O_i	E_i	$\chi_P^2(i)$	Day (i)	O_i	E_i	$\chi_P^2(i)$
1	20	13.76	2.83	8	5	4.88	0.003
2	14	11.86	0.39	9, 10	5	7.84	1.029
3	10	10.23	0.005	11, 12	5	5.83	0.118
4	8	8.82	0.076	13, 14, 15	6	6.07	0.001
5	6	7.61	0.341	$16 \le X \le 20$	6	5.68	0.018
6	5	6.56	0.371	$X \ge 21$	6	5.17	0.133
7	4	5.66	0.487	—	—	—	—

$$\chi_P^2 = \sum_i \chi_P^2(i) = 5.798.$$

If the data are, in fact, a random sample from an exponential distribution this test statistic has an approximate χ^2 distribution with $(13 - 1 - 1) = 11$ degrees of freedom. Using ECOSTAT, we find that the p-value associated with our test statistic is 0.8865, indicating that the exponential distribution is a good fit for these data.

The grouping in the above example is not unique. The optimal grouping is not evident. However, the grouping we used is reasonable given the guidelines associated with the use of the chi-squared test.

When the population distribution is continuous, an alternative to the chi-squared test is the Kolmogorov–Smirnov test. Instead of forcing the data into arbitrary groupings, the continuous nature of the data is exploited in the Kolmogorov–Smirnov test. To perform the test, we must first compute the empirical and hypothesized distribution functions. We begin by defining these functions.

The empirical distribution function, $S(x)$, provides an estimate of the population distribution function without making any distributional assumptions. $S(x)$, a function of x, equals the fraction of the sample values that are less than or equal to x for any real value of x.

Example 2.6

Consider again the data given in Example 2.5. One hundred guppies are placed in the aquarium at the onset of the experiment. The time until each dies is recorded. Let X_i be the day on which the ith guppy was observed dead. Then $S(1)$ equals the fraction of guppies that have been observed dead on or before day 1; therefore, $S(1) = 10/100 = 0.1$. Similarly, $S(10)$ equals the fraction of guppies dead on or before day 10; that is, $S(10) = 77/100 = 0.77$. $S(x)$ may be computed for each day of the experiment and is given in Table 2.8.

Table 2.8 Empirical Distribution Function for the Number of Dead Guppies

Day	No. of dead guppies	$S(x)$	Day (x)	No. of dead guppies	$S(x)$
1	20	0.20	14	2	0.86
2	14	0.34	15	2	0.88
3	10	0.44	16	1	0.89
4	8	0.52	17	2	0.91
5	6	0.58	18	1	0.92
6	5	0.63	19	1	0.93
7	4	0.67	20	1	0.94
8	5	0.72	21	2	0.96
9	3	0.75	22	1	0.97
10	2	0.77	23	1	0.98
11	3	0.80	24	0	0.98
12	2	0.82	25	1	0.99
13	2	0.84	26	1	1.00

Notice that the empirical distribution function did not change values from day 23 to day 24 because no guppy was observed to have died on day 24. Also, the value of the empirical distribution function does not change between sample values. For example, $S(5.5) = S(5)$ because the same number of guppies has been observed dead after 5.5 days as were observed dead after 5 days. True, more may be dead and not yet observed, but we have no way of knowing that. A graph of the empirical distribution function is shown in Figure 2.1.

The empirical distribution function is compared to the hypothesized distribution function in the Kolmogorov–Smirnov test. Often the form of the distribution function is hypothesized, but the parameters must be estimated. These estimated parameters are used to estimate the hypothesized distribution function.

Example 2.7

Again consider the data in Examples 2.5 and 2.6. Suppose the hypothesis to be tested is that the number of days until death for a newborn guppy, exposed to this pathogen, is exponentially distributed. The average time the 100 guppies in the sample survived was 6.67 days. Then the parameter λ was estimated to be 0.148. Therefore, the hypothesized distribution function has the form

$$F(x) = 1 - e^{-0.148x}, \qquad x > 0.$$

This hypothesized distribution function is shown in Figure 2.2.

Kolmogorov proposed using the greatest distance between the empirical and hypothesized distribution functions in the vertical direction as the test statistic, T. The null hypothesis is rejected at the α significance level if T exceeds the $1 -$

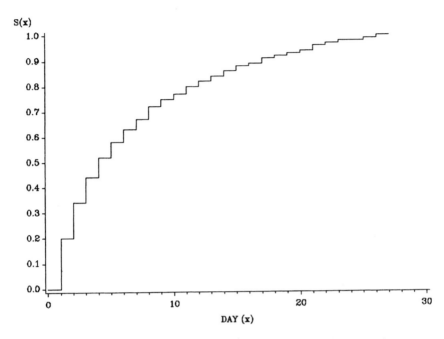

Figure 2.1. Empirical distribution function $S(x)$ for the number of dead guppies.

α quantile $w_{1-\alpha}$ as given in many references (*see* Conover 1972). The tabulated values are exact for $n \leq 20$ and are good approximations for $20 < n \leq 40$. For $n > 40$, the approximations are based on the asymptotic distribution of the test statistic and are not very accurate until n becomes large.

Example 2.8

One last time, let us consider the data in Examples 2.5, 2.6, and 2.7. For these data, we have the following values of the empirical and hypothesized distributions for each day of the experiment (*see* Table 2.9).

Both the empirical distribution function $S(x)$ and the hypothesized distribution function $F^*(x)$ are depicted in Figure 2.3. Note that the greatest distance between $S(x)$ and $F^*(x)$ in the vertical distance will occur immediately after a change in $S(x)$ if $S(x)$ lies above $F^*(x)$ or immediately before a change in $S(x)$ if $S(x)$ lies below $F^*(x)$. For this example, $S(x)$ and $F^*(x)$ differ by $T = 0.138$ for $X = 1$. This is the point at which the difference in the two functions is the greatest, and thus this is the value of the test statistic, T. From the table we can only conclude that the p-value exceeds 0.20 because 0.138 is less than the approximate tabled value of 0.152. Therefore, we do not reject the null hypothesis and conclude that the exponential distribution fits the data within error.

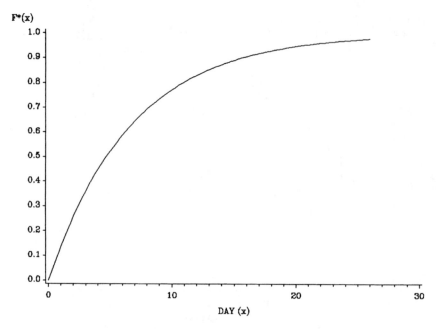

Figure 2.2. Hypothesized distribution function for the number of dead guppies.

Table 2.9 Empirical and Hypothesized Distribution Functions for the Number of Dead Guppies

Day	S(x)	F*(x)	Day (x)	S(x)	F*(x)
1	0.20	0.138	14	0.86	0.874
2	0.34	0.256	15	0.88	0.891
3	0.44	0.358	16	0.89	0.906
4	0.52	0.447	17	0.91	0.919
5	0.58	0.523	18	0.92	0.930
6	0.63	0.589	19	0.93	0.940
7	0.67	0.645	20	0.94	0.948
8	0.72	0.694	21	0.96	0.955
9	0.75	0.736	22	0.97	0.961
10	0.77	0.772	23	0.98	0.967
11	0.80	0.804	24	0.98	0.971
12	0.82	0.831	25	0.99	0.975
13	0.84	0.854	26	1.00	0.979

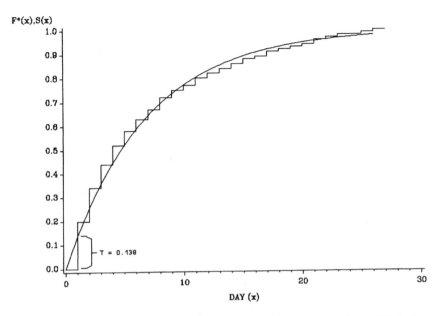

Figure 2.3. The greatest distance *T* between the hypothesized and empirical distribution functions for the number of dead guppies.

The Kolmogorov–Smirnov test is exact when the population distribution is continuous. It is conservative when the population distribution is discrete; that is, the stated *p*-value associated with the test tends to be greater than the true significance level of the test (Conover, 1972). This results in the hypothesized distribution not being rejected as frequently as it should be. We tend to use the Kolmogorov–Smirnov test when the underlying distribution is continuous.

Summary

Goodness-of-fit tests are used to determine whether a particular population distribution fits the data. More than one distribution may adequately describe a data set. Often, the distribution that fits the best (i.e., has a smaller value of the test statistic) is chosen. The difficulty with this approach is that as other data sets are collected at different times and/or locations, other distributions may provide a better fit. This may reflect a change in the population distribution, but not necessarily. Instead of seeking the distribution that best fits each set of data, it is better to search for a distribution that consistently fits data from a given ecosystem. At times, the distribution may fit better than any other. For other data sets, the distribution may describe the data well, but another distribution fits better.

Occasionally, we conclude that the distribution does not fit the data. This last conclusion is expected for one in every 20 data sets if we consistently use a 5% significance level in testing *and the null hypothesis correctly specifies the distribution*. Through this broader look at data from the ecosystem, we can gain insight into the population distribution.

Sample size is another important consideration in goodness-of-fit tests. It has been said that if the sample is too small, all distributions will fit the data; if the sample size is too large, no distribution will fit the data. As the sample size increases, the sample distribution must get closer to the population distribution for the null hypothesis not to be rejected. For very large samples, even small differences in the sample and hypothesized distributions will lead to rejecting the null hypothesis. Caution should be used in such cases as these small differences may not be sufficient to keep the hypothesized distribution from serving as a useful model.

Exercises

ECOSTAT software can be used for these exercises.

1. A large study on the Southern pine beetle, *Dendroctonus frontalis* Zimmerman (Coleoptera: Scolytidae), in Southeast Texas was conducted over a 5-year period: 1975–77, 1982–83 (Lin, 1985, as reported by Gates, 1989). The Southern pine beetle infests all species of southern yellow pine. The adults mine under the bark of the pine trees and eventually girdle the trees. For the brood to develop, the pines must be killed. Suppose a tree is colonized. Then, depending on site, stand, and weather conditions, many adjacent pines may also be colonized. The subsequent damage is great. In 1976, the economic loss was $14 million in stumpage alone in state and private forests. The data set below is the number of "spots" of infestations in 5' × 5' geographic "rectangles," which are roughly 8 km square. Initial counts are from aerial photographs, but the counts were verified by ground crews before a new spot was considered infested. The data were collected in September 1976.

No. of spots	Frequency
0	1,169
1	144
2	92
3	54
4	29
5	18
6	10
7	12
8	6

No. of spots	Frequency
9	9
10	3
11	2
12	0
13	0
14	1
15	0
16	0
17	0
18	0
19	1

The researcher wants to know the distribution of spots in the sampled forest. Investigate this for him. When you have completed your work, report what you did and your conclusions.

2. This data set is from study of the sorghum midge, *Contarinia sorghicola* (Coquillett), (Diptera: Cecidomyiidae) (Gates, 1989). This insect, a pest of sorghum, causes the greatest economic damage of all sorghum pests in the United States. The sorghum midge is a small fly that lays eggs in the flowers of sorghum plants. Larvae feed on the developing seed, causing yield loss. A panicle (seed head) was the sampling unit. The number of sorghum midges/panicle was recorded. Systematic sampling was used; every 50 paces a single plant was sampled. The data were taken from a field near Corpus Christi, Texas, during the summer of 1987.

No. of sorghum midges/panicle	Frequency
0	46
1	25
2	4
3	0
4	1

The researcher wants to know the distribution of sorghum midges on the panicles. Investigate this for him. When you have completed your work, report what you did and your conclusions. Do you have any suggestions?

3. A study of the pecan weevil, *Curculio caryae* (Horn), (Celeoptera: Curculionidae), was conducted (Ring, 1978, as reported in Gates, 1989). Random samples of pecan nut clusters composed of approximately 100 nuts were taken twice a week from August 5 to August 29 and once a week from September 1 to November 1 in 1975. With rare exceptions, the number of weevils per nut appear to have resulted from a single oviposition event by

one adult female. The pecans were from native, rather than cultivated, trees located near Hamilton, Texas. The results of the sampling are presented in the table below:

No. of weevils per nut	Frequency
0	9,313
1	240
2	409
3	606
4	496
5	153
6	28
7	14
8	1
9	1
10	1

A. Express any concerns you have about finding a distribution to describe these data (Yes, you should have some concerns.)

B. Try to find a feasible distribution from which this sample was drawn. Explain any aspects of the data set that might make this difficult to do.

4. During June 1991, a Nebraska soybean field was sampled for the presence of pigweed. The sample unit was 1 meter of linear row (G.A. Johnson, personal communication). The number of pigweeds within each sample unit was counted with sample units occurring on a grid. The data are displayed in the table below.

No. of Pigweeds	Frequency
0	84
1	42
2	37
3	26
4	14
5	12
6	6
7	5
8	5
9	3
10	3
11	2
12	0
13	1
14	0
15	1

No. of Pigweeds	Frequency
16	0
17	0
18	1
19	0
20	0

The researcher was interested in the distribution of pigweeds in the field. Investigate this on his behalf. Report your work and conclusions.

5. An alfalfa field had an increasing population of alfalfa weevil larvae (Young and Willson, 1987). On three different occasions samples of the number of alfalfa weevil larvae on an alfalfa terminal were collected. The first sample was taken in the early stages of the infestation. The second sample was drawn after the population had been allowed to build for a period of time. The last sample was taken when the population had become large, and the crop was suffering heavy damage.

Observed no. of larvae	Frequency for data set		
per terminal	I	II	III
0	100	36	16
1	37	35	42
2	15	24	37
3	5	16	43
4	2	9	38
5	0	6	12
6	1	3	1
7	0	2	0
8	0	0	0
9	0	1	1

Investigate the possible population distributions from which these samples were drawn. Does it appear that the form of the distribution is changing through time, or will a single distribution adequately fit all three data sets? Give a possible biological interpretation for what you have observed.

6. Find a journal article in your field of study that discusses the application of a goodness-of-fit test. Copy the paper. Describe the purpose of sampling, the goodness-of-fit test used, and the conclusions.

7. Reanalyze the data in problem 1 using

   A. the likelihood ratio test

   B. the Freeman–Tukey test

   C. the Nass test for the data grouped as in problem 1

    D.   the Nass test using less grouping (do not let an expectation drop below 1)

8. Reanalyze the data in problem 2 using
    A.   the likelihood ratio test
    B.   the Freeman–Tukey test
    C.   the Nass test for the data grouped as in problem 2
    D.   the Nass test using less grouping (do not let an expectation drop below 1)

9. Reanalyze the data in problem 3 using
    A.   the likelihood ratio test
    B.   the Freeman–Tukey test
    C.   the Nass test for the data grouped as in problem 3
    D.   the Nass test using less grouping (do not let an expectation drop below 1)

10. Reanalyze the data in problem 4 using
    A.   the likelihood ratio test
    B.   the Freeman–Tukey test
    C.   the Nass test for the data grouped as in problem 4
    D.   the Nass test using less grouping (do not let an expectation drop below 1)

11. Reanalyze the data in problem 5 using
    A.   the likelihood ratio test
    B.   the Freeman–Tukey test
    C.   the Nass test for the data grouped as in problem 5
    D.   the Nass test using less grouping (do not let an expectation drop below 1)

Appendix

DISCRETE is a FORTRAN program developed by Gates and Ethridge (1972) and Gates et al. (1987). It is designed to fit the following distributions: Poisson, negative binomial, binomial, Thomas double Poisson, Neyman type A, Poisson-binomial, Poisson with zeroes, and logarithmic with zeroes.

For the Poisson, the probability mass function is taken to be

$$P(X = x) = \frac{e^{-\mu}\mu^x}{x!}, \qquad x = 0, 1, 2, \ldots$$

Notice that this is the same as in equation (1.34), with μ used here instead of λ.

The negative binomial distribution is parameterized in terms of k and $P\,(=\mu/k)$ as in equation (1.21). The binomial distribution has parameterization as in equation (1.31).

The Thomas double Poisson distribution (Thomas, 1949) is defined as

$$P(X = x) = e^{-m}, \qquad x = 0$$
$$= \sum_{r=1}^{x} \frac{m^r e^{-m}}{r!} \left(\frac{(r\lambda)^{x-r} e^{-r\lambda}}{(x-r)!} \right), \qquad x = 1, 2, 3, \ldots$$

where m and λ are positive. Note that if $\lambda = 0$, the distribution degenerates to the Poisson.

The Neyman type A distribution (Douglas, 1955) is

$$P(X = x) = \frac{e^{-m_1} m_2^x}{x!} \sum_{j=0}^{\infty} \frac{(m_1 e^{-m_2}) j^x}{j!}, \qquad x = 0, 1, 2, \ldots$$

where $m_1 > 0$ and $m_2 > 0$.

The probability mass function for the Poisson-binomial distribution is

$$P(X = x) = e^{-\alpha} \sum_{j \geq x/n} \frac{\alpha^j}{j!} \binom{nj}{x} p^x q^{nj-x}$$

where $\alpha > 0$, $n > 0$, $0 < p < 1$, and $q = 1 - p$.

The Poisson distribution with zeroes (Cohen, 1960) is defined as

$$P(X = x) = 1 - \theta, \qquad x = 0$$
$$= \frac{\theta e^{-\lambda} \lambda^x}{(1 - e^{-\lambda}) x!}, \qquad x = 1, 2, 3, \ldots$$

Notice that if $\theta = 0$, the distribution is degenerate at $x = 0$. If $\theta = 1$, the distribution is Poisson with the zeroes truncated. If $\theta = 1 - e^{-\lambda}$, the distribution is Poisson.

The logarithmic distribution with zeroes (Chakravarti 1967) is defined as

$$P(X = x) = 1 - \lambda, \qquad x = 0$$
$$= \frac{\lambda \theta^x}{(-x \ln(1 - \theta))}, \qquad x = 1, 2, 3, \ldots$$

where $0 \leq \lambda \leq 1$ and $0 < \theta < 1$.

To run DISCRETE, one must first construct a data file. The first line of the data file has the number of problems, N, in columns 1–4, right-justified, to be analyzed. The second line is for a title that can be up to 80 characters long. The third line is comprised of 20 fields of length four, all numbers are right-justified. An entry is required in the first four columns, but all others are optional.

NO, the number of cells per problem beginning with the zero class, is entered in the first four columns. *There must be a line for every cell between 0 and the last class filled, inclusive, including all vacant cells.* Thus if *M* were the largest number of muskrats observed in a quadrat, $NO = M + 1$. DISCRETE tests for the data input sequence and prints an error message for each error.

MAXCLS (columns 5–8) specifies the maximum class value to be used. This value may be either greater than the largest class in the data set or smaller than the largest class in the data set. In the latter case, if the TRUN (truncate) option is not used, all classes larger than MAXCLS are pooled into that class; if TRUN is used, all values greater than MAXCLS are truncated. MAXCLS does not affect other computations. It is a rough approximation to fitting a truncated distribution. *Default:* MAXCLS = NO − 1.

MMM (columns 9–12) represents the maximum number of iterations in estimating parameters wherever iterative techniques are required. *Default:* MMM = 50.

IEXP (columns 13–16) is a convergence criterion used in iterative methods. The algorithm will terminate when $|\phi_i - \phi_{i-1}| \leq 10^{(-\text{IEXP})}\phi$. *Default:* IEXP = 6.

ISW (columns 17–20) is 1 if the data are grouped. *Default:* data not grouped. Grouped data, as meant here, occurs when counts are so large and hence sparse that it is necessary to group the data into intervals (otherwise most cells would have zero frequencies). The data must then be entered as follows: mid-class value (read as an *F*-field, inclusive upper class limit, and frequency. Thus we could have (F4.1,2I4). At present, the option for grouped data works only for the Poisson distribution. If the option for grouped data is used, all other distributions are automatically skipped.

The next 8 fields are used to specify which distributions are fit to the data. If the field is left blank the distribution is automatically fit. If a 1 is placed in the field, the distribution is not fit. The distributions and their corresponding columns are as follows: Poisson (columns 21–24), negative binomial (columns 25–28), binomial (columns 29–32), Thomas double Poisson (columns 33–36), Neyman type A (columns 37–40), Poisson-binomial (columns 41–44), Poisson with zeroes (columns 45–48), and logarithmic with zeroes (columns 49–52).

MO (columns 53–56) is the number of calculations of the Poisson-binomial minus 1. MO must be an integer of 2 or larger. The values of *N* in the Poisson-binomial take on the integer values 2, 3, . . . , MO). *Default:* MO = 5.

JSW (column 60) determines whether plots of observed (*O*) and expected (*E*) values for each distribution are printed. A Kolmogorov–Smirnov test of the goodness-of-fit is completed automatically. Printing plots increases the execution time and the amount of computer output. The default is to print plots. If no plots are to be printed, set JSW = 1.

KSW (column 64) is used when a variable format card is desired. The default is KSW = 0 and implies that the format is (2I4). In this case, no variable format card is permitted. If KSW $\neq$ 0, a variable format card, such as (18X, I4, 10X, I4), is required. Note that the variable format card is required if raw data are being read as opposed to a frequency table. If the class number and observed frequency are entered in columns 1–4 and 5–8; respectively, no format card is required. If the class number and observed frequency are in any other format, the format card is required. The word "FORMAT" is not permitted.

POOL (columns 65–68) indicates the amount of pooling that is to be done. The default, POOL = 0, indicates that adjacent classes with small expected frequencies are to be pooled until the cumulative frequency exceeds 1. If POOL $\neq$ 0, the adjacent classes with small expected frequencies are pooled until the cumulative frequency exceeds POOL.

ITRUN (column 72) has a default value of 0, indicating no truncation. If ITRUN = 1, classes greater than MAXCLS are deleted instead of being pooled into the maximum class. ITRUN is operational only if a nondefault value of MAXCLS is specified.

ISKIP (column 76) has a default value of 0, indicating that intermediate computations in iterative methods are not to be printed. If ISKIP = 1, these intermediate computations are printed.

If KSW $\neq$ 0 in column 64, then a variable format card is required; otherwise, it is not needed. Integer format is required for the data.

The next MAXCLS + 1 lines have the data. For each data line, the class value is in columns 1 to 4, and the class frequency is in columns 5 to 8.

Although the instructions are lengthy, setting up the data file can be quite simple. For the greenbug data given in Example 2.1, the data file has the following form:

> *1*
> *GREENBUGS ON SEEDLING OATS*
> *7*
> *0 43*
> *1 29*
> *2 16*
> *3 6*
> *4 3*
> *5 1*
> *6 1*
> *7 1*

In this case, we only have one set of data as indicated on the first line. The problem is titled "Greenbugs on Seedling Oats." The largest observed value is (third line, column 4). We have decided to accept all default settings so all distributions will be fit, grouping will be done until the cumulative expected frequencies are at least one, and plots of observed and expected values will be produced.

DISCRETE can be modified to read raw data instead of frequency tables. For further discussion, *see* Gates et al. (1987).

Before running the program, the data file must be constructed. Remember that DISCRETE is a DOS program. To begin the program, select DISCRETE from ECO-STAT. You will be asked for the name of the data file in single quotes. Then you will be asked for the name of the output file in single quotes. Be careful at this point. If the output name is that of an existing file, the original file will be totally lost. In its stead will be the new output file.

If you have difficulty running the program, do not despair. We have found the most common errors are in entering the original data file. Check to be sure that the entries are in the appropriate columns and that no lines are missing. Another common error is failing to enter the input and output data file names in single quotes. Finally, be sure that the directory being accessed is the one that contains the program and its files.

3

Models and Sampling

Introduction

Population ecology has struggled with the problems of distribution throughout its history. Just how do you view or model the movement of organisms in the environment? In this chapter, several ways to approach this problem are given. In addition, we review some basic sampling concepts and terminology.

Population ecology is central to the understanding of ecosystems, problems of pollution, and contamination, and long-term effects of human activities on the environment. The environmental questions are many and varied, but they will not be solved until population dynamics are better defined and reliable databases are established. Often we build models based on existing data sets and find that output is meaningless because of the wide variation in the data sets used. When several data sets are needed to build a model, the errors are confounded to a point of unreality. Most of the plant growth, insect pests, and insect predator models have limitations, and most of their limitations are parameter-estimation related. The study of chaos and statistical mechanics offers new insights into these problems. As has been pointed out earlier, the binomial, Poisson, negative binomial, and geometric distributions are central to population dynamics studies. In this chapter, we begin by discussing historical models that give rise to these distributions. These are extended, using principles from physics, to include models of immigration, emigration, and movement. Finally, some concepts of sampling are introduced. It should be noted that the models and simulations in this chapter are focusing on the probability distribution of the organisms and not the change in population density through time.

Binomial Models

The binomial is perhaps the most frequently used discrete distribution because of the following model. Let X be the number of successes in n independent trials, each with the probability p of success. Then X has a binomial distribution with parameters n and p. This model has been used extensively in presence–absence sampling. Is the corn plant damaged or undamaged? Is there at least one insect, or is there no insect on the plant? Is the organism alive or dead, fertile or not fertile, male or female? The exact conditions of the model, independence of trials and a constant probability p of success, are rarely satisfied. Yet, the binomial distribution often gives a sufficiently accurate representation. Other models give rise to the binomial distribution (*see* Johnson et al. 1992), but they have not been broadly applied in ecology.

Poisson Models

Poisson published the derivation of the Poisson distribution as a limiting form of the binomial distribution in 1837 (*see* also Johnson et al., 1992). An example of a biological model in keeping with this derivation follows. Suppose that each sampling unit contains a large number n of microhabitats, each of which can be occupied by a single organism. Each microhabitat has the same probability p of being occupied. Then the probability of exactly $X = x$ microhabitats in any one sampling unit being occupied is binomial with parameters n and p. Now assume that n is large and p is small, but the mean of the distribution $\lambda = np$ is of moderate magnitude. Then, as $n \to \infty$, the probability of observing $X = x$ organisms within a sampling unit is Poisson with parameter λ. As Pielou (1977) pointed out, this model assumes that the maximum number of organisms within a sampling unit and the expected number of organisms per sampling unit is the same for all sampling units. These restrictions seldom hold, and indeed, the Poisson distribution rarely fits the observed frequency distributions of the number of organisms per sampling unit. However, because spatial randomness is inherent in the description of this model, the Poisson distribution and spatial randomness are assumed to be equivalent in much of the biological literature (*see* Neyman, 1939; Gurland, 1959; Waters, 1959; Iwao, 1970b; Southwood, 1978; Taylor et al., 1978).

Patil and Stiteler (1974) pointed out that a Poisson distribution does not imply a random spatial pattern. Suppose that the number of organisms within a sampling unit has a binomial distribution with parameters n and p. Further assume that due to heterogeneity in the environment, n varies over the region according to the Poisson distribution with mean λ. Then the numerical distribution of the organisms is Poisson with parameter $\lambda^* = p\lambda$, but the spatial structure should reflect the heterogeneity of the ecosystem and will not be random unless the heterogeneity occurs randomly.

Negative Binomial Models

In 1907, Student encountered the negative binomial distribution while studying the distributions of yeast cells counted with a hemocytometer. He reasoned that if the liquid in which the cells were suspended was properly mixed, a given particle had an equal chance of falling on any unit area of the hemocytometer. Thus, he was working with the binomial distribution and the fact that the probability of a binomial random variable X assumes a value x is equal to the $(x + 1)$st term in the expansion of $(p + q)^n$, where p, q, and $n > 0$, and $p + q = 1$. Student estimated p, q, and n from the first two sample moments. In two of his four series, the second moment exceeded the mean, resulting in negative estimates of p and n. Nevertheless, these "negative" binomials fit his data well. He noted that this may have occurred due to a tendency of the yeast cells "to stick together in groups which was not altogether abolished even by vigorous shaking" (p. 357). Several biological models have been presented in an effort to explain the phenomena observed by Student and so often witnessed in nature.

Suppose that a female insect lays eggs in clusters and that the clusters are randomly distributed among plants in a given area; that is, the number of clusters per plant has a Poisson distribution with parameter λ. Further assume that the number of eggs per cluster has a logarithmic distribution with parameter θ. Then the number of eggs per plant has the negative binomial distribution with a mean of

$$\mu = -\frac{\lambda\theta}{(1 - \theta)\ln(1 - \theta)} \tag{3.1}$$

and

$$k = -\frac{\lambda}{\ln(1 - \theta)} \tag{3.2}$$

(*see* Patil and Stiteler, 1974; Pielou, 1977). Notice that if more than one plant is included in a sampling unit, the number of egg clusters per sampling unit continues to be Poisson but with the parameter λ proportional to the number of plants per sampling unit. The distribution of the number of eggs per cluster remains unchanged. This is an example of a true contagion, or generalized Poisson, process because the organisms occur in small clusters that are randomly scattered throughout the field (*see* Gurland, 1959). Because aggregation occurs on a scale smaller than the sampling unit, the observed spatial structure is random.

Now suppose that the region over which the organisms are dispersed varies in its attractiveness. Some plants provide more favorable habitats than others. To model this, assume that the organisms are independent of one another and that all plants are identical. The number of organisms per plant is distributed according to the Poisson distribution with parameter λ. However, as a result of the differential attractability, the expected number of organisms per plant λ varies according to the gamma distribution with parameters r and θ (*see* equation (1.55) with θ replacing λ). Then the distribution of the number of organisms per plant is negative binomial, with the parameter k being equal to the r of the gamma distribution and a mean $\mu = k/\theta$. This is an example of an apparent contagion, or compound Poisson, process because the organisms are randomly scattered within each sampling unit, but the average number of organisms within a sampling unit is a variable that takes different values from sample to sample (*see* Gurland, 1959). Here aggregation is present on a larger scale than the sampling unit. If the average number of insects within a sampling unit is spatially influenced, the spatial structure will not be random.

Models giving rise to both the Poisson and negative binomial distributions have been presented that result in a random spatial pattern, and others have been given that would lead to spatial aggregation. Yet, it is commonly assumed that the negative binomial results in spatial aggregation and that the aggregation becomes more pronounced with decreasing values of k. Southwood (1978) expresses this view when he says, "The dispersion of a population, the description of the pattern of the distribution or disposition of the animals in space, is of considerable ecological significance." This statement is followed by a discussion of the role of the binomial, Poisson, and negative binomial distributions in describing spatial patterns. Something more than fitting a specific probability distribution to data is needed to address the question of spatial randomness. Approaches that specifically address the issue of spatial structure are considered in Chapters 7 and 8.

Our ability to discern which, of perhaps many, possible models gave rise to an observed frequency distribution is extremely limited. Because each generalized, and compound, distribution is based on at least two assumptions, a single set of observations does not permit all assumptions to be evaluated. Therefore, the fact that any given probability distribution consistently fits observational data within error does not permit us to explain the biological mechanisms that give rise to the distribution. More thought and research should be directed toward discerning which of many possible models may truly reflect the biology of organisms.

The models described to this point provide insight into what might have occurred at a particular instant in time. Yet, ecosystems are not static. Insects and other organisms continue to enter through birth or immigration, move within the ecosystem, and leave through death or emigration. These models are unable to reflect these dynamics of the ecosystem through time. Models of population size have been developed. Although they model population size, they do not provide insight into the distribution of the organisms within the system, and the development of sampling programs is strongly influenced by distribution. We want to model the population dynamics of organisms and their impact on the within ecosystem distribution through time. Some progress toward that goal is considered next.

Bose–Einstein versus Maxwell–Boltzmann Statistics

For the moment, we restrict our discussion to arthropods within agroecosystems. By so doing, we have greatly simplified the ecological system under study.

The plants represent an even-aged monoculture from isogenic or near-isogenic lines. This, together with the fact they receive uniform cultivation, irrigation, fertilizer, etc., results in broad expanses with little variation. In addition, a limited number of arthropod species enter these ecosystems, providing further simplification. Modeling these ecosystems is a reasonable first step toward more complex systems.

The approach to modeling insect dynamics we consider parallels the statistical mechanics study of particle distribution in a phase space (Willson et al., 1987; Young and Willson, 1987). To illustrate some of the ideas, consider the distribution of two arthropods on two plants. The classical, or Maxwell–Boltzmann, approach assumes that every permutation of the arthropods is equiprobable (*see* Figure 3.1).

We have four possible arrangements of the insects on the plants. The two insects are shaded differently so that we may distinguish arrangements (A) and (B), but in fact, the insects are indistinguishable. If we are then interested in the probability distribution of the number of insects on a plant, we have two possible distributions, each of which would occur with probability one-half. Now if we allow the number of insects and the number of plants to approach infinity while holding the average number of insects on a plant fixed, we obtain the Poisson as the limiting distribution. This derivation of the Poisson is consistent with referring to it as the "random" distribution because it does agree with intuitive notions of randomness. The fact that Poisson population counts are rarely observed is par-

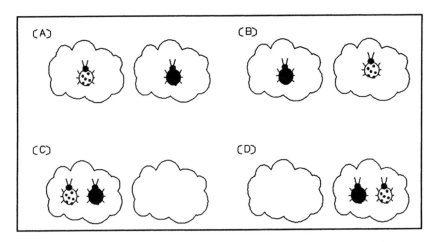

Figure 3.1. Four equally likely ways of placing two insects on two plants using Maxwell–Boltzmann statistics. The insects are shaded differently so they may be distinguished. Reprinted from Young and Young, 1994, p. 738, with kind permission from Elsevier Science-NL, Sara Burgerhartstraat 25, 1055 KV Amsterdam, The Netherlands.

alleled by the fact that no known body of matter is known to behave according to Maxwell–Boltzmann statistics.

During the 1920s, Bose and Einstein proposed an alternate approach to particle distribution in a phase space (Pathria, 1972). Returning again to our example of two insects on two plants, we apply their argument in the following manner. The plants may be distinguished by their position in the field. However, the insects are indistinguishable. It does not matter whether insect 1 or insect 2 is on the first plant. It only matters that there is one insect on the plant. Therefore, equal probability is given to distinguishable spatial patterns. Thus the case of one insect on each plant has only one possibility, and not two as under Maxwell–Boltzmann statistics (*see* Figure 3.2). The limiting distribution in this case is the geometric distribution. This is also equivalent to maximizing the entropy subject to the constraints that the average number of insects on a plant is μ and the sum of the proportions of plants with 0, 1, 2, . . . , insects is 1. In statistical mechanics, some physical bodies, such as photons, nuclei, and atoms with an even number of elementary particles, have been found that behave according to Bose-Einstein statistics.

Intuition led to the classical use of Maxwell-Boltzmann statistics in statistical mechanics. However, failure of this approach to explain observed phenomena led physicists to employ Bose–Einstein statistics. Although Bose–Einstein statistics are not as intuitive, they do produce results that agree with experience. Biologists

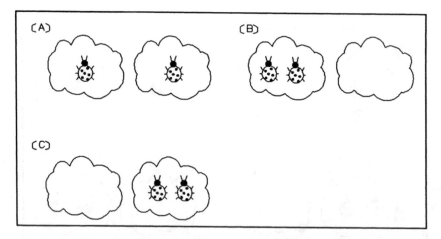

Figure 3.2. Three equally likely ways of placing two insects on two plants using Bose–Einstein statistics. Both insects have identical shading, emphasizing that they are indistinguishable. Reprinted from Young and Young, 1994, p. 740, with kind permission from Elsevier Science-NL. Sara Burgerhartstraat 25, 1055 KV, Amsterdam, The Netherlands.

have been experiencing the same difficulties. Although intuition may dictate that the Poisson distribution is the expected distribution of arthropods in agroecosystems, this distribution rarely occurs (Taylor 1965, 1984; Taylor et al., 1978). The geometric distribution is found in nature as shown in Table 3.1.

For a generalized model of arthropod distribution, consider the case where C arthropods are on N plants. Let

$$\mu = \frac{C}{N} \tag{3.3}$$

represent the average number of arthropods per plant. Modeling may proceed in one of at least two ways. The first approach considers each plant individually. The probability of exactly j arthropods on the plant is computed, and a uniform $(0, 1)$ random variate is used to assign arthropods to the plant according to those probabilities. Of course, the probability of exactly k arthropods on a given plant differs under Maxwell–Boltzmann statistics (Feller, 1968, p. 59)

$$Pr(k) = \binom{C}{k} \frac{(N - 1)^{C-k}}{N^C} \tag{3.4}$$

and Bose–Einstein statistics (Feller 1968, p. 61)

Table 3.1. Arthropod Distribution in Agroecosystems[a]

Arthropod	Crop	Sampling unit	No. of fields sampled	Sample size range	Range of estimates of k
Greenbug	Seedling oats	Single seedling	57	300–1,600	0.96–1.04
Collops spp.	Cotton	Single plant	182	106–$\frac{1}{4}$ acre	1.1–1.3
Convergent lady beetle	Cotton	Single plant	186	92–$\frac{1}{4}$ acre	0.7–1.2
Notuxus monodon F.	Cotton	Single plant	186	106–$\frac{1}{4}$ acre	0.92–1.6
Cotton fleahopper nymphs	Cotton	Single terminal	28	198–222	0.8–1.1
Cotton fleahopper adults	Cotton	Single terminal	28	240–242	1.1–1.4

[a]Reprinted from *Ecological Modelling* **36,** Use of Bose–Einstein Statistics in population dynamics models of arthropods, p. 94, 1994, with kind permission of Elsevier Science-NL, Sara Burgerhartstraat 25, 1055 KV Amsterdam, The Netherlands.

$$Pr(k) = \frac{\binom{N + C - k - 2}{C - k}}{\binom{N + C - 1}{C}}. \tag{3.5}$$

The difficulty with this approach is that the combinatorics rapidly become intractable when extending the models to include limiting cases.

An alternate approach is to develop stochastic models that allow the arthropods to enter the model sequentially. This approach provides insight into both Maxwell–Boltzmann and Bose–Einstein statistics. It is also easily extended to embrace various limiting conditions.

Stochastic Immigration Model

Under Maxwell–Boltzmann statistics, every plant has equal probability of being occupied by each arthropod that enters the model (or sample area); that is, the probability of the next arthropod selecting plant i is

$$P_i = \frac{1}{N}$$

for all i. The probability remains constant, regardless of the number of arthropods already on the plant. The placement of each arthropod is completely independent of any other arthropod. As C and N become large but

$$\mu = \frac{C}{N}$$

remains constant, this model leads to the distribution of arthropods being that of a Poisson distribution (Feller 1968, p. 59). With as few as a hundred plants and a hundred insects, a distribution resembling the Poisson is consistently observed. When considering 10,000 insects on 10,000 plants, the fluctuations about the Poisson are small. As pointed out by Feller (1968, pp. 20–21), this process is intuitive and explains a great many of the applications for collection and analysis of data. Because it is intuitively appealing, we see again why many biologists term the Poisson distribution as the random distribution (Waters, 1959). However, Kac (1983) points out that no *random* process has been described and that the term random may be defined in a number of ways.

Now consider a stochastic model based on Bose–Einstein statistics. The first arthropod is equally likely to choose each plant as in the Maxwell–Boltzmann example. Once the arthropod selects a plant for occupancy, that plant has an additional attraction to other arthropods. Therefore, the plant with an arthropod has twice the probability of receiving a second arthropod as an unoccupied plant. When there are N plants, the probability of the next arthropod selecting plant i is

$$P_i = \frac{1 + \text{number of arthropods on plant } i}{N + \text{total number of arthropods already placed}}. \tag{3.8}$$

As the number of insects C and the number of plants N become large but

$$\mu = \frac{C}{N} \tag{3.9}$$

remains fixed, this model results in the arthropods being in a geometric distribution (Gibbs, 1902; Willson et al., 1987). Distributions similar to the geometric distribution are observed for as few as 100 arthropods on 100 plants. When modeling 10,000 arthropods on 10,000 plants, the variation about the geometric distribution is slight. See Chapter 1 for a discussion of the geometric distribution.

Studying the models based on Maxwell–Boltzmann and Bose–Einstein statistics, it is evident that both are special cases of a more general model. Consider a model that sequentially places C arthropods on N plants. Suppose that the probability of the next insect selecting plant i for occupancy is

$$P_i = \frac{1 + w(\text{number of arthropods currently on plant } i)}{N + w(\text{total number of arthropods already placed})} \tag{3.10}$$

where w is a measure of the additional attraction of an arthropod toward a plant due to the occupancy of that plant by another arthropod. Maxwell–Boltzmann statistics occurs when $w = 0$. Bose–Einstein statistics occur when $w = 1$. This model is an example of the Polya process (Fisz, 1963). If C and N become large but

$$\mu = \frac{C}{N} \tag{3.11}$$

is constant, it leads to the negative binomial distribution with

$$k = \frac{1}{w} \tag{3.12}$$

for all positive values of w (*see* Gurland, 1959). If w is less than 0, k is negative and the binomial distribution is produced. When $w = -1$, each plant will have exactly 1 insect if the mean is one. Simulations for a broad range of values for μ and w indicated 10,000 plants are sufficient to show the corresponding distributions with only slight fluctuations, and the basic tendencies of the model can be observed for as few as 100 arthropods on 100 plants. The corresponding relationships of w to distributions is given in Table 3.2.

As can be seen from these examples, this method of emulating arthropod populations allows a wide range of distributions to be simulated. ECOSTAT, the accompanying software, permits the user to investigate the infestation model. To do so, select *Models* from the *Probability Distributions* menu. Then choose *Infest*. You will then be asked to provide the value for w, the number of arthropods infesting the field, whether each plant has a maximum carrying capacity and, if so, what that maximum carrying capacity is. The infestation is simulated based on these inputs. The number of plants with 0, 1, 2, . . . , arthropods is continuously updated. Upon completing the infestation, more arthropods may be added to the field by pressing the *More Arthropods* button. The graph of the observed distribution of arthropods can be compared to those of the Poisson distribution and the geometric distribution with the same mean by pressing the *Display Graphs* button.

Modeling Within Field Movement

The application of Bose-Einstein statistics in biology is not complete without considering an additional parallel with physics. In statistical mechanics, it is known that particles move toward the most probable distribution regardless of what their initial distribution is, thereby maximizing entropy. Populations of arthropods that are mobile or reproducing should move toward a geometric distribution unless a limiting factor such as restricted carrying capacity stops them. Young and Young (1989) have shown that cotton fleahoppers disrupted by insecticide applications to random spots in the field quickly return to a geometric distribution as the insecticide effects diminish. Any proposed model of insect movement should have this feature.

Consider the following approach to modeling movement. Suppose that after entering the field, each arthropod is equally likely to be the next one to move, die, or emigrate. The distribution of the arthropods is invariant to the random removal of an arthropod from the system so we continue to have a Bose–Einstein distribution (Skellam, 1952). If the arthropod remains within the field, then the

Table 3.2. *Relationships of* w *to Distributions*

Value of w	Distribution
0	Poisson
1	Geometric
>0	Negative binomial
<0	Binomial

selection of the next habitat for occupation is according to the same process as if it was initially entering the field. Because this process led to the Bose–Einstein distribution, the distribution continues to be Bose–Einstein. Further, if the initial distribution of the arthropods is not Bose–Einstein, the probability distribution converges to that of the Bose-Einstein distribution. If $w = 1$, we have the Bose–Einstein distribution with the geometric as the limiting distribution, and the process is equivalent to maximizing entropy.

This model has several appealing features. First, it provides a coherent approach to immigration, movement and emigration. Second we can model any distribution in the negative binomial-Poisson family of distributions. Further after the system is disrupted, the population begins to move immediately toward the most probable distribution. To model the movement of arthropods on 100 plants using ECOSTAT, select *Models* from the *Probability Distributions* menu. Next choose *Move*. You will be asked to provide the number of arthropods in the field, the initial distribution of the arthropods, and the number of arthropod movements to be simulated. After the simulation is complete, you may request that additional movements be simulated by pressing the *More Moves* button. The graph of the observed arthropod distribution can be compared to the graphs of the Poisson and geometric distributions by pressing the *Display Graphs* button.

A remaining modeling issue for consideration is how to integrate the rate of immigration, movement, and emigration. If we allow 1000 insects to enter a field, 500 hundred movements to occur and 100 to leave the field through death or emigration, then we can build this model. Yet, are these relative rates reasonable? By selecting *Models* from the *Probability Distributions* menu of ECOSTAT and then *Both* from the *Models* menu, you can simulate both arthropod infestation and movement within the field.

Restrictions on Carrying Capacity

One of the biological limitations of Bose–Einstein statistics is a limitation on the number of arthropods that each sampling unit can carry. This becomes a factor when arthropods begin to behave differently due to an excessive number of arthropods already inhabiting a particular plant. The simplest case occurs when each plant has the ability to hold the same maximum number of arthropods. Adding this constraint to the model results in a truncated Poisson for Maxwell–Boltzmann statistics and a truncated geometric under Bose–Einstein statistics. We have observed a similar phenomenon in alfalfa weevils in alfalfa (*see* Young and Willson, 1987). Initially, the geometric describes the distribution of alfalfa weevils (Figure 3.3). As the population increases, the negative binomial fits (Figure 3.4). As the destruction of the crop increases, disagreement with the model occurs in each tail (*see* Figure 3.5). We believe this can be explained in the following manner. These data were collected when the weevil population was increasing rapidly, becoming chaotic, and on the verge of destroying the crop. At this time,

percentage

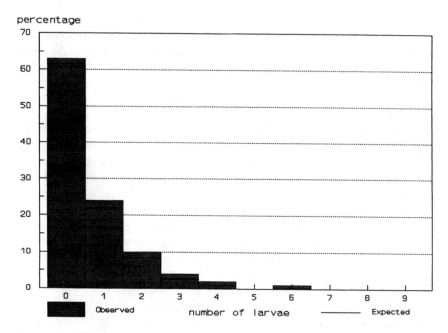

Figure 3.3. Relative frequency histograms of the observed number of alfalfa weevil larvae on an alfalfa terminal where the sample mean is 0.6. The alfalfa weevil population exhibits a geometric distribution. Redrawn from *Ecological Modelling* **36,** Young and Willson, Use of Bose–Einstein statistics in population dynamics models of arthropods, p. 96, 1994, with kind permission of Elsevier Science-NL. Sara Burgerhartstraat 25, 1055 KV Amsterdam, The Netherlands.

it is unlikely that all plants had an equal carrying capacity, but in fact the carrying capacities varied depending on the amount of destruction that had already occurred. Again, how do we maximize entropy under the restriction that carrying capacity varies in a prescribed manner?

It is easy to think of many situations in which none of these models would fit, but the general equations should work for most cases. For example, unequal attractiveness due to sexual preference, soil types, etc., could be modeled by giving different parts of the population different values of *w*. Thus we have a modeling tool that allows populations of arthropods to be modeled from one arthropod on each plant to all the arthropods occurring on one plant (an unlikely case).

It seems that the geometric distribution is the basic underlying distribution for many arthropods and, for that matter, many dispersing objects. The biological and physical principles giving rise to Bose–Einstein statistics are not completely known. The most prevalent explanation is that interacting objects or organisms

percentage

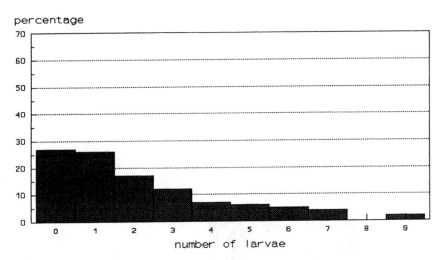

Figure 3.4. Relative frequency histograms of the observed number of alfalfa weevil larvae on an alfalfa terminal where the sample mean is 1.8. The negative binomial distributions fits the data. Redrawn from *Ecological Modelling* **36,** Young and Willson, Use of Bose–Einstein statistics in population dynamics models of arthropods, p. 96, 1994, with kind permission of Elsevier Science-NL. Sara Burgerhartstraat 25, 1055 KV Amsterdam, The Netherlands.

are subject to Bose–Einstein statistics and non-interacting objects, like playing cards, are subject to Maxwell–Boltzmann statistics.

Once a potential model has been selected based on one's biological understanding of the system, data must be gathered either to support or to refute the model. Although we worked with data sets in the last chapter, we have not addressed the issues surrounding data collection. We give a brief overview now. For more detailed accounts, see books on sampling, such as those by Cochran (1977), Kish (1965), and Thompson (1992).

Sampling Concepts

Suppose all biological organisms in a given geographical region can be enumerated. In this case, sampling is not needed because we are able to list each organism present. Unfortunately, based on today's technology, complete enumeration is rarely possible. Instead, we must be satisfied with looking at only a portion of the population. Some of the principles for selecting that portion are the focus of this section.

Sampling is the major tool used to gather information about populations of organisms. Population density estimation is a focal point of most sampling pro-

percentage

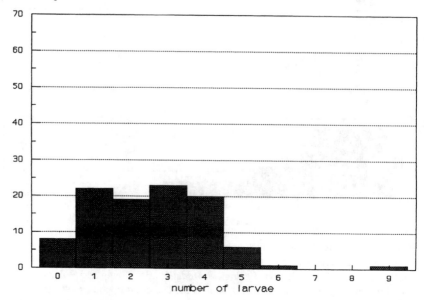

Figure 3.5. Relative frequency histograms of the observed number of alfalfa weevil larvae on an alfalfa terminal where the sample mean is 2.5. The distribution is uniform-like. This would be expected if carrying capacity is limited. Redrawn from *Ecological Modelling* **36,** Young and Willson, Use of Bose–Einstein statistics in population dynamics models of arthropods, p. 96, 1994, with kind permission of Elsevier Science-NL. Sara Burgerhartstraat 25, 1055 KV Amsterdam, The Netherlands.

grams. Estimates of population density may be divided into two basic groups: absolute population density estimates, and relative population density estimates. In absolute density estimation, data are collected in such a manner that the number of organisms per unit area of the habitat can be estimated. Fixed-size sampling quadrats, such as a square-foot quadrat, a plant, or a linear foot of row, are often used for this purpose. Capture–recapture methods, line transects, and nearest neighbor techniques are other approaches to obtaining absolute density estimates. All these methods provide an estimate of the population density in the sampling area that must be clearly defined and delineated. For many organisms, these estimates are expressed as the number of organisms per acre, per hectare, etc.

Relative population density estimates are based on measurements that cannot be directly related to a unit of area within the sampling region; that is, the numbers of organisms are measured in unknown units. Sweep nets and traps, such as light traps or pheromone traps, are used to obtain relative density estimates. This type of sampling is generally used in migration, relative abundance, or species diversity studies. Relative density measurements provide insight into trends in the abundance of the organisms. As the population density increases, more organisms are measured using the relative techniques; as the population density decreases, fewer organisms are measured. If the correlation is strong, it may be possible to determine a relationship, usually through regression, between the relative density estimates obtained and the desired absolute density estimates. In some insect pest management studies, sweep net (or trap) counts are correlated with absolute population density estimates from the sampling area. If the two sampling procedures are correlated closely enough, these counts can be used to initiate pest control activities, thereby emulating an absolute measure. As an example, treatment may begin when four insects are found per 10 sweeps; however, studies using absolute population density estimates must be used to determine that four insects in 10 sweeps surpasses the economic threshold.

Regardless of the sampling procedure used, sampling remains one of the weakest links in population studies. Many of the principles of population biology are based on empirical data sets. The collection of these data sets should be the aim of most population studies. One of the reasons that data sets are often inadequate is a lack of resources and a lack of time. Also there is often a misunderstanding of the basic principles of sampling. The aim of most population sampling is to obtain estimates of densities and, short of an actual census, these density estimates have variability associated with them. In general, it takes less effort to obtain precise estimates of plentiful populations than it does to estimate sparse or low-density ones with precision. Precisely estimating low population densities of organisms may be too costly in terms of time and manpower. Yet, these low densities may be very important biologically and/or economically.

Sample sizes are often inadequate, and the precision of the estimates obtained are sometimes unknown. Biologists take to the field, gather information, and hope that they have the resources to identify and characterize the parameters of the population. When sample sizes are inadequate, interpretations of the data are frequently incoherent with other information available. Year after year, biologists spend all year, or in some cases, years, collecting data that give meaningless or poorly validated results. It is hoped that the sampling procedures outlined in this and subsequent chapters will aid in this problem. Many journals that publish ecological research are requiring improved data validation. Increasingly, a measure of the precision of population estimates is required in addition to description of the sampling procedures.

In some areas of population studies, standard or quasi-standard sampling methods have been established. These methods are not always the most efficient but serve as bench marks. Any addition to the sampling routines should explain how the systems compare and how the new routine improves upon the one it replaces. Basic ecological information about the system being studied is often a result of accumulated evidence, and although the sampling procedure is poor, the accumulated trends make powerful arguments. Some of the best samples are taken by workers who have sampled in an area for long periods. Knowledge about the flora and fauna, soil types, climate, and in some cases cropping systems, assist in determining sampling systems. This fact has led many workers to avoid using some of the more resource laden and advanced sampling methods. The general rationale is that they can sample better without them than an inexperienced worker can using them. No one would deny the truth of this statement, but solid advancements are more frequently made by careful examination of the data, and this requires an understanding of data validation techniques. In population studies, sampling methods should be refined as increased information becomes available about the system. Generally, descriptive data about the ecosystem may be obtained with relatively crude sampling methods, but as the subtle interactions in the systems are investigated, measurements require refinement. The human brain has trouble coping with secondary and tertiary effects on multilevel dimensions. Computer-assisted science can identify subtle effects, validate them, and present them in some graphic representation that is more understandable.

Modern ecosystem studies generally require a team or interdisciplinary approach where experienced biologists that are computer and mathematically literate work closely with other areas. In addition, they should be backed by sufficient resources to obtain the necessary information and to assess this information as well as possible. One of the problems with many ecological investigations is that they are overly ambitious and try to explain too much with insufficient resources. As stated by Fisher, the aim of any investigation is to extract all the useful information from a data set and to determine the part that is not useful. A single sample reflects the state of a system during the sampling period. It does not necessarily reflect the state of the system all, or even most, of the time. Events in nature, even chemical events, are seldom unique for all time.

The first prerequisite in choosing a sampling method is to make sure that the research question will be answered. The researcher should have or obtain enough information to enact a sampling system that estimates parameters, such as population density, with the desired precision, or tests hypotheses with specified probabilities of error. The sampling generally needs to be repeated several times in the initial periods of the investigation. Few biologists accept any one time sampling scheme from a new unchartered system. Always try to err on the side of sampling more than is required. The theme of the field biologist is to take data, take data, and take more data.

Numerous aids are available for sampling. Diverse types of traps have proved useful. Light traps, sweep nets, pheromone traps, pit traps, sticky traps, live traps, snap traps, and leg traps may be used. Southwood (1978) describes many types used for arthropods.

Sampling may be used (1) to obtain precise estimate(s) of the parameter(s) of interest, or (2) to test hypotheses concerning the parameter(s) of interest with specified error probabilities. If the purpose of sampling is to obtain a precise estimate of population density, (1) is the primary consideration. However, if the purpose of sampling is to determine whether a population is above or below a specified level, such as above or below an economic threshold, (2) is the appropriate approach. Note that after performing a test of hypotheses, the researcher may want to obtain an estimate of the population. This may certainly be done, but the precision with which this estimate is made may be less than it would have been if estimation was the primary objective of sampling. Here we define statistical terms commonly used to describe aspects of sampling. We also consider the general aspects of simple random sampling and stratified sampling as well as ratio estimates. Subsequent chapters discuss various approaches to estimating population density and testing hypotheses.

To sample a biological population effectively, it is essential that the population be well defined. For example, suppose a researcher wants to determine the current size of the tadpole population. Are we interested in the tadpole population in the state, in the county, or in some other geographical region? Are all tadpoles of interest, or are we restricting our attention to ditches, wetlands, ponds, or some other habitat or combination of these habitats? The biological population of interest is called the *target population*. The population sampled is termed the *sampled population*. Ideally, the target and sampled population should be the same. This is rarely the case. If we are interested in the tadpole population in a county, but sample only the ponds, then depending on the county, we may have failed to sample a significant portion of the population. Is that a problem? It can be if the target population members that are not in the sampled population differ significantly from those in the sampled population. Part of the difficulty is that the biological population that is our target population may not have well-defined spatial boundaries. Further, these boundaries may change with time.

Careful thought should be given to the differences in the sampled and target population and what impact these differences may have on inferences drawn about the population. In some cases, the differences may not be significant. For example, in his study of spruce budworm [*Choristoneura fumiferana* (Clem.)] on balsam fir, Wilson (1959) noted that most of the egg masses were laid on the branch tips; he determined that the number of eggs masses on the tips was highly correlated with the number on the branch. Therefore, unless the branches were severely or completely defoliated, a precise estimate of the number of egg masses on a branch could be obtained by increasing the number on the tip by 10%. This sampling approach gave an accurate measure of the level of infestation while reducing the amount of time required for sampling based on the branches by 25% to 40%. However, the number of examples of where there is no significant impact on inferences to the target population based on the sampled population is small compared to the problems that have been encountered by failing to consider the differences.

The sampled population is made up of members, each of which has a numerical value for a variable X of interest. For example, an item may be a plant, and X could be the observed number of damaged fruit on that plant. Here the *sampling unit* is the plant. It is important that each plant in the population is associated with one and only one sampling unit. A sampling unit may be simple, such as a single rabbit or a single elm tree, or it may be more complex such as a core of soil, a 1-m^2 quadrat, or a branch of an apple tree. In most areas of ecology, practical experience is the guide to help one decide on an appropriate sampling unit.

A *sampling frame* lists all the sampling units in the population. If the sampling unit is a plot of ground, then a map of the region covered by the plots provides a sampling frame. For a small population, finding and in some way identifying each member of that population provides the basis for a frame that lists each member. Because sampling frames are generally expensive to develop and maintain, they are rarely used in ecological studies.

Simple Random Sampling

Once the sampled population is well defined, a decision must be made as to the best way to choose sampling units for inclusion in the sample. A *simple random sample* is drawn from the population in such a way that every member of the sampled population has an equal probability for inclusion in the sample. Simple random sampling is easy to conceptualize, easy to describe, and difficult to implement. Sampling can be *with replacement,* if sampling units are replaced in the population after selection and may be included in the sample more than once, or *without replacement,* if sampling units are not replaced in the population after selection.

Typically, a random sample is selected by numbering the sampling units from 1 to N. In the past, random numbers between 1 and N have been selected from a table of random numbers or from a set of numbers in a container, such as a bowl. Now computers can be used to draw the numbers using random number generators. A total of n such numbers is selected. The n units corresponding to the selected numbers are measured and constitute the sample. Although simple to describe, it is generally difficult to implement a random sampling plan in the field. Assuming that sampling units have fixed locations, software can be developed that selects the position of the sampling unit in the study region, making it unnecessary to enumerate all of the sampling units.

The estimates of the mean ($\bar{X}$) and variance (s^2) developed in Chapter 1 are appropriate, whether sampling is conducted with or without replacement. In estimating the standard error of the sample mean in Chapter 1, we assumed that sampling was with replacement and thus

$$s_{\bar{X}}^2 = \frac{s^2}{n}. \tag{3.13}$$

However, if sampling is without replacement, the estimated standard error of the sample mean is

$$s_{\bar{W}}^2 = (1 - f)\frac{s^2}{n} \tag{3.14}$$

where $f = n/N$ is called the *finite population correction (fpc) factor*. We should notice at least two things in viewing equation (3.14). First, when sampling without replacement, the variability associated with estimating the population mean is smaller than it would be if sampling was with replacement. This decrease in variance becomes more pronounced as the sampling fraction n/N becomes larger. Second, if N is large relative to n, the finite population correction is negligible and can be ignored, giving the same result as when sampling with replacement.

In ecology, several other methods of sampling are often confused with simple random sampling (Krebs, 1989). *Accessibility sampling* is conducted by taking observations that are readily accessible. For example, in the North American Breeding Bird Survey, data collectors drive along secondary roads to observe birds. Therefore, the sampled population is the breeding birds along secondary roads. In *haphazard sampling,* the sample is collected in a haphazard manner. The first rabbit one grabs from a cage may be the next one to enter the sample, or soil core samples are collected from a location that "looks good" to the investigator. *Judgment sampling* is perhaps the most tempting alternative to random sampling. Here the researcher selects a series of "typical" sampling units based on experience. In simple random sampling, the distribution of the sample should be approximately that of the distribution. In judgment sampling, atypical units are seldom included in the sample, resulting in the variability of the estimates being underestimated. *Volunteer sampling* occurs when individuals are able to self-select for inclusion in the study. For example, hunters may choose to complete survey forms on kill statistics. Some hunters would never choose to complete the forms. Whether the hunters willing to fill out the forms differ significantly in their response from those who refuse to fill out the form is a critical question that must be answered before we can determine whether the inferences can be extended to the population of hunters or only to the population of hunters willing to fill out the form. A general rule of thumb is that, if at all possible, conduct random sampling. This assures a probabilistic foundation for inferences based on the sample values.

Stratified Random Sampling

At first glance, simple random sampling is intuitively appealing. Suppose a plot of land within a field is the sampling unit. Every plot (sampling unit) has an equal chance for inclusion in the sample. Consequently, most of the observations may be concentrated in a relatively small portion of the study region. This is fine as long as the variable being measured has a fairly uniform distribution over the study region. However, the study region may not be homogeneous, resulting in some good and other poor habitats. Intuitively, biologists want to be sure to sample both types of habitats whether they are in a random sample of the full region or not. *Stratified random sampling* is extremely useful in these circumstances.

In stratified sampling, the study region is divided into L strata. Each stratum is comprised of homogeneous sampling units. Each sampling unit must be in one and only one stratum. Each stratum is sampled separately. Assuming that a random sample is taken within each stratum, we have stratified random sampling. Cochran (1977) gives four general reasons for stratifying a sample:

1. Estimated means and confidence intervals may be needed for each stratum.

2. Sampling problems may depend greatly on the stratum.

3. Stratification can result in more precise estimates of population parameters. Consequently, confidence intervals can be much narrower when strata are chosen well.

4. If sampling is being done by more than one field crew, stratification may be useful.

Define the following:

N	Number of sampling units in the population.
N_i	Number of sampling units in stratum i, $i = 1, 2, \ldots, L$.
n_i	Sample size taken from stratum i, $i = 1, 2, \ldots, L$.
$f_i = n_i/N_i$	Sampling fraction in stratum i.
$\bar{X}_i$	Sample mean for stratum i, $i = 1, 2, \ldots, L$.
s_i^2	Sample variance for stratum i, $i = 1, 2, \ldots, L$.
$W_i = N_i/N$	Weight for stratum i.
$\bar{X}_{ST}$	Estimated population mean based on stratified random sampling.

Notice that the weight for stratum i represents the proportion of population sampling units in the ith stratum. Therefore, the sum of the stratum weights must be one.

Based on stratified random sampling, the population mean is estimated as

$$\bar{X}_{ST} = \frac{\sum\limits_{i=1}^{L} N_i \bar{X}_i}{N}. \tag{3.15}$$

The estimated variance of the stratified sample mean is

$$s_{\bar{X}_{ST}}^2 = \sum_{i=1}^{L} \left(\frac{W_i^2 s_i^2}{n_i} (1 - f_i) \right) \tag{3.16}$$

if sampling is without replacement, and

$$s_{\bar{X}_{ST}}^2 = \sum_{i=1}^{L} \frac{W_i^2 s_i^2}{n_i} \tag{3.17}$$

if sampling is with replacement.

Notice that the estimated variance in equation (3.17) depends only on the estimated variances *within* each stratum. Therefore, if the strata comprise homogeneous units, the variability within each stratum is reduced as compared to the variability within the population, leading to a more precise estimate of the population mean for stratified random sampling as opposed to random sampling.

Confidence intervals have the form

$$\bar{X}_{ST} \pm t_{\alpha/2} s_{\bar{X}_{ST}} \tag{3.18}$$

The only problem is the choice of the appropriate degrees of freedom for t. The appropriate degrees of freedom lie somewhere between the smallest of the values $(n_h - 1)$ and the total degrees of freedom $(\Sigma n_i - 1)$. The Satterthwaite (1946) approximation for the degrees of freedom is

$$df = \frac{\left(\sum\limits_{i=1}^{L} a_i s_i^2 \right)^2}{\sum\limits_{i=1}^{L} \frac{(a_i s_i^2)^2}{n_i - 1}} \tag{3.19}$$

where

$$a_i = \frac{N_i (N_i - n_i)}{n_i}. \tag{3.20}$$

If all the stratum sizes N_i are equal and all the stratum sample sizes n_i are equal, the Satterthwaite approximation for the degrees of freedom is $N - L$.

Suppose we have decided to take a stratified random sample of size n. How many samples should be taken from within each stratum? If one does not have any information about the population, a natural choice is *proportional allocation*. That is, the proportion of the sample taken within stratum i is chosen to be equal to the proportion of the population sampling units in stratum i:

$$n_i = \frac{nN_i}{N}.$$ (3.21)

If the variability is the same within each stratum, proportional allocation also results in the smallest variability in the estimated population mean.

More frequently, the variability differs from stratum to stratum. In this case, *optimum allocation* results in the lowest variability of the estimated population mean and gives

$$n_i = \frac{nN_i\sigma_i}{\sum\limits_{j=1}^{L} N_j\sigma_j}$$ (3.22)

where σ_i is the stratum i population standard deviation. In practice, σ_i is not known, but it can be estimated using the sample standard deviations from past data. Notice that under optimum allocation, more observations are taken from the strata with high variability than from strata with low variability.

Systematic Sampling

Field biologists often use systematic sampling. For example, sticky traps may be placed on a square grid at 25-m intervals within a corn field. Systematic sampling may give a better coverage of a population than a simple random sample. However, it provides no basis for estimating sampling errors. Often, a systematic sample is treated as if it is a random sample, and estimation proceeds as discussed earlier. This approach is based on the assumption that the values for the response variable are randomly ordered in the population.

From an ecological perspective, the primary concern when choosing a systematic sample instead of a random sample is the possible existence of periodic variation in the system under analysis (Krebs, 1989). If periodic variation is present, then systematic sampling runs the risk of sampling at the same periodicity. This leads to a biased estimate of the population parameter. Milne (1959) investigated how often this might be a problem in field situations by looking at systematic samples taken on biological populations that had been completely enumerated. After analyzing 50 populations, he found no error introduced by assuming that a systematic sample is a simple random sample and using all the appropriate formulas from random sampling theory. Caughley (1977b) conducted a computer simulation of a kangaroo population based on observed aerial counts. He found that although the mean kangaroo density was estimated without bias, the estimated standard error was below the true value, raising some concern when using systematic samples for aerial surveys. In summary, if at all possible, a random sample should be used instead of a systematic one. If the cost and inconvenience of randomization are too great, then little may be lost by conducting a systematic sample and treating the data as if they were taken from a random sample. However, watch out for possible periodic trend.

Ratio Estimation

Sometimes ecologists want to estimate the ratio of two variables based on a simple random sample, but both variables vary from sampling unit to sampling unit. For example, the ratio of predators to prey within a cotton field may be of interest. The intuitive estimate of the mean ratio is

$$\hat{R} = \frac{\bar{X}}{\bar{Y}} \tag{3.23}$$

where $\hat{R}$ is the estimated mean ratio of X to Y, $\bar{X}$ is the sample mean of X, and $\bar{X}$ is the observed mean of Y. The estimated variance of this estimated ratio is

$$s_{\hat{R}}^2 = \frac{1 - f}{n \bar{Y}^2} \left(\frac{\sum_{i=1}^{n} X_i^2 - 2\hat{R} \sum_{i=1}^{n} X_i Y_i + \hat{R}^2 \sum_{i=1}^{n} Y_i^2}{n - 1} \right) \tag{3.24}$$

when sampling without replacement and

$$s_{\hat{R}}^2 = \frac{1}{n \bar{Y}^2} \left(\frac{\sum_{i=1}^{n} X_i^2 - 2\hat{R} \sum_{i=1}^{n} X_i Y_i + \hat{R}^2 \sum_{i=1}^{n} Y_i^2}{n - 1} \right) \tag{3.25}$$

when sampling with replacement.

Confidence intervals are generally based on the usual normal approximation. However, ratios tend to have distributions that are skewed to the right, particularly when the coefficient of variation of the denominator is high (Atchley et al., 1976). Therefore, the true confidence level tends to be below the stated level unless the sample size is large (Sukhatme and Sukhatme 1970). Therefore, the stated level of confidence should only be considered as an approximation unless the sample size is large.

Summary

Understanding the biological processes giving rise to a particular probability distribution is challenging. Because each probability model can result from more than one type of biological setting, knowing that a probability distribution fits the data well provides only marginal help in this endeavor. If we are to discern among models, goodness-of-fit tests are not sufficient. More basic biological studies are required.

Effective sampling is a key to increasing our understanding of biological populations. Simple random sampling is intuitively appealing. It is sometimes difficult to fully implement in the field. Yet, it provides a probabilistic foundation for making inferences about biological populations. When some areas (strata) of the study region are better environments for the population than other regions, we want to be sure that these different strata are all sampled. Stratified random sampling allows for random sampling within each stratum. Then the information from the different strata can be combined to give estimates of the population parameters. Systematic sampling does not provide a foundation for estimation of the variability of parameter estimates. Often the observations from a systematic sample are treated as a random sample. This may lead to biased estimators if periodic trends are present. Ratio estimates are sometimes useful.

4

Sequential Estimation

Introduction

Determining the number of members in a biological population is basic in the study of any population. Chapter 3 discussed some principles of sampling. The importance of defining the biological population and choosing an appropriate sampling unit is clear. We have a basic idea of what to do and how to do it. Yet, one extremely important question remains unanswered. How many observations are needed to obtain either estimates with the desired precision, or tests of hypotheses with specified error rates? Much of this book looks at various sampling approaches that help us address this question. In this chapter, we consider both fixed-sample size and sequential methods for estimation of population density for well-defined discrete sampling units. Subsequent chapters discuss alternate methods of density estimation as well as hypothesis testing.

Suppose we are to help prepare a sampling plan. The goal is to estimate the population density. The target and sampled populations are clearly defined. An appropriate sampling unit has been selected, and we have decided to use simple random sampling. Everything seems to be well planned, and then someone asks us about the sample size. How many observations should we take? We know that as the number of observations increases, the precision of the estimate increases. One approach that is commonly adopted is to take as many observations as time and resources permit. Yet, this is not very satisfying from two perspectives. First, after exhausting our resources, we may determine that we have an extremely poor estimate of population density. Perhaps it would have been better to narrow our research question or to redirect our efforts because the estimates obtained have little value. A second possibility is that our estimates are more precise than needed to answer the research question. If we had known, we could have spent time and resources on the next phase of the study. Seldom does the sample size come out "just right" by chance alone.

Before initiating sampling, we need to decide how we will determine sample size. In this chapter, we discuss two basic approaches, fixed and sequential sampling. First, suppose we are going to use a fixed sample size. If we have experience with the population, then we may have an idea of the possible range of means and variances that may be present. We discuss how to use these to determine appropriate sample sizes. Next suppose we are willing to sample until the population density is estimated with the desired precision. Observations are taken sequentially. When enough data have been collected, we stop. One of the consequences of this sequential sampling is that sample sizes are no longer fixed but random. However, before implementing such a plan, some idea of the possible sample size range is essential. Knowledge of the possible range of population means and variances can be used to determine this range and to help assess whether a sequential sampling plan is reasonable based on the available resources.

Whether using fixed or sequential sampling, the goal of sampling is to estimate the population parameter of interest, population density in this case, with the desired precision. Methods have been developed to obtain density estimates based on the measure of precision that is to be used. Some methods are designed to obtain an estimate with a specified coefficient of variation. Other methods provide interval estimates with specified confidence and either (1) a fixed width, or (2) a width proportional to the true parameter of interest. We address these three approaches in both the fixed and sequential settings.

Sample Sizes Required to Control $CV(\bar{X})$

Suppose that the purpose of sampling is to obtain a precise estimate of population density with a specified coefficient of variation; that is, we want to estimate the mean of the population with a specified coefficient of variation of the sample mean $[CV(\bar{X})]$. The $CV(\bar{X})$ is the standard deviation of the sample mean over the population mean:

$$CV(\bar{X}) = \frac{\sigma_{\bar{x}}}{\mu} = \frac{\left(\frac{\sigma}{\sqrt{n}}\right)}{\mu} \tag{4.1}$$

where μ and σ are the population mean and standard deviation, respectively. As the sample size increases, the variability of the estimated sample mean decreases, and consequently, the $CV(\bar{X})$ decreases. Note that μ and σ are fixed, but unknown, population parameters. Therefore, the $CV(\bar{X})$ decreases as the sample size increases. The sample size required to obtain the desired $CV(\bar{X})$, say $CV(\bar{X}) = C$, may be found by solving equation (4.1) for n and is

$$n = \left(\frac{\sigma}{\mu C}\right)^2. \tag{4.2}$$

Rojás (1964), Karandinos (1976), and others have noted that equations (4.1) and (4.2) take varying forms if available information concerning the form of the population distribution is used. These are summarized in Table 4.1 for the binomial, negative binomial, and Poisson distributions. To understand the derivation of these formulae, suppose we are sampling from a negative binomial population. The variance of the negative binomial distribution is $\sigma^2 = \mu + \mu^2/k$. Thus, the $CV(\bar{X})$ may be written as

$$CV(\bar{X}) = \sqrt{\frac{1}{n}\left(\frac{1}{\mu} + \frac{1}{k}\right)}. \tag{4.3}$$

The sample size needed to achieve a $CV(\bar{X}) = C$ is then

$$n = \frac{1}{C^2}\left(\frac{1}{\mu} + \frac{1}{k}\right). \tag{4.4}$$

Example 4.1

A primary pest of cotton in the United States is the cotton bollworm (*Heliothis zea* (Boddie)). A standard sampling unit is the plants in 3 row feet (with 40-inch rows, there are 13,068 row feet per acre). An observation consists of all bollworms observed on the plants in that 3 row feet. Experience has shown that the distribution tends to be negative binomial with a k value of about 3. If the mean is 2 bollworms/3 row feet, how many observations are needed to estimate the mean with a $CV(\bar{X})$ of no more than 10%?

$$n = \frac{1}{C^2}\left(\frac{1}{k} + \frac{1}{\mu}\right) = \frac{1}{(0.1)^2}\left(\frac{1}{3} + \frac{1}{2}\right) = 83.3.$$

Because it is not possible to take a fractional part of an observation, we would take 84 observations. This would result in a $CV(\bar{X})$ slightly less than 10%. Suppose a $CV(\bar{X})$ of 25% was considered sufficient for some preliminary scouting. Then the required sample size is

$$n = \frac{1}{C^2}\left(\frac{1}{k} + \frac{1}{\mu}\right) = \frac{1}{(0.25)^2}\left(\frac{1}{3} + \frac{1}{2}\right) = 13.3.$$

In this case, only 14 observations are needed. One problem exists with this and the next few examples: The mean is *not* known. If we knew the mean (population density), then it would not be necessary to sample. However, these examples will help you become acquainted with the formulae. The problem of what to do in practice will be addressed shortly.

Table 4.1. Optimal Sample Sizes When Estimating the Population Density With a Specified $CV(\bar{X}) = C(C)$, Within $D\mu$ With Confidence $(1 - \alpha)$ 100% (D), and Within H With Confidence $(1 - \alpha)$ 100% Confidence (H)

	C	D	H
Binomial	$\dfrac{1-p}{pC^2}$	$\dfrac{(1-p)z_{\alpha/2}^2}{pD^2}$	$p(1-p)\left(\dfrac{z_{\alpha/2}}{H}\right)^2$
Poisson	$\dfrac{1}{\mu C^2}$	$\dfrac{1}{\mu}\left(\dfrac{z_{\alpha/2}}{D}\right)^2$	$\mu\left(\dfrac{z_{\alpha/2}}{H}\right)^2$
Negative binomial	$\dfrac{1}{C^2}\left(\dfrac{1}{\mu}+\dfrac{1}{k}\right)$	$\left(\dfrac{z_{\alpha/2}}{D}\right)^2\left(\dfrac{1}{\mu}+\dfrac{1}{k}\right)$	$\left(\mu+\dfrac{\mu^2}{k}\right)\left(\dfrac{z_{\alpha/2}}{H}\right)^2$
Iwao's Patchiness Regression	$\dfrac{1}{C^2}\left(\dfrac{\alpha+1}{\mu}+(\beta-1)\right)$	$\left(\dfrac{z_{\alpha/2}}{D}\right)^2\left(\dfrac{\alpha+1}{\mu}+(\beta-1)\right)$	$[(\alpha+1)\mu+(\beta-1)\mu^2]\left(\dfrac{z_{\alpha/2}}{H}\right)^2$
Taylor's Power Law	$\dfrac{a\mu^{b-2}}{C^2}$	$\dfrac{az_{\alpha/2}^2\mu^{b-2}}{D^2}$	$a\mu^b\left(\dfrac{z_{\alpha/2}}{H}\right)^2$

Reprinted from Young and Young, 1994, p. 744; with kind permission from Elsevier Science-NL, Sara Burgerhartstraats 25, 1055 KV Amsterdam, The Netherlands.

ECOSTAT can be used to determine the optimal sample size. From the *Sequential Sampling* menu, select *Estimation*. First, indicate whether the sample size is fixed. Next choose the appropriate distribution from the menu and the measure of precision. Use the scroll bars to indicate values for the parameters and the level of precision. For fixed sample sizes, the optimal sample size is displayed after *Sample Size* is pressed. For the binomial distribution, the parameter n should usually be set to one. If groups of size n are taken for each observation, n is the group size and may differ from one.

The geometric distribution results if the negative binomial parameter $k = 1$; that is, $\sigma^2 = \mu + \mu^2$. Substituting into (4.4), we have

$$CV(\bar{X}) = \sqrt{\frac{1}{n}\left(\frac{1}{\mu} + 1\right)}. \tag{4.5}$$

The number of observations that must be taken to obtain a $CV(\bar{X}) = C$ is

$$n = \frac{1}{C^2}\left(\frac{1}{\mu} + 1\right). \tag{4.6}$$

Example 4.2

A field of seedling oats is infested by greenbugs. Experience has shown that these insects are usually in the geometric distribution when the sampling unit is a seedling oat. The purpose of sampling is to estimate the mean with a $CV(\bar{X})$ of 10%. If the average number of greenbugs on a plant is 8, how many observations must be taken?

$$n = \frac{1}{(0.1)^2}\left(\frac{1}{8} + 1\right) = 112.5$$

Therefore, 113 observations are needed.

The Poisson distribution has a variance equal to the mean, $\sigma^2 = \mu$. Thus the $CV(\bar{X})$ may be written as

$$CV(\bar{X}) = \sqrt{\frac{1}{n\mu}}. \tag{4.7}$$

The sample size required to obtain a $CV(\bar{X}) = C$ is

$$n = \frac{1}{\mu C^2}. \tag{4.8}$$

Example 4.3

Some spider species have been reported to be in the Poisson distribution. Suppose that the sampling unit is 1/5,000th of an acre. Assume the average number of spiders in 1/5,000th of an acre is 3. How many observations are needed to estimate the mean with a $CV(\bar{X})$ of no more than 10%?

$$n = \frac{1}{\mu C^2} = \frac{1}{3(.1)^2} = 33.3$$

Thus we need to take at least 34 observations.

The binomial distribution has $\sigma^2 = np(1 - p)$. Generally, this arises when doing presence–absence or success–failure sampling. In this case, p is the parameter of interest, and n is 1 for each experimental unit; that is, the Bernoulli distribution is used and $\sigma^2 = p(1 - p)$. Then the $CV(\bar{X})$ may be written as

$$CV(\bar{X}) = \sqrt{\frac{q}{pn}}. \qquad (4.9)$$

where $q = 1 - p$ and now n represents the number of observations taken. The sample size required to achieve a $CV(\bar{X}) = C$ when sampling from the binomial distribution is then

$$n = \frac{q}{pC^2}. \qquad (4.10)$$

Example 4.4

Suppose that 40% of a cow herd is infested with cattle grubs. How many observations are needed to check this assumption with a $CV(\bar{X})$ of 20%? First, notice that since 40% of the herd is infested, $p = 0.4$ and $q = 1 - 0.4 = 0.6$. Then we have

$$n = \frac{q}{pC^2} = \frac{0.6}{(0.4)(0.2)^2} = 37.5.$$

That is, a sample size of 38 cows is needed.

Note that in each case, a larger sample size is required when the population mean is decreased. A decrease in the specified value of $CV(\bar{X})$ also requires a larger sample size.

Sample Sizes Required to Set Confidence Intervals

Suppose we take a random sample of size n from a population with mean μ and variance σ^2. By invoking the Central Limit Theorem, we have

$$P\left(\bar{X} - z_{\alpha/2}\frac{\sigma}{\sqrt{n}} < \mu < \bar{X} + z_{\alpha/2}\frac{\sigma}{\sqrt{n}}\right) \approx 1 - \alpha \qquad (4.11)$$

where $(1 - \alpha)$ is the specified confidence level and $z_{\alpha/2}$ is the $(1 - \alpha/2)$ quantile of the standard normal distribution. For $(1 - \alpha) = 0.90, 0.95$, and 0.99, we have $z_{\alpha/2} = 1.645, 1.96$, and 2.58, respectively. From equation (4.11), we have that the probability of the interval

$$\bar{X} \pm z_{\alpha/2}\frac{\sigma}{\sqrt{n}} \qquad (4.12)$$

covering the mean μ is approximately $(1 - \alpha)$, regardless of the form of the parent population, provided the population has a finite variance. As n increases, the approximation improves for any nonnormal distribution. However, for highly skewed distributions, such as the geometric, the actual confidence level may tend to stay below the stated level of $(1 - \alpha)$ for sample sizes of several hundred, emphasizing that the level of approximation depends on the population distribution being sampled.

The half-length of the confidence interval in equation (4.12) is

$$z_{\alpha/2}\frac{\sigma}{\sqrt{n}}. \qquad (4.13)$$

From this, we can observe that the length of the confidence interval depends on three things: the confidence level $(1 - \alpha)$, the standard deviation of the population σ, and the number of observations drawn from the population. If two of these values are known or set at desired levels and the half-length of the confidence interval is specified, we can solve for the remaining value. Therefore, given σ, the desired level of confidence $(1 - \alpha)$, and the specified half-length (half of the desired length) of the interval, we can solve for the optimal sample size n. Here optimal means that it is the smallest value of n that will result in an interval with no more than the specified half-length (or length) and the desired level of confidence. We consider specifying the half-length of the confidence interval (a) as proportional to the true parameter and (b) as a fixed length. The results are summarized in Table 4.1.

Length Proportional to the Parameter of Interest

Suppose our goal in sampling is to estimate the mean μ of the population using a confidence interval with half-length a fixed proportion D of the mean; that is,

$$z_{\alpha/2} \frac{\sigma}{\sqrt{n}} = D\mu. \tag{4.14}$$

Solving for n, we find the optimal sample size n to be

$$n = \left(\frac{z_{\alpha/2}\sigma}{D\mu}\right)^2. \tag{4.15}$$

Suppose we are sampling from the negative binomial distribution. Then by substituting $\sigma^2 = \mu + \mu^2/k$, we find that the optimal sample size for estimating the mean within a specified proportion D of μ with $(1 - \alpha)$ confidence is

$$n = \left(\frac{1}{\mu} + \frac{1}{k}\right)\left(\frac{z_{\alpha/2}}{D}\right)^2. \tag{4.16}$$

Example 4.5

As in Example 4.1, we are interested in estimating the mean number of bollworms in 3 row feet of cotton. Our goal is to estimate the mean within 10% of the mean with 95% confidence. If the mean is 2 bollworms/3 row feet and $k = 3$, how many observations are needed?

$$n = \left(\frac{1}{\mu} + \frac{1}{k}\right)\left(\frac{z_{\alpha/2}}{D}\right)^2 = \left(\frac{1}{3} + \frac{1}{2}\right)\left(\frac{1.96}{0.10}\right)^2 = 320.1$$

Thus 321 observations are needed.

By setting $k = 1$ in equation (4.16), we obtain the following for the geometric distribution:

$$n = \left(\frac{1}{\mu} + 1\right)\left(\frac{z_{\alpha/2}}{D}\right)^2. \tag{4.17}$$

Example 4.6

Greenbugs in seedling oats are often found to be in the geometric distribution (*see* Example 4.2). Suppose we want to estimate the mean within 10% with 90% confidence. If the average number of greenbugs on a seedling oat is 8, how many observations are required?

$$n = \left(\frac{1}{\mu} + 1\right)\left(\frac{z_{\alpha/2}}{D}\right)^2 = \left(\frac{1}{8} + 1\right)\left(\frac{1.645}{0.10}\right)^2 = 304.4$$

That is, 305 observations are needed.

If the population being sampled is Poisson, then by using $\sigma^2 = \mu$, equation (4.15) becomes

$$n = \frac{1}{\mu}\left(\frac{z_{\alpha/2}}{D}\right)^2. \tag{4.18}$$

Example 4.7

In Example 4.3, the average number of spiders in 1/5,000th of an acre was of interest. Assume that the distribution of the spiders is Poisson with a mean of 3 spiders per 1/5,000th of an acre. How many observations are needed to estimate the mean within 10% of the mean with 95% confidence?

$$n = \frac{1}{\mu}\left(\frac{z_{\alpha/2}}{D}\right)^2 = \frac{1}{3}\left(\frac{1.96}{0.10}\right)^2 = 128.1$$

Thus 129 observations must be made.

When sampling from a binomial population for the purpose of estimating the proportion p ($p = \mu$ when $n = 1$), (4.14) may be written as

$$z_{\alpha/2}\frac{p(1 - p)}{\sqrt{n}} = Dp. \tag{4.19}$$

Solving for n to obtain the optimal sample sizes gives

$$n = \left(\frac{1 - p}{p}\right)\left(\frac{z_{\alpha/2}}{D}\right)^2. \tag{4.20}$$

Example 4.8

As in Example 4.4, consider a cow herd that has 40% of the cattle infested with grubs. If we want to estimate the proportion of infested cattle within 30% of the population proportion with 95% confidence, how many cattle must be checked?

$$n = \left(\frac{1 - p}{p}\right)\left(\frac{z_{\alpha/2}}{D}\right)^2 = \left(\frac{0.6}{0.4}\right)\left(\frac{1.96}{0.30}\right)^2 = 64.02$$

Therefore, 65 observations are required.

Length Fixed

Now suppose that we want the half-length of the $(1 - \alpha)$ 100% confidence interval on the mean to be a specified positive value of H; that is,

$$z_{\alpha/2} \frac{\sigma}{\sqrt{n}} = H. \tag{4.21}$$

By solving for n in equation (4.21), we find the optimal size to be

$$n = \left(\frac{z_{\alpha/2}\sigma}{H}\right)^2. \tag{4.22}$$

When sampling from the negative binomial distribution, the relationship in mean and variance permits us to rewrite equation (4.22) as

$$n = \left(\mu + \frac{\mu^2}{k}\right)\left(\frac{z_{\alpha/2}}{H}\right)^2. \tag{4.23}$$

Example 4.9

Consider again the problem in Examples 4.1 and 4.5 of estimating the mean number of bollworms in 3 row feet of cotton. Suppose now that we want to estimate the mean within 0.2 bollworm per 3 row feet with 95% confidence. If the distribution is negative binomial with a mean of 2 bollworms/3 row feet and a k of 3, how many observations are needed?

$$n = \left(\mu + \frac{\mu^2}{k}\right)\left(\frac{z_{\alpha/2}}{H}\right)^2 = \left(2 + \frac{2^2}{3}\right)\left(\frac{1.96}{0.20}\right)^2 = 320.1$$

That is, 321 observations should be taken.

Setting $k = 1$ in equation (4.23), we obtain the required sample size for estimating the meaning within H with $(1 - \alpha)$% confidence based on the geometric distribution:

$$n = (\mu + \mu^2)\left(\frac{z_{\alpha/2}}{H}\right)^2. \tag{4.24}$$

Example 4.10

Refer again to Example 4.2 and 4.6, where the average number of greenbugs on a seedling oat was 8, and the frequency distribution when the sampling unit is a seedling is geometric. Suppose we wanted to estimate the mean within 0.25 with 90% confidence:

$$n = (\mu + \mu^2)\left(\frac{z_{\alpha/2}}{H}\right)^2 = (8 + 8^2)\left(\frac{1.645}{0.25}\right)^2 = 3117.3$$

The sample size should be 3118.

For the Poisson distribution, the variance is equal to the mean. This permits equation (4.22) to be written as

$$n = \mu\left(\frac{z_{\alpha/2}}{H}\right)^2. \tag{4.25}$$

Example 4.11

Consider again the spiders in Examples 4.3 and 4.7. Assume the average number of spiders in 1/5,000th of an acre is 3. The purpose of sampling is to estimate the mean within 0.1 with 95% confidence.

$$n = \mu\left(\frac{z_{\alpha/2}}{H}\right)^2 = 3\left(\frac{1.96}{0.10}\right)^2 = 1152.4$$

The sample size needs to be 1153.

Now consider sampling from a binomial population. Again assume that $n = 1$ in which case $\mu = p$. Because the variance is $p(1 - p)$, equation (4.22) may be written as

$$n = p(1 - p)\left(\frac{z_{\alpha/2}}{H}\right)^2. \tag{4.26}$$

Example 4.12

Suppose that, as in Examples 4.4 and 4.8, 40% of a cow herd is infested with cattle grubs. How many observations would be required to estimate this proportion within 10% with 95% confidence? Because 40% of the herd is infested, $p = 0.4$ and $q = 0.6$. To obtain an estimate within 10% of the true population proportion p, set $D = 0.10$. The required sample is

$$n = p(1 - p)\left(\frac{z_{\alpha/2}}{H}\right)^2 = 0.4(0.6)\left(\frac{1.96}{0.10}\right)^2 = 92.2.$$

In this example, 93 observations are needed.

Sequential Estimation

The primary difficulty in determining the optimal sample size n for any of the cases given above is that, in each case, the optimal sample size depends upon one or more unknown parameters. A worker may put in some "reasonable" values for these unknown parameters to determine approximate sample sizes. However, one can never be sure whether a predetermined fixed sample size will result in estimates of the population density μ with the desired level of precision.

When attempting to estimate the mean using a confidence interval no longer than a predetermined length, Stein (1945) suggested replacing μ and σ^2 in equation (4.22) with some preliminary estimates of the mean and variance, say $\bar{X}_0$ and s_0^2, based on an initial sample of size n_0. Thus,

$$n = \left(\frac{z_{\alpha/2} s_0}{H} \right)^2. \tag{4.27}$$

However, the choice of n_0 may be difficult. In addition, we are not assured that

$$n \geq \left(\frac{z_{\alpha/2} s_n}{H} \right)^2 \tag{4.28}$$

where s_n^2 is the sample estimate of the variance based on n observations.

Sequential sampling permits the sampler to update estimates of population parameters after each field observation. These estimates then permit an informed decision as to whether the population density can be estimated with the desired level of precision. Therefore, in sequential sampling, the sample size is not fixed. It is random. Sampling continues until a stopping criterion is satisfied. The purpose of the stopping criterion is to indicate the smallest sample size for which the estimate of the mean has met the specified level of precision.

Chow and Robbins (1965) suggested updating the estimates of mean and variance after each observation and sampling until equation (4.28) was satisfied in order to estimate the mean within H units with $(1 - \alpha)$ 100% confidence; that is, sample until

$$n \geq \left(\frac{z_{\alpha/2} s_n}{H} \right)^2. \tag{4.29}$$

They were able to show that this method is both asymptotically consistent and efficient for any population distribution with finite variance. However, Monte Carlo simulation studies have shown that for highly skewed distributions, the probability of coverage is below the stated level unless H is small (the sample size is large).

Nadás (1969) suggested a similar approach when the goal of sampling is to estimate the mean within a specified proportion D of the mean with a stated level of confidence. Using equation (4.15) and updating the estimates of the mean and variance after each observation, he proposed to sample until

$$n \geq \left(\frac{z_{\alpha/2} S_n}{D \bar{X}_n}\right)^2. \tag{4.30}$$

Nadás showed that the above procedure is asymptotically consistent and efficient. However, again simulation studies have shown that for the highly skewed distributions frequently encountered in population studies, the actual confidence level is below the stated one until the sample size becomes large (in the hundreds).

Rojás (1964) introduced sequential *CV* control sampling for soil insects. This approach is widely used in entomology. By estimating the parameters in equation (4.2) and updating these estimates after each observation, he was led to propose sampling until the first time

$$n \geq \left(\frac{S_n}{\bar{X}_n C}\right)^2 \tag{4.31}$$

where $\bar{X}_n$ and s_n^2 are sample estimates of mean and variance, respectively, based on the n observations in the sample. Willson (1981), building on the work of Mukhopadhyay (1978) and Ghosh and Mukhopadhyay (1979), showed that this procedure was asymptotically consistent and efficient. Again large simulation studies have shown that for highly skewed distributions, the actual $CV(\bar{X})$ is greater than the stated level of C, unless C becomes small (the sample size is large).

The procedures outlined in equations (4.27) to (4.31) are nonparametric in nature; that is, they may be used with any population distribution that has a finite variance. They are intuitive extensions of fixed-sample-size results. However, we are only assured that the goal of estimating the mean with a specified level of precision is accomplished in the limit. In each case, Monte Carlo studies have shown that for the highly skewed distributions encountered in population studies, the sample size must be in the hundreds before the precision of the estimate is at or near the specified level. Also, the need to update estimates of the mean and variance after each observation requires the use of some form of computing equipment in the field. Although this is certainly feasible, it may not be as desirable as a method that does not require such computations.

An alternative approach is to use the properties of the population distribution being sampled in devising a method of sampling. For the negative binomial, geometric, Poisson, and binomial distributions, as well as many others, the relationship in mean and variance is well understood. By using that known relationship, we are able to develop a sampling program for which the precision of the estimate is more nearly at the specified level. This is done for each of the four distributions; the stopping rules are summarized in Table 4.2.

Sequential Estimation for the Negative Binomial

Assume that we have sufficient experience with an ecosystem that we are willing to assume that, for the sampling unit being used, the distribution of the number of organisms per sampling unit is distributed as a negative binomial random variable or as a variable with a distribution very close to a negative binomial distribution. The knowledge of k, or lack of that knowledge, is a determining factor in the development of a sampling program for negative binomial populations. Sometimes, a biological model of the ecosystem may imply a value for k. If the model has been verified previously, then this value of k may be used in the development of a sampling scheme. In other cases, the value of k has been found to be stable through time and for varying population densities. This value of k could then be appropriately used in a sampling program. If knowledge of k, or a precise estimate of k, is not available, k must be estimated at the time of sampling. We first consider estimation of the mean when k is not known. Next, we consider methods of obtaining an estimate of k for use in these sampling schemes. Finally, we consider estimation of the mean when k is assumed known.

Parameter k *Unknown*

Intuitive sequential sampling schemes for the negative binomial distribution could be developed by updating the estimates of μ and k after each observation and sampling until one of the equations (4.4), (4.16), or (4.23) is satisfied, depending on the measure of precision being used. Let $\bar{X}_n$ and $\hat{k}_n$ be the estimates of μ and k, respectively, based on n observations. Then, if the goal of sampling is to estimate the mean with a specified $CV(\bar{X}) = C$, we sample until the first time

$$n \geq \frac{1}{C^2}\left(\frac{1}{\bar{X}_n} + \frac{1}{\hat{k}_n}\right). \tag{4.32}$$

Now assume that the mean is to be estimated within a proportion D of the mean with $(1 - \alpha)$ 100% confidence. Based on equation (4.16), we continue sampling until the first time

$$n = \left(\frac{1}{\bar{X}_n} + \frac{1}{\hat{k}_n}\right)\left(\frac{z_{\alpha/2}}{D}\right)^2. \tag{4.33}$$

Finally, suppose the mean is to be estimated within H units with $(1 - \alpha)$ 100% confidence. Following equation (4.23), we propose sampling until the first time

Table 4.2. Stopping Rules When Estimating the Population Density With a Specified $CV(\bar{X}) = C(C)$, Within $D\mu$ With Confidence $(1 - \alpha)$ 100% Confidence (D), and Within H With Confidence $(1 - \alpha)$ 100% Confidence (H)

	C	D	H
Binomial	$0 < T_n < n$ and $T_n \geq \dfrac{n}{nC^2 + 1}$	$0 < T_n < n$ and $T_n \geq \dfrac{n z_{\alpha/2}^2}{nD^2 + z_{\alpha/2}^2}$	$0 < T_n < n$ and either $T_n \geq \dfrac{n}{2} + n\sqrt{\dfrac{1}{4} - \dfrac{nH^2}{z_{\alpha/2}^2}}$ or $T_n \leq \dfrac{n}{2} - n\sqrt{\dfrac{1}{4} - \dfrac{nH^2}{z_{\alpha/2}^2}}$
Poisson	$n \geq 2$ and $T_n \geq \dfrac{1}{C^2}$	$n \geq 2$ and $T_n \geq \left(\dfrac{z_{\alpha/2}}{D}\right)^2$	$T_n > 0$ and $T_n \leq \left(\dfrac{nH}{z_{\alpha/2}}\right)^2$
Negative binomial	$n > \dfrac{1}{C^2 k} - \dfrac{1}{k}$ and $T_n \geq \dfrac{nk}{(nk+1)C^2 - 1}$	$n > \dfrac{z_{\alpha/2}^2}{kD^2} - \dfrac{1}{k}$ and $T_n \geq \dfrac{nk z_{\alpha/2}^2}{(nk+1)D^2 - z_{\alpha/2}^2}$	$T_n > 0$ and $T_n \leq -\dfrac{nk}{2} + n\sqrt{\dfrac{k^2}{4} + \dfrac{nkH^2}{z_{\alpha/2}^2}}$
Iwao's Patchiness Regression	$n > \dfrac{\beta - 1}{C^2}$ and $T_n \geq \dfrac{n(\alpha+1)}{nC^2 - \beta + 1}$	$n > \dfrac{(\beta-1)z_{\alpha/2}^2}{D^2}$ and $T_n \geq \dfrac{n(\alpha+1)z_{\alpha/2}^2}{nD^2 - (\beta-1)z_{\alpha/2}^2}$	$T_n > 0$ and $T_n \leq -\dfrac{n(\alpha+1)}{2(\beta-1)} + n\sqrt{\dfrac{nH^2}{(\beta-1)z_{\alpha/2}^2} + \left(\dfrac{\alpha+1}{2(\beta-1)}\right)^2}$

Reprinted from Young and Young, 1994, p. 747, with kind permission from Elsevier Science-NL, Sara Burgerhartstroat 25, 1055 KV, Amsterdam, The Netherlands.

$$n \geq \left(\bar{X}_n + \frac{\bar{X}_n^2}{\hat{k}_n} \right) \left(\frac{z_{\alpha/2}}{H} \right)^2.$$ (4.34)

Suppose the method-of-moments estimator of k is used in equations (4.32), (4.33), and (4.34). It is left as an exercise to show that equations (4.32), (4.33), and (4.34) reduce to equations (4.31), (4.30), and (4.29), respectively. That is, the sampling procedures based on the method-of-moments estimator of k become equivalent to the nonparametric methods. Although these procedures have been shown to behave well asymptotically, they do not perform well for moderate sample sizes when the negative binomial is the underlying distribution. The difference in the stated and actual levels of precision differs depending on μ and k. For example, suppose we want to estimate the mean with a $CV(\bar{X}) = 0.25$. When $(\mu, k) = (1, 1)$ and $(5, 5)$, simulations indicate that the actual $CV(\bar{X})$ is 0.42 and 0.32, respectively (Willson 1981). Hence, the true $CV(\bar{X})$ is larger than the specified 0.25, indicating that the mean is not estimated as precisely as the sampling procedure would lead us to believe. Similar results hold for the other two procedures.

No sequential sampling method for negative binomial populations has been proposed that performs better than the nonparametric methods when the parameter k is unknown. Monte Carlo studies can provide insight into the behavior of a particular method in the region of the parameter space of interest. These permit a realistic assessment of the precision of the estimated mean for the proposed sampling scheme. Perhaps a lower value of $CV(\bar{X})$ could be used in equation (4.32), for example, in order to attain the desired $CV(\bar{X})$. Thus, in spite of their weaknesses, they serve a useful role in data collection. However, if a precise estimate of k can be obtained, we can develop sampling procedures involving only the unknown parameter μ. Methods to obtain that precise estimate of k will be considered next.

Estimating Parameter k

The parameter k may be called a nuisance parameter when the purpose of sampling is to estimate μ; that is, it is not of primary interest, but its presence makes estimation of the mean more difficult. Sometimes, biological models or experience may lead us to anticipate the distribution will be negative binomial with a specified value of k. Goodness-of-fit tests can be used to verify that the negative binomial with that value of k does, in fact, describe the data adequately. More frequently, the estimates of k from several data sets tend to be *close* to each other. From these, we may want to obtain an estimate of a common k value for the ecosystem and to test whether that common value of k seems to hold over all data sets. Several methods of estimating a common k have been proposed (Beall, 1942; Kleezkowski, 1949; Anscombe, 1949, 1950; Bliss and Fisher, 1953). The method proposed by Bliss and Owen (1958) is the most widely used and is discussed first. The likelihood ratio test for a common k is also discussed. Finally, a multistage method of estimating k is presented.

Suppose t data sets have been collected, and we want to test for a common k. Let n_i, s_i, and $\bar{X}_i$ denote the sample size, estimated standard deviation, and estimated mean, respectively, for the tth data set, $i = 1, 2, \ldots, t$. The estimation process begins by computing

$$x_i' = \bar{X}_i^2 - \frac{s_i^2}{n_i} \tag{4.35}$$

and

$$y_t' = s_i^2 - \bar{X}_i \tag{4.36}$$

for each of the t data sets. The regression line of y' on x' passes through the origin and has slope $1/\hat{k}_c$. The precision of the estimate may be increased by weighting each population inversely to the variance; that is,

$$w_i = \frac{0.5(n_i - 1)k_c^4}{\mu_i^2(\mu_i + k_c)^2 \left(k_c(k_c + 1) - \dfrac{(2k_c - 1)}{n_i} - \dfrac{3}{n_i^2} \right)}. \tag{4.37}$$

Because the weight involves unknown parameters, including k_c, the estimation process must be iterative. An unweighted regression provides an initial estimate of k_c:

$$\frac{1}{\hat{k}_c} = \frac{\sum\limits_i x'y'}{\sum\limits_i x'^2}. \tag{4.38}$$

Then estimates of the parameters are used to obtain the weight w_i for each population i, $i = 1, 2, \ldots, t$:

$$w_i = \frac{0.5(n_i - 1)\hat{k}_c^4}{\bar{X}_i^2(\bar{X}_i + \hat{k}_c)^2 \left(\hat{k}_c(\hat{k}_c + 1) - \dfrac{(2\hat{k}_c - 1)}{n_i} - \dfrac{3}{n_i^2} \right)}. \tag{4.39}$$

The weighted estimate of k_c is now found using

$$\frac{1}{\hat{k}_c} = \frac{\sum\limits_i w_i x_i' y_i'}{\sum\limits_i w_i x_i'^2}. \tag{4.40}$$

If the weighted estimate of k_c differs appreciably from its initial unweighted estimate, then the weights should be recomputed using the weighted estimate of k_c, and a new weighted estimate of k_c determined. This process is repeated until the change in the estimate of k_c is negligible. The estimate of a common k may change very little during these iterations. However, the iterations are essential for a valid test of homogeneity.

The test of homogeneity is a test of the null hypothesis that there is a common k for all t populations. Two tests have been developed for the test of this hypothesis. The first test is based on the fact that the error sum of squares from the weighted regression through the origin is approximately χ^2 with $(t - 1)$ degrees of freedom; that is, if a common k is present,

$$\sum_i (w_i y_i^2) - \frac{\sum_i^2 w_i x_i' y_i'}{\sum_i w_i x_i'^2} \sim \chi^2_{t-2} \tag{4.41}$$

where $\sum_i^2 x_i$ represents the square of the sum of the x_i's; that is, $(\Sigma x_i)^2$.

If the value of the test statistic exceeds the $(1 - \alpha)$ quantile of a χ^2 distribution with $(t - 2)$ degrees of freedom, then we reject the null and conclude that the use of a common k is not appropriate for this set of t populations. It should be noted that the first term in (4.41) is the uncorrected sum of squares and the second term is the sum of squares due to the regression of y' on x'. Thus the expression in equation (4.41) is the error sum of squares associated with a linear regression through the origin. This fact is useful when using statistical software to estimate k_c.

The second test requires four quantities that may be obtained from the weighted linear regression constrained to pass through the origin and the weighted linear regression with an intercept. First, the sum of squares associated with the regression constrained to pass through the origin is

$$B_0^2 = \frac{\sum_i^2 (w_i x_i' y_i')}{\sum_i w_i x_i'^2}. \tag{4.42}$$

For the regression that passes through the origin, the corrected total sum of squares is

$$CSS = \sum_i (w_i y'^2) - C \tag{4.43}$$

where

$$C = \frac{\sum\limits_i^2 w_i y'}{\sum\limits_i w_i} \qquad (4.44)$$

is the sum of squares associated with the intercept. The regression sum of squares when the line is not constrained to pass through the origin is

$$B^2 = \frac{\left(\sum\limits_i w_i x_i' y_i'\right) - \dfrac{\sum\limits_i (w_i x_i') \sum\limits_i (w_i y_i')}{\sum\limits_i w_i}}{\sum\limits_i (w_i x_i'^2) - \dfrac{\sum\limits_i^2 (w_i x_i')}{\sum\limits_i w_i}}. \qquad (4.45)$$

Then the test may be arranged as an analysis of variance as in Table 4.3.

If a common value of k, k_c, is justified, the F-value in the first row should be significant and that in the second row should not be significant. A significant F in the second row indicates a progressive change in k.

The two tests proposed by Bliss and Owen do not always result in the same conclusion as to the appropriateness of a common k. The second test seems to be preferred, but a definitive comparison of these and the other techniques discussed later in this chapter has yet to be conducted.

Example 4.13

The number of adult Colorado potato beetles (*Leptinotarsa decemlineata* Say) in 10 feet of row in a heavily infested potato field was recorded (Beall, 1939; Bliss and Owen, 1958). Eight such observations were made within each of 16 blocks. The experimenter wanted to determine whether a common value of k was appropriate for all 16 blocks. For each block, the sample mean and sample variance based on the eight observations were recorded.

In Table 4.4, x' and y' are computed using equations (4.35) and (4.36). For example,

Table 4.3. *Test for a Common* k *Presented in Form of Analysis of Variance*

Effect of	df	Sum of squares	Mean square	F
Slope, $1/k_c$	1	B_0^2	B_0^2	B_0^2/s^2
Computed intercept against 0	1	$C + B^2 - B_0^2$	I_0	I_0/s^2
Error	$t - 3$	$CSS - B^2$	s^2	—

Table 4.4. *Components Computed as Part of Test for a Common* k *Based on Beall's (1939) Colorado Potato Beetle Data*

Block	$\bar{X}$	s^2	x'	y'
1	75.75	539.07	5670.7	463.32
2	84.00	627.14	6977.6	543.14
3	42.25	257.93	1752.8	215.68
4	32.50	77.43	1046.6	44.93
5	48.00	172.57	2282.4	124.57
6	48.62	91.12	2352.5	42.50
7	66.50	737.43	4330.1	670.93
8	56.75	71.64	3211.6	14.89
9	64.88	101.27	4196.8	36.39
10	52.12	417.55	2664.3	365.43
11	40.12	137.55	1592.4	97.43
12	26.62	48.55	702.6	21.93
13	34.12	252.41	1132.6	218.29
14	34.12	14.70	1162.3	− 19.42
15	13.88	45.84	186.9	31.96
16	37.62	167.70	1394.3	130.08

$$x_1' = \bar{X}_1^2 - \frac{s_1^2}{n_1} = 75.75^2 - \frac{539.07}{8} = 5670.7$$

and

$$y_1' = s_i^2 - \bar{X}_1 = 539.07 - 75.75 = 463.32.$$

An initial estimate of the reciprocal of k_c can now be determined using equation (4.38); that is,

$$\frac{1}{k_c} = \frac{\sum\limits_i x'y'}{\sum\limits_i x'^2} = \frac{11888444.7}{157172903.6} = 0.07564.$$

This provides a starting estimate of k_c:

$$\hat{k}_c = \frac{1}{0.07564} = 13.221.$$

This estimate of k_c may now be used to estimate the weight for each sample using (4.39). For example, for the first block, we find

$$w_1 = \frac{0.5(8 - 1)(13.221)^4}{75.75^2(75.75 + 13.221)^2\left(13.221(13.221 + 1) - \dfrac{2(13.221) - 1}{8} - \dfrac{3}{8^2}\right)}$$

$$= 0.0000127.$$

Using these weights, a weighted regression passing through the origin is computed to improve the estimate of k_c:

$$\frac{1}{\hat{k}_c} = \frac{\sum\limits_i w_i x_i' y_i'}{\sum\limits_i w_i x_i'^2} = \frac{388.030}{5270.030} = 0.07362.$$

The updated estimate of k_c is then

$$\hat{k}_c = \frac{1}{0.07362} = 13.583.$$

Because the estimate of k_c changed in the first decimal place, we recompute the weights and perform another weighted regression. In this iteration, we obtain $\hat{k}_c = 13.588$. The estimate of the common k only changed by 0.005 in this case; therefore, we stop and test for a common k. Figure 4.1 depicts the final weighted regression of y' on x' with the sample values. The modeled relationship between mean and variance as shown in Figure 4.2.

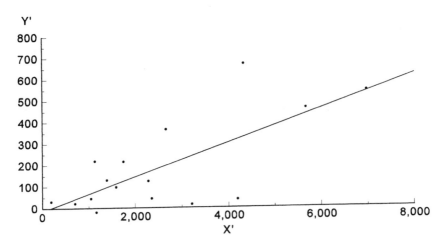

Figure 4.1. The regression of y' on x' used to obtain a common k with Beall's (1939) Colorado potato beetle data.

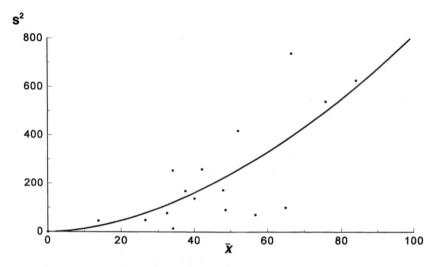

Figure 4.2. The relationship between mean and variance for Beall's (1939) Colorado potato beetle data based on the common k with the observed data.

To conduct the χ^2 test for a common k, we use equation (4.41)

$$\sum_i (w_i y_i'^2) - \frac{\left(\sum_i w_i x_i' y_i'\right)^2}{\sum_i w_i x_i'^2} = 48.86 - 29.83 = 19.03.$$

If the null hypothesis is true and there is a k common to all blocks, 19.03 is an observation from an approximate χ^2 distribution with 14 degrees of freedom (df). The probability of obtaining a value larger than 19.03 from this distribution is 0.16; therefore, we do not reject the hypothesis of a k common to all blocks. It should be noted that 48.86 is the uncorrected sum of squares, 19.83 is the sum of squares due to regression, and 19.03 is the error sum of squares from a weighted regression of y' on x' passing through the origin.

For the F-test to test the equality of mean and variance, we must compute two additional quantities. First, from equation (4.44), we have

$$C = \frac{\left(\sum_i w_i y'\right)^2}{\sum_i w_i} = 16.59.$$

This may also be computed as the difference in the uncorrected sum of squares

(48.85592) and the corrected sum of squares (32.26809). Next, using equation (4.45), we find the weighted regression sum of squares when the line is not constrained to pass through the origin, which is 14.187. These quantities may be used to establish the analysis of variance (ANOVA) table shown in Table 4.5. Because the F-test associated with the slope is significant but that for the intercept is not, the hypothesis of a common k is not rejected.

Maximum likelihood estimation (MLE) of a k common to several negative binomial populations that may have different means is an alternative to the Bliss and Owen regression technique. The MLE of k_c is the root of the following equation in $\hat{k}_c$:

$$\sum_{i=1}^{t} n_i \ln(\bar{X}_i + \hat{k}_c) - n \ln(\hat{k}_c) = \sum_{i=1}^{t} \sum_{j=0}^{\infty} m_{ij} \sum_{s=0}^{j} \frac{1}{k_c + s} \tag{4.46}$$

where m_{ij} is the number of observations in the sample from population i with the value j.

The likelihood ratio test may then be used to test the hypothesis that there is a k common to the t negative binomial populations under consideration. To perform this test, the likelihood functions for the restricted and unrestricted parameter spaces must be computed. In the restricted parameter space, the mean may differ with populations but there is a k common to the t populations; therefore, the maximum of the likelihood function in the restricted parameter space is

$$L(\hat{\omega}) = \prod_{i=1}^{t} \left[\left(\prod_{b=1}^{n_i} \frac{(\hat{k}_c + x_{ib} - 1)!}{x_{ib}!\hat{k}_c!} \right) \left(\frac{\bar{X}_i}{\bar{X}_i + \hat{k}_c} \right)^{\sum_{b=1}^{n_i} x_{ib}} \left(\frac{\hat{k}_c}{\bar{X}_i + \hat{k}_c} \right)^{n_i \hat{k}_c} \right]$$

where x_{ij} is the jth observation in the ith population. When k is not restricted to be the same for the t populations, the maximum of the likelihood function is

$$L(\hat{\Omega}) = \prod_{i=1}^{t} \left[\left(\prod_{b=1}^{n_i} \frac{(\hat{k}_i + x_{ib} - 1)!}{x_{ib}!\hat{k}_i!} \right) \left(\frac{\bar{X}_i}{\bar{X}_i + \hat{k}_i} \right)^{\sum_{b=1}^{n_i} x_{ib}} \left(\frac{\hat{k}_i}{\bar{X}_i + \hat{k}_i} \right)^{n_i \hat{k}_i} \right]$$

where $\hat{k}_i$ is the maximum likelihood estimate of k based on the observations from the ith population. The likelihood ratio is then denoted by

Table 4.5. *Test for a Common k Based on Beall's (1939) Colorado Potato Beetle Data*

Effect of	df	Sum of squares	Mean square	F
Slope, $1/k_c$	1	29.82786	29.82786	21.45
Computed intercept against 0	1	0.94661	0.94661	0.68
Error	13	18.08145	1.39088	—

$$\lambda = \frac{L(\hat{\omega})}{L(\hat{\Omega})}. \tag{4.47}$$

If the null hypothesis is true and there is a k common to the t populations, -2 $\ln(\lambda)$ has an approximate χ^2 distribution with $(t - 1)$ degrees of freedom. We reject the null hypothesis and conclude that at least one population has a different k from the others at the α significance level if $-2 \ln(\lambda)$ exceeds the $(1 - \alpha)$ quantile value of a χ^2 value with $(t - 1)$ degrees of freedom.

Maximum likelihood estimation and the likelihood ratio test for a common k has not been used much in biological applications. The reason for this probably lies in the extensive computations involved. However, with today's computing equipment, this is no longer a major concern. A limited Monte Carlo study (Willson, 1981) indicated that the likelihood ratio test controlled the size of the test better than the regression technique, while at the same time giving more power. Perhaps this method should be considered more carefully in biological applications.

ECOSTAT may be used to estimate a common k for t populations. To begin, select *Common k* from the *Sequential Sampling* menu. Then indicate whether the complete data sets are available or summary data are to be used. By complete data sets we mean that all observations for each of the t populations are available. In this case, the first line of the ASCII data file has the number of populations t that have been sampled. The second line has a numeric population identification and the number (n_1) of observations from that population, separated by a space. The next n_1 lines each have an observation from the first population. The population identification and the number of observations for the second population follows the observations from the first population. The observations from the second population are next. The sequence is repeated for each of the t populations. The maximum likelihood estimate of the common k is automatically computed, and the likelihood ratio test conducted.

If summary data are present, the ASCII data file should have the number of populations t on the first line. The next t lines should each have a numeric population identification, the sample size, sample mean, and sample variance in that order. Bliss and Owen's (1958) method of estimating the common k is used in this case, and both tests are conducted. The ECOSTAT data file beetle.dat has the data for this example.

A third method of estimating k is a multistage method given by Willson and Young (1983). This method estimates k in the field at the time data are taken. In this method, five observations are taken, k is estimated using the method of moments estimator. Five more observations are taken, and k is estimated on the basis of all 10 observations. This procedure is continued until the last two estimates of k differ by no more than 0.05. Computer simulations and field studies have shown this to be equal or superior to the method-of-moments or maximum likelihood estimators based on comparable fixed sample sizes.

Example 4.14

In April 1981, data were collected on greenbugs on seedling oats (Willson et al., 1984). A seedling oat was the sampling unit. The multistage procedure was used to obtain a precise estimate of k. Table 4.6 shows the number of greenbugs per plant for the first five plants, then the next five, and so forth. The MME of k is recalculated after each successive set of five counts. The stopping criterion is met after 20 observations, and $\hat{k} = 1.0$.

Parameter k Known

Consider now the case for which either k is known or a precise estimate of k is available. Further assume that the purpose of sampling is to obtain a precise estimate of the mean μ. Let T_n denote the sum of the first n observations; that is,

$$T_n = \sum_{i=1}^{n} X_i. \tag{4.48}$$

Table 4.6. Multistage Estimation of a Common k for Greenbugs on Seedling Oats

Observation No.	No. of greenbugs per seedling oat
1	1
2	18
3	2
4	0
5	40
$\hat{k} = 0.5241$	
6	3
7	8
8	5
9	12
10	11
$\hat{k} = 0.7488$	
11	8
12	7
13	6
14	3
15	15
$\hat{k} = 0.9562$	
16	27
17	8
18	4
19	3
20	2
$\hat{k} = 0.9648 \leftarrow$ stop	

If the goal of sampling is to estimate μ with a specified $CV(\bar{X}) = C$, from (4.4), we could propose to sample until

$$n \geq \frac{1}{C^2}\left(\frac{1}{\bar{X}_n} + \frac{1}{k}\right). \tag{4.49}$$

However, this results in larger sample sizes than are needed to attain the desired precision because the reciprocal of the sample mean has a positive bias (tends to be too large) when estimating the reciprocal of the population mean. Equivalently, it may be thought of as resulting from using the positively biased estimate of the variance given in equation (1.16). Willson and Young (1983) recommended using

$$n \geq \frac{V_n}{\bar{X}_n^2 C^2} \tag{4.50}$$

where V_n is the unbiased estimate of the variance based on n observations as given in equation (1.17).

The effect of the stopping rule may be seen better by writing (4.50) as

$$n \geq \frac{n}{nk + 1}\left(\frac{1}{C^2}\right)\left(\frac{1}{\bar{X}_n} + \frac{1}{k}\right). \tag{4.51}$$

The impact of the correction factor

$$\frac{n}{nk + 1}$$

is greater for smaller k values. As n becomes large, the effect of the correction factor becomes negligible. Equation (4.51) may be rewritten as sample until

$$n > \frac{1}{C^2 k} - \frac{1}{k} \tag{4.52}$$

and

$$T_n \geq \frac{nk}{(nk + 1)C^2 - 1}. \tag{4.53}$$

When equations (4.52) and (4.53) are applied, computations may be completed before sampling begins. Often tables are constructed giving the total needed to stop at several levels of precision, say $C = 0.1, 0.15, 0.20,$ and 0.25. Simulation studies indicate this procedure results in an estimate of the mean with the desired precision for a C as moderate as 0.3.

Example 4.15

Suppose we are sampling all life stages of cotton fleahoppers on cotton. Experience has shown that the distribution of counts are negative binomial with a k of about 2 when the terminal of a cotton plant is the sampling unit. If we want to estimate the mean number of fleahoppers on a terminal of a cotton plant with a $CV(\bar{X})$ of 0.25 or less, the minimum sample size is

$$n > \frac{1}{C^2 k} - \frac{1}{k} = \frac{1}{(0.25)^2(2)} - \frac{1}{2} = 7.5.$$

Therefore, at least eight observations must be taken. Beginning with the 8th observation, we compare the total of the observations to the number required by the stopping rule. When $n = 12$,

$$T_n \geq \frac{nk}{(nk + 1)C^2 - 1} = \frac{12(2)}{(12(2) + 1)(0.25)^2 - 1} = 42.7.$$

Thus, we need a total of at least 43 fleahoppers in the first 12 observations to stop. If the total does not exceed 43, we continue sampling.

The purpose of rewriting equations (4.51) in the form of equations (4.52) and (4.53) is to allow all computations to be completed before beginning data collection. These are often summarized in a table. For the above example, the data sheet could appear as in Table 4.7.

Table 4.7. Relationship Between Mean and Variance for Beall's (1939) Colorado Potato Beetle Data Based on the Common k *With the Observed Data*

Sample No.	Count	Total	Total needed to stop	Sample No.	Count	Total	Total needed to stop
1	2	2	.	9	5	36	96
2	4	6	.	10	8	43	64
3	0	6	.	11	0	43	51
4	9	15	.	12	5	48	43
5	5	20	.	13	—	—	38
6	1	21	.	14	—	—	35
7	7	28	.	15	—	—	32
8	3	31	256	16	—	—	31

Prior to sampling, the Sample Number and Total Needed to Stop columns are completed. The periods in the Total Needed to Stop column indicate that the minimum sample size has not been attained. The Count and Total columns are completed during the sampling process. The number of cotton fleahoppers in the ith sampling unit is entered on the ith row of the Count column. In the example above, there were two cotton fleahoppers on the first terminal, 4 on the second, and so on. The total number of organisms observed in the first j sampling units are recorded on the jth line of Total columns. Given either the Count or Total column, the other column can be determined, so it is not necessary to have both columns. If only one of the two are included on the data sheet, it is usually the Total column because the stopping rule is in terms of the total.

In this example, sampling stops after the 12th observation because the total number of cotton fleahoppers counted on the first 12 cotton plant terminals was 48, more than the total required to satisfy the stopping rule. Upon completing sampling, our estimate of the mean number of cotton fleahoppers per terminal is $48/12 = 4$, and the coefficient of variation of this estimate is 25%.

ECOSTAT can be used to develop the sequential estimation plans based on probability distributions presented in this chapter. To begin, choose *Estimation* from the *Sequential Sampling* menu. Indicate that the type of sampling is to be *Sequential*. Choose a distribution and then specify the measure of precision. For Example 4.15, we would choose *Negative Binomial* and *C, Coefficient of Variation of Sample Mean*. Use the scroll bars to set the level of precision. By pressing *Sampling Plan*, the totals needed to stop for sample sizes from one to a hundred are displayed on the screen. These may be saved to a data file by pressing, *Data Sheet*. To print the file, use a 10-point fixed font with 1-inch margins. A scalable font will prevent the columns from aligning.

Suppose now that the purpose of sampling was to estimate the mean within a specified proportion D of μ with $(1 - \alpha)$ 100% confidence. Based on (4.16), we propose sampling until

$$n \geq \left(\frac{1}{\bar{X}_n} + \frac{1}{k} \right) \left(\frac{z_{\alpha/2}}{D} \right)^2.$$

Again, however, this results in larger sample sizes than needed to attain the desired precision. Using V_n as the estimate of the variance, a proposed stopping rule is sample until

$$n \geq \left(\frac{n}{nk + 1} \right) \left(\frac{1}{\bar{X}_n} + \frac{1}{k} \right) \left(\frac{z_{\alpha/2}}{D} \right)^2. \tag{4.54}$$

This may be rewritten as sample until

$$n > \frac{z_{\alpha/2}^2}{kD^2} - \frac{1}{k} \tag{4.55}$$

and

$$T_n \geq \frac{nkz^2}{(nk + 1)D^2 - z^2}. \tag{4.56}$$

Simulation studies indicate that with a D as moderate as 0.30 and 95% confidence, the goal of estimating the mean within 30% of the mean with 95% confidence is achieved.

Example 4.16

Suppose, as in Example 4.15, we are sampling cotton fleahoppers and that from experience we know the distribution of counts can be modeled with a negative binomial distribution with a value of $k = 2$. Further, assume we want to estimate the mean number of cotton fleahoppers on a terminal of a cotton plant within 30% of the mean with 95% confidence. The minimum sample size is

$$n > \frac{z_{\alpha/2}^2}{kD^2} - \frac{1}{k} = \frac{(1.96)^2}{2(0.30)^2} - \frac{1}{2} = 20.84.$$

Therefore, at least 21 observations must be made. For $n = 50$, the total must exceed

$$T_n \geq \frac{nkz^2}{(nk + 1)D^2 - z^2} = \frac{50(2)(1.96)^2}{(50(2) + 1)(0.3)^2 - (1.96)^2} = 73.20.$$

That is, a total of 74 cotton fleahoppers must be accumulated in the first 50 observations or sampling will continue.

Now assume that the goal of sampling is to estimate the mean within H with a specified level of confidence. Based on equation (4.23), we propose sampling until

$$n \geq \left(\bar{X}_n + \frac{\bar{X}_n^2}{k} \right) \left(\frac{z_{\alpha/2}}{H} \right)^2. \tag{4.57}$$

Here we do *not* propose using V_n to estimate the variance, thereby reducing the sample size. Even using the larger sample sizes from equation (4.57), the confidence level tends to be below the stated level unless D is small. Notice that if the first observation is 0, sampling stops because $\bar{X}_1$, and thus the right-hand side of equation (4.57), is 0. To avoid this difficulty, two approaches are possible. First, we could require at least one nonzero observation before stopping. Alternatively, we could establish a minimum sample size based on our knowledge of the ecosystem being sampled. Rewriting equation (4.57), we propose to sample until at least one non-zero value has been observed ($T_n > 0$) and

$$T_n \leq -\frac{nk}{2} + n\sqrt{\frac{k^2}{4} + \frac{nkH^2}{z_{\alpha/2}^2}}. \tag{4.58}$$

Notice that now sampling continues until the sum of the observations is below a threshold. If the mean is one and k is small, D should not exceed 0.2; otherwise, the confidence level tends to be well below the stated level. More study is needed to determine how large D can be in other regions of the parameter space without reducing the confidence below the stated level.

Example 4.17

Assume, as in Examples 4.15 and 4.16, that we are sampling cotton fleahoppers and that we know the distribution is negative binomial with $k = 2$. Suppose the purpose of sampling is to estimate the mean number of cotton fleahoppers on a terminal of a cotton plant within 0.5 with 90% confidence. For $n = 10$,

$$T_n \leq -\frac{nk}{2} + n\sqrt{\frac{k^2}{4} + \frac{nkH^2}{z_{\alpha/2}^2}} = -\frac{10(2)}{2} + 10\sqrt{\frac{(2)^2}{4} + \frac{10(2)(0.5)^2}{(1.645)^2}} = 6.9.$$

That is, in order to stop with 10 observations, the total cotton fleahoppers in those 10 observations could not exceed 6. Notice that because we are now stopping when the total falls below a certain level, we round down instead of up to ensure that the desired precision of the estimator is attained.

The sampling methods developed in this section depend on precise knowledge of k. In each case, using a value smaller than the true k results in an estimate of the population mean that is more precise than the stated level. If a range of possible values of k is available, a conservative approach is to use one of the smallest, if not the smallest, value in the range as the k-value. With this approach, more sampling is required when compared to using larger values of k but, upon stopping, the estimate of the mean is at least as precise as stated. If the value of k used in developing the sampling program is within one of the true parameter value, these procedures continue to produce estimates with a level of precision close to the stated one. Therefore, if a fairly reliable estimate of k is available, it is better to use that than to turn too quickly to the nonparametric approaches presented earlier.

Sequential Estimation for the Geometric

The sampling plans for the geometric distribution may be obtained by using a k of one in each of the sampling plans for the negative binomial distribution. We consider estimating the mean of a geometric population using each of the three measures of precision discussed earlier: attaining a specified $CV(\bar{X}) = C$, estimating the mean within $D\mu$ with $(1 - \alpha)$ 100% confidence, and estimating the mean within H units with $(1 - \alpha)$ 100% confidence.

Assume that the purpose of sampling is to obtain an estimate of the mean μ with a specified $CV(\bar{X}) = C$. Based on equation (4.51), we propose to sample until

$$n \geq \frac{n}{n + 1} \left(\frac{1}{C^2}\right)\left(\frac{1}{\bar{X}_n} + 1\right). \tag{4.59}$$

This may be rewritten as sample until

$$n > \frac{1}{C^2} - 1 \tag{4.60}$$

and

$$T_n \geq \frac{n}{(n + 1)C^2 - 1}. \tag{4.61}$$

Based on extensive simulations, this procedure results in an estimate of the mean with the desired precision for a C as moderate as 0.3.

Example 4.18

The geometric distribution consistently fits the counts of greenbugs on seedling oats. If we want to obtain an estimate of the mean number of greenbugs/plant in a field of seedling oats with a $CV(\bar{X})$ value of 0.25 or less, the minimum sample size is

$$n > \frac{1}{C^2} - 1 = \frac{1}{(0.25)^2} - 1 = 15.$$

Therefore, at least 16 observations must be taken. Beginning with the 16th observation, we compare the total of the observations to the number required by the stopping rule. When $n = 22$,

$$T_n \geq \frac{n}{(n + 1)C^2 - 1} = \frac{22}{(22 + 1)(0.25)^2 - 1} = 50.3.$$

If, after 22 observations, we have counted 51 or more greenbugs, we stop sampling; otherwise, we continue sampling. Upon stopping, the sample mean is our estimate of the mean number of greenbugs/seedling oat, and this estimate has a 25% coefficient of variation.

Suppose now that the purpose of sampling is to estimate the mean within a specified proportion D of μ with $(1 - \alpha) 100\%$ confidence. Based on (4.54), we propose sampling until

$$n \geq \frac{n}{n + 1} \left(\frac{1}{\bar{X}_n} + 1 \right) \left(\frac{z_{\alpha/2}}{D} \right)^2. \tag{4.62}$$

As in (4.55) and (4.56), this may be rewritten as sample until

$$n > \left(\frac{z_{\alpha/2}}{D} \right)^2 - 1 \tag{4.63}$$

and

$$T_n \geq \frac{n z_{\alpha/2}^2}{(n + 1)D^2 - z_{\alpha/2}^2}. \tag{4.64}$$

Simulation studies indicate that with a D as moderate as 0.30 and 95% confidence, the goal of estimating the mean within 30% of the mean with 95% confidence is achieved.

Example 4.19

Suppose, as in Example 4.18, we are sampling greenbugs on seedling oats and that the number of greenbugs/oat is known to have a geometric distribution. Further, assume we want to estimate the mean within 25% of the mean with 90% confidence. The minimum sample size is

$$n > \left(\frac{z_{\alpha/2}}{D} \right)^2 - 1 = \left(\frac{1.645}{0.25} \right)^2 - 1 = 42.3.$$

Therefore, a minimum of 43 observations must be made before sampling ceases. For $n = 50$, the total must exceed

$$T_n \geq \frac{n z_{\alpha/2}^2}{(n + 1)D^2 - z_{\alpha/2}^2} = \frac{50(1.645)^2}{(50 + 1)(0.25)^2 - (1.645)^2} = 281.0.$$

That is, a total of 281 must be accumulated in the first 50 observations or sampling will continue.

Now assume that the goal of sampling is to estimate the mean within H with a specified level of confidence. Based on equation (4.57), we propose sampling until

$$n \geq (\bar{X}_n + \bar{X}_n^2)\left(\frac{z_{\alpha/2}}{H}\right)^2. \qquad (4.65)$$

As with the negative binomial, notice that if the first observation is a 0, sampling stops because $\bar{X}_1$ would be 0. To resolve this problem, we require at least one nonzero observation before stopping. Rewriting equation (4.65), we propose to sample until at least one nonzero value has been observed ($T_n > 0$) and

$$T_n \leq -\frac{n}{2} + n\sqrt{\frac{1}{4} + \frac{nH^2}{z_{\alpha/2}^2}}. \qquad (4.66)$$

If the mean is one, D should not exceed 0.2; otherwise, the confidence level will be well below the stated level. More study is needed to determine how large D can be in other regions of the parameter space without reducing the confidence below the stated level.

Example 4.20

Assume, as in Examples 4.18 and 4.19, that the counts of greenbugs on seedling oats have a geometric distribution. Suppose the purpose of sampling is to estimate the mean within 0.3 with 95% confidence. For $n = 15$,

$$T_n \leq -\frac{n}{2} + n\sqrt{\frac{1}{4} + \frac{nH^2}{z_{\alpha/2}^2}} = -\frac{15}{2} + 15\sqrt{\frac{1}{4} + \frac{15(0.3)^2}{(1.96)^2}} = 4.1.$$

That is, in order to stop with 15 observations, the total of those 15 observations could not exceed 4.

Sequential Estimation for the Poisson

Suppose we are sampling an ecosystem for the purpose of estimating the population mean. The precision of the estimated mean will be determined using one of three measures: estimating μ with a specified coefficient of variation, estimating the mean within $D\mu$ with $(1 - \alpha)$ 100% confidence, or estimating the mean within H with $(1 - \alpha)$ 100% confidence. Having sampled this ecosystem before, we are willing to assume that the distribution of counts will be Poisson.

First, assume that the purpose of sampling is to estimate the population mean μ with a specified $CV(\bar{X}) = C$. An intuitive approach is to update the estimate of μ after each observation and sample until equation (4.8) (with the sample estimate replacing the parameter) is satisfied; that is, sample until

$$n \geq \frac{1}{\bar{X}_n C^2}. \tag{4.67}$$

This may be rewritten as sample until

$$T_n \geq \frac{1}{C^2}. $$

Notice that in this case, the total needed to satisfy the stopping rule does not depend on the sample size and therefore remains constant.

Example 4.21

The number of grasshoppers in a m^2 quadrat has been found to be well modeled by the Poisson distribution during certain life stages of the grasshopper. Suppose that we want to estimate the average number of grasshoppers in a m^2 quadrat with a coefficient of variation of the estimated mean of 15%. We sample until

$$T_n \geq \frac{1}{C^2} = \frac{1}{(0.15)^2} = 44.4.$$

Therefore, we sample until a total of 45 grasshoppers have been observed. The sample mean is the estimate of the population mean.

Now assume that the purpose of sampling is to estimate the mean of a Poisson population within $D\mu$ with $(1 - \alpha)$ 100% confidence. By updating the estimate of the population mean after each observation and sampling until equation (4.18) is satisfied, we have sample until

$$n \geq \frac{1}{\bar{X}_n} \left(\frac{z_{\alpha/2}}{D} \right)^2. \tag{4.69}$$

This may be rewritten as sample until

$$T_n \geq \left(\frac{z_{\alpha/2}}{D} \right)^2. \tag{4.70}$$

Example 4.22

Consider, as in Example 4.21, the problem of estimating the mean number of grasshoppers in a m^2 quadrat. The distribution of grasshopper counts is believed to be Poisson. The purpose of sampling is to estimate the population mean within 10% of the mean with 95% confidence. To achieve this degree of precision, we sample until

$$T_n \geq \left(\frac{z_{\alpha/2}}{D}\right)^2 = \left(\frac{1.96}{0.1}\right)^2 = 384.2.$$

Sampling continues until a total of 385 grasshoppers have been counted.

Again assume that we are sampling from a Poisson population. However, now the goal of sampling is to estimate the mean within H with $(1 - \alpha)$ 100% confidence. Using sample estimates in equation (4.25), we propose sampling until

$$n \geq \bar{X}_n \left(\frac{z_{\alpha/2}}{H}\right)^2. \tag{4.71}$$

This may be written equivalently as sample until at least one nonzero value has been observed and

$$T_n \leq \left(\frac{nH}{z_{\alpha/2}}\right)^2. \tag{4.72}$$

Example 4.23

Again, recall the grasshoppers discussed in Examples 4.21 and 4.22. The purpose of sampling is to obtain a precise estimate of the average number of grasshoppers in a m² area within the sampling region. Now suppose that we want this estimate to be within 0.5 of the true mean with 90% confidence. Then if we have observed at least one grasshopper, we stop with 20 observations if

$$T_n \leq \left(\frac{nH}{z_{\alpha/2}}\right)^2 = \left(\frac{20(0.5)}{1.645}\right)^2 = 36.95.$$

Therefore, if a total of 36 or fewer grasshoppers have been observed, we stop and use the sample mean as our estimate of the population mean.

Sequential Estimation for the Binomial

Suppose we want to estimate the proportion of the sample units within an ecosystem having a particular characteristic. We may want to estimate the proportion of fruit with damage, the proportion of infested plants, or the proportion of deer with ticks. An observation will consist of randomly selecting a sampling unit and observing whether that unit has the characteristic of interest. If it does, we record a 1. If it does not, we record a zero. When sampling in this manner, the binomial parameter n is 1. Therefore, in this section, the binomial parameter n is taken to be 1, and n in the equations represents the sample size. The purpose of sampling is to estimate the binomial proportion p with precision. The measures of precision considered in this section are attaining a specified coefficient of variation, C, of the estimated proportion p, estimating the proportion p within Dp with $(1 - \alpha)$ 100% confidence, and estimating the proportion p within H with $(1 - \alpha)$ 100% confidence.

First consider estimating the proportion p with a coefficient of variation of C. After n observations, we estimate p using $\bar{X}_n$. Since $q = 1 - p$, the estimate of q based on n observations is $(1 - \bar{X}_n)$. Using these estimates in equation (4.10), we propose sampling until

$$n \geq \frac{1 - \bar{X}_n}{\bar{X}_n C^2}. \tag{4.73}$$

Note that if the characteristic has not been observed, $\bar{X}_n$ is zero and (4.73) is not well defined. Also, if all units in the sample after n observations possessed the characteristic of interest, $\bar{X}_n = 1$. This results in the numerator in (4.73) being zero, and the inequality is satisfied. To avoid both of these extremes, we propose sampling until at least one sampling unit has been observed with the trait of interest and another sampling unit had been observed without that same trait. Therefore, we propose sampling until at least one zero and at least one has been observed and the following inequality, derived from (4.73), is satisfied:

$$T_n \geq \frac{n}{nC^2 + 1}. \tag{4.74}$$

Example 4.24

Suppose we want to estimate the proportion of tomatoes with damage in a large commercial tomato field. Further, we want the coefficient of variation of this estimate to be 10%. Because we want to estimate the proportion of damaged fruit, we record a 1 if a tomato is damaged and a zero if it is not damaged. If we have observed at least one damaged tomato and at least one undamaged tomato, we can stop after 20 observations if

$$T_n \geq \frac{n}{nC^2 + 1} = \frac{20}{20(0.1)^2 + 1} = 16.7.$$

That is, if in the first 20 tomatoes observed, 17 or more are damaged and at least one is undamaged, we stop sampling. The estimate of the proportion of damaged fruit is the sample mean, and the coefficient of variation of this estimate is 10%.

Now assume the goal of sampling is to estimate the proportion p of the population possessing the characteristic of interest within Dp with $(1 - \alpha)$ 100% confidence. Again, we propose sampling until at least one unit with the trait and another without the trait have been observed. Then, based on equation (4.20) with sample estimates replacing the population parameters, we propose sampling until

$$n \geq \left(\frac{1 - \bar{X}_n}{\bar{X}_n}\right)\left(\frac{z_{\alpha/2}}{D}\right)^2. \tag{4.75}$$

This may be presented in a form that permits computations to be completed prior to the initiation of sampling; that is, we sample until at least one unit with the characteristic of interest and another without it have been observed and

$$T_n \geq \frac{n z_{\alpha/2}^2}{n D^2 + z_{\alpha/2}^2}. \tag{4.76}$$

Example 4.25

Again consider, as in Example 4.24, the problem of estimating the proportion of damaged tomatoes in a large commercial field. Suppose the producer wants an estimate within 20% of the true proportion of damaged tomatoes with 90% confidence. We could stop after 15 observations if

$$T_n \geq \frac{n z_{\alpha/2}^2}{n D^2 + z_{\alpha/2}^2} = \frac{15(1.645)^2}{15(0.2)^2 + (1.645)^2} = 12.3.$$

That is, we stop if at least 13 damaged and 1 undamaged tomato have been observed in the first 15 randomly selected tomatoes. The estimate of the proportion of damaged fruit is the sample mean, and we are 90% confident that this estimate is within 20% of the true proportion of damaged tomatoes in the field.

Now suppose that the goal of sampling is to estimate the proportion of sampling units in a population with a given trait within H of the true proportion with $(1 - \alpha)$ 100% confidence. Again assuming that sampling continues until at least one unit with that characteristic and another without it have been observed, we propose, based on equation (4.26), sampling until

$$n \geq \bar{X}_n(1 - \bar{X}_n)\left(\frac{z_{\alpha/2}}{H}\right)^2. \tag{4.77}$$

Equivalently, we propose sampling until at least one unit with and another without the trait of interest have been observed and either

$$T_n \geq \frac{n}{2} + n\sqrt{\frac{1}{4} - \frac{nH^2}{z_{\alpha/2}^2}}$$

(4.78)

or

$$T_n \leq \frac{n}{2} - n\sqrt{\frac{1}{4} - \frac{nH^2}{z_{\alpha/2}^2}}.$$

(4.79)

Example 4.26

Consider again the problem of estimating the proportion of damaged tomatoes in a large commercial field as discussed in Examples 4.24 and 4.25. Now assume that the grower wants to estimate the proportion of damaged tomatoes within 0.05 with 90% confidence. Stopping ceases with the 50th observation if

$$T_n \geq \frac{n}{2} + n\sqrt{\frac{1}{4} - \frac{nH^2}{z_{\alpha/2}^2}} = \frac{50}{2} + 50\sqrt{\frac{1}{4} - \frac{50(0.05)^2}{(1.645)^2}} = 47.6$$

or

$$T_n \leq \frac{n}{2} - n\sqrt{\frac{1}{4} - \frac{nH^2}{z_{\alpha/2}^2}} = \frac{50}{2} - 50\sqrt{\frac{1}{4} - \frac{50(0.05)^2}{(1.645)^2}} = 2.4.$$

That is, we stop if at least one damaged and one undamaged tomato have been observed and if at least 48 damaged or no more than two undamaged tomatoes have been observed. When the stopping criterion has been met, the sample mean provides the estimate of the proportion of damaged tomatoes, and this estimate is within 0.05 of the true proportion with 90% confidence.

Sequential Estimation Based on Iwao's Patchiness Regression

The sequential methods discussed thus far are either nonparametric and make no assumption about the relationship in mean and variance or parametric and assumed the distribution, and hence the relationship in mean and variance, were known. Another approach that is seen frequently in the literature attempts to relate the mean and variance empirically without assuming a parametric form of the model. Iwao's Patchiness Regression and Taylor's Power Law are two such approaches. Their use in developing sequential sampling plans is discussed in the final two sections of this chapter.

Lloyd (1967) proposed to define mean crowding as the "mean number per individual of other individuals in the same quadrat." He further determined that this index of mean crowding (x^*) could be approximated by

$$x^* \approx \mu + \frac{\sigma^2}{\mu} - 1. \tag{4.80}$$

Given a parametric model, mean crowding can be evaluated. For example, when the population distribution is negative binomial, mean crowding is

$$x^* \approx \mu + \frac{\mu}{k}.$$

Mean crowding for a Poisson distributed population is

$$x^* \approx \mu.$$

If the population distribution is binomial, mean crowding is

$$x^* \approx (n - 1)p.$$

Based on empirical studies, Iwao (1968) proposed that mean crowding has the following relationship to mean density:

$$x^* = \alpha + \beta\mu. \tag{4.81}$$

If the relationship in equation (4.81) holds, then from equation (4.80), we conclude that

$$\sigma^2 = (\alpha + 1)\mu + (\beta - 1)\mu^2. \tag{4.82}$$

The negative binomial, geometric, Poisson, and binomial are special cases of this more general relationship in mean and variance. For example, if $\alpha = -1$ and $\beta > 0$, we have the form of the negative binomial distribution with $k = 1/(\beta - 1)$. However, other distributions could also have this same relationship in mean and variance in at least some region of the parameter space. Therefore, although the relationship in mean and variance is assumed to be of a certain form, the parametric form of the distribution is not assumed to be known.

Given a series of data sets collected from the type of ecosystem of interest, a regression of estimated mean crowding on the sample mean provides estimates of α and β. To perform these computations, first let $\hat{x}_i^*$ denote the estimate of mean crowding based on the ith sample; that is,

$$\hat{x}_i^* = \bar{X}_i + \frac{s_i^2}{\bar{X}_i} - 1. \tag{4.83}$$

A linear regression of $\hat{x}_i^*$ on $\bar{X}_i$ based on the t populations sampled is performed. The estimated intercept is α and the estimated slope is β.

The r^2 from the regression has historically been used as the measure of the adequacy of the model. Rarely do independent samples result in the same value of $\bar{X}$, and thus a test for lack-of-fit is seldom possible. An alternative method of testing the adequacy of this model is to regress the predicted value of σ^2 based on the model on the observed estimate of σ^2 from the samples. If the model is accurate, a line passing through the origin with a slope of one should fit within error. For this approach to be fully valid, care must be taken to account for the variability in the estimated σ^2 from the samples.

Often most of the data sets used to establish the relationship in mean and variance produce small estimated means. The few data sets with large estimated means tend to greatly influence the model. A study of the influence of each point on the regression of $\hat{x}^*$ on $\bar{X}$ may also be appropriate. In addition, the adequacy of the model for different ranges of the mean may be investigated by performing the regression of the model-based variance estimates on the sample-based variance estimates, as mentioned above, within each of the ranges of interest.

Example 4.27

Consider again the 16 samples of Colorado potato beetles discussed in Example 4.13 during the process of determining a common k. The estimated means, variances, and mean crowding measures are given in Table 4.8. Here equation (4.83) is used to compute $\hat{x}^*$. For example,

$$\hat{x}_1^* = \bar{x}_1 + \frac{s_1^2}{\bar{x}_1} - 1 = 75.75 + \frac{539.07}{75.75} - 1 = 81.87.$$

Most statistical software packages can be used to perform a regression of estimated mean crowding on the estimated sample mean. For this example, the estimated intercept is -0.01 and thus

$$\hat{\alpha} = -0.01.$$

The estimated slope is 1.07 and hence

Table 4.8. Estimates of Means, Variances, and Mean Crowding for Beall's (1939) Colorado Potato Beetle Data

Block	$\bar{X}$	s^2	$\hat{x}*$
1	75.75	539.07	81.87
2	84.00	627.14	90.47
3	42.25	257.93	47.35
4	32.50	77.43	33.88
5	48.00	172.57	50.60
6	48.62	91.12	49.49
7	66.50	737.43	76.59
8	56.75	71.64	57.01
9	64.88	101.27	65.44
10	52.12	417.55	59.13
11	40.12	137.55	42.55
12	26.62	48.55	27.44
13	34.12	252.41	40.52
14	34.12	14.70	33.55
15	13.88	45.84	16.18
16	37.62	167.70	41.08

$$\hat{\beta} = 1.07.$$

The r^2 is 0.98, indicating a strong linear relationship. Based on the estimated parameters from the regression, the estimate of the variance is

$$\hat{\sigma}^2 = 0.99\bar{X}_n + 0.07\bar{X}_n^2.$$

Notice that in this case, there is little difference in this and the estimated variance from the negative binomial of

$$\hat{\sigma}^2 = \bar{X}_n + \frac{\bar{X}_n^2}{\hat{k}_c} = \bar{X}_n + \frac{\bar{X}_n^2}{13.58} = \bar{X}_n + 0.07\bar{X}_n^2$$

that is based on a common k.

To test the adequacy of the model, we regressed the model-based variance estimates of the sample variance estimates and obtained

$$\hat{\sigma}_I^2 = 107.27 + 0.51s^2$$

where $\hat{\sigma}_I^2$ is the estimate of the variance based on Iwao's patchiness regression and s^2 is the estimated variance based on the observed sample values. The intercept is significantly different from zero ($t = 2.809, p = 0.0139$), and the slope is significantly different from one ($F = 17.0, p = 0.001$). After accounting for the variability in the estimated sample variances, our conclusions remain unchanged. Therefore, even though the r^2 is very high, we question the adequacy of the model for these data.

Estimated mean crowding is plotted against the estimated mean in Figure 4.3. The line in Figure 4.3 is the estimated regression line relating mean crowding and mean density. The purpose of the regression is to obtain an estimated relationship between mean and variance. Therefore, the estimated variances are plotted against the estimated means in Figure 4.4. The curve represents the estimated relationship in variance and mean.

Once the relationship in mean and variance has been estimated. This estimate can be used to develop a sequential sampling plan to estimate the mean density of a population. Three measures of precision are again considered: attaining a specified coefficient of variation of the estimated mean density, estimating the mean density within $D\mu$ with $(1 - \alpha)$ 100% confidence, and estimating the mean density within H with $(1 - \alpha)$ 100% confidence.

First, suppose we want to estimate the mean density of a population with a $CV(\bar{X})$ value of C. Using equation (4.2), the relationship in mean and variance, and estimates of unknown parameters, we propose sampling until

$$ n \geq \frac{1}{C^2}\left(\frac{\alpha + 1}{\bar{X}_n} + (\beta - 1)\right). \tag{4.84} $$

Alternatively, this could be written as sample until

$$ n > \frac{\beta - 1}{C^2} \tag{4.85} $$

and

$$ T_n \geq \frac{n(\alpha + 1)}{nC^2 - \beta + 1}. \tag{4.86} $$

When the stopping criterion is met, the sample mean is the estimate of mean density, and the coefficient of this estimate is C. Simulation studies have not been conducted to provide insight into the behavior of this sampling method.

Example 4.28

Suppose that we wanted to develop a sampling plan for the Colorado potato beetle after estimating the relationship in mean and variance in Example 2.27.

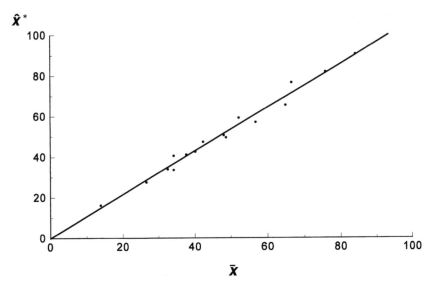

Figure 4.3. The regression of estimated mean crowding on the sample mean with the sample estimates for Beall's (1939) Colorado potato beetle data.

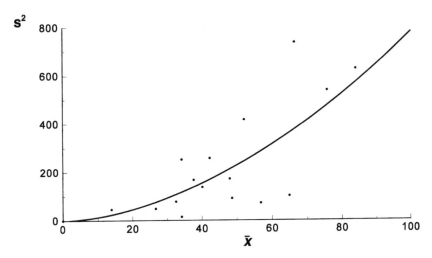

Figure 4.4. The relationship between the mean and variance implied by Iwao's patchiness regression with the sample estimates for Beall's (1939) Colorado potato beetle data.

The goal is to estimate the mean density with a $CV(\bar{X})$ of 20%. From Example 2.27, we have

$$\hat{\alpha} = -0.01 \qquad \hat{\beta} = 1.07.$$

Therefore, from equation (4.85), we determine the minimum sample size to be

$$n > \frac{\hat{\beta} - 1}{C^2} = \frac{1.07 - 1}{(0.2)^2} = 1.8.$$

That is, we must take at least two observations before sampling ceases. Sampling continues until the stopping criterion in equation (4.86) is satisfied. For 20 observations, the total needed to stop is

$$T_n \geq \frac{n(\hat{\alpha} + 1)}{nC^2 - \hat{\beta} + 1} = \frac{20(-0.01 + 1)}{20(0.2)^2 - 1.07 + 1} = 27.12.$$

Therefore, if the total number of Colorado potato beetles in the first 20 observations was at least 28, we could stop sampling and use the sample mean as an estimate of the mean density. The coefficient of variation of this estimate is 20%. If the total is less than 28, sampling continues.

Now assume that the purpose of sampling is to estimate the mean density within $D\mu$ with $(1 - \alpha)$ 100% confidence. Extending equation (4.15) by using the estimated relationship in mean and variance and updating estimates of population parameters after each observation, we propose sampling until

$$n \geq \left(\frac{\alpha + 1}{\bar{X}_n} + \beta - 1\right)\left(\frac{z_{\alpha/2}}{D}\right)^2. \tag{4.87}$$

This may be written in a manner that will permit computations to be completed prior to the initiation of sampling; that is, we sample until

$$n \geq \frac{(\beta - 1)z_{\alpha/2}^2}{D^2} \tag{4.88}$$

and

$$T_n \geq \frac{n(\alpha + 1)z_{\alpha/2}^2}{nD^2 - (\beta - 1)z_{\alpha/2}^2}. \tag{4.89}$$

Example 4.29

Consider again estimating the mean density of Colorado potato beetles as in Examples 4.27 and 4.28. Now suppose that the purpose of sampling is to estimate the mean density within 10% of the mean with 90% confidence. In this case, the minimum sample size is

$$n \geq \frac{(\hat{\beta} - 1)z_{\alpha/2}^2}{D^2} = \frac{(1.07 - 1)(1.645)^2}{(0.1)^2} = 18.9.$$

At least 19 observations must be taken to obtain an estimate of mean density with the desired precision. On the 30th observation, sampling ceases if

$$T_n \geq \frac{n(\hat{\alpha} + 1)z_{\alpha/2}^2}{nD^2 - (\hat{\beta} - 1)z_{\alpha/2}^2} = \frac{30(-0.01 + 1)(1.645)^2}{30(0.1)^2 - (1.07 - 1)(1.645)^2} = 726.8.$$

That is, at least 727 Colorado potato beetles must be counted in the first 30 randomly selected sampling units for sampling to cease with 30 observations. If the total number is less than 727, sampling continues until the sampling criterion is met for the sample mean to provide the desired precision in estimating the population mean density.

Assume the goal of sampling is to estimate the population mean density within H with $(1 - \alpha)$ 100% confidence based on Iwao's Patchiness Regression. Using the estimated relationship in mean and variance and updating estimates of the population mean after each observation in equation (4.22), we propose sampling until

$$n \geq ((\alpha + 1)\bar{X}_n + (\beta - 1)\bar{X}_n^2)\left(\frac{z_{\alpha/2}}{H}\right)^2. \qquad (4.90)$$

Rewriting equation (4.90) in a form that permits the completion of calculations before sampling begins, we have sample until at least one beetle has been observed and

$$T_n \leq -\frac{n(\alpha + 1)}{2(\beta - 1)} + n\sqrt{\frac{nH^2}{(\beta - 1)z_{\alpha/2}^2} + \left(\frac{\alpha + 1}{2(\beta - 1)}\right)^2}. \qquad (4.91)$$

Example 4.30

Recall the Colorado potato beetles discussed in Examples 4.27, 4.28, and 4.29. Suppose that the purpose of sampling is to estimate the mean population density of these beetles within 0.5 with 95% confidence. Further, assume that we want to take advantage of the estimated relationship in mean and variance expressed by Iwao's Patchiness Regression. We sample until at least one beetle has been observed. Then we stop sampling as soon as equation (4.91) has been satisfied. On the 15th observation, we stop if

$$T_n \leq -\frac{n(\hat{\alpha} + 1)}{2(\hat{\beta} - 1)} + n\sqrt{\frac{nH^2}{(\beta - 1)z_{\alpha/2}^2} + \left(\frac{\hat{\alpha} + 1}{2(\hat{\beta} - 1)}\right)^2} = -\frac{15(-0.01 + 1)}{2(1.07 - 1)}$$

$$+ 15\sqrt{\frac{15(0.5)^2}{(1.07 - 1)(1.96)^2} + \left(\frac{-0.01 + 1}{2(1.07 - 1)}\right)^2} = 13.9.$$

If no more than 13 potato beetles have been counted in the first 15 sampling units, sampling ceases. The sample mean is within 0.5 of the true population mean density with 95% confidence.

Sequential Sampling Based on Taylor's Power Law

Based on empirical data, Taylor (1961, 1965, 1971) suggested that the relationship in mean and variance could be expressed by

$$\sigma^2 = a\mu^b \tag{4.92}$$

for numerous species. The relationship expressed in equation (4.92) has come to be described as Taylor's Power Law. Parameter b is referred to as an index of aggregation characteristic of the species under consideration (Taylor et al., 1978). The parameter a is considered a sampling factor. While a may change with environment or sampling unit, b is considered fixed for any species, although Banerjee (1976) disputes this. The parameter b is often used as a measure of aggregation. A b of 1 is indicative of randomness, a $b > 1$ implies aggregation, and a $b < 1$ indicates more regularity. However, as with the numerical distributions, these terms can only be applied to the numerical distribution of the organisms and not to the spatial patterns of those organisms.

Several samples are required to obtain estimates of a and b in Taylor's Power Law, much as they were required in determining the presence of a common k or in performing Iwao's Patchiness Regression. The estimation process begins by regressing the natural logarithm of the estimated variance on the natural logarithm of the estimated mean. The resulting regression equation is of the form:

$$\ln(s_i^2) = \ln \hat{a} + \hat{b} \ln(\bar{X}_i) + \hat{\epsilon}_i. \tag{4.93}$$

Therefore, parameter a may be estimated by taking the exponentiation of the estimated intercept, and parameter b is the estimated slope of the line. The r^2 from regression is used to assess the strength of the relationship, with high r^2 values being common. Graphs of this relationship are usually on the log-log scale.

This approach to evaluating the adequacy of Taylor's power law in describing the relationship in mean and variance has a primary weakness. Once a and b are estimated, inference concerning the population is based on the estimated mean and variance and not on the natural logarithms of those estimates. The r^2 from regression is for the log-log scale and may be (and usually is) a poor indicator of the fit on the untransformed scale. At the very least, the relationship expressed by Taylor's Power Law should be plotted with the sample means and variances on the untransformed scale. This permits a more realistic assessment of the adequacy of the model. However, we also suggest formally testing the adequacy of the model as we did with Iwao's patchiness regression. This entails regressing the estimated variance from Taylor's Power Law on the sample-based variance estimates, adjusting for the variability in the estimated sample variance. If Taylor's power law is a good model for the data, a regression line passing through the origin with a slope of 1 should fit within error. Based on these alternate approaches, our experience indicates that Taylor's Power Law is not as strong as is often believed. This is especially true when a high proportion of the samples have small means and variances.

Example 4.31

Once again consider the problem of estimating the relationship in mean and variance of the adult Colorado potato beetles as discussed in Examples 4.13 and 4.27. However, now we explore the use of Taylor's Power Law in describing this relationship. Because regression is on the log-log scale, the natural logarithms of the estimated mean and variance are shown in addition to these sample estimates in Table 4.9.

A linear regression of $\ln(s^2)$ on $\ln(\bar{X})$ produces the following:

$$\ln(s_i^2) = -0.87 + 1.55 \ln(\bar{X}_i) + \hat{\epsilon}_i.$$

A plot of the estimated regression line with the observed logarithms of the estimated means and variances is shown in Figure 4.5. The r^2 from the regression is 0.42, a value much lower than is commonly observed from Taylor's Power Law. Therefore, for these data, Taylor's Power Law does not appear as appropriate as either using a common k for the negative binomial or Iwao's Patchiness Regression. However, for illustrative purposes, we continue. From the estimated intercept, we can obtain an estimate of a:

$$\hat{a} = e^{-0.87} = 0.42.$$

This permits Taylor's Power Law to be represented as

Table 4.9. Estimates of Means, Variances, and Their Logarithms for Beall's (1939) Colorado Potato Beetle Data

Block	$\bar{X}$	s^2	$\ln(X)$	$\ln(s^2)$
1	75.75	539.07	4.33	6.29
2	84.00	627.14	4.43	6.44
3	42.25	257.93	3.74	5.55
4	32.50	77.43	3.48	4.35
5	48.00	172.57	3.87	5.15
6	48.62	91.12	3.88	4.51
7	66.50	737.43	4.20	6.60
8	56.75	71.64	4.04	4.27
9	64.88	101.27	4.17	4.62
10	52.12	417.55	3.95	6.03
11	40.12	137.55	3.69	4.92
12	26.62	48.55	3.28	3.88
13	34.12	252.41	3.53	5.53
14	34.12	14.70	3.53	2.69
15	13.88	45.84	2.63	3.83
16	37.62	167.70	3.63	5.12

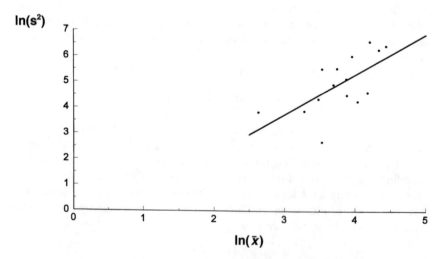

Figure 4.5. The regression of $\ln(s^2)$ on $\ln(\bar{X})$ with the sample estimates for Beall's (1939) Colorado potato beetle data.

$$\hat{\sigma}^2 \approx 0.42\bar{X}_n^{1.55}.$$

This estimated relationship in mean and variance is shown with the sample estimates of mean and variance on the untransformed scale in Figure 4.6, which permits a more realistic assessment of the validity of using Taylor's Power Law for these data than does Figure 4.5 because later applications will be in terms of the sample means and variances and not their logarithms.

The variance estimates based on Taylor's Power Law were regressed on the sample variance estimates to assess the adequacy of the model. The resulting equation is

$$\hat{\sigma}_T^2 = 94.39 + 0.34s^2 + \hat{\varepsilon}_i$$

where $\hat{\sigma}_T^2$ is the estimated variance from Taylor's Power Law and s^2 is the estimated variance based on the observed sample values. The intercept is significantly different from zero ($t = 3.617, p = 0.0028$), and the slope is significantly different from one ($F = 65.34, p = 0.0001$). After adjusting for the variability in the sample variances, the test statistics remain significant. Therefore, based on both the r^2 and the regression, we question the adequacy of the model for these data.

Now suppose that we are satisfied with the relationship in mean and variance expressed by Taylor's Power Law. Next we want to estimate the population mean

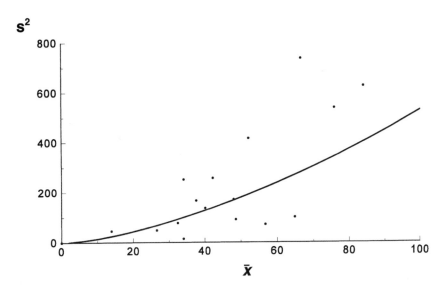

Figure 4.6. The relationship between the mean and variance implied by Taylor's power law with the sample estimates for Beall's (1939) Colorado potato beetle data.

density with a $CV(\bar{X}) = C$. Using the estimated relationship in mean and variance and equation (4.2), we propose sampling until

$$n \geq \frac{a\bar{X}_n^{b-2}}{C^2}. \tag{4.94}$$

This may be rewritten in a form that permits computations to be completed before the initiation of sampling.

Assume that the purpose of sampling is to estimate the mean within $D\mu$ with $(1 - \alpha)$ 100% confidence. By extending equation (4.15) and using the relationship in mean and variance from Taylor's Power Law, we propose sampling until

$$n \geq \frac{a\bar{X}_n^{b-2}z_{\alpha/2}^2}{C^2}. \tag{4.95}$$

Again the stopping rule may be rewritten in terms of the total.

When the goal is to estimate the population mean density within H with $(1 - \alpha)$ 100% confidence, we propose sampling until

$$n \geq a\bar{X}_n^b\left(\frac{z_{\alpha/2}}{H}\right)^2 \tag{4.96}$$

when working with Taylor's Power Law.

Summary

Developing a sampling program is an important aspect of most population biology studies. In this chapter, we considered fixed and sequential sampling plans. Based on knowledge of the population, the size of sample required to obtain an estimate of population density with the desired precision can be determined. In practice, our knowledge of the population is imprecise, and the sample sizes computed are approximately those required to obtain the desired precision of the estimate. Yet, these approximations are much better than initiating sampling without any thought given to the sample size needed to obtain the desired precision of the estimate. Alternatively, we can use sequential sampling methods and sample until the parameter is estimated with a stated level of precision. However, here too we must evaluate the plan to be sure that sufficient resources are available.

Three measures of precision were considered. First, estimating the population density with a specified coefficient of variation of the estimate was developed. Next, obtaining an interval estimate for population density with a specified level of confidence and length a given proportion of the population density was discussed. Finally, estimating the population density within a specified amount with a stated level of confidence was addressed. This last approach should be more fully studied before broad-based application.

Models of the relationship of mean and variance have been developed. Iwao's patchiness regression and Taylor's Power Law are two popular approaches. More appropriate methods for assessing the fit of these models are needed. Further, the statistical properties of the sampling programs based on these variance-to-mean relationships are not known and would benefit from further study.

Exercises

1. Mazagan, a large horse bean was planted in small beds (Taylor, 1970). During or immediately after the primary aphid (*Aphis fabae* Scop.) migration in May, the plants were removed. The number of aphids on each stem were counted. Thirty-four samples were collected in this manner. The data are presented in the table below:

Number of Aphids (Aphis fabae Scop.) on Mazagan

Sample No.	n	$\bar{X}$	s^2	Sample No.	n	$\bar{X}$	s^2
1	61	1.54	82.0	18	22	10.18	547.0
2	58	1.14	34.0	19	7	5.00	175.0
3	74	3.41	721.0	20	93	0.05	0.1
4	90	0.48	3.2	21	70	0.14	0.6
5	70	5.23	404.0	22	31	1.10	28.0
6	46	2.22	83.0	23	36	0.42	4.1
7	69	1.97	124.0	24	23	1.04	19.0
8	76	4.49	367.0	25	17	0.12	0.2
9	22	14.05	1037.0	26	18	1.44	27.0
10	29	16.83	2270.0	27	28	0.14	0.6
11	43	20.77	4957.0	28	76	0.17	1.1
12	100	0.17	1.5	29	97	0.14	0.8
13	92	0.95	27.0	30	12	0.75	3.2
14	47	0.36	2.2	31	17	0.53	3.0
15	57	1.60	30.0	32	16	1.94	46.0
16	64	0.58	16.0	33	23	0.17	0.2
17	38	10.53	1108.0	34	29	5.48	755.0

From Taylor (1970).

A. Assuming a negative binomial consistently fits data collected in this manner estimate a common value of k for these 34 samples.

B. Use two different methods to test whether a common value of k is appropriate for these data.

2. In problem 1, both apterae (unwinged) and alatae (winged) aphids were included in the counts. However, the populations probably tended to be primarily alate because the data were collected during or immediately after the primary aphid migration. Another researcher (reported in Taylor 1970) considered the apterae and alatae populations in yearling field beans separately. He sampled in 8 consecutive weeks, during which time the aphid population rose and fell. The data are presented in the table below:

Number of Apterae and Alatae Aphids in Yearling Bean Fields for Eight Consecutive Weeks

Sampling date	No. of stems	Apterae		Alatae	
		$\bar{X}$	s^2	$\bar{X}$	s^2
4 June	50	18.5	2162	0.26	0.28
9 June	50	136.0	65025	0.44	0.82
16 June	49	732.3	839056	0.45	4.16
23 June	50	2202.0	3841600	8.06	581.00
1 July	50	6761.0	19856000	56.00	3808.00
8 July	51	6935.0	14357000	157.00	16039.00
15 July	49	2514.0	468300	66.00	7365.00
22 July	48	126.0	55225	0.63	2.28

A. Assuming a negative binomial consistently fits data collected in this manner estimate a common value of k for the apterae and estimate another common value of k for the alatae.

B. Use two different methods to test whether a common value of k is appropriate for the apterae or for the alatae.

C. Estimate a common k for both apterae and alatae aphids.

D. Use two different methods to test whether a common value of k is appropriate for both apterae and alatae.

3. A study was conducted to determine the distribution of mites on the fruit of lime trees. From this work, it was concluded that the distribution of mites was geometric. Develop a sequential sampling plan to estimate the mean number of mites on a lime within an orchard. Use a $CV(\bar{X})$ of 0.1, 0.15, 0.2, and 0.25.

4. The number of mites on a lime can be large and time consuming to determine. Therefore, a worker proposed simply determining whether or not each lime had at least one mite on it. Develop a sequential sampling plan to estimate the proportion of limes with at least one mite. Use a $CV(\bar{X})$ of 0.1, 0.15, 0.2, and 0.25.

5. Contrast the two approaches in exercises 3 and 4. Discuss the strengths and weaknesses of each.

6. Suppose now the workers assumed that the number of mites on a lime was Poisson because "all count data are Poisson." Develop a sequential sampling plan to estimate the mean number of mites on a lime. Again use a $CV(\bar{X})$ of 0.1, 0.15, 0.2, and 0.25.

7. Contrast the plans developed in exercises 4 and 6. How would you respond to the workers?

8. The worker now tells you that he wonders whether the distribution was truly geometric. The distribution is negative binomial, but the k might not be 1. What effect would an inaccurate value of k have on the sampling plan developed in exercise 3?

9. A researcher was interested in estimating the mean number of pigweeds in a 1-meter length of row in a soybean field. The rows were 40 inches apart. Therefore, 20-inches on either side of the row were included in the count for each 1-meter length. From experience, the researcher expects the distribution to be negative binomial with a k of about 0.6. Develop a sequential sampling plan to estimate the mean population pigweed density within 15% of the mean with 90% confidence.

10. The researcher described in exercises 9 now decides to record only whether one or more pigweeds is in each sampling unit.

   A. Develop a sequential sampling plan to estimate the proportion of sampling units in the population with one or more pigweeds within 10% of the true proportion with 95% confidence.

   B. Suppose upon the completion of sampling, we estimate the population proportion to be 0.40 based on 50 observations. Use this and the information in exercise 9 to obtain an estimate of the population mean density of pigweeds. With 90% confidence, within what percentage of the mean do you expect this estimate to be?

11. Suppose we are sampling from a Poisson distribution. We are having trouble deciding whether to estimate the mean within 10% of the mean with 95% confidence or within 1 with 95% confidence. For which values of the mean would the optimal sample size be smaller using each method? For which value of the mean would the optimal sample sizes be equal? Discuss the value of this information as it would relate to a sampling problem in your discipline.

12. Find a journal article within your field of study that discusses or is based on a sequential sampling approach. Describe the purpose of sampling and the methods used.

13. Rework exercise 1 using Iwao's Patchiness Regression instead of a common k. Fully discuss the results.

14. Rework exercise 2 using Iwao's Patchiness Regression instead of a common k. Fully discuss the results.

15. Rework exercise 1 using Taylor's Power Law instead of a common k. Fully discuss the results.

16. Rework exercise 2 using Taylor's Power Law instead of a common k. Fully discuss the results.

17. Show that when the method-of-moments estimator of k is used in equations (4.32), (4.33), and (4.34), the stopping rules are equivalent to equations (4.31), (4.30), and (4.29), respectively.

5

Sequential Hypothesis Testing

Introduction

Sequential sampling is a fast efficient tool for many sampling problems. Sequential sampling may be used (1) to obtain precise estimate(s) of the parameter(s), or (2) to test hypotheses concerning the parameters. Sequential estimation is used when the purpose of sampling is to obtain precise parameter estimates. Several sequential estimation procedures are discussed in Chapter 4. The focus of this chapter is sequential hypothesis testing. This approach is appropriate when we are interested in determining whether the population density is above or below a stated threshold. As in sequential estimation, sequential hypothesis testing requires taking observations sequentially until some stopping criterion is satisfied. The observations are taken at random over the sampling area. Generally, the accumulated total of the observations relative to the number of observations taken determines when sampling is stopped. The sequential hypothesis testing we consider requires some prior knowledge of the population distribution. This permits most computations to be completed in advance of sampling and to be stored in handheld calculators, laptop computers, or printed on cards or sheets. Wald's sequential probability ratio test was the earliest sequential test and is described first. Lorden's 2-SPRT is a more recent development that has some exciting possibilities for tests of hypotheses concerning population density and is discussed in the latter parts of this chapter.

Wald's Sequential Probability Ratio Test

Wald (1947) introduced a sequential method of testing a simple null versus a simple alternative hypothesis with specified error probabilities. He called this sequential process the sequential probability ratio test (SPRT). The SPRT was used as a basis for inspecting wartime weapons and evaluating war research problems. Oakland (1950) and Morgan et al. (1951) presented early applications of the SPRT in the biological sciences. The SPRT was used increasingly in biology after Waters (1955) presented details and examples of its use in forest insect studies. By the 1960s, the SPRT was widely applied in integrated pest management for a wide variety of pests and cultivars. Today, the SPRT is most often used to determine whether a specified region needs to be treated with pesticides (*see* Onsager, 1976; Pieters and Sterling, 1974; Oakland, 1950; Allen et al., 1972; Rudd 1980).

As in any sampling program, the first step is to identify clearly the sampling unit and the associated random variable of interest, such as the number of lady beetles (random variable) on a cotton plant (sampling unit) or the number of pigweeds (random variable) in a one-row meter (sampling unit) of a corn field. The SPRT assumes that the parametric form of the random variable's distribution is known. In presence–absence sampling, the binomial distribution is the parametric form of the distribution. The negative binomial and Poisson distributions are likely parametric forms when the number of organisms within a sampling unit is the random variable of interest. For the negative binomial, the parameter k is assumed known; if it is unknown, it must be estimated before the test of hypotheses can begin. The equations associated with the normal distribution with a known variance will be given because these are used in some biological applications. Because Iwao's Patchiness Regression and Taylor's Power Law are based on a relationship in mean and variance, and not on a parametric distribution, these approaches are not appropriate for use in Wald's SPRT.

Once the sampling unit and associated random variable are clearly defined, the hypotheses to be tested must be stated:

$$H_0: \quad \theta = \theta_0$$
$$H_1: \quad \theta = \theta_1. \tag{5.1}$$

Two types of error are associated with the tests of these hypotheses. A type I error results when H_0 is true, but it is rejected in favor of H_1. A type II error occurs when H_0 is false, and it is not rejected. The probabilities of type I and type II errors are α and β, respectively.

Usually we actually want to test two composite hypotheses:

$$H_0: \quad \theta \leq \theta_0$$
$$H_1: \quad \theta \geq \theta_1. \tag{5.2}$$

In (5.2), the type I and type II error rates are associated with θ_0 and θ_1, respectively, because the maximum probability of error occurs at these points. Consequently, the test of the composite hypotheses in (5.2) is the same as the one for the simple hypotheses in (5.1). In many biological applications, the hypotheses are being tested to determine whether the population density is above the economic threshold, implying treatment is needed, or below a safety level in which case no treatment is necessary. For these cases, (5.2) may be stated as

H_0: The population is below a stated safety level

H_1: The population is above the economic threshold.   (5.3)

In (5.3), we have implicitly assumed that $\theta_0 < \theta_1$. We use this convention throughout the remainder of our discussion for mathematical reasons. However, from an applied view, the order of the hypotheses makes no difference because the probabilities of type I and II errors are controlled.

We briefly discuss the foundation of hypothesis testing for fixed-sample sizes and describe how these ideas are extended to sequential testing. The Neyman-Pearson lemma states that, with a fixed sample size n and type I error probability α, the test that minimizes β is the likelihood ratio test. The likelihood ratio is

$$\lambda = \frac{\prod_{i=1}^{n} f_X(x_i;\theta_1)}{\prod_{i=1}^{n} f_X(x_i;\theta_0)}.$$

where f_X is the probability mass (density) function associated with the random variable X of interest. The critical region that leads to rejection of H_0 is given by $\lambda > C_n$. C_n is obtained using the predetermined values of α and n. Consideration of β may impact the choice of n and/or α, but once α and n are set, the type II error rate β is determined.

Wald presented an alternative approach by setting α and β and then minimizing the sample size n. Thus, the final sample size N becomes a random variable of interest. A sequence of random observations $(X_1, X_2, X_3, \ldots)$ is drawn from one of the two hypothesized distributions resulting in a sequence of likelihood ratios:

$$\lambda_n = \frac{\prod_{i=1}^{n} f_X(x_i;\theta_1)}{\prod_{i=1}^{n} f_X(x_i;\theta_0)}.$$ (5.4)

A decision is made and the test terminated after the nth observation if $\lambda_n \geq A$ (accept H_1) or $\lambda_n \leq B$ (accept H_0), where A and B are constants, depending on the predetermined α and β. Exact computations for A and B are extremely difficult. Thus, the approximations

$$A \approx \frac{1 - \beta}{\alpha} \tag{5.5}$$

and

$$B \approx \frac{\beta}{1 - \alpha} \tag{5.6}$$

are used. The approximations for A and B result in error rates α' and β' that differ from the desired error rates α and β. However, it may be shown that the total of the actual error rates, α' and β', is no larger than the sum of the specified error rates, α and β; that is, $\alpha' + \beta' \leq \alpha + \beta$.

Sampling continues as long as $B < \lambda_n < A$. In this form, computations are required after each observation in order to determine whether the stopping criterion is met. However, this relationship can be simplified. To do so, first, take the natural logarithm to obtain

$$b = \ln(B) < \ln(\lambda_n) < \ln(A) = a. \tag{5.7}$$

For the distributions we consider, equation (5.7) may be rewritten as bounds on the cumulative sum of the observations. Let T_n denote the sum of the first n observations. Then, the decision procedure may be stated as

1. Terminate the test and accept the alternative hypothesis if

$$T_n \geq h_2 + Sn,$$

2. Terminate the test and accept the null hypothesis if

$$T_n \leq h_1 + Sn,$$

3. Continue sampling if

$$h_1 + Sn < T_n < h_2 + Sn.$$

Notice that the decision boundaries are straight lines. They are parallel because both have slope S. The intercepts are h_1 and h_2 for the lower and upper boundaries, respectively (see Figure 5.1). S, h_1, and h_2 depend on the distribution, the hypothesized parameters, and the specified error rates (see Table 5.1).

The extensive use of Wald's SPRT in pest management programs gives added meaning to the three zones in Figure 5.1 as may be seen in Figure 5.2. Figure 5.2 is based on the assumption that a pest species is being sampled to determine whether treatment is necessary to control the population. Three zones are pro-

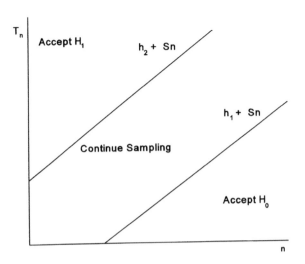

Figure 5.1. Decision lines for Wald's SPRT.

Table 5.1. Intercepts and Slopes for Wald's SPRT for the Binomial (B), *Negative Binomial* (NB), *Normal* (N), *and Poisson* (P) *Distributions*

Distribution	h_1	h_2	S
B	$\dfrac{b}{\ln\left[\dfrac{p_1 q_0}{p_0 q_1}\right]}$	$\dfrac{a}{\ln\left[\dfrac{p_1 q_0}{p_0 q_1}\right]}$	$\dfrac{\ln\left[\dfrac{q_0}{q_1}\right]}{\ln\left[\dfrac{p_1 q_0}{p_0 q_1}\right]}$
NB	$\dfrac{b}{\ln\left[\dfrac{\mu_1(\mu_0 + k)}{\mu_0(\mu_1 + k)}\right]}$	$\dfrac{a}{\ln\left[\dfrac{\mu_1(\mu_0 + k)}{\mu_0(\mu_1 + k)}\right]}$	$k\,\dfrac{\ln\left[\dfrac{\mu_1 + k}{\mu_0 + k}\right]}{\ln\left[\dfrac{\mu_1(\mu_0 + k)}{\mu_0(\mu_1 + k)}\right]}$
N	$\dfrac{b\sigma^2}{\mu_1 - \mu_0}$	$\dfrac{a\sigma^2}{\mu_1 - \mu_0}$	$\dfrac{\mu_1 + \mu_0}{2}$
P	$\dfrac{b}{\ln\left[\dfrac{\lambda_1}{\lambda_0}\right]}$	$\dfrac{a}{\ln\left[\dfrac{\lambda_1}{\lambda_0}\right]}$	$\dfrac{\lambda_1 - \lambda_0}{\ln\left[\dfrac{\lambda_1}{\lambda_0}\right]}$

where $\quad a = \ln\left(\dfrac{1 - \beta}{\alpha}\right)\quad$ and $\quad b = \ln\left(\dfrac{\beta}{1 - \alpha}\right)$

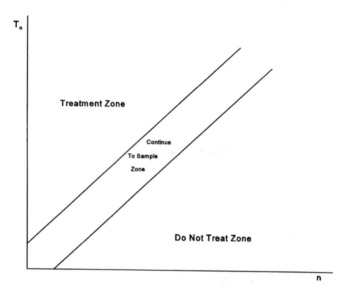

Figure 5.2. Wald's SPRT applied to integrated pest management.

duced: the zone of decision to recommend control or treatment, the zone of decision to recommend no treatment, and the indecision zone where sampling continues. Wald's SPRT is used extensively in professional scouting programs, such as those for cotton, fruits, vegetables, and forests, and in disease control. Extensive field evaluation has shown that the use of Wald's SPRT saves in time spent sampling when compared to fixed-sample-size methods. The greatest time savings result from decisions being made quickly when the populations are either considerably above the economic threshold or considerably below the safety level. Sampling takes more time when the population mean density is near or between the economic threshold and the safety level. Thus, sequential hypothesis testing results in a greater allocation of resources when decision-making is most critical.

The following information is needed to design a Wald's sequential sampling plan from a pest management view point:

1. *Economic threshold:* This is the population density or damage level where economic loss will occur if control procedures are not initiated

2. *Safety level:* This is the population density that will ensure that economic damage will not occur

3. *The distribution of the pest or damage being sampled:* If the distribution is negative binomial, the common k_c, or a precise estimate of k, is needed. The variance, or a precise estimate of it, is needed when hypotheses concerning the mean of a normal distribution are being tested

4. *The probabilities of type I and type II errors, α and β, respectively:* Based on the hypotheses in (5.3), a type I error occurs when the population density is declared to be above the economic threshold when it is in fact below the safety level. A type II error is committed when the population is stated to be below the safety level when it is above the economic threshold.

The safety level is often the most difficult to determine. In many scouting programs, a region is monitored throughout the growing season. Periodically, tests are conducted to determine whether treatment is needed. If the decision is made that no treatment is necessary, the scout wants to be assured that the population is not likely to exceed the economic threshold for some period of time; that is, the population density should be enough below the economic threshold that the likelihood of economic damage before the region is sampled again is small. The largest population density that offers this assurance is the safety level. Time and labor constraints and the relationship between pest density and damage are two of the primary factors considered when establishing this safety level. As the difference in the economic threshold and safety level increases, the distance between the decision lines decreases and the number of observations needed to reach a decision decreases.

The probabilities of type I and type II errors, α and β, respectively, also influence the distance between the lines. As the probability of error that the sampler is willing to accept increases, the distance between the lines decreases and the required sampling effort decreases. Often these error rates are set at the same level. Historically, many scouts and other pest management personnel have believed that it is more serious to make a type II error than a type I error. This largely results from the fact that when a producer suffers economic damage from a pest due to lack of treatment, the error is often more obvious than the error of having a producer apply a treatment when it was not really needed. Further, many producers who are paying for pest management advice may believe they are getting a greater return on their money if they are told a particular action is needed instead of being told that no action is required. With the increase in environmental concerns and the development of resistance to pesticides, this approach is debatable and is being reevaluated. If the sampling plan is based on sound ecological studies, each risk factor is equally important.

The equations for Wald's SPRT can be developed using the equations in Table 5.1 after these factors are determined. These provide the decision boundaries as in Figures 5.1 and 5.2. The totals needed to stop are generally presented in tabular, instead of graphic, form when implementing the SPRT.

SPRT for the Negative Binomial Distribution

The negative binomial distribution has been used extensively in developing sequential tests of hypotheses because it adequately models the distribution of many organisms. Therefore, sampling plans for the negative binomial distribution are widely available (*see* Pieters and Sterling, 1974; Allen et al., 1972; Oakland, 1950). For the negative binomial distribution, the hypotheses of interest are

$$H_0: \quad \mu = \mu_0$$
$$H_1: \quad \mu = \mu_1 (>\mu_0). \tag{5.8}$$

Knowledge, or a precise estimate, of k is required to determine the boundaries of the SPRT. The robustness of the SPRT to misspecification of k has been studied (Hubbard and Allen, 1991). Improperly specifying k does not yield error rates that are seriously different from those attained at the true value of k. If k is underestimated, the test is conservative; that is, the error rates are less than stated. If the estimate of k is larger than the true k, the error rates are larger than specified. As k becomes larger, the effect of misspecification is less pronounced. Another concern is that k may also differ, depending on the mean. A reasonable approach is to consider the estimates of k for means near and between the two hypothesized values. Use a common k if one can be found within that region. If not, a conservative test can be developed by using an estimate of k at the lower end of the range of the estimates from historical data. Hefferman (1996) has suggested adjusting α and β so that a test constructed based on a common k has the desired error rates if k varies with μ in a known manner or if the test is truncated.

Example 5.1

Suppose we want to determine whether a pest has sufficient population density to warrant treatment. Earlier work has determined that the economic threshold and safety levels are 10 and 7 pests/sampling unit, respectively. The distribution of this pest is known to be modeled well by the negative binomial distribution with $k = 3$. Further, assume that we have specified $\alpha = 0.1$ and $\beta = 0.1$. Then the hypotheses of interest are

$$H_0: \quad \mu = \mu_0 = 7$$
$$H_1: \quad \mu = \mu_1 = 10.$$

The slope of the decision boundaries is

$$S = k \frac{\ln\left(\dfrac{\mu_1 + k}{\mu_0 + k}\right)}{\ln\left[\dfrac{\mu_1(\mu_0 + k)}{\mu_0(\mu_1 + k)}\right]} = 3 \frac{\ln\left(\dfrac{10 + 3}{7 + 3}\right)}{\ln\left[\dfrac{10(7 + 3)}{7(10 + 3)}\right]} = 8.35.$$

The upper intercept is

$$h_2 = \frac{\ln\left(\dfrac{1 - \beta}{\alpha}\right)}{\ln\left[\dfrac{\mu_1(\mu_0 + k)}{\mu_0(\mu_1 + k)}\right]} = \frac{\ln\left(\dfrac{1 - 0.1}{0.1}\right)}{\ln\left[\dfrac{10(7 + 3)}{7(10 + 3)}\right]} = 23.30.$$

and the lower intercept is

$$h_1 = \frac{\ln\left(\dfrac{\beta}{1 - \alpha}\right)}{\ln\left[\dfrac{\mu_1(\mu_0 + k)}{\mu_0(\mu_1 + k)}\right]} = \frac{\ln\left(\dfrac{0.1}{1 - 0.1}\right)}{\ln\left[\dfrac{10(7 + 3)}{7(10 + 3)}\right]} = -23.30.$$

The upper decision boundary U in Figures 5.1 and 5.2 is given by

$$U = h_2 + Sn = 23.30 + 8.35n,$$

and the lower decision boundary L is

$$L = h_1 + Sn = -23.30 + 8.35n.$$

For the first observation,

$$L = -23.30 + (8.35)(1) = -15.0.$$

It is not possible to have minus pests, so a minimum sample size (MSS) is required before a decision can be made in favor of H_0, and this is

$$\text{MSS} = \frac{|h_1|}{S} = \frac{23.30}{8.35} = 2.8.$$

This is always rounded up, giving 3 as the MSS for the lower decision boundary. In other words, the MSS is the point where the smallest observation number multiplied by the slope is greater than the absolute value of the line intercept. It is customary to note the L for observation numbers below the MSS with *nd* for no decision. The upper line is calculated in the same manner. For example, at sample 11, we have

$$U = 23.30 + (8.35)(11) = 115.2.$$

In order to attain the desired error rates, U, the upper limit, should always be rounded up (giving a value of 116 for this example) and L, the lower limit, should always be rounded down. It is customary to program these systems into handheld calculators or laptop computers or to construct charts or graphs as in Table 5.2.

ECOSTAT can be used to construct an SPRT for the binomial, geometric, negative binomial, and Poisson distributions. Begin by selecting *SPRT* from the *Sequential Sampling* menu. Select the appropriate distribution. Then enter the hypothesized values and the error rates for the test. By pressing *Set Boundaries,* the equations for the upper and lower decision boundaries are displayed. A data sheet as in Table 5.2 can be stored by pressing *Data Sheet.* The data sheet can be printed using a 10-point fixed font, such as Courier, with one-inch margins.

In practice, sequential sampling is simple. An observation is made and recorded. Another observation is made, added to the previous sample total, and the total count is recorded. This procedure is repeated until a stopping rule is satisfied. The stopping rule has the following form:

Stop if the total number or organisms counted equals or exceeds the upper limit,

OR

Stop if the total number of organisms counted equals or is lower than the lower limit.

For instance, in Table 5.2, if the running total at observation 11 is 68 or less, stop sampling and recommend that no treatment be used. If the total number of organism is 116 or more, stop sampling and recommend treatment. If the total number of organisms is between 68 and 116, continue to sample until the total number of organisms crosses a decision boundary and satisfies the stopping rule. Excessive sampling may be needed to meet the stopping criterion. This may be due to the choices of hypothesized values and error probabilities. How to evaluate the reasonableness of a proposed sampling program is discussed later in this chapter. Also, if the true mean population density is between the two hypothesized means, sampling may continue beyond reasonable limits. If this case is frequently encountered, a 2-SPRT test (*see* 2-SPRT in this chapter) could be used. Many pest management scouts simply return to the field at a shorter interval. Knowledge of the pest and cropping system that is being sampled aids in this decision.

ECOSTAT can be used to construct an SPRT for the binomial, geometric, negative binomial, and Poisson distributions. Begin by selecting *SPRT* from the *Sequential Sampling* menu. Select the appropriate distribution. Then enter the hypothesized values and the error rates for the test. By pressing *Set Boundaries,* the equations for the upper and lower decision boundaries are displayed. A data sheet as in Table 5.2 can be stored by pressing *Data Sheet.* The data sheet can be printed using a 10-point fixed font, such as Courier, with one-inch margins.

SPRT for the Poisson Distribution

The Poisson distribution has one parameter, λ. The mean is λ, and the mean and variance are equal. Expressing (5.3) in terms of λ, we have

Table 5.2. Sequential Sampling for the Negative Binomial Distribution[a]

Sample number	Lower limit	Running total	Upper limit	Sample number	Lower limit	Running total	Upper limit
1	nd	—	32	52	402	—	449
2	nd	—	40	54	419	—	466
3	1	—	49	56	435	—	483
4	10	—	57	58	452	—	500
5	18	—	66	60	469	—	516
6	26	—	74	62	485	—	533
7	35	—	82	64	502	—	550
8	43	—	91	66	519	—	566
9	51	—	99	68	535	—	583
10	60	—	107	70	552	—	600
11	68	—	116	72	569	—	616
12	76	—	124	74	585	—	633
13	85	—	132	76	602	—	650
14	93	—	141	78	619	—	666
15	101	—	149	80	636	—	683
16	110	—	157	82	652	—	700
17	118	—	166	84	669	—	716
18	126	—	174	86	686	—	733
19	135	—	182	88	702	—	750
20	143	—	191	90	719	—	767
21	151	—	199	92	736	—	783
22	160	—	207	94	752	—	800
23	168	—	216	96	769	—	817
24	177	—	224	98	786	—	833
25	185	—	232	100	802	—	850
26	193	—	241	102	819	—	867
27	202	—	249	104	836	—	883
28	210	—	257	106	853	—	900
29	218	—	266	108	869	—	917
30	227	—	274	110	886	—	933
31	235	—	283	112	903	—	950
32	243	—	291	114	919	—	967
33	252	—	299	116	936	—	984
34	260	—	308	118	953	—	1000
35	268	—	316	120	969	—	1017
36	277	—	324	122	986	—	1034
37	285	—	333	124	1003	—	1050
38	293	—	341	126	1019	—	1067
39	302	—	349	128	1036	—	1084
40	310	—	358	130	1053	—	1100
41	318	—	366	132	1069	—	1117
42	327	—	374	134	1086	—	1134
43	335	—	383	136	1103	—	1150

Table 5.2.   Continued

Sample number	Lower limit	Running total	Upper limit	Sample number	Lower limit	Running total	Upper limit
44	343	—	391	138	1120	—	1167
45	352	—	399	140	1136	—	1184
46	360	—	408	142	1153	—	1201
47	368	—	416	144	1170	—	1217
48	377	—	424	146	1186	—	1234
49	385	—	433	148	1203	—	1251
50	393	—	441	150	1220	—	1267

$^{a}\mu_0 = 7, \mu_1 = 10, k = 3, \alpha = 0.1, \beta = 0.1.$

$$H_0: \quad \mu = \mu_0 = \lambda_0$$
$$H_1: \quad \mu = \mu_1 = \lambda_1(>\lambda_0) \tag{5.9}$$

After the type I and type II error rates for the test have been specified, the decision boundaries may be constructed using the formulae in Table 5.1

Example 5.2

Suppose that if 30,000 spiders per acre are present in a field, treatment to control a pest population within that field is not needed. Further, assume that this field has 40,000 plants per acre and that 20,000 spiders per acre leave the field vulnerable to pest attack. Experience has shown that the distribution of the number of spiders tends to be Poisson in most fields. This example differs from earlier ones in that since the spiders are predators, no treatment is required if the population is large enough. Also, the example is more realistic, in that the sampling unit, the hypotheses, and the error rates have not been specified. In this case, it seems reasonable to use a plant as the sampling unit. The hypotheses of interest are

$$H_0: \quad \mu = \lambda_0 = \frac{20,000}{40,000} = 0.5$$

$$H_1: \quad \mu = \lambda_1 = \frac{30,000}{40,000} = 0.75.$$

Notice that the hypotheses are stated in terms of the mean density on the sampling unit basis. Assume that we specify $\alpha = \beta = 0.1$. Then the slope is

$$S = \frac{\lambda_1 - \lambda_0}{\ln\left(\frac{\lambda_1}{\lambda_0}\right)} = \frac{0.75 - 0.5}{\ln\left(\frac{0.75}{0.5}\right)} = 0.617.$$

The upper intercept is

$$h_2 = \frac{\ln\left(\frac{1 - \beta}{\alpha}\right)}{\ln\left(\frac{\lambda_1}{\lambda_0}\right)} = \frac{\ln\left(\frac{1 - 0.1}{0.1}\right)}{\ln\left(\frac{0.75}{0.50}\right)} = 5.42,$$

and the lower intercept is

$$h_1 = \frac{\ln\left(\frac{\beta}{1 - \alpha}\right)}{\ln\left(\frac{\lambda_1}{\lambda_0}\right)} = \frac{\ln\left(\frac{0.1}{1 - 0.1}\right)}{\ln\left(\frac{0.75}{0.50}\right)} = -5.42.$$

The upper decision line is

$$U = 5.42 + 0.617n,$$

and the lower decision line is

$$L = -5.42 + 0.617n.$$

The minimum sample size is

$$\text{MSS} = \frac{|h_1|}{S} = \frac{|-5.42|}{0.617} = 8.78.$$

Therefore, at least nine observations must be made before deciding in favor of H_0 that the population density of the spiders is not sufficient to protect the field from the pest species.

SPRT for the Binomial Distribution

Binomial sequential sampling plans are sometimes used in pest management programs. This usually arises when each observation may be considered an outcome of a Bernoulli trial; that is, the fruit is damaged or not damaged, the organism is male or female, the plant is infested or not infested, and so on. When counting organisms is time consuming, the binomial may also be used as an alternative to the negative binomial and Poisson distributions. In this case, each sampling unit is judged to have no organisms or at least one organism. Then the probability of 0 for the distribution of the number of organisms per sampling unit is equated with q of the binomial distribution.

The hypotheses to be tested are

$$H_0: \quad p = p_0$$
$$H_1: \quad p = p_1(>p_0). \tag{5.10}$$

The decision boundaries may be constructed using the formulae in Table 5.1.

Example 5.3

Dry cowpeas in storage are to be tested for weevil damage. If 50% or more of the cowpeas are infested, they must be destroyed. If 30% or less are infested, the cowpeas can be saved by treating them with a fumigant. The sampling unit is one cowpea. Suppose that the type I and type II error rates are specified to be 0.15 and 0.20, respectively.

The hypotheses to be tested are

$$H_0: \quad p = 0.3$$
$$H_1: \quad p = 0.5.$$

For the binomial distribution, $q = 1 - p$. A value of q is associated with each hypothesized value of p; that is,

$$q_0 = 1 - 0.3 = 0.7$$

and

$$q_1 = 1 - 0.5 = 0.5.$$

The slope of the decision lines is then

$$S = \frac{\ln\left(\dfrac{q_0}{q_1}\right)}{\ln\left[\dfrac{p_1(q_0)}{p_0(q_1)}\right]} = \frac{\ln\left(\dfrac{0.7}{0.5}\right)}{\ln\left[\dfrac{0.5(0.7)}{0.3(0.5)}\right]} = 0.397.$$

The upper intercept is

$$h_2 = \frac{\ln\left(\dfrac{1-\beta}{\alpha}\right)}{\ln\left(\dfrac{p_1 q_0}{p_0 q_1}\right)} = \frac{\ln\left(\dfrac{1-0.2}{0.15}\right)}{\ln\left[\dfrac{0.5(0.7)}{0.3(0.5)}\right]} = 1.98,$$

and the lower intercept is

$$h_1 = \frac{\ln\left(\dfrac{\beta}{1-\alpha}\right)}{\ln\left[\dfrac{p_1 q_0}{p_0 q_1}\right]} = \frac{\ln\left(\dfrac{0.2}{1-0.15}\right)}{\ln\left[\dfrac{0.5(0.7)}{0.3(0.5)}\right]} = -1.71.$$

Notice that in the example above, $h_1 \neq -h_2$. When $\alpha = \beta$, the intercept of the lower decision boundary is the negative of the intercept of the upper decision boundary. This relationship no longer holds if $\alpha \neq \beta$.

The minimum sample size for deciding in favor of H_0 is

$$\text{MSS} = \frac{|h_1|}{S} = \frac{|-1.71|}{0.397} = 4.31.$$

Therefore, at least five observations must be taken before one can conclude that the proportion of infested cowpeas is no more than 30%.

Operating Characteristic and Average Sample Number Functions

The properties of any proposed sampling system should be known as completely as possible before that program is implemented. The evaluation of Wald's SPRT is generally based on the Operating Characteristic (OC) and Average Sample Number (ASN) functions.

The OC function, $L(\theta)$, is the probability that the null hypothesis is accepted given any parameter θ in the parameter space. In constructing the test, the user specifies $L(\theta_0) = 1 - \alpha$ and $L(\theta_1) = \beta$. Wald (1947) proved that, for general θ,

$$L(\theta) \approx \frac{A^{h(\theta)} - 1}{A^{h(\theta)} - B^{h(\theta)}} \tag{5.11}$$

where A and B are given in equations (5.5) and (5.6), respectively, and $h(\theta)$ is the solution of

$$\sum_x \left(\frac{f(x;\theta_1)}{f(x;\theta_0)}\right)^{h(\theta)} f(x;\theta) = 1 \qquad (5.12)$$

or

$$\int_{-\infty}^{\infty} \left(\frac{f(x;\theta_1)}{f(x;\theta_0)}\right)^{h(\theta)} f(x;\theta)dx = 1, \qquad (5.13)$$

depending on whether $f(x;\theta)$ is discrete or continuous, respectively. With the exception of $h(\theta_0) = 1$ and $h(\theta_1) = -1$, it is difficult to solve for $h(\theta)$ given θ in equations (5.12) and (5.13). Therefore, Wald (1947) suggested solving for θ given h values, rather than vice versa. For $h(\theta) \neq 0$, the solution for θ is given in Table 5.3 (see also Waters, 1955; Fowler and Lynch, 1987). Then equation (5.11) is used to compute $L(\theta)$. This is also given in Table 5.3 for completeness. When $h(\theta) = 0$, θ is the solution of

$$E\left[\ln\left(\frac{f(x;\theta_1)}{f(x;\theta_0)}\right)\right] = 0 \qquad (5.14)$$

and

$$L(\theta) \approx \frac{a}{a - b} \qquad (5.15)$$

where a and b, as defined in equation (5.7), are $\ln(A)$ and $\ln(B)$, respectively. The θ for $h(\theta) = 0$ and the corresponding $L(\theta)$ are given in Table 5.4. By computing $L(\theta)$ for h values ranging from -2 to 2 in increments of 0.2, the OC function can usually be adequately described. These computations may be done using ECOSTAT. Simply press Wald's Approximations after the decision lines have been found as described in Example 5.1.

One of the disturbing features of the SPRT for many first-time users is that the indecision zone lies between two parallel lines. The fear of staying in the indecision zone forever is common. Wald (1945, 1947) did prove that the probability of making a terminating decision was one, regardless of the true value of θ. However, this does not guarantee that the sample size needed to reach a decision will be within the time and labor constraints of the sampler. Therefore, another important measure of the feasibility of a proposed sampling plan is the average sample size that is required to satisfy the stopping criterion. The ASN function $[E_\theta(n)]$ is the average number of observations needed to make a terminating decision for each θ in the parameter space Θ. Wald (1947) approximated the ASN function using

Table 5.3. Operating Characteristic Function, L(θ); Average Sample Number Function, E_θ(n); and the Relationship Between θ and h(θ), when h(θ) ≠ 0, for the Binomial (B), Negative Binomial (NB), Normal (N), and Poisson (P) Distributions

Distribution	θ	$L(\theta)$	$E_\theta(n)$	$\theta = f[h(\theta)]$
B	p	$\dfrac{A^{h(p)} - 1}{A^{h(p)} - B^{h(p)}}$	$\dfrac{bL(p) + a[1 - L(p)]}{p \ln\left[\dfrac{p_1(1 - p_0)}{p_0(1 - p_1)}\right] + \ln\left[\dfrac{1 - p_1}{1 - p_0}\right]}$	$p = \dfrac{1 - \left[\dfrac{1 - p_1}{1 - p_0}\right]^{h(p)}}{\left[\dfrac{p_1}{p_0}\right]^{h(p)} - \left[\dfrac{1 - p_1}{1 - p_0}\right]^{h(p)}}$
NB	μ	$\dfrac{A^{h(\mu)} - 1}{A^{h(\mu)} - B^{h(\mu)}}$	$\dfrac{bL(\mu) + a[1 - L(\mu)]}{k \ln\left[\dfrac{\mu_0 + k}{\mu_1 + k}\right] + \mu \ln\left[\dfrac{\mu_1(\mu_0 + k)}{\mu_0(\mu_1 + k)}\right]}$	$\mu = k\left\{\dfrac{1 - \left[\dfrac{\mu_0 + k}{\mu_1 + k}\right]^{h(\mu)}}{\left[\dfrac{\mu_1(\mu_0 + k)}{\mu_0(\mu_1 + k)}\right]^{h(\mu)} - 1}\right\}$
N	μ	$\dfrac{A^{h(\mu)} - 1}{A^{h(\mu)} - B^{h(\mu)}}$	$\dfrac{bL(\mu) + a[1 - L(\mu)]}{\left[\dfrac{\mu_1 - \mu_0}{\sigma^2}\right]\left[\mu - \dfrac{\mu_1 - \mu_0}{2}\right]}$	$\mu = \dfrac{\mu_1 + \mu_0}{2} - \dfrac{h(\mu)(\mu_1 - \mu_0)}{2}$
P	λ	$\dfrac{A^{h(\lambda)} - 1}{A^{h(\lambda)} - B^{h(\lambda)}}$	$\dfrac{bL(\lambda) + a[1 - L(\lambda)]}{\lambda \cdot \ln\left[\dfrac{\lambda_1}{\lambda_0}\right] + \lambda_0 - \lambda_1}$	$\lambda = \dfrac{(\lambda_1 - \lambda_0)h(\lambda)}{\left[\dfrac{\lambda_1}{\lambda_0}\right]^{h(\lambda)} - 1}$

where $a = \ln\left(\dfrac{1 - \beta}{\alpha}\right)$ and $b = \ln\left(\dfrac{\beta}{1 - \alpha}\right)$

Reprinted from Young and Young, 1994, p. 758, with kind permission from Elsevier Science-NL, Sara Burgerhartstraat 25, 1055 KV Amsterdam, The Netherlands.

Table 5.4. *Operating Characteristic Value,* $L(\theta)$; *Average Sample Number Value,* $E_\theta(n)$; *and* θ, *When* $h(\theta) = 0$ *for the Binomial* (B), *Negative Binomial* (NB), *Normal* (N), *and Poisson* (P) *Distributions*

Distribution	θ	$L(\theta)$	$E_\theta(n)$	θ when $h(\theta) = 0$
B	p	$\dfrac{a}{a - b}$	$\dfrac{-ab}{\ln\left[\dfrac{p_1}{p_0}\right]\ln\left[\dfrac{1 - p_0}{1 - p_1}\right]}$	$p = \dfrac{\ln\left[\dfrac{1 - p_0}{1 - p_1}\right]}{\ln\left[\dfrac{p_1(1 - p_0)}{p_0(1 - p_1)}\right]}$
NB	μ	$\dfrac{a}{a - b}$	$\dfrac{-ab}{\left[\dfrac{\mu(\mu + k)}{k}\right]\left[\ln\left(\dfrac{\mu_1(\mu_0 + k)}{\mu_0(\mu_1 + k)}\right)\right]^2}$	$\mu = k\,\dfrac{\ln\left[\dfrac{\mu_1 + k}{\mu_0 + k}\right]}{\ln\left[\dfrac{\mu_1(\mu_0 + k)}{\mu_0(\mu_1 + k)}\right]}$
N	μ	$\dfrac{a}{a - b}$	$\dfrac{-ab}{\dfrac{(\mu_1 - \mu_0)^2}{\sigma^2}}$	$\mu = \dfrac{\mu_1 + \mu_0}{2}$
P	λ	$\dfrac{a}{a - b}$	$\dfrac{-ab}{\lambda\left[\ln\left(\dfrac{\lambda_1}{\lambda_0}\right)\right]^2}$	$\lambda = \dfrac{(\lambda_1 - \lambda_0)}{\ln\left[\dfrac{\lambda_1}{\lambda_0}\right]}$

where $a = \ln\left(\dfrac{1 - \beta}{\alpha}\right)$ and $b = \ln\left(\dfrac{\beta}{1 - \alpha}\right)$

$$E_\theta(n) \approx \frac{bL(\theta) + a(1 - L(\theta)}{E\left[\ln\left(\dfrac{f_X(x;\theta_1)}{f_X(x;\theta_0)}\right)\right]} \qquad (5.16)$$

for $h(\theta) \neq 0$ and

$$E_\theta(n) \approx \frac{-ab}{E\left[\ln\left(\dfrac{f_X(x;\theta_1)}{f_X(x;\theta_0)}\right)\right]} \qquad (5.17)$$

for $h(\theta) = 0$. The forms (5.16) and (5.17) take for the binomial, negative binomial, Poisson, and normal distribution are shown in Tables 5.3 and 5.4 (*see* Waters, 1955; Fowler and Lynch, 1987).

Wald's equations for the OC and ASN functions are based on the assumption that sampling stops on the boundary. This rarely occurs. Usually, there is some *overshooting* of the boundary with the size of the overshoot being the absolute difference in the final total T_N and the boundary at that N. The approximations tend to overstate the error probabilities and understate the average sample size. The difference in the true and approximate error probabilities is not of practical significance in most cases. However, Corneliussen and Ladd (1970) found that Wald's formula under-represented the ASN at the maximum by about 20% when sampling from binomial distributions. Similar results were obtained by Fowler and Lynch (1987) and Seebeck (1989) for the Poisson and negative binomial distributions. In addition, the distribution of the sample size tends to be highly skewed. This means that sample sizes substantially larger than the average may be required to reach a decision. If all available resources are needed to gather a sample of average size, the sampler will often have to cease sampling before meeting the stopping criterion.

A realistic assessment of the ASN function and the possible sample size range is needed. Corneliussen and Ladd (1970) developed an algorithm for computing the exact OC and ASN functions for the binomial distribution. This work was extended by Young (1994) to cover discrete distributions that are members of the exponential family with special emphasis on the Poisson and negative binomial distributions. These algorithms, along with Wald's approximations, are in the accompanying software ECOSTAT. The exact computations take much longer than Wald's approximations. Therefore, we recommend investigating the properties of a proposed SPRT using Wald's approximations. Once a reasonable test seems to have been developed, examine the exact properties of the test. In addition to the OC and ASN functions for various parameter values, the 95th percentile of sample size $N_{.95}$ is given; 95% of the samples collected will have sample sizes $\leq N_{.95}$, and 5% of the sample sizes will be $> N_{.95}$.

Example 5.4

Consider again the SPRT developed in Example 5.1. Computing Wald's approximations for the OC and ASN functions as $h(P)$ ranges from -2 to 2 in increments of 0.2, we obtain Table 5.5. To verify these numbers, suppose $h(P) = 0.4$. Then the parameter μ corresponding to $h(P) = 0.4$ is

$$\mu = k \frac{1 - \left[\dfrac{\mu_0 + k}{\mu_1 + k}\right]^{h(\mu)}}{\left[\dfrac{\mu_1(\mu_0 + k)}{\mu_0(\mu_1 + k)}\right]^{h(\mu)} - 1} = 3 \frac{1 - \left[\dfrac{7 + 3}{10 + 3}\right]^{0.4}}{\left[\dfrac{10(7 + 3)}{7(10 + 3)}\right]^{0.4} - 1} = 7.77.$$

For $\alpha = 0.10$ and $\beta = 0.10$, we can calculate

Table 5.5. Wald's Approximations of the OC and ASN Functions for Various Means with $k = 3$.[a]

Mean	OC	ASN
12.04	0.012	6.1
11.60	0.019	6.9
11.17	0.029	7.8
10.77	0.044	8.8
10.37	0.067	9.9
10.00	0.100	11.3
9.64	0.147	12.7
9.30	0.211	14.2
8.97	0.293	15.5
8.65	0.392	16.6
8.35	0.500	17.2
8.05	0.608	17.3
7.77	0.707	16.8
7.51	0.789	16.0
7.25	0.853	15.0
7.00	0.900	13.8
6.76	0.933	12.7
6.53	0.956	11.7
6.31	0.971	10.8
6.10	0.981	10.0
5.90	0.988	9.3

[a]The economic threshold and safety levels are 10 and 7 pests per sampling unit, respectively.

$$A = \frac{1 - \beta}{\alpha} = \frac{1 - 0.1}{0.1} = 9.0$$

and

$$B = \frac{\beta}{1 - \alpha} = \frac{0.1}{1 - 0.1} = 0.1111.$$

A, B, and $h(\mu)$ can now be used to evaluate the OC function for $\mu = 7.77$:

$$\frac{A^{h(\mu)} - 1}{A^{h(\mu)} - B^{h(\mu)}} = \frac{(9.0)^{0.4} - 1}{(9.0)^{0.4} - (0.1111)^{0.4}} = 0.707.$$

Recall that $a = \ln(A) = 2.197$ and $b = \ln(B) = -2.197$. Wald's approximation for the ASN function may now be computed:

$$\frac{bL(\mu) + a[1 - L(\mu)]}{k \ln\left[\frac{\mu_0 + k}{\mu_1 + k}\right] + \mu \ln\left[\frac{\mu_1(\mu_0 + k)}{\mu_0(\mu_1 + k)}\right]} = \frac{(-2.197)(0.707) + 2.197(1 - 0.707)}{3 \ln\left[\frac{7 + 3}{10 + 3}\right] + 7.77 \ln\left[\frac{10(7 + 3)}{7(10 + 3)}\right]}$$

$$= 16.8.$$

The exact properties of the test were computed using ECOSTAT and are given in Table 5.6. Wald's approximate and the exact OC and ASN functions are shown in Figures 5.3 and 5.4, respectively. In ECOSTAT, the exact values of the OC, ASN, and $N_{.95}$ functions may be found by pressing *Exact Computations* after the decision boundaries have been determined as described in Example 5.1. They can be stored as an ASCII file by pressing *Save Properties*.

The 2-SPRT

No other test of a simple versus a simple hypothesis with type I and type II error rates at most α and β, respectively, results in a smaller average sample size at the hypothesized values than Wald's sequential probability ratio test (Wald and

Table 5.6. *Exact Values of the OC and ASN Functions and $N_{.95}$ for Wald's SPRT with* $k = 3$.[a]

Mean	OC	ASN	$N_{.95}$
62.00	0.000	1.22	2
29.50	0.000	1.95	4
18.67	0.000	3.31	7
13.25	0.002	6.26	14
10.00	0.083	14.70	37
9.62	0.132	16.81	43
9.26	0.203	18.98	49
8.93	0.298	20.96	55
8.61	0.413	22.38	58
8.30	0.538	22.97	60
8.02	0.658	22.62	58
7.74	0.760	21.51	54
7.48	0.839	19.95	49
7.24	0.896	18.22	44
7.00	0.933	16.54	39
3.82	1.000	5.78	9
2.17	1.000	4.31	5
1.17	1.000	3.84	4
0.49	1.000	3.42	4

[a]The economic threshold and safety levels are 10 and 7 pests per sampling unit, respectively.

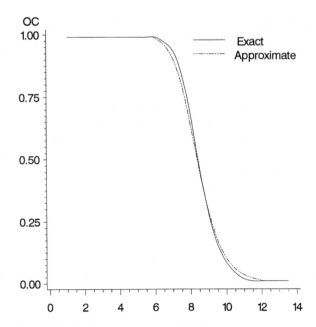

Figure 5.3. The SPRT OC function based on Wald's approximation and exact compu- tations based on the negative binomial distribution with $k = 3$. Economic threshold and safety level are 10 and 7, respectively. $\alpha = \beta = 0.1$.

Wolfowitz, 1948; Lehmann, 1959). However, since θ may not always assume one of the hypothesized values, the behavior of the ASN over the full parameter space is often of interest. As shown in Figure 5.4, the ASN function peaks at a point in the parameter space intermediate to the two hypothesized values, and this peak may represent a substantially larger average sample size than that at either hypothesized value. In addition, the sample size distribution tends to be highly skewed, and the sampling effort may greatly exceed the average as can be seen by the line denoting the 95th percentile of sample size, $N_{.95}$, in Figure 5.4

Suppose that sampling is conducted according to the SPRT and that no decision has been made by the time the maximum possible sample size has been attained. The sampler could assume that the true parameter value is intermediate to the two hypothesized values. If sampling is being conducted to determine whether a field should be treated, a value of the parameter between the economic threshold and safety level indicates that a pest population may be building in numbers, and the field should be closely monitored. The sampler would return to the field sooner than the normal scouting schedule would dictate, as a delay could result in eco- nomic damage. Alternatively, if no decision has been made after attaining the maximum feasible sample size, a rule can be established for deciding between

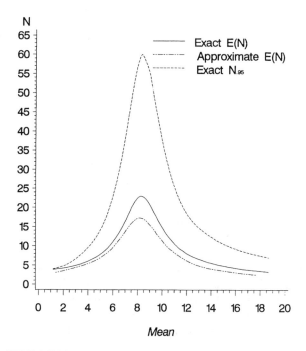

Figure 5.4. SPRT ASN function based on Wald's approximation and exact computations as well as $N_{.95}$ based on the negative binomial distribution with $k = 3$. Economic threshold and safety level are 10 and 7, respectively. $\alpha = \beta = 0.1$.

the two hypotheses. This approach generally decreases the expected sample size at the hypothesized values of the parameter and increases the actual error probabilities.

Several different schemes have been proposed to modify or replace the SPRT with a closed sequential test, a test with a bounded zone of indecision resulting in a decision on or before a known maximum sample size. Weiss (1953) introduced the generalized SPRT. For the generalized SPRT, the predetermined constants A and B change at each stage of sampling instead of remaining constant as they do in the SPRT. Armitage (1957) proposed restricted SPRT's for testing the mean of the normal distribution. The 2-SPRT, proposed by Lorden (1976, 1980), is another closed test and is presented here.

An alternative approach to minimizing the expected sample size at the hypothesized values as in the SPRT is to minimize the average sample size at an intermediate parameter value θ^*. The problem of finding such a test is known as the modified Kiefer–Weiss problem. Minimizing the ASN at the value of θ^* for which the ASN is a maximum provides a solution to the Kiefer–Weiss problem.

An asymptotic solution to the modified Kiefer–Weiss problem was given by Lorden (1976, 1980). He developed the 2-SPRT test, which simultaneously performs two one-sided SPRTs. Again consider testing

$$H_0: \quad \theta = \theta_0$$
$$H_1: \quad \theta = \theta_1 (>\theta_0) \tag{5.18}$$

Let θ^* be a value intermediate to θ_0 and θ_1 for which the ASN is to be minimized. Define a third hypothesis

$$H_2: \quad \theta = \theta^* \tag{5.19}$$

A one-sided hypothesis of H_2 against H_0 is conducted for possible rejection of H_0. Simultaneously, another one-sided SPRT of H_2 against H_1 is conducted for the possible rejection of H_1. The decision boundaries for this test are two converging lines that produce a triangular continuation region (*see* Figure 5.5). To gain an understanding of the 2-SPRT, we first develop the 2-SPRT for a given θ^*. Then we show how to choose θ^* to minimize the ASN asymptotically. However, in practice, we use ECOSTAT to perform all the computations.

Let $X_1, X_2, X_3, \ldots$ be a random sample from a density of the Koopman–Darmois form; that is

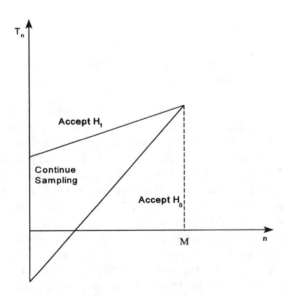

Figure 5.5. Decision boundaries for the 2-SPRT.

$$f(x;\theta) = \exp\{k(x) + \theta x - b(\theta)\}$$

where $k(x)$ is a function of x alone and $b(\theta)$ is a function of θ alone. θ and $b(\theta)$ for the binomial, Poisson, and negative binomial distributions are given in Table 5.7. Note that in each case, the original set of hypotheses of interest must be restated in terms of computational hypotheses involving θ to develop the test just as we did with the negative binomial when developing the SPRT. It is desired to test the hypotheses in (5.18) with type I and type II error probabilities equal to α and β, respectively. As before, let

$$T_n = \sum_{i=1}^{n} x_i.$$

Sampling will continue until

Table 5.7. Quantities Used for Lorden's 2-SPRT for the Binomial, Negative Binomial, and Poisson Distributions

	Binomial	Poisson	Negative binomial
θ	$\ln\left(\dfrac{p}{q}\right)$	$\ln(\lambda)$	$\ln\left(\dfrac{\mu}{\mu + k}\right)$
$b(\theta)$	$-n \ln(1 - p)$	λ	$-k \ln\left(\dfrac{k}{\mu + k}\right)$
h_0	$\dfrac{\ln\left(\dfrac{B(\theta^*)}{1 - A(\theta^*)}\right)}{\ln\left(\dfrac{p_1 q^*}{q_1 p^*}\right)}$	$\dfrac{\ln\left(\dfrac{B(\theta^*)}{1 - A(\theta^*)}\right)}{\ln\left(\dfrac{\lambda_1}{\lambda^*}\right)}$	$\dfrac{\ln\left(\dfrac{B(\theta^*)}{1 - A(\theta^*)}\right)}{\ln\left(\dfrac{\mu_1(\mu^* + k)}{\mu^*(\mu_1 + k)}\right)}$
h_1	$\dfrac{\ln\left(\dfrac{1 - B(\theta^*)}{A(\theta^*)}\right)}{\ln\left(\dfrac{p^* q_0}{q^* p_0}\right)}$	$\dfrac{\ln\left(\dfrac{1 - B(\theta^*)}{A(\theta^*)}\right)}{\ln\left(\dfrac{\lambda^*}{\lambda_0}\right)}$	$\dfrac{\ln\left(\dfrac{1 - B(\theta^*)}{A(\theta^*)}\right)}{\ln\left(\dfrac{\mu^*(\mu_0 + k)}{\mu_0(\mu^* + k)}\right)}$
s_0	$\dfrac{\ln\left(\dfrac{q^*}{q_1}\right)}{\ln\left(\dfrac{p_1 q^*}{q_1 p^*}\right)}$	$\dfrac{\lambda_1 - \lambda^*}{\ln\left(\dfrac{\lambda_1}{\lambda^*}\right)}$	$\dfrac{k \ln\left(\dfrac{\mu_1 + k}{\mu^* + k}\right)}{\ln\left(\dfrac{\mu_1(\mu^* + k)}{\mu^*(\mu_1 + k)}\right)}$
s_1	$\dfrac{\ln\left(\dfrac{q_0}{q^*}\right)}{\ln\left(\dfrac{p^* q_0}{q^* p_0}\right)}$	$\dfrac{\lambda^* - \lambda_0}{\ln\left(\dfrac{\lambda^*}{\lambda_0}\right)}$	$\dfrac{k \ln\left(\dfrac{\mu^* + k}{\mu_0 + k}\right)}{\ln\left(\dfrac{\mu^*(\mu_0 + k)}{\mu_0(\mu^* + k)}\right)}$

$$(1)\ T_n \geq h_1 n + S_1 \qquad (\text{accept } H_1),$$

OR

$$(2)\ T_n \leq h_0 n + S_0 \qquad (\text{accept } H_0). \qquad (5.20)$$

It remains to determine h_0, h_1, S_0, and S_1. The values h_1 and S_1 are determined by the one-sided test of H_2 versus H_0. Similarly, h_0 and S_0 are obtained from a one-sided test of H_1 and H_2.

We begin by defining some quantities that are used repeatedly in determining θ^* and the bounds of the 2-SPRT. The Kullback–Leibler information numbers are

$$I_i(\theta) = (\theta - \theta_i)b'(\theta) - [b(\theta) - b(\theta_i)] \qquad (5.21)$$

for $i = 0, 1$. Further define

$$a_i(\theta') = \frac{\theta' - \theta_i}{I_i(\theta')} \qquad (5.22)$$

for $i = 0, 1$.

Now recall that the sequential likelihood ratio for the SPRT is

$$\lambda_n = \frac{\prod\limits_{i=1}^{n} f_X(x_i;\theta_1)}{\prod\limits_{i=1}^{n} f_X(x_i;\theta_0)}.$$

Since two SPRTs are to be conducted simultaneously, two likelihood ratios need to be formed. Let

$$\lambda_{0n} = \frac{\prod\limits_{i=1}^{n} f_X(x_i;\theta_0)}{\prod\limits_{i=1}^{n} f_X(x_i;\theta^*)} \qquad (5.23)$$

and

$$\lambda_{1n} = \frac{\prod\limits_{i=1}^{n} f_X(x_i;\theta_1)}{\prod\limits_{i=1}^{n} f_X(x_i;\theta^*)} \qquad (5.24)$$

The 2-SPRT takes the form:

(1) Terminate the test, reject H_0, and accept H_1 if $\lambda_{0n} \leq A$;
(2) Terminate the test, reject H_1, and accept H_0 if $\lambda_{1n} \leq B$;
(3) Otherwise, continue sampling. (5.25)

The quantities A and B are chosen so that the desired errors rates α and β are attained when both one-sided tests are conducted simultaneously. Ignoring the excess over the boundary, Huffman (1983) recommended for practical uses that

$$A(\theta) = \frac{a_0(\theta) - a_1(\theta)}{a_0(\theta)} \alpha \qquad (5.26)$$

and

$$B(\theta) = \frac{a_1(\theta) - a_0(\theta)}{a_1(\theta)} \beta. \qquad (5.27)$$

The slopes and intercepts of the decision boundaries in (5.20) may now be obtained. The form of the distribution permits the quantities in the decision rule in (5.25) to be computed as follows:

$$h_1 = \frac{\ln\left[\dfrac{1 - B(\theta^*)}{A(\theta^*)}\right]}{\theta^* - \theta_0}, \qquad (5.28)$$

$$h_0 = \frac{\ln\left[\dfrac{B(\theta^*)}{1 - A(\theta^*)}\right]}{\theta_1 - \theta^*}, \qquad (5.29)$$

$$S_1 = \frac{b(\theta^*) - b(\theta_0)}{\theta^* - \theta_0}, \qquad (5.30)$$

and

$$S_0 = \frac{b(\theta_1) - b(\theta^*)}{\theta_1 - \theta^*}. \qquad (5.31)$$

This gives a solution to the modified Kiefer–Weiss problem. The maximum sample size for the SPRT occurs at the point of intersection of the two decision boundaries:

$$M = \frac{S_1 - S_0}{h_0 - h_1}. \tag{5.32}$$

Intuitively, one might think that the maximum of the ASN function occurs equidistant from θ_0 and θ_1. When the hypotheses concerning the binomial parameter p are symmetric about 0.5, this is indeed the case. The midpoint between θ_0 and θ_1 is $p = 0.5$, yielding a symmetric distribution. However, when the distribution is skewed at the midpoint, the maximum of the ASN function does not occur at the midpoint between θ_0 and θ_1. Hence, we now focus our attention on determining the θ^* for which the ASN is a maximum. By constructing the 2-SPRT at that point, an asymptotic solution to the Kiefer–Weiss problem is obtained.

In order to obtain θ^*, θ' is first determined such that

$$\frac{\ln[A(\theta')]^{-1}}{I_1(\theta')} = \frac{\ln[B(\theta')]^{-1}}{I_0(\theta')}. \tag{5.33}$$

Let n' be the common value of the two sides in equation (5.33). Denote $a_i(\theta')$ as a_i'. Find r' such that

$$\Phi(r') = \frac{a_1'}{a_1' - a_0'} \tag{5.34}$$

where Φ is the distribution function for the standard normal random variable. Also, for $\theta = \theta'$,

$$\sigma' = \sqrt{Var_{\theta'}(X)}. \tag{5.35}$$

The point θ^* for which the ASN is to be minimized may be expressed as

$$\theta^* = \theta' + \frac{r'}{\sigma'\sqrt{n'}}. \tag{5.36}$$

The adjusted error rates based on H_2: $\theta = \theta^*$ are

$$A(\theta^*) = \frac{a_0(\theta^*) - a_1(\theta^*)}{a_0(\theta^*)} \alpha \tag{5.37}$$

and

$$B(\theta^*) = \frac{a_1(\theta^*) - a_0(\theta^*)}{a_1(\theta^*)} \beta. \tag{5.38}$$

Finally, the values needed to construct the decision boundaries for Huffman's extension of the 2-SPRT are as in equations (5.28) to (5.31), with the value of θ^* found in equation (5.36) and the corresponding corrected error rates from equation (5.37) and (5.38).

Huffman's (1983) extension of Lorden's work determines the value of θ^*, that minimizes the maximum expected sample size to within $o((\log \alpha^{-1})^{1/2})$ as α and β tend to zero. This provides an asymptotic solution to the Kiefer–Weiss problem.

The expressions for θ, $b(\theta)$, h_0, h_1, S_0, and S_1 for the binomial, Poisson, and negative binomial distributions are given in Table 5.7.

Example 5.5

Consider again the test of hypotheses discussed in Examples 5.1 and 5.4:

$$H_0: \quad \mu = \mu_0 = 7$$

$$H_1: \quad \mu = \mu_1 = 10.$$

The distribution of pests in the field is known to be negative binomial with $k = 3$. α and β are each specified to be 0.1. From Table 5.7, we have the following computational hypotheses:

$$H_0: \quad \theta_0 = \ln\left(\frac{\mu_0}{\mu_0 + k}\right) = \ln\left(\frac{7}{7 + 3}\right) = -0.357$$

$$H_1: \quad \theta_1 = \ln\left(\frac{\mu_1}{\mu_1 + k}\right) = \ln\left(\frac{10}{10 + 3}\right) = -0.262 \; (>\theta_0).$$

Also from Table 5.7, we have

$$b(\theta) = -k \ln\left(\frac{k}{\mu + k}\right).$$

The derivative of $b(\theta)$ is the mean of the distribution for members in the Koopman–Darmois family of distributions. We can demonstrate that property for this distribution by using the chain rule for taking derivatives to obtain

$$b'(\theta) = \mu.$$

Later, we will need the value of $b(\theta)$ at the hypothesized values so we find these now:

$$b(\theta_0) = -k \ln\left(\frac{k}{\mu + k}\right) = -3 \ln\left(\frac{3}{7 + 3}\right) = 3.612$$

and

$$b(\theta_1) = -k \ln\left(\frac{k}{\mu + k}\right) = -3 \ln\left(\frac{3}{10 + 3}\right) = 4.399.$$

The Kullback–Leibler information number for this distribution may be expressed in terms of μ and k (*see* Table 5.7) as

$$I_i(\theta) = \mu \ln\left[\frac{\mu(\mu_i + k)}{\mu_i(\mu + k)}\right] + k \ln\left[\frac{\mu_i + k}{\mu + k}\right]$$

for $i = 0, 1$. Further, we have

$$a_i(\theta) = \frac{\theta - \theta_i}{I_i(\theta)}$$

for $i = 0, 1$. $a_0(\theta)$ and $a_1(\theta)$ are used later to compute the adjustments to the error rates.

To find θ^*, the value of θ for which the ASN function is to be minimized, we first determine θ' such that

$$\frac{\ln\left[\frac{1}{A(\theta')}\right]}{I_0(\theta')} = \frac{\ln\left[\frac{1}{B(\theta')}\right]}{I_1(\theta')}.$$

Solving for θ' iteratively, we find $\theta' = -0.307$, corresponding to $\mu' = 8.36$. This can be verified by showing that equality holds in the above equation. Several steps are required to accomplish this. We begin by computing the information numbers

$$I_0(8.36) = 8.36 \ln\left[\frac{8.36(7 + 3)}{7(8.36 + 3)}\right] + 3 \ln\left(\frac{7 + 3}{8.36 + 3}\right) = 0.0359$$

and

$$I_1(8.36) = 8.36 \ln\left[\frac{8.36(10 + 3)}{10(8.36 + 3)}\right] + 3 \ln\left(\frac{10 + 3}{8.36 + 3}\right) = 0.0343.$$

To determine the corrected error rates, we must first find

$$a_0(\theta') = \frac{\theta' - \theta_i}{I_0(\theta')} = \frac{-0.307 - (-0.357)}{(0.0359)} = 1.396$$

and

$$a_1(\theta') = \frac{\theta' - \theta_1}{I_1(\theta')} = \frac{-0.307 - (-0.262)}{(0.0343)} = -1.289$$

The adjusted error rates may now be found as follows:

$$A(\theta') = \frac{a_1(\theta') - a_0(\theta')}{a_1(\theta')} \alpha = \frac{-1.289 - 1.396}{-1.289} (0.1) = 0.208$$

and

$$B(\theta') = \frac{a_0(\theta') - a_1(\theta')}{a_0(\theta')} \alpha = \frac{1.396 - (-1.289)}{1.396} (0.1) = 0.192$$

Finally, we can verify that

$$n' = \frac{\ln\left(\dfrac{1}{A(\theta')}\right)}{I_1(\theta')} = \frac{\ln\left(\dfrac{1}{B(\theta')}\right)}{I_0(\theta')} = 45.8.$$

The next step is to adjust θ' to obtain θ^*. This requires first finding r' and σ'. r' is chosen such that

$$\Phi(r') = \frac{a_1(\theta')}{a_1(\theta') - a_0(\theta')} = \frac{-1.308}{-1.308 - 1.396} = 0.48.$$

Finding the inverse of the standard normal distribution function, we have

$$r' = -0.050.$$

The value of σ' is the standard deviation of the distribution evaluated at θ', or equivalently μ':

$$\sigma' = \sqrt{Var_{\theta'}(X)} = \sqrt{8.36 + \frac{8.36^2}{3}} = 5.6.$$

The value of θ, θ^*, for which the ASN function is to be minimized is

$$\theta^* = \theta' + \frac{r'}{\sigma'\sqrt{n'}} = -0.307 + \frac{-0.050}{5.6\sqrt{65.8}} = -0.308.$$

This is equivalent to $\mu^* = 8.32$.

The adjusted error rates for θ^* need to be calculated. This requires first finding the information numbers for this value of θ:

$$I_0(8.32) = 8.32 \ln\left(\frac{8.32(7 + 3)}{7(8.32 + 3)}\right) + 3 \ln\left(\frac{7 + 3}{8.32 + 3}\right) = 0.0361$$

and

$$I_1(8.32) = 8.32 \ln\left(\frac{8.32(10 + 3)}{10(8.32 + 3)}\right) + 3 \ln\left(\frac{10 + 3}{8.32 + 3}\right) = 0.0339.$$

The information numbers are then used to compute

$$a_0(\theta^*) = \frac{\theta^* - \theta_0}{I_0(\theta^*)} = \frac{-0.308 - (-0.357)}{(0.0361)} = 1.44$$

and

$$a_1(\theta^*) = \frac{\theta^* - \theta_1}{I_1(\theta^*)} = \frac{-0.308 - (-0.262)}{(0.0339)} = -1.26.$$

Based on these values, the corrected error rates for use in constructing the 2-SPRT may be calculated as follows:

$$A(\theta^*) = \frac{a_1(\theta^*) - a_0(\theta^*)}{a_1(\theta^*)} \alpha = \frac{-1.26 - 1.44}{-1.26}(0.1) = 0.214$$

and

$$B(\theta^*) = \frac{a_0(\theta^*) - a_1(\theta^*)}{a_0(\theta^*)} \alpha = \frac{1.44 - (-1.26)}{1.44}(0.1) = 0.187.$$

To determine the decision boundaries, we need to evaluate $b(\theta)$ at $\theta = \theta^*$:

$$b(\theta^*) = -k \ln\left(\frac{k}{\mu + k}\right) = -3 \ln\left(\frac{3}{8.32 + 3}\right) = 3.984$$

Now we are ready to determine the intercepts and slopes of the decision boundaries:

$$h_1 = \frac{\ln\left(\dfrac{1 - B(\theta^*)}{A(\theta^*)}\right)}{\theta^* - \theta_0} = \frac{\ln\left(\dfrac{1 - 0.187}{0.214}\right)}{-0.308 - (-0.357)} = 34.29,$$

$$h_0 = \frac{\ln\left(\dfrac{B(\theta^*)}{1 - A(\theta^*)}\right)}{\theta_1 - \theta^*} = \frac{\ln\left(\dfrac{0.187}{1 - 0.214}\right)}{-0.262 - (-0.308)} = -33.86,$$

$$S_1 = \frac{b(\theta^*) - b(\theta_0)}{\theta^* - \theta_0} = \frac{3.98 - 3.61}{-0.308 - (-.357)} = 7.63,$$

and

$$S_0 = \frac{b(\theta_1) - b(\theta^*)}{\theta_1 - \theta^*} = \frac{4.40 - 3.98}{-0.262 - (-0.308)} = 9.12.$$

Therefore, after n observations, we

1. Stop sampling and conclude the population density exceeds the economic threshold if the total number of observed insects exceeded $34.29 + 7.63n$;
2. Stop sampling and conclude the population density is less than the safety level if the total number of observed insects is less than $-33.86 + 9.12n$;
3. Continue sampling otherwise.

The maximum sample size is

$$M = \frac{S_1 - S_0}{h_0 - h_1} = \frac{7.63 - 9.12}{-33.86 - 34.29} = 45.81.$$

Therefore, at most, 46 observations will be taken.

The example above clearly demonstrates why the 2-SPRT has not been used extensively in biological, and other, applications; the computations are complex. In addition, if the reader works through some of the above calculations, it soon becomes evident that round-off error can cause severe problems. Fortunately, ECOSTAT may be used to construct these decision boundaries using double-precision to reduce the impact of round-off error. The process is the same as that for the SPRT except *2-SPRT* is chosen from the *Sequential Sampling* menu. In addition, ECOSTAT determines the values of the operating characteristic function, the average sample number, and the 0.95 quantile of sample size. For Example 5.5, these are given in Table 5.8. Plots with the exact characteristics associated with the SPRT and the 2-SPRT are shown in Figures 5.6, 5.7, and 5.8, so that the properties of the two tests can be compared. Notice that the error rates are comparable (Figure 5.6). If the mean is equal to one of the hypothesized values, then the 2-SPRT requires a few more observations than the SPRT whether observing the ASN function (Figure 5.7) or the 95th percentile of sample size (Figure 5.8). However, for means intermediate to the hypothesized values, the 2-SPRT can result in significantly lower sample sizes, especially if one considers the 95th percentile of sample size.

Summary

Wald's (1945, 1947) SPRT has been used extensively to make management decisions in integrated pest management. For the SPRT, the type I and type II error rates are specified in advance of sampling, and the sample size is random.

Table 5.8. Exact Values of the OC and ASN Functions and $N_{.95}$ for Lorden's 2-SPRT With k = 3[a]

Mean	OC	ASN	$N_{.95}$
62.00	0.000	1.38	2
29.50	0.000	2.39	4
18.67	0.000	4.11	8
13.25	0.001	7.44	15
10.00	0.106	14.71	29
9.62	0.166	16.02	31
9.26	0.246	17.20	32
8.93	0.342	18.14	33
8.61	0.450	18.75	34
8.30	0.560	18.99	34
8.02	0.666	18.84	33
7.74	0.758	18.37	33
7.48	0.833	17.64	32
7.24	0.891	16.75	31
7.00	0.931	15.78	29
3.82	1.000	7.05	10
2.17	1.000	5.45	7
1.17	1.000	4.83	5
0.49	1.000	4.31	5

[a]The economic threshold and safety levels are 10 and 7 pests per sampling unit, respectively.

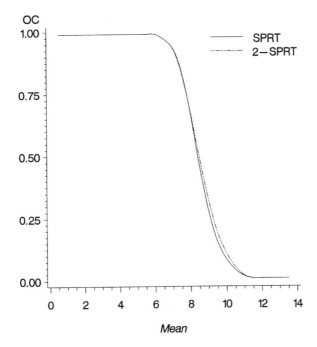

Figure 5.6. Exact OC functions for the SPRT and 2-SPRT based on the negative binomial distribution with $k = 3$. Economic threshold and safety level are 10 and 7, respectively. $\alpha = \beta = 0.1$.

If an infestation is high or low, few observations are required to make a decision. If the population is at or near the economic threshold or safety level, more observations are required. Therefore, sampling effort is increased during the critical period of decision-making. The properties of a proposed sampling program can be evaluated using Wald's approximation or using ECOSTAT to determine exact properties. It is recommended that the approximations be used for initial screening of a proposed test. Then the exact properties should be investigated. Use of the 95th percentile of sample size provides insight into whether sufficient resources exist so that sampling can probably continue until a decision is made.

Lorden (1976, 1980) and Huffman (1983) have developed the 2-SPRT, which is based on simultaneously conducting two one-sided SPRTs. Instead of minimizing the sample size at the hypothesized values as in the SPRT, an effort is made to minimize the maximum sample size. This results in a triangular no-decision zone, ensuring that sampling stops with a known maximum sample size. Exact properties of the test can be investigated using ECOSTAT. Compared to the SPRT, the 2-SPRT has comparable error rates, slightly higher sample sizes at the

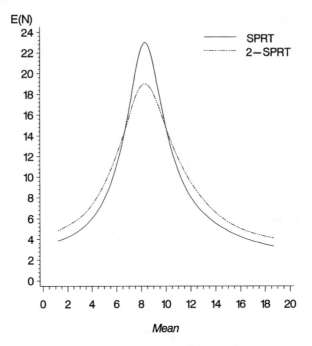

Figure 5.7. Exact ASN functions for the SPRT and 2-SPRT based on the negative binomial distribution with $k = 3$. Economic threshold and safety level are 10 and 7, respectively. $\alpha = \beta = 0.1$.

hypothesized values, and substantially lower sample sizes at values intermediate to the hypothesized values. Now that software is available for ready implementation, we recommend this approach be strongly considered for IPM use.

Exercises

For Problems 1 through 5 below,

A. Develop Wald's SPRT to test the specified hypotheses.

B. Estimate the OC and ASN functions for Wald's SPRT.

C. Based on the results in B, adjust the error rates if this is needed to obtain a reasonable plan in terms of sample size.

D. Compute the exact OC and ASN functions.

E. Based on the results in D, adjust the error rates to attain the specified error rates using the plan.

F. Print out a sampling sheet for the final sampling plan.

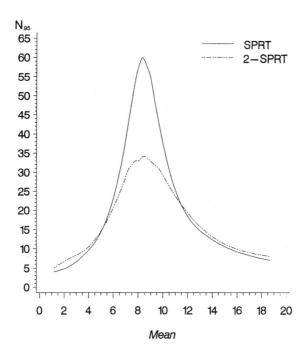

Figure 5.8. $N_{.95}$ for the SPRT and 2-SPRT based on the negative binomial distribution with $k = 3$. Economic threshold and safety level are 10 and 7, respectively. $\alpha = \beta = 0.1$.

1. The economic threshold for cotton fleahoppers is 40 fleahoppers/100 terminals; 25 fleahoppers/100 terminals is considered safe. Experience has shown that the distribution of cotton fleahoppers tends to be negative binomial with a k value of 2. The purpose of sampling is to determine whether the population is above the economic threshold or below the safety level with type I and type II error rates of 0.10 and 0.15, respectively.

2. The distribution of greenbugs in seedling oats is consistently geometric. For an oat field, the farmer wants to know whether the mean population density is above the economic threshold of 25/linear row foot or below the safety level of 10/linear row foot with type I and type 2 error rates of 0.10 and 0.05, respectively.

3. The distribution of snails in streams is negative binomial with a k value of 0.5, when the sample unit is a 0.1-meter dredge with a standardized net. Develop a test of the hypotheses that the mean population density is above the economic threshold of 2 or below the safety level of 1 snail/0.1-meter dredge with type I and type II error probabilities of 0.1 and 0.1, respectively.

4. The distribution of grasshoppers in rangeland is often Poisson. Develop a test of the hypotheses that the mean population density is above the economic

threshold of 3.5 or below the safety level of 1.5 grasshoppers/square yard with type I and type II error probabilities of 0.15 and 0.10, respectively.

5. If the percentage of damaged oranges in an orchard exceeds 10%, the orchard will usually not be harvested. Develop a test of the hypotheses that the percent damaged fruit is above this threshold versus it is below the safety level of 5% with type I and type II error rates of 0.1 and 0.05, respectively.

Consider problems 1 to 5 again. However, this time

A. Specify the hypotheses.

B. Develop the 2-SPRT to test the specified hypotheses.

C. Compute the OC and ASN functions for the SPRT.

D. Based on the results in C, adjust the error rates if this is needed to obtain a reasonable plan in terms of sample size or to have the test with error rates more nearly equal to the specified levels.

E. If the test was modified in D, compute the exact OC and ASN functions for the modified test.

6

Sequentially Testing Three Hypotheses

Introduction

The sequential probability ratio test (SPRT) and 2-SPRT, discussed in Chapter 5, are sequential methods for deciding between two hypotheses. It is assumed here that the reader is familiar with the notation and concepts presented there. Increasingly ecologists are attempting to choose among three or more hypotheses. Morris (1954) and Waters (1955) appear to be the first in the biological literature to consider the problem of deciding among three or more hypotheses. Each developed sequential plans for determining whether the infestation of trees by spruce budworm egg masses was light, medium, or heavy. Waters also considered the infestation of spruce budworm larvae. The standard survey procedure in the northeastern United States had been to examine five 15-inch twigs from each of five balsam fir trees at every collection point (Waters, 1955). The twigs were cut by pole pruner from approximately the mid-crown height of the trees. In the study reported by Morris (1954), very tall trees for which the mid-crown could not be reached with pole pruners were felled and sample branches selected as follows: one branch from the apical quarter of the crown, one from the second quarter, one-half of a branch from the third quarter, and one-half from the basal quarter. These four values were averaged to give a mean for the tree. Alternately, Waters avoided including very tall trees in the sample. For sequential sampling, the tree was taken as the sampling unit. One observation per tree was recorded. Waters compared the results from sequential sampling to those from the standard procedure for each plan. A 40% savings in twig-examination time was observed with no loss of reliability in correctly identifying the degree of infestation for each.

Ecologists and statisticians have each developed sequential sampling plans for this circumstance. Unfortunately, the dialogue between these two groups has been limited. This is regrettable as each has a perspective of the sampling problem that can contribute to developing a strong sampling system. It is hoped that, by considering the sampling plans proposed by both ecologists and statisticians, this chapter will help increase communication.

Ecologists' Sequential Test

Ecologists generally view the above problem as one of classification. The parameter space can be divided into three regions based on whether the population density is light, medium, or heavy, as expressed in the following hypotheses:

$$
\begin{aligned}
H_0^+&: \quad \mu < a_1 \\
H_1^+&: \quad a_1 \leq \mu \leq a_2 \\
H_2^+&: \quad \mu > a_2.
\end{aligned}
\tag{6.1}
$$

In integrated pest management (IPM) studies of pest species, a light infestation might indicate that the population is unlikely to become a problem, a medium infestation might imply that close monitoring is needed in the event that the population begins to increase rapidly, and a heavy infestation might show the need for control measures to be immediately implemented. In order to develop a sequential procedure, it is assumed that there is an interval (μ_0, μ_1) about a_1 in which we have no strong preference for H_0 or H_1 but rejection of H_2 is strongly desired. Similarly, an interval (μ_1, μ_2) about a_2 may be found in which we strongly prefer to reject H_0 but have no preference for H_1 or H_2. The intervals (μ_0, μ_1) and (μ_1, μ_2) are sometimes referred to as indifference zones because we are indifferent as to the choice between two hypotheses when the mean lies within these intervals. We have now effectively partitioned the sample space into five regions as depicted in Figure 6.1. The hypotheses to be tested are

$$
\begin{aligned}
H_1^*&: \quad \mu \leq \mu_0 \\
H_2^*&: \quad \mu_1 \leq \mu \leq \mu_2 \\
H_3^*&: \quad \mu \geq \mu_3.
\end{aligned}
\tag{6.2}
$$

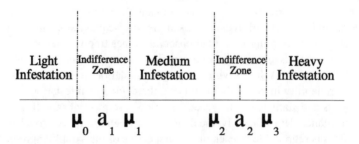

Figure 6.1. Partitioned parameter space resulting from the following hypotheses: H_0^+: $\mu < a_1$; H_1^+: $a_1 \leq \mu \leq a_2$; H_2^+: $\mu > a_2$.

where $\mu_0 < \mu_1 < \mu_2 < \mu_3$. Based on (6.2), two sets of hypotheses are considered. The first set is used to determine whether the population is light or medium. The other set distinguishes between medium and heavy. The hypotheses could be stated as follows:

$$H_0: \quad \mu \leq \mu_0 \qquad \text{versus} \qquad H_1: \quad \mu \geq \mu_1 \qquad (6.3)$$

and

$$H_0': \quad \mu \leq \mu_2 \qquad \text{versus} \qquad H_1': \quad \mu \geq \mu_3$$

When the mean number of budworm larvae per twig is between μ_0 and μ_1, the level of infestation could be said to be either light or medium. Similarly, it would be difficult to say definitively whether the infestation is medium or heavy if the mean number of budworm larvae per twig is between μ_2 and μ_3. The test of the hypotheses in (6.2) is conducted using two SPRTs, one for each set of hypotheses in (6.3), simultaneously. The first SPRT, denoted by S, distinguishes between light and medium infestation levels, and the second SPRT, denoted by S', between medium and heavy levels. Type I and II error rates are denoted by α and β, respectively, for test S and α' and β', respectively, for test S'. The combined test has three decision zones as may be seen in Figure 6.2. Notice that to

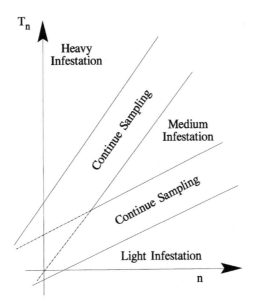

Figure 6.2. The general form of the continuation and decision zones when ecologists simultaneously conduct two SPRTs.

conclude that the population density level is medium, we must simultaneously decide in favor of H_1 based on the test S and in favor of H_0' for the test S'. This clearly indicates the interdependence of the two tests.

The operating characteristic (OC) and average sample number (ASN) functions for the individual SPRTs provide insight into, but are not the same as, the OC and ASN functions for the combined test because of the lack of independence of the two tests. Based on a simulation study, Fowler (1985) and Fowler and Lynch (1987) have noted that the ASN function for the combined test has two peaks, corresponding to the peaks of the individual SPRTs. These authors suggested simulation studies to investigate the properties of the combined test. As with the SPRT, we recommend using ECOSTAT for an initial evaluation of the feasibility of conducting the individual SPRTs based on Wald's approximations. If sufficient resources appear to be available for the individual tests, an evaluation of the exact OC and ASN functions of the combined test can be obtained from ECOSTAT.

Example 6.1

Waters (1955) considered both spruce budworm larvae and eggs. Here we consider the properties of the larvae sampling program. Based on prior experience, Waters defined light, medium, and heavy larval infestations as those with means less than 3 larvae/twig, those with means between 6 and 9 larvae per twig, and those with an average of 12 or more larvae per twig, respectively. For these limits, the hypotheses in equation (6.2) take the following form:

$$H_1^*: \quad \mu \leq 3$$
$$H_2^*: \quad 6 \leq \mu \leq 9$$
$$H_2^*: \quad \mu \geq 12.$$

To construct a sequential sampling program to discriminate among these three levels of infestation, Waters considered the following two sets of hypotheses:

$$H_0: \quad \mu \leq 3 \qquad \text{versus} \qquad H_1: \quad \mu \geq 6$$

and

$$H_0': \quad \mu \leq 9 \qquad \text{versus} \qquad H_1': \quad \mu \leq 12.$$

The purpose of the first set is to determine whether the population is light or medium. The other set distinguishes between medium and heavy infestations. For each set of hypotheses, the type I and type II errors were set at 0.1. Based on historical data, the distribution of the larvae was assumed to be negative binomial with a k value of 7.228.

Wald's SPRT was developed for each set of hypotheses. First, consider H_0 versus H_1. The slope of the decision boundaries is

$$S = k \frac{\ln\left(\dfrac{\mu_1 + k}{\mu_0 + k}\right)}{\ln\left[\dfrac{\mu_1(\mu_0 + k)}{\mu_0(\mu_1 + k)}\right]} = 7.228 \frac{\ln\left(\dfrac{6 + 7.228}{3 + 7.228}\right)}{\ln\left[\dfrac{6(3 + 7.228)}{3(6 + 7.228)}\right]} = 4.2646$$

The upper intercept for this set of hypotheses is

$$h_2 = \frac{\ln\left(\dfrac{1 - \beta}{\alpha}\right)}{\ln\left[\dfrac{\mu_1(\mu_0 + k)}{\mu_0(\mu_1 + k)}\right]} = \frac{\ln\left(\dfrac{1 - 0.1}{0.1}\right)}{\ln\left[\dfrac{6(3 + 7.228)}{3(6 + 7.228)}\right]} = 5.0402,$$

and the lower intercept is

$$h_1 = \frac{\ln\left(\dfrac{\beta}{1 - \alpha}\right)}{\ln\left[\dfrac{\mu_1(\mu_0 + k)}{\mu_0(\mu_1 + k)}\right]} = \frac{\ln\left(\dfrac{0.1}{1 - 0.1}\right)}{\ln\left[\dfrac{6(3 + 7.228)}{3(6 + 7.228)}\right]} = -5.0402.$$

Next consider H_0' versus H_1'. The slope for this set of decision boundaries is

$$S = k \frac{\ln\left(\dfrac{\mu_3 + k}{\mu_2 + k}\right)}{\ln\left[\dfrac{\mu_3(\mu_2 + k)}{\mu_2(\mu_3 + k)}\right]} = 7.228 \frac{\ln\left(\dfrac{12 + 7.228}{9 + 7.228}\right)}{\ln\left[\dfrac{12(9 + 7.228)}{9(12 + 7.228)}\right]} = 10.3859.$$

The upper intercept for this set of hypotheses is

$$h_2 = \frac{\ln\left(\dfrac{1 - \beta}{\alpha}\right)}{\ln\left[\dfrac{\mu_3(\mu_2 + k)}{\mu_2(\mu_3 + k)}\right]} = \frac{\ln\left(\dfrac{1 - 0.1}{0.1}\right)}{\ln\left[\dfrac{12(9 + 7.228)}{9(12 + 7.228)}\right]} = 18.6122,$$

and the lower intercept is

$$h_1 = \frac{\ln\left(\dfrac{\beta}{1 - \alpha}\right)}{\ln\left[\dfrac{\mu_3(\mu_2 + k)}{\mu_2(\mu_3 + k)}\right]} = \frac{\ln\left(\dfrac{0.1}{1 - 0.1}\right)}{\ln\left[\dfrac{12(9 + 7.228)}{9(12 + 7.228)}\right]} = -18.6122,$$

Both SPRTs are considered simultaneously, resulting in the decision zones shown in Figure 6.3.

Before implementing a sampling program based on these boundaries, we want to determine some of the properties of the tests. To do this, we use ECOSTAT. To do so, select *3 Hypotheses* from the *Sequential Sampling* menu. After that, proceed as you did with the SPRT described in Examples 5.1 and 5.4. We obtain the output in Table 6.1. Notice that we no longer have a single OC function; instead, an OC function can be defined for each set of hypotheses. However, in view of our application, we are most interested in determining the probability of

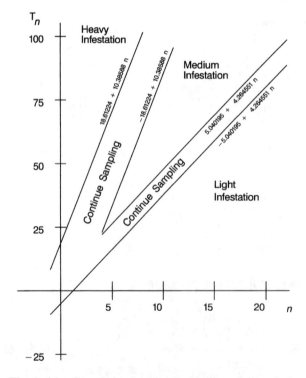

Figure 6.3. The decision boundaries associated with determining whether spruce budworm egg infestation is light, medium, or heavy for the ecologists' or Sobel and Wald's methods.

Table 6.1 Based on the Ecologist Approach, Probabilities of Deciding in Favor of Light, Medium, and Heavy Infestation, Average Sample Size, and 95th Percentile of Sample Size for Various Values of the Mean When Sampling Spruce Budworm Eggs using the Decision Boundaries in Figure 6.3

Mean μ	p_L	p_M	p_H	$E(N)$	$N_{.95}$
0.450	1.000	0.000	0.000	2.017	2
0.961	1.000	0.000	0.000	2.154	3
1.544	1.000	0.000	0.000	2.501	4
2.216	0.999	0.001	0.000	3.207	6
3.000	0.974	0.025	0.000	4.781	11
3.486	0.889	0.110	0.000	6.187	14
4.020	0.679	0.321	0.000	7.430	17
4.611	0.389	0.611	0.000	7.574	17
5.267	0.172	0.828	0.000	6.888	14
6.000	0.066	0.934	0.001	6.435	11
6.508	0.034	0.965	0.001	6.544	10
7.056	0.018	0.981	0.001	7.060	12
7.650	0.009	0.987	0.004	8.137	15
8.296	0.005	0.979	0.016	10.081	21
9.000	0.002	0.931	0.067	13.344	31
9.523	0.001	0.832	0.167	16.260	40
10.080	0.001	0.642	0.357	18.525	48
10.676	0.001	0.393	0.606	18.277	47
11.314	0.000	0.189	0.810	15.506	40
12.000	0.000	0.077	0.922	12.132	30
16.807	0.000	0.000	1.000	3.931	8
24.819	0.000	0.000	1.000	2.025	4
40.842	0.000	0.000	1.000	1.257	2
88.912	0.000	0.000	1.000	1.011	1

concluding that the infestation is light, medium, or heavy given the true mean infestation level. Therefore, we give each of these probabilities, using p_L, p_M, and p_H, to denote the probability of deciding the infestation is light, medium, and heavy, respectively. The average sample size $E(N)$ and 95th percentile of sample size $N_{.95}$ are also given.

Before using ECOSTAT to investigate the properties of the combined tests, one should evaluate the individual tests using Wald's approximations. This permits the user to assess the feasibility of the individual tests. If enough resources are not available for these, then adjustments should be made at this point. Once the practicality of the individual tests has been established, ECOSTAT can be used to assess the properties of the combined test. However, due to the computational intensity of this assessment, one should plan to allow a significant amount of time for the program to run.

The values in Table 6.1 become more meaningful when they are viewed graphically. Figures 6.4, 6.5, and 6.6 were developed by splining the values in Table 6.1 to produce smooth curves. Figure 6.4 compares the OC functions for the individual SPRTs to that of the combined test. Notice that the OC functions associated with the tests of H_0 versus H_1 and of H_0' versus H_1' are denoted by p_L and p_M, respectively, for the combined test values in Table 6.1. Further, ECOSTAT was used to obtain the exact OC functions for the individual SPRTs instead of using Wald's approximation. Because the computations for the combined test are exact, this permits us to evaluate the impact of combining the SPRTs on the OC functions. There is good agreement in the two over much of the parameter space. The dependence of the two tests is illustrated by the drop in the value of p_M for the combined, but not the individual test, as the mean decreased. In this range of the parameter space, the test of H_0 versus H_1 began to have an impact on the combined test. As with the individual SPRTs, the error rates for the combined test are below the stated level.

As we saw in Chapter 5, the exact ASN function differs significantly from Wald's approximate ASN function. We now want to concentrate on the effect of combining tests. Therefore, we compared the exact ASN functions of the indi-

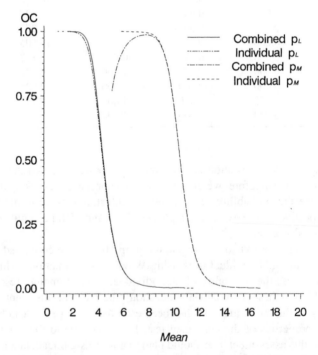

Figure 6.4. The OC function for the combined and individual SPRTs using the ecologists' method.

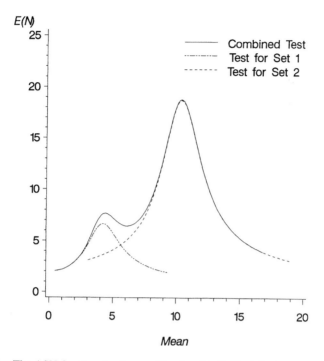

Figure 6.5. The ASN function for the combined and individual SPRTs using the ecologists' method.

vidual tests to the ASN function of the combined test. H_0 versus H_1 is the first set of test hypotheses and H_0' versus H_1' is the second set. When compared to the individual SPRTs, the combined test results in a larger average sample size over the range of means considered as seen in Figure 6.5. Over much of the parameter space, the ASN for the combined test is only slightly larger than that of an individual test. The exception lies when the infestation level is medium; that is, the mean is between 6 and 9. Both tests are impacting sampling in this region causing the sample size to be larger than it would be the case for either individual test. Similar results are observed for the 95th percentile of sample size as may be seen in Figure 6.6. These results should be expected. By combining tests, we have enlarged our zone of indecision when compared to each individual test. For any sequence of observations, we cannot stop sooner than we would if we had only one SPRT, and we may take longer to reach a decision.

Sobel and Wald's Method

Sobel and Wald (1949) considered the simultaneous test of three hypotheses concerning the mean of a normal distribution with known variance σ^2. They were

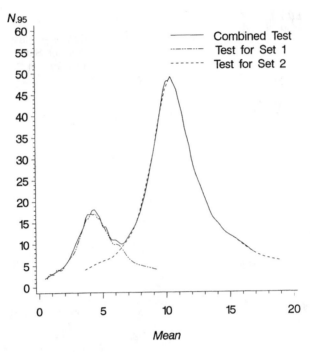

Figure 6.6. The 95th percentile of sample size for the combined and individual SPRTs for the ecologists' method.

the first to view the problem as one of choosing among the three mutually exclusive and exhaustive hypotheses given in equation (6.1). They partitioned the parameter space as in Figure 6.1 and proposed the simultaneous test of the hypotheses given in equation (6.2). Therefore, the Sobel and Wald approach is conceptually the same as that used by ecologists.

Now consider the test proposed by Sobel and Wald. First, an SPRT is developed for each set of hypotheses in equation (6.2). However, the tests are terminated independently of one another; that is, both tests are conducted simultaneously with calculations stopping on each as soon as its inequalities are satisfied. The three possible terminal decisions are illustrated in Table 6.2.

Notice that the possibility of S accepting H_0 and S' accepting H_1' is not included in Table 6.2. This case would be particularly disturbing because we would have no way of reaching a logical conclusion when one test indicates a light infestation and the other test indicates a heavy infestation. Fortunately, this case does not arise in most practical settings. It is impossible when $(A, B) = (A', B')$ where (as in Chapter 5), $A = (1 - \beta)/\alpha$, $B = \beta/(1 - \alpha)$, and A' and B' are the corresponding functions of α' and β'. (If $\alpha = \alpha'$ and $\beta = \beta'$, these equalities

Table 6.2. Possible Terminal Decisions Resulting from Sobel and Wald's (1949) Test

Test S accepts	Test S' accepts	Terminal decision
H_0	H'_0	H^*_1
H_1	H'_0	H^*_2
H_1	H'_1	H^*_3

are satisfied.) Further, Sobel and Wald showed that this case was not possible as long as $A/A' \leq 1$ and $B/B' \leq 1$. In the section on Armitage's (1950) method, we consider a method of avoiding this difficulty. However, here we assume that the error rates are such that this case does not arise.

By terminating the two SPRTs independently of one another, Sobel and Wald were able to obtain a good approximation to the OC function and some bounds on the ASN function. They noted that the procedure could not be optimal because final inference need not be based on the sample mean, the complete sufficient statistic for the mean μ. However, they believed the loss in efficiency was not great. Further, they observed that the sample size for the combined test must be at least as large as that required by the individual tests.

Example 6.2

Consider again the set of hypotheses in Example 6.1. The decision boundaries are computed exactly as in that example. The difference in the two tests lies in the stopping rules. During the sampling period, there may be times that the stopping criterion for one test, but not the other, is satisfied. Ecologists using the preceding method would continue to sample until the sampling criterion for both S and S' are simultaneously satisfied. Alternatively, Sobel and Wald cease to consider a test when its stopping criterion is met. This results in a smaller ASN function.

To clearly illustrate the impact that the stopping rules have on sampling, consider the partial sampling sheet given in Table 6.3. This sheet is appropriate for either the methods due to Waters or Sobel and Wald because the decision boundaries are the same. Suppose that observations are taken sequentially and that the first eight are as follows: 6, 8, 3, 9, 0, 1, 2, 10. This gives corresponding running totals of 6, 14, 17, 26, 26, 27, 29, and 39.

First consider the stopping rule by Sobel and Wald. After the first two observations, the running total equals the upper bound for the test S of the light vs. medium infestation, indicating the population is not light but is instead medium or heavy. Therefore, this test is no longer considered. However, the stopping criterion for the test of medium versus heavy infestation has not been satisfied so sampling continues. On the fourth observation, the running total of 26 falls below the lower limit for the test S' of medium versus heavy infestation, indicating the

Table 6.3   A Data Sheet for Use With Ecologists or Sobel and Wald's Method

Sample No.	Running total	Light lower limit	vs	Medium upper limit	Medium lower limit	vs	Heavy upper limit
1	—	nd		10	nd		29
2	—	3		14	2		40
3	—	7		18	12		50
4	—	12		23	22		61
5	—	16		27	33		71
6	—	20		31	43		81
7	—	24		35	54		92
8	—	29		40	64		102
9	—	33		44	74		113
10	—	37		48	85		123
11	—	41		52	95		133
12	—	46		57	106		144
13	—	50		61	116		154
14	—	54		65	126		165
15	—	58		70	137		175
16	—	63		74	147		185
17	—	67		78	157		196
18	—	71		82	168		206
19	—	75		87	178		216
20	—	80		91	189		227

infestation is medium and not heavy. Because the stopping criterion for the second test has now been satisfied, sampling ceases. The conclusion is that the level of infestation is medium. Now consider the ecologists' approach. After the first two observations, the stopping criterion for the first test has been met, but that for the other test has not. Therefore, both tests are continued. Again, after the fourth observation, the stopping criterion for the second test has been satisfied, but that of the first test is no longer satisfied. Hence, sampling continues. After the eighth observation, the stopping criterion of both tests are satisfied. Therefore, sampling ceases and the conclusion is that the infestation level is medium. We should note here that it is doubtful that we would be comfortable making any management decision based on two, four, or eight observations. In constructing the ASN and OC functions for a proposed plan, we could determine whether the anticipated sample sizes are large enough to cover the sampling region. If not, α and/or β can be reduced. If this is encountered during the course of sampling, sampling generally continues long enough to be certain the sampling region has been covered. By taking an increased number of samples, we are effectively reducing the error rates, but we will not know by how much.

Because each test may be terminated at different times, we are not able to compute the exact ASN and OC functions as we did for the ecologists' method. Monte Carlo studies can provide insights into the test. Such a study was conducted for the means given in Table 6.1. The results are shown in Table 6.4.

Graphs of the OC and ASN functions and $N_{.95}$ are shown in Figure 6.7 and 6.8 and 6.9 respectively. Compare these graphs to the corresponding ones based on the ecologists' method (Figures 6.4, 6.5, and 6.6). Notice that Sobel and Wald's test lies closer to the individual SPRTs than does the ecologists' test. Because both stopping criterion must be satisfied simultaneously, the ecologists' method will always take at least as many observations as under Sobel and Wald's method, and usually more. Therefore, the error rates tend to be slightly smaller, and the average sample size and 95th percentile of sample size tend to be slightly larger

Table 6.4. Based on Sobel and Wald's method, probabilities of Deciding in Favor of Light, Medium, and Heavy Infestation, Average Sample Size, and 95th Percentile of Sample Size for Various Values of the Mean When Sampling Spruce Budworm Eggs Using the Decision Boundaries in Figure 6.3

Mean μ	p_L	p_M	p_H	$E(N)$	N_{05}
0.450	1.000	0.000	0.000	2.068	3
0.961	0.994	0.006	0.000	2.311	3
1.544	0.961	0.039	0.000	2.665	4
2.216	0.864	0.136	0.000	3.124	5
3.000	0.654	0.346	0.000	3.662	6
3.486	0.496	0.504	0.000	3.900	6
4.020	0.337	0.663	0.000	4.124	7
4.611	0.196	0.804	0.000	4.317	7
5.267	0.099	0.901	0.000	4.623	7
6.000	0.045	0.955	0.000	5.136	8
6.508	0.024	0.975	0.000	5.710	10
7.056	0.016	0.983	0.001	6.497	12
7.650	0.008	0.989	0.003	7.782	15
8.296	0.004	0.980	0.016	9.884	21
9.000	0.002	0.931	0.067	13.349	32
9.523	0.001	0.837	0.162	16.089	39
10.080	0.001	0.633	0.366	18.262	47
10.676	0.000	0.388	0.612	18.291	47
11.314	0.000	0.189	0.810	15.506	40
12.000	0.000	0.076	0.924	11.988	30
16.807	0.000	0.001	0.999	3.918	8
24.819	0.000	0.000	1.000	2.019	4
40.842	0.000	0.000	1.000	1.256	2
88.912	0.000	0.000	1.000	1.012	1

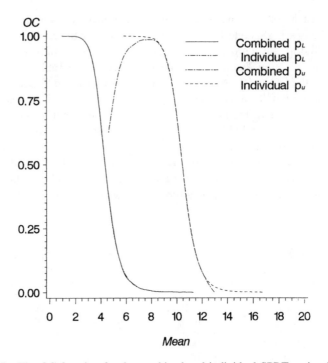

Figure 6.7. The OC function for the combined and individual SPRTs using Sobel and Wald's (1949) method.

for the ecologists' method, especially for the mean values intermediate to the two sets of hypotheses.

Armitage's Method

Welch (1939) and Smith (1947) noted that Fisher's linear discriminant function for discriminating among two multivariate normal populations with the same covariance matrix but differing means is equivalent to the likelihood ratio between the two populations. Rao (1947) used this fact to develop a method of discriminating among more than two populations. Armitage (1950) extended this work to a sequential choice among more than two hypotheses. The problem of discriminating among the light, medium, and heavy infestations that we have been discussing was presented as in (6.1). To choose among these three hypotheses, Armitage suggested choosing μ_0, μ_1^*, and μ_3 such that $\mu_0 < a_1 < \mu_1^* < a_2 < \mu_3$. Sampling is conducted to choose among the three hypotheses:

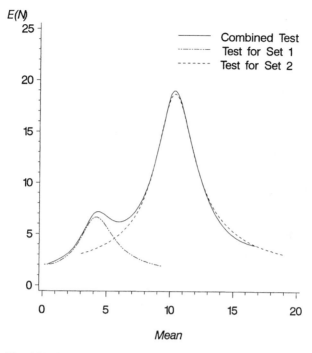

Figure 6.8. The ASN function for the combined and individual SPRTs using Sobel and Wald's (1949) method.

$$H_0: \quad \mu = \mu_0$$
$$H_1: \quad \mu = \mu_1^*$$
$$H_2: \quad \mu = \mu_3.$$

If $\mu_1^* = \mu_1 = \mu_2$ in (6.2), then we have the case considered by Armitage. Here we used μ_1^* to denote the hypothesized parameter in H_1 to emphasize the fact that this value may be, and probably is, different from either μ_1 or μ_2 in (6.2), but that μ_0 and μ_3 may well be the same values as in (6.2). As in the ecologists' approach, SPRTs of H_0 versus H_1 and H_1 versus H_2 are performed simultaneously until both reach a decision. To avoid the possibility that a decision could be made for both H_0 and H_2, a third SPRT of H_0 versus H_2 is conducted for any small values of n, for which this case broadens the zone for continued sampling. The resulting decision zones are shown in Figure 6.10. Theoretical knowledge of the properties of these boundaries is scanty (Wetherill and Glazebrook 1986). However, the properties of a proposed test may be evaluated using ECOSTAT.

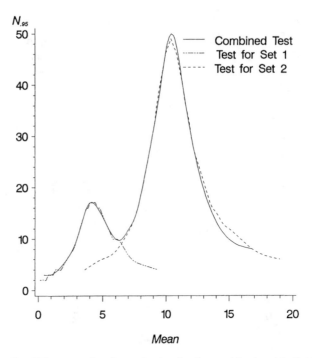

Figure 6.9. The 95th percentile of sample size for the combined and individual SPRTs based on Sobel and Wald's (1949) method.

Example 6.3

Once again consider Example 6.1. The purpose of sampling is to determine whether the infestation of spruce budworm larvae is light, medium, or heavy. As opposed to the earlier two examples, we must choose a mean, as opposed to a range of means, to represent a medium infestation. Consequently, we choose the midpoint of the range, 7.5, as the mean associated with a medium infestation level. Therefore the hypotheses to be tested are

$$H_0: \quad \mu = 3$$
$$H_1: \quad \mu = 7.5$$
$$H_2: \quad \mu = 12$$

S is the SPRT of H_0 versus H_1. S' is the SPRT of H_1 versus H_2. S^* is the SPRT of H_0 versus H_2. Type I and II error rates are 0.1 for all tests. The decision zones are illustrated in Figure 6.10. Notice that S^* does not impact the decision bound-

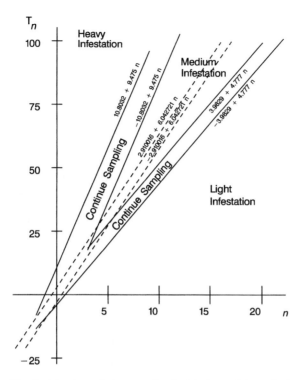

Figure 6.10. The decision boundaries associated with determining whether spruce bud-worm egg infestation is light, medium, or heavy for Armitage's (1950) method.

aries in the positive quadrant and thus has no effect on the combined test. This was expected because the error rates for S and S' are equal; consequently, it is not possible to decide in favor of H_0 and H_2 simultaneously (as discussed in the section describing Sobel and Wald's test). Therefore, in this case, this test is equivalent to the one proposed by ecologists for the case that $\mu_1 = \mu_2$. ECOSTAT was used to determine the OC and ASN functions and $N_{.95}$ for this test. These are reported Table 6.5 and depicted in Figures 6.11, 6.12, and 6.13, respectively.

Notice that for $\mu = 7.5$, the error rate is 0.12. The specified error rate for S was 0.1, that for S' was 0.1. By combining the tests, Armitage reasoned that the error rate at the intermediate value would be approximately $\alpha + \beta'$, or 0.2 in this example. The actual error rate is significantly less than this. The reason for this is twofold. First, the true error rates tend to be less than the stated ones for the SPRT. Further, the two tests are not independent, and therefore the error rate is not simply the sum of the individual ones.

Table 6.5. Armitage's (1950) OC and ASN Functions and $N_{.95}$ for a Range of Means for the Negative Binomial With k = 3[a]

Mean μ	p_L	p_M	p_H	E(N)	$N_{.95}$
0.450	1.000	0.000	0.000	1.355	2
0.961	1.000	0.000	0.000	1.615	2
1.544	1.000	0.000	0.000	1.878	3
2.216	1.000	0.001	0.000	2.282	4
3.000	0.987	0.013	0.000	3.076	7
3.666	0.918	0.082	0.000	4.065	9
4.424	0.704	0.295	0.000	5.081	11
5.296	0.386	0.613	0.001	5.451	11
6.309	0.155	0.839	0.007	5.586	10
7.500	0.052	0.895	0.052	6.745	14
8.223	0.029	0.826	0.145	7.953	18
9.021	0.015	0.642	0.343	8.974	22
9.906	0.008	0.378	0.614	8.717	22
10.893	0.004	0.165	0.831	7.164	18
12.000	0.002	0.058	0.940	5.424	13
16.807	0.000	0.001	0.999	2.424	5
24.819	0.000	0.000	1.000	1.460	3
40.842	0.000	0.000	1.000	1.087	2
88.912	0.000	0.000	1.000	1.002	1

[a]Hypothesized means are 3, 7.5, and 12. Error rates are 0.1.

The method due to ecologists is the most appealing from an ecological viewpoint. It permits a range of values for each of the levels of infestation, unlike Armitage's method. Unlike Sobel and Wald's method, it also requires a simultaneous decision for both SPRTs. By using ECOSTAT to first evaluate the feasibility of the individual SPRTs through Wald's approximations, and then to determine the properties of the combined test, the properties of the sampling program can be assessed. If the actual error rates depart significantly from the specified ones, adjustments can be made in the specified error rates until the true ones are at an acceptable level.

The examples to this point in the chapter have been based on the negative binomial distribution. The method proposed by ecologists should hold for any distribution in the exponential family for which an SPRT can be developed. The properties of the combined test are not known in general. However, using ECOSTAT, the properties can be determined when the test concerns the mean of the Poisson or negative binomial distributions or the proportion p of the binomial distribution. Sobel and Wald were working with the normal distribution and Armitage with the binomial distribution; however, their methods can be easily extended to other distributions.

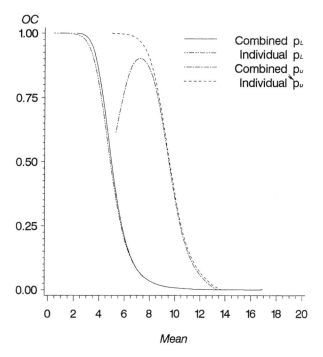

Figure 6.11. The OC function for the combined and individual SPRTs using Armitage's (1950) method.

Testing Composite Hypotheses

Consider using the SPRT to determine whether a population is above the economic threshold, indicating a need for control. The primary interest is in whether the population mean is above or below the economic threshold. Therefore, the hypotheses of interest are in fact

$$H_0: \mu = \mu_0 \quad \text{versus} \quad H_1: \mu \neq \mu_0. \quad (6.4)$$

We consider a couple of approaches to sequentially testing these hypotheses.

Iwao's Method

To address this problem, Iwao (1975) first assumed that the m^*–m relationship held (*see* Chapter 4). This permits the variance to be written as a quadratic function of the mean; that is,

$$\sigma^2 = (\alpha + 1)\mu + (\beta - 1)\mu^2. \quad (6.5)$$

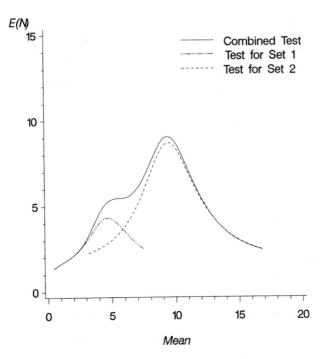

Figure 6.12. The ASN function for the combined and individual SPRTs using Armitage's (1950) method.

Then the half-width of confidence interval of the estimated mean density, d, based on the normal distribution and assuming the mean is μ_0 can be written as

$$d = t \sqrt{\frac{(\alpha + 1)\mu_0 + (\beta - 1)\mu_0^2}{n}} \qquad (6.6)$$

where n is the number of observations. Iwao proposed that $(1 - \alpha)$ 100% confidence intervals be set about μ_0 for $n = 1, 2, 3, \ldots$. Therefore, the interval is of the form

$$\mu_0 \pm t \sqrt{\frac{(\alpha + 1)\mu_0 + (\beta - 1)\mu_0^2}{n}}. \qquad (6.7)$$

The continuation zone may be rewritten in terms of the total of the first n observations, T_n, $n = 1, 2, 3, \ldots$ as follows:

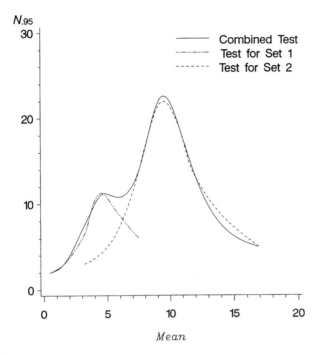

Figure 6.13. The 95th percentile of sample size for the combined and individual SPRTs for Armitage's (1950) method.

$$n\mu_0 - t\sqrt{n[(\alpha + 1)\mu_0 + (\beta - 1)\mu_0^2]} \leq T_n$$
$$\leq n\mu_0 + t\sqrt{n[(\alpha + 1)\mu_0 + (\beta - 1)\mu_0^2]}. \quad (6.8)$$

Sampling continues as long as T_n lies within the interval in equation (6.8). If T_n falls above the upper limit, sampling ceases and the population level is declared to be above the economic threshold. If T_n falls below the lower limit, sampling ceases and the population level is declared to be below the economic threshold. Usually, a maximum number of observations is set in advance and sampling ceases with the conclusion that $\mu = \mu_0$ if the stopping criterion has not been reached by that time.

In spite of its intuitive appeal, Nyrop and Simmons (1984) showed that the error levels for this test are significantly higher than the stated ones using this approach. The reason for this lies in that the confidence interval set about μ_0 is based on a fixed number of observations. For a given sample size, the probability of error is α. However, if sampling continues beyond this point, the error rate must necessarily increase above the stated level. Nyrop and Simmons (1984) suggested using simulations to study the behavior of this method before implementing it in the field. However, alternative approaches to this same problem have been developed that do have the desired properties. These will be considered next.

Armitage's Methods

In 1947, Armitage proposed a method of sequential t-tests for the purpose of testing the hypotheses in (6.4). To do this, he considered a set of three hypotheses:

$$H_0^*: \quad \mu = \mu_0 - a$$
$$H_1^*: \quad \mu = \mu_0$$
$$H_2^*: \quad \mu = \mu_0 + a \qquad\qquad (6.9)$$

Armitage suggested testing

$$H_0: \quad \mu = \mu_0 - a \qquad \text{versus} \qquad H_1: \quad \mu = \mu_0 \qquad (6.10)$$

and

$$H_0': \quad \mu = \mu_0 \qquad \text{versus} \qquad H_1': \quad \mu = \mu_0 + a.$$

Note that this is in fact a special case of the hypotheses in (6.2), where μ_1 and μ_2 in (6.2) are both equal to μ_0 in equation (6.5). The test Armitage proposed is exactly that proposed later by Morris (1954); that is, two SPRTs are to be conducted simultaneously. The first SPRT, S, tests H_0 versus H_1; the second, S', tests H_0' versus H_1'.

The test Armitage proposed in 1947 could result in the simultaneous decisions that the population density was both above and below the economic threshold. As discussed earlier in the chapter when considering tests to determine whether an infestation is light, medium, or heavy, this case usually does not arise for the sampling programs considered in ecology. However, should it occur, then Armitage's (1950) method avoids this difficulty by simultaneously conducting three SPRTs. In addition to S and S', a test of H_0 versus H_2 is performed. Armitage presented an analogy of this test with linear discriminant function analysis.

The difficulty with using Armitage's method to determine whether the population density is above or below the economic threshold is that we are again faced with three decision zones. In the end, we make one of the following decisions: (1) $\mu \leq \mu_0 - a$; (2) $\mu = \mu_0$; and (3) $\mu \geq \mu_0 + a$. Yet, we really want to know whether $\mu < \mu_0$ or $\mu \geq \mu_0$. Iwao's method gives an intuitively pleasing way of addressing the question we want to answer, but the error rates are significantly greater than those specified. No current sequential sampling program gives us an answer to this question with actual error probabilities close to the specified ones.

Perhaps a modification of Iwao's method would accomplish this. Until that occurs or another test is proposed, it appears that Wald's SPRT will remain the primary tool for making these management decisions. However, more recent work for testing the hypotheses in (6.9) may be used increasingly as software for the methods becomes more readily available.

Billard and Vagholkar (1969) and Billard (1977) used a computer optimization method to determine the procedure that minimizes the ASN function at a specified value of the normal mean μ and the binomial proportion p, respectively. The decision boundaries for the test are based on a geometric set of seven parameters, and the values of these are determined during the optimization process. Because determination of the boundaries require specialized software that has only been developed for the normal and binomial distributions, this test has had limited use.

Payton (1991) developed a method of adjusting the decision boundaries in Armitage's (1947) test so that the observed error rates are near the nominal ones. The procedure was developed for members of the Koopman–Darmois family of densities but has only been applied to the exponential, normal, and binomial distributions.

Summary

For more than 40 years, ecologists' have sought sequential methods for deciding among three or more hypotheses. The intuitive approach adopted by ecologists' simultaneously tests two sets of hypotheses using the SPRT. The stopping criterion must be met simultaneously for each test before stopping. In contrast, Sobel and Wald (1949) also simultaneously conducts SPRTs of two sets of hypotheses, but the stopping criterion for each is evaluated independently of the other. That is, as soon as a conclusion can be drawn for one set of hypotheses, the SPRT associated with that set is terminated and sampling continues until the stopping criterion of the other SPRT is satisfied. Armitage (1950) considered the simultaneous test of three hypotheses as opposed to two sets of hypotheses. ECO-STAT may be used to evaluate the properties of the methods commonly used by ecologists.

Some work has been done on testing a null hypothesis versus a two-sided alternative. Iwao's (1975) method should be evaluated through simulation before implementation. Armitage's (1950) approach is easily conducted. Other methods will undoubtedly be used more extensively as computer software needed for their implementation becomes more readily available.

Exercises

1. Tostowaryk and McLeod (1972) proposed a sampling plan for determining whether the populations of Swaine jack pine sawfly (*Neodiprion swainei*

Middleton) in large tracts of pure jack pine (*Pinus banksiana* Lamb.) in Quebec represented light, moderate, or severe infestations. The sampling unit was an individual tree, and the number of egg clusters per tree was the observed random variable. The number of egg clusters per tree was found to have an approximate negative binomial distribution with an estimated k value of 1.103. Infestations less than an average of 3.3 egg clusters per tree was considered a light infestation. Moderate infestations were described as having between 8.3 and 14.0 egg masses on average. If the average number of egg masses exceeded 26 per tree, the infestation was said to be heavy.

A. Develop a sequential sampling plan that will classify an infestation as light, medium, or heavy, using 0.1 error rates.

B. Evaluate the OC function, the ASN function, and the 95th percentile of sample size for the proposed plan.

C. Adjust the test developed in A so that the actual error rates are closer to the stated rates. What impact does this have on the OC function, the ASN function, and the 95th percentile of sample size?

7

Aggregation and Spatial Correlation

Introduction

In the first three chapters, we considered some distributions commonly encountered in ecological studies, models that give rise to those distributions, and methods of determining the fit of a proposed distribution to the data. One of the models giving rise to the Poisson distribution assumes that each member of the species behaves independently of every other member of that species and that the probability that a species member is in a sampling unit is the same for all sampling units. This requires that each member of a species reacts with complete indifference to other members of that species or to the environment. This model is probably more appropriate for animal than for plant populations. For plant populations, a possible Poisson model assumes that every sampling unit contains a large number, n, of locations, each of which has the probability p of being occupied by a species member. Both models have very restrictive assumptions. Therefore, it is not surprising that the Poisson rarely provides a good fit to ecological data sets. At the same time, these models represent two cases that are not very biologically interesting. It is the interaction of plant and animals with members of their own species, with members of other species, and with their environment that makes ecology exciting. From this perspective, it makes sense first to determine whether the population departs from this most sterile case and then to describe the nature of those departures.

We have continually stressed the distinction between probability distribution and spatial pattern. Terms such as aggregated, random, or regular are often used in ecological studies when the variance is greater than, less than, or equal to the mean, respectively. Unfortunately, these terms bring with them ideas of corresponding spatial patterns that cannot be justified through the probability distribution alone. In this chapter, we begin by reviewing the traditional measures of aggregation and discuss their limited value in drawing inference about spatial structure. Then we turn our attention to methods that address questions relating to the spatial structure of the data. These are divided into two broad categories. The first includes methods useful for investigating spatial correlation, and the other consists of approaches to the study of point patterns. This chapter considers measures of spatial correlation. Chapter 8 describes nearest neighbor techniques and analysis of spatial point patterns. However, before beginning, let us consider the circumstances under which each is the most appropriate.

Often in sampling biological populations, units closer together tend to be more alike than units farther apart. Suppose we are taking soil samples in order to determine the level of a certain pollutant, say atrazine. If the measure of atrazine is high in one sample, we expect measures from samples in close proximity to the first one to also tend to be high. If the measure of atrazine is low, we anticipate surrounding samples to also tend to have low levels. Similarly, if the number of insects on a plant is recorded, counts on plants close together may tend to be more alike than counts farther apart. To think about this is a statistical manner, we can consider the study region as a whole. Each sample point within the study region has associated with it an (x, y)-coordinate that identifies its location uniquely. Further, the random variable Z (level of the pollutant or number of insects in our examples) may be observed at any sample point. Therefore, each observation has a unique spatial location associated with it. The distance between sample points can be used to assess whether in fact observations are spatially correlated. In this chapter, geostatistics is the set of methods that are used for this purpose.

Biological populations are often distributed in space and how individual members came to be in a particular location may be of interest. For example, are trees in a forest randomly distributed or are they distributed in a more aggregated, or perhaps a more regular, pattern than would be associated with randomness? In this case, the (x, y)-coordinate of each tree is recorded. Here we are not interested in observing a random variable at each spatial location. Instead, we are interested in the spatial pattern that results by noting each tree's location. Chapter 8 introduces methods that provide a foundation for determining whether the spatial pattern is aggregated, random, or regular.

Measures of Aggregation

The traditional measures of aggregation are functions of the mean and variance. The value of a measure when the distribution is Poisson is generally taken as the standard, and departures from this value are used to provide insight into the behavior of the population under consideration. We consider the variance-to-mean ratio, Fisher et al.'s (1922) index of dispersion (ID), David and Moore's (1954) index of clumping (IC), Lloyd's (1967) index of mean crowding (IMC), and Lloyd's index of patchiness (IP).

Variance-to-Mean Ratio and Index of Dispersion

The feature that all contagious distributions have in common is that the variance exceeds the mean. For the Poisson, the variance equals the mean. For underdispersed distributions, such as the binomial, the variance is less than the mean. Based on these relationships, an intuitive measure of aggregation is the ratio of variance to the mean: σ^2/μ. Values of less than 1, 1, and greater than 1 are often said to represent underdispersion, randomness, and aggregation, respectively. The term of randomness for a ratio of one has served to confuse researchers. In the scientific literature, the term randomness is often extended to mean randomness in the spatial pattern. However, as we saw in Chapter 3, spatial randomness is not always associated with the Poisson. Because of this, we are led to repeatedly emphasize the narrowness with which the term randomness can be used in this context. More accurately, a ratio of one is associated with the Poisson distribution, not randomness. A variance-to-mean ratio of one does *NOT* imply that a model of spatial randomness or a model of independence among species members holds. Moreover, when the variance-to-mean ratio is different from one, the question of spatial randomness remains unanswered.

Because σ^2 and μ are unknown, the variance-to-mean ratio is estimated using $s^2/\bar{X}$. Because of the variability of the estimator, we rarely observe a value of 1, even if the distribution is Poisson. How far from one must this ratio be before we decide that the distribution is not Poisson? Fisher et al. (1922) proposed using the estimated index of dispersion (ID),

$$\text{ID} = \frac{(n-1)s^2}{\bar{X}} \tag{7.1}$$

as the test statistic to test the hypothesis that the variance is equal to the mean. This is the same test statistic given in equation (1.7). If the variance equals the mean, ID has an approximate χ^2 distribution with $(n-1)$ degrees of freedom. Examples of the application of the test statistic are given in Chapter 1.

Index of Clumping

David and Moore (1954) proposed an index of clumping (ICS):

$$\text{ICS} = \left(\frac{\sigma^2}{\mu}\right) - 1. \tag{7.2}$$

This is estimated using

$$\widehat{\text{ICS}} = \left(\frac{s^2}{\bar{X}}\right) - 1. \tag{7.3}$$

ICS is simply the variance-to-mean ratio less 1. The rationale behind the index is that the Poisson case of the variance equaling the mean has value 0. Values less than zero are associated with underdispersion, and values greater than zero represent aggregation. The test of the hypothesis that the variance equals the mean (or that ICS = 0), is the same as that testing ID = 1 given in equation (7.1).

Douglas (1975) used ICS as an Index of Cluster Size, leading to the acronym. If ICS measures cluster size, ICF = μ/ICS is a measure of the average number of clusters per sampling unit. Douglas derived standard errors for ICS and ICF and presented an APL computer program for their computation.

Mean Crowding and Mean Patchiness

Lloyd (1967) defined the index of mean crowding x^* as the "mean number per individual of other individuals in the same quadrat." Biologically, x^* is most meaningful if the sampling unit, or quadrat, corresponds to the microhabitat of the organism. For example, suppose the microhabitat of a bird is a tree. Then by taking the tree as a sampling unit, x^* represents the crowding due to other birds on the same tree. If instead, a group of five trees is the sampling unit, then x^* measures crowding for a given bird as the number of other birds on the five trees, even though birds on a different tree from that of the specified bird may not be crowding that bird. Therefore, the measure of mean crowding would not accurately reflect the crowding being experienced by the bird.

Lloyd determined that x^* is approximately

$$x^* \approx \mu + \frac{\sigma^2}{\mu} - 1 \qquad (7.4)$$

and can be estimated using

$$\hat{x}^* = \bar{X} + \frac{s^2}{\bar{X}} - 1. \qquad (7.5)$$

Based on x^*, Lloyd developed the index of patchiness (IP), the ratio of mean crowding to mean density:

$$IP = \frac{x^*}{\mu}. \qquad (7.6)$$

IP is estimated by

$$\widehat{IP} = \frac{\hat{x}^*}{\bar{X}} = 1 + \frac{s^2}{\bar{X}^2} - \frac{1}{\bar{X}}. \qquad (7.7)$$

As mean crowding increases relative to the mean, the population aggregation increases. This rising aggregation is reflected in increasing values of *IP*. Values of *IP* that are less than, equal to, and greater than 1 correspond to underdispersion, randomness, and overdispersion, respectively. Fisher's test of *ID* = 1 is used to test *IP* = 1.

Morista's (1959) index of dispersion, I_δ, is almost identical to Lloyd's index of crowding; however, the rationale leading to the index differs. Let there be x_i organisms in the *i*th sampling unit and assume that the *q* sampling units do not overlap. Then $\Sigma x_i = N$, the number of organisms in the population. The probability that any two organisms chosen at random from the population of *N* organisms are in the same sampling unit is

$$\delta = \frac{\sum_i x_i(x_i - 1)}{N(N - 1)} \tag{7.8}$$

If the organisms are crowded into relatively few of the quadrats, then δ is large. It is small if the organisms are uniformly spaced so that they are somewhat evenly apportioned among the quadrats. The expected value of δ when organisms are randomly dispersed is $1/q$. Taking the ratio of the estimated δ to that expected under randomness, we obtain

$$I_\delta = \frac{\sum_{i=1}^{s} x_i(x_i - 1)/(N(N - 1))}{1/q} = \frac{q}{N - 1}x^* = \frac{N}{N - 1}IP \tag{7.9}$$

where IP is the index of patchiness.

Comparison of Indices

Several measures of aggregation have been presented. By substituting the appropriate parameters for the mean and variance, we can obtain a general form for each index based on any given distribution. The results for the binomial, Poisson, and negative binomial distributions are presented in Table 7.1.

For the negative binomial, a decrease in the parameter *k* results in an increase in each measure of aggregation. This has led to *k* often being called the aggregation parameter. Small values of *k* are associated with strong aggregation and aggregation is said to decrease as *k* increases. It is interesting to note that neither parameter of the binomial distribution has this property. Increasing values of *p* are associated with decreasing aggregation (increasing underdispersion) for the variance-to-mean ratio, index of dispersion, and the index of clumping. Mean crowding, however, increases with *p*, and *p* is not even present in the index of mean patchiness. For patchiness, increasing values of *n* result in a decrease in the measure of aggregation.

220 / Statistical Ecology

Table 7.1.   General Expressions of Aggregation Measures for the Binomial, Poisson, and Negative Binomial Distributions

	Distribution		
Measure	Binomial	Poisson	Negative binomial
Variance-to-mean ratio (σ^2/μ)	$q = 1 - p$	1	$1 + \mu/k$
Index of dispersion	$q(n - 1)$	$n - 1$	$(n - 1)(1 + \mu/k)$
Index of clumping ($\sigma^2/\mu - 1$)	$-p$	0	μ/k
Mean crowding			
$(x^* = \mu + \sigma^2/\mu - 1)$	$p(n - 1)$	μ	$\mu + \mu/k$
Index of patchiness (x^*/μ)	$1 - 1/n$	1	$1 + 1/k$

For each measure of aggregation, the test of the null hypothesis that the relationship expressed by the Poisson distribution holds is tested using the index of dispersion test presented in Chapter 1 (*see* Cormack, 1979). Kathirgamatamby (1953) has shown that the χ^2 approximation is good provided that the sample size n is greater than 6 and the mean density μ is greater than 1. The power of the test increases with n, but it also depends on the size and shape of the individual quadrats. The results of power studies (Perry and Mead, 1979; Stiteler and Patil, 1971; Diggle, 1979) suggest that the test is powerful in detecting aggregation, but may be weak in determining underdispersion.

Example 7.1

To illustrate the inability of these measures of aggregation to assess spatial pattern, we consider a field that has been divided into 100 plots as may be seen in Figures 7.3 to 7.8 (*see* Nicot et al., 1984, for a similar example). For each plot, the soil inoculum level of a plant pathogen is observed. We consider the two data sets presented in Table 7.2.

A frequency diagram for the first data set is plotted with expected probabilities of a Poisson random variable with a mean of one in Figure 7.1. A χ^2 goodness-of-fit test (*see* Chapter 2) indicates that the Poisson distribution describes this first

Table 7.2.   Two Hypothetical Data Sets, Each With a Mean of 1

	X						
	0	1	2	3	4	5	6
Data set 1	37	37	17	7	2	0	0
Data set 2	50	25	11	7	4	2	1

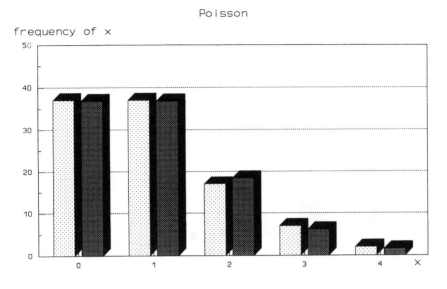

Figure 7.1. Observed and expected frequencies for a sample from a Poisson distribution with a mean of 1.

set of data well ($\chi_P^2 = 0.23$, $P(\chi^2 > .23) = 0.99$). Similarly, a frequency diagram of the second data set is plotted with the expected probabilities of a geometric (negative binomial with $k = 1$) random variable with a mean of one in Figure 7.2. Based on a χ^2 goodness-of-fit test, the geometric distribution describes this data set well ($\chi_P^2 = 0.84$, $P(\chi_P^2 > 0.84) = 0.99$).

Each data set was used to form three fields. As shown in Figures 7.3 to 7.8, the spatial patterns for fields based on the same data set can be quite different. However, each of the aggregation measures studied in this section is the same for all three spatial patterns associated with a given data set. The variance-to-mean ratio for the Poisson data set is 1.01, and it is 1.84 for the geometric data set. (The accompanying software ECOSTAT can be used to compute the variance-to-mean ratio as well as the other traditional measures of aggregation. To do so, select *Variance-to-Mean* from the *Spatial Statistics* menu. Enter the sample size, estimated mean, and estimated standard deviation. Pressing *Measures of Aggregation* results in all of the measures being computed. The test statistics and their *p*-values are displayed when *Variance-to-Mean Test* is pressed. See Chapter 1 for further details.) Based on these numerical distributions, we declare the first data set random and the second aggregated. However, for each set, one field demon-

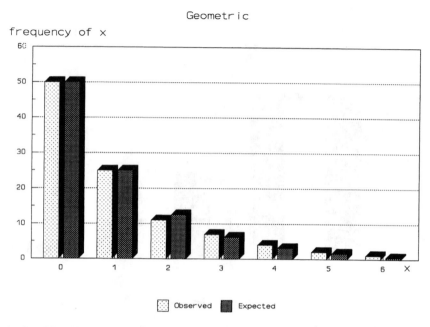

Figure 7.2. Observed and expected frequencies for a sample from a geometric distribution with a mean of 1.

strates aggregation, another shows randomness, and the third field has a regular spatial pattern. Clearly, something more is needed if we are to be able to say anything sensible about spatial aggregation.

By considering the effect of changes in quadrat size, some limited insight into spatial patterns can be gained. Pielou (1977) discusses the effect of quadrat size on aggregation measures for differing spatial patterns. Limited information is available. In addition, extensive sampling effort is required to assess spatial pattern in this manner. If the quadrats are randomly placed within the region of interest, a random sample is taken for each different quadrat size. Not only is this time consuming, it may also result in significant trampling or even destruction of the vegetation. The fact that the size of the quadrat affects some of the indices is a further warning of the care with which the sampling unit should be chosen.

Spatial Correlation

Gurland (1959) discussed the difference between apparent and true contagion. To illustrate this difference, consider the distribution of insect eggs on plants within an even-aged monoculture. The sampling unit is a plant. A true contagion,

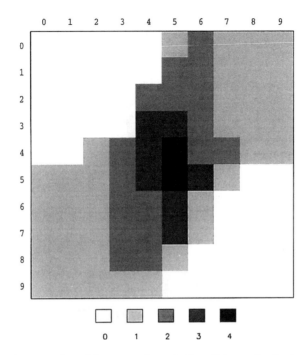

Figure 7.3. Aggregated spatial pattern of sample from Poisson distribution with $\mu = 1$.

or generalized Poisson, process describes a pattern in which the eggs are laid in small clusters that are randomly scattered among the plants. The observed spatial structure is random because aggregation is occurring on a scale smaller than the sampling unit (the plant). Suppose now that the eggs are randomly scattered within each plant, but the average number of eggs on a plant varies from plant to plant. This is an apparent contagion, or compound Poisson, process because aggregation is present on a larger scale than the sampling unit. If the average number of eggs on a plant is spatially influenced, the spatial structure is not random. The measures of aggregation discussed in the preceding section cannot distinguish between these two types of contagion. Any attempt to determine whether the spatial structure is random must in some way take into account the position and/or the relative location of the organisms. Returning to our present example, we want to know whether we are more or less likely to observe insect eggs on a plant if insect eggs are on an adjacent plant. Moran's I (1950) and Geary's c (1954) were the first measures that in some way measured this spatial correlation (or spatial autocorrelation). During the past few years, geostatistical methods have begun to impact ecological studies of spatial structure.

Numerous software packages have been developed for the analysis of spatial data. However, most of the standard statistical programs have not incorporated

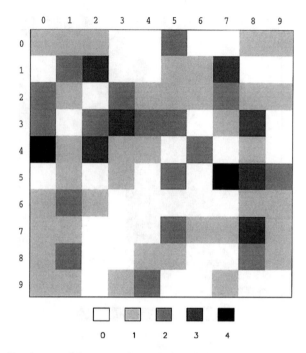

Figure 7.4. Random spatial pattern of sample from Poisson distribution with $\mu = 1$.

spatial methods yet. We have included some spatial methods in ECOSTAT. However, the data structure that ECOSTAT accepts is limited. Observations must be collected on a rectangular grid. Therefore, data are to be entered by identifying the row and column associated with the observed value of the random variable. It is assumed that rows and columns are each numbered consecutively, beginning with one. It is not necessary for the sampling units to be physically touching, as in Figures 7.3 to 7.8; instead, they may be separated in space. Yet, the form of rows and columns must be maintained. This data structure is a special case of a lattice, a mathematical term that is used at times in this chapter. Certainly not all lattices take this form, and the methods do not require such a restricted form. Other software should be used if your sampling points are not on a rectangular grid.

Moran's *I* and Geary's *c*

Moran (1950) suggested considering, in some manner, the correlation between adjoining sampling units. Two units are said to be adjoining, or linked by a join,

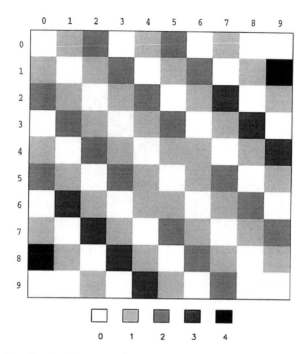

Figure 7.5. Regular spatial pattern of sample from Poisson distribution with $\mu = 1$.

if they share a common boundary of positive (nonzero) length. The total number of joins in the system is denoted by A. Let δ_{ij} be an indicator function for which $\delta_{ij} = 1$ if sampling units i and j are joined, and $\delta_{ij} = 0$ otherwise. Further, define x_i to be the number of organisms within the ith sampling unit and $\bar{x}$ to be the mean of the n x_i's for the n sampling units. The coefficient proposed by Moran is given by

$$I = \frac{n}{2A} \frac{\sum\limits_{i \neq j}^{n} \sum\limits_{j=1}^{n} \delta_{ij}(x_i - \bar{x})(x_j - \bar{x})}{\sum\limits_{i=1}^{n} (x_i - \bar{x})^2}. \tag{7.10}$$

Notice that the numerator is the traditional measure of covariance among the x_i's from adjoining plots, and the denominator is a measure of the variance, giving I the classical form of a correlation coefficient.

Geary (1954) proposed the following statistic c as a measure of aggregation:

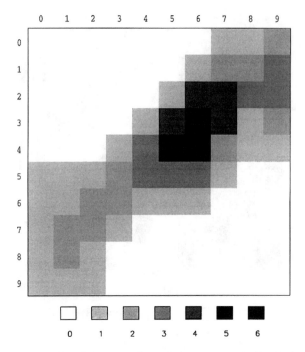

Figure 7.6. Aggregated spatial pattern of sample from geometric distribution with $\mu = 1$.

$$c = \frac{(n-1)\sum\limits_{i \ne j}^{n}\sum\limits_{j=1}^{n}\delta_{ij}(x_i - x_j)^2}{4A\sum\limits_{i=1}^{n}(x_i - \bar{x})^2}. \tag{7.11}$$

Here the measure of variance in the denominator is the same as in equation (7.10), but the numerator has the form of that in the Durbin–Watson d-statistic. It can be shown that $c \approx 1 - I$. Therefore, as I increases, c decreases, and vice versa.

Both I and c are asymptotically normally distributed as n increases. Generally, it is assumed that normality holds approximately for small lattices; therefore, I and c are standardized and tested for significance as standard normal deviates. To standardize I and c, we must first be able to compute their means and variances. These may be evaluated under either of two assumptions: normality or randomization. Under the assumption N of normality, we assume that the observed values $x_i, i = 1, 2, 3, \ldots, n$, are n independent observations from a normal population. Under the assumption R of randomization, we consider all of the $n!$ possible

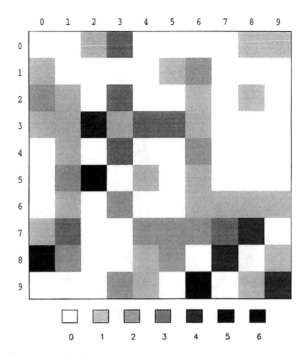

Figure 7.7. Random spatial pattern of sample from geometric distribution with $\mu = 1$.

permutations of the values x_i, $i = 1, 2, 3, \ldots, n$. We consider the observed values of I and c relative to the set of all possible values I and c could assume for each of the $n!$ permutations.

Using the subscripts N and R to denote the assumptions of normality and randomization, respectively, it can be shown (*see* Cliff and Ord, 1973) that the mean of I, *given no spatial correlation,* is the same under both assumptions:

$$E_N(I) = E_R(I) = -\frac{1}{n-1}. \tag{7.12}$$

I values significantly greater than $-1/(n-1)$ indicate aggregation while those significantly less than $-1/(n-1)$ point to a regular spatial pattern. To determine whether the statistic I is significantly different from $-1/(n-1)$, indicating a departure from randomness, we must first determine the variance of I. We begin by computing the second moment of I, which differs under the two assumptions. Denote the number of units joined to the ith unit by L_i. Then the total number of joins in the system is

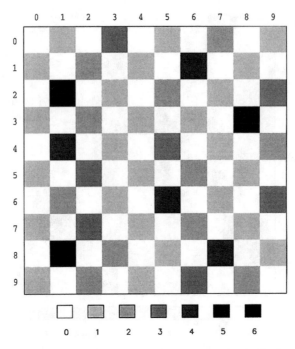

Figure 7.8. Regular spatial pattern of sample from geometric distribution with $\mu = 1$.

$$A = \frac{1}{2}\sum_{i=1}^{n} L_i. \tag{7.13}$$

Further define

$$D = \frac{1}{2}\sum_{i=1}^{n} L_i(L_i - 1). \tag{7.14}$$

Let b_2 be the sample kurtosis coefficient,

$$b_2 = \frac{\dfrac{1}{n}\sum_{i=1}^{n}(x_i - \bar{x})^4}{\left(\dfrac{1}{n}\sum_{i=1}^{n}(x_i - \bar{x})^2\right)^2}. \tag{7.15}$$

Assuming the observations are normally distributed, the second moment of I may be written as

$$E_N(I^2) = \frac{4An^2 - 8(A + D)n + 12A^2}{4A^2(n^2 - 1)}. \tag{7.16}$$

Under the randomization assumption, the second moment of I is

$$E_R(I^2) =$$
$$\frac{n[4A(n^2 - 3n + 3) - 8(A + D)n + 12A^2] - b_2[4A(n^2 - n) - 16(A + D)n + 24A^2]}{4A^2(n - 1)(n - 2)(n - 3)}. \tag{7.17}$$

Based on the first two moments, the variance is computed as

$$Var(I) = E(I^2) - [E(I)]^2 \tag{7.18}$$

where subscripts N and R continue to represent the assumptions of normality and randomization, respectively.

As in the case of Moran's I, the mean of Geary's c, *given no spatial correlation,* is the same under both the assumptions of normality and randomization:

$$E_N(c) = E_R(c) = 1. \tag{7.19}$$

Values of c greater than 1 are indicative of a regular spatial pattern and those less than one are associated with aggregation. To test whether c differs significantly from 1, we must first compute its variance. The variance under the assumption N of normality is

$$Var_N(c) = \frac{(2A + D)(n - 1) - 2A^2}{(n + 1)A^2} \tag{7.20}$$

and that under the assumption R of randomization is

$$\begin{aligned} Var_R(c) = \frac{1}{2n(n - 2)(n - 3)A^2} & \{2A^2[-(n - 1)^2 b_2 + (n^2 - 3)] \\ & + 2A(n - 1)[-(n - 1)b_2 + n^2 - 3n + 3] \\ & + (D + A)(n - 1)[(n^2 - n + 2)b_2 - (n^2 + 3n - 6)]\}. \end{aligned} \tag{7.21}$$

Suppose we want to test the null hypothesis H_0: No spatial correlation is present. Whether using I or c, the test statistic is obtained by subtracting the mean and dividing by the standard deviation (square root of the variance). This value is compared to a standard normal distribution to determine its significance.

Example 7.2

Consider again the data in Example 7.1. We look only at the spatial arrangement in Figure 7.3. The number of joins is 180. The accompanying software ECOSTAT was used to compute I, c, and their means and variances. First select *Moran's I and Geary's c* from the *Spatial Statistics* menu. Press *Enter Data*. The data must be in an ASCII data file. The first line of the data set specifies the number of observations N in the set. Each of the next N lines has the x-coordinate, the y-coordinate, and the observed value at that point. The ECOSTAT data file morpagg.dat was developed for this example. Press *Sample Moments* to observe the sample moments. After pressing *Moran*, *Moran's I* and *Geary's c* are displayed, in addition, the tests for randomness under both normality and randomization are conducted. The test statistics and p-values are displayed. We obtained

$$I = 0.79 \qquad E(I) = 0.01 \qquad Var_N(I) = 0.0053 \qquad Var_R(I) = 0.0053$$
$$c = 0.26 \qquad E(c) = 1 \qquad Var_N(c) = 0.0057 \qquad Var_R(c) = 0.0058$$

To test the hypothesis of no spatial correlation, we must first determine whether the assumption of normality or randomization is most appropriate. Because the data are discrete, we know that the assumption of normality cannot be valid. Therefore, we use the test based on randomization which makes no distributional assumptions.

For the statistic I, we compute the test statistic:

$$\frac{I - E_R(I)}{\sqrt{VAR_R(I)}} = \frac{.7889 - 0.0101}{\sqrt{0.0053}} = 10.8106.$$

The significance level for the test is $P(|z| > 10.8106) < 0.0001$. Therefore, we reject the null hypothesis and conclude that there is evidence of spatial aggregation.

For the statistic c, the test statistic is

$$\frac{c - E_R(c)}{\sqrt{Var_R(c)}} = \frac{0.2612 - 1}{\sqrt{0.0058}} = -9.709.$$

The significance level for the test is $P(|z| > |-9.709|) < 0.0001$. Again, we reject the null hypothesis and conclude that there is spatial aggregation.

The I and c statistics have two weaknesses as measures of spatial structure (Cliff and Ord, 1981). First, suppose that the existence of a join between two sampling units has been established. The strength of the link as well as the size and shape of the sampling units are ignored. Therefore, the measures are invariant under certain transformations of the underlying sampling unit structure, a problem called "topological invariance" by Dacey (1965, p. 28). As an illustration, consider a system with four sampling units. The three systems illustrated in Figure 7.9 have the same join structure and, consequently, the same values of I and c.

Second, the statistics measure spatial correlation only between sampling units that are physically contiguous (first nearest neighbors). Thus, it is not possible to determine how the strength of the correlation changes over space. Modifications of the statistic that include second and higher order neighbors have been considered (Cliff and Ord, 1981). An alternative approach to evaluating the presence of spatial correlation comes from the study of geostatistics. We consider this approach next.

Geostatistics

Geostatistics came into prominence in the 1980's as a set of statistical methods useful in estimating ore reserves. In recent years, it has been used increasingly in numerous other disciplines. Applications of geostatistics to ecological data have provided new tools for assessing spatial patterns. In this section, we introduce some of the concepts and methods associated with this rapidly developing field of statistics. Isaaks and Srivastava's (1989) *An Introduction to Applied Geostatistics* is an introductory text on geostatistical theory and practice. Cressie's (1993) *Statistics for Spatial Data* is a comprehensive work that is more theoretically inclined. An overview paper of geostatistical methods in ecology is given by Rossi et al. (1992). We have drawn heavily on these sources in developing this section.

In previous chapters, we have discussed drawing samples from a population. In each case, we assumed that the observations were *independent* and *identically distributed*. However, in numerous biological settings, observations taken from sampling units in close proximity to each other tend to be more alike than values taken from sampling units farther apart; that is, observations taken from sampling

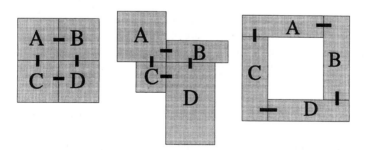

Figure 7.9. Three arrangements of four sampling units, each with the same join structure. Based on Cliff and Ord (1981) by permission of Pion Limited, London.

units in close proximity to one another tend to be positively correlated. Alternately, if a species expresses territoriality, observations close together may tend to differ more than observations further apart, meaning observations close together are negatively correlated. In geostatistics, efforts are made to model this correlation and, based on the model, to predict values at locations where no data were taken. Because our interest is in exploring the relationships of organisms with their environment and with each other, we develop models of spatial correlation, but we do not consider prediction.

Let Z be the random variable of interest. For example, Z could represent the number of insects on a plant, the yield of a crop, or the presence or absence of damage to a plant. Because we believe that observations close together are correlated, we denote the value of Z at a *fixed* location s in the sampling region as $Z(s)$. If a row of plants is the sampling area, then s is one-dimensional. When sampling a field (with numerous rows of plants) or a meadow, s is two-dimensional; s could also be three-dimensional if we are taking observations at different heights from the Earth's surface. In geostatistics, s is viewed as varying continuously over the sampling region. Because of the region being sampled, observations may be taken on a lattice of locations, leading us to think of neighbors, second nearest neighbors, and so on. Although s does not vary continuously in this case, geostatistical methods are still useful. We tend not to draw the distinction in the two types of data in the discussion that follows.

Before beginning an analysis of spatial pattern, however, an exhaustive exploratory data analysis (EDA) should be performed (*see* Tukey, 1977). An EDA consists of univariate and bivariate statistics, histograms, regression plots and scattergrams. Cluster analysis, principal component analysis, and analysis of variance may also be needed if the data are multivariate. We explain some of the possible components of an EDA as we work through a classical example.

Example 7.3

Mercer and Hall (1911) conducted mangold and wheat uniformity trials at Rothamsted Experimental Station in 1910. A uniformity trial is an experiment in which all the plots receive the same treatment, a control. The plot yields for the control treatment, wheat in this case, are recorded. Adjacent plots are aggregated into new plots. If x is the size of the original plot, then plots of size $2x, 3x, 4x, \ldots$, can be formed through this aggregation. The mean-squared error is computed for each plot size. By viewing the mean-squared error as a function of plot size, the optimal plot size can be determined.

The layout of the wheat uniformity trial was a 20 × 25 lattice of plots, covering approximately one acre. As indicated in Figure 7.10, the 25 columns run in the north-south direction, and the 20 rows lie in the east–west direction. The spatial analysis of these data has been considered by numerous investigators, including Fairfield Smith (1938), Whittle (1954), Besag (1974), Ripley (1981), McBratney

NORTH

WEST

3.63	4.15	4.06	5.13	3.04	4.48	4.75	4.04	4.14	4.00	4.37	4.02	4.58	3.92	3.64	3.66	3.57	3.51	4.27	3.72	3.36	3.17	2.97	4.23	4.53
4.07	4.21	4.15	4.64	4.03	3.74	4.56	4.27	4.03	4.50	3.97	4.19	4.05	3.97	3.61	3.82	3.44	3.92	4.26	4.36	3.69	3.53	3.14	4.09	3.94
4.51	4.29	4.40	4.69	3.77	4.46	4.76	3.76	3.30	3.67	3.94	4.07	3.73	4.58	3.64	4.07	3.44	3.53	4.20	4.31	4.33	3.66	3.59	3.97	4.38
3.90	4.64	4.05	4.04	3.49	3.91	4.52	3.05	4.59	5.07	4.01	3.34	4.06	3.19	3.75	4.54	3.97	3.77	4.30	4.10	3.81	3.89	3.32	3.46	3.64
3.63	4.27	4.92	4.64	3.76	4.10	4.40	4.17	3.67	3.83	3.79	3.63	3.74	4.14	3.70	3.92	3.79	4.29	4.22	3.74	3.55	3.67	3.57	3.96	4.31
3.16	3.55	4.08	4.73	3.61	3.66	4.39	3.84	4.26	4.36	3.79	4.09	3.72	3.76	3.37	4.01	3.87	4.35	4.24	3.58	4.20	3.94	4.24	3.75	4.29
3.18	3.50	4.23	4.39	3.28	3.56	4.94	4.06	4.32	4.86	3.96	3.74	4.33	3.77	3.71	4.59	3.97	4.38	3.81	4.06	3.42	3.05	3.44	2.78	3.44
3.42	3.35	4.07	4.66	3.72	3.84	4.44	3.40	4.07	4.93	3.93	3.04	3.72	3.93	3.71	4.76	3.83	3.71	3.54	3.66	3.95	3.84	3.76	3.47	4.24
3.97	3.61	4.67	4.49	3.75	4.11	4.64	2.99	4.37	5.02	3.56	3.59	4.05	3.96	3.75	4.73	4.24	4.21	3.85	4.41	4.21	3.63	4.17	3.44	4.55
3.40	3.71	4.27	4.42	4.13	4.20	4.66	3.61	3.99	4.44	3.86	3.99	3.37	3.47	3.09	4.20	4.09	4.07	3.95	4.08	4.24	4.03	3.97	2.84	3.91
3.39	3.64	3.84	4.51	4.01	4.21	4.77	3.95	4.17	4.39	4.10	3.07	4.09	3.29	3.37	3.74	3.41	3.86	4.54	4.36	4.11	3.97	3.89	3.47	3.29
4.43	3.70	3.82	4.45	3.59	4.37	4.45	4.08	3.72	4.56	4.10	3.41	4.09	3.14	4.86	4.36	3.51	3.47	3.94	4.54	4.11	4.08	4.07	3.56	3.83
4.52	3.79	4.41	4.57	3.94	4.47	4.42	3.92	3.86	4.77	4.99	3.91	4.09	3.05	3.39	3.60	3.89	3.89	3.67	4.54	4.11	4.58	4.02	3.93	4.33
4.46	4.09	4.39	4.31	4.29	4.47	4.37	3.44	3.82	4.63	4.36	3.79	3.56	3.29	3.64	3.60	3.19	3.80	3.72	3.91	3.35	4.39	4.39	3.47	3.93
3.46	4.42	4.29	4.08	3.96	3.96	3.89	4.11	3.73	4.03	4.09	3.82	3.57	3.43	3.73	3.39	3.08	3.48	3.05	3.71	3.71	3.69	3.43	3.43	3.38
5.13	3.89	4.26	4.32	3.78	3.54	4.27	4.12	4.13	4.47	3.41	3.55	3.16	3.47	3.30	3.39	2.92	3.23	3.25	3.86	3.22	3.69	3.80	3.79	3.63
4.23	3.87	4.23	4.58	3.19	3.49	3.91	4.41	4.21	4.61	4.27	4.06	3.75	3.91	3.51	3.45	3.05	3.68	3.52	3.91	3.87	3.76	4.21	3.68	4.06
4.38	4.12	4.39	3.92	4.84	3.94	4.38	4.24	3.96	4.29	4.52	4.19	4.49	3.82	3.60	3.14	2.73	3.09	3.66	3.77	3.48	3.69	3.69	3.84	3.67
3.85	4.28	4.69	5.16	4.46	4.41	4.68	4.37	4.15	4.91	4.68	5.13	4.19	4.41	3.54	3.01	2.85	3.36	3.85	4.15	3.93	3.91	4.33	4.21	4.19
3.61	4.22	4.42	5.09	3.66	4.22	4.06	3.97	3.89	4.46	4.44	4.52	3.70	4.28	3.24	3.29	3.48	3.49	3.68	3.36	3.71	3.54	3.59	3.76	3.36

Figure 7.10. Plot wheat yields from a uniformity trial. From Mercer and Hall, (1911).

and Webster (1981), Wilkinson et al. (1983), and Cressie (1993). We use these data to explore various techniques that are useful in geostatistics. The analysis presented here is most heavily influenced by that of Cressie.

The dimensions of the plots used in the Mercer and Hall study are not known precisely (*see* Cressie, 1993, for a discussion). However, we use the dimensions of Wilkinson et al. (1983); they are 3.30 meters (east–west) × 2.51 meters (north–south). In this discussion, the rows in Figure 7.10 are numbered 1 to 20 from south to north, and the columns are numbered from 1 to 25 from west to east. In a spatial analysis, the yield for each plot must be associated with a point. Here the northeast corner of each plot is the point representing the entire plot. In order to identify the location of each plot, an X–Y coordinate system with origin in the southwest corner is superimposed on the field. Then the location of the plot in row i and column j has an x-coordinate of $j(3.30)$ and a y-coordinate of $i(2.51)$. The yield is the response of interest and is denoted by $z(j(3.30), i(2.51))$. (Because we anticipated the presence of the X–Y coordinate system, the rows were numbered from south to north instead of the more common approach of numbering from north to south.)

To conduct an EDA, we begin by graphing the data in several ways and by obtaining some summary statistics. A three-dimensional plot of the data as shown in Figure 7.11 permits a view of the variation of the yields in space. The bar chart in Figure 7.12 provides a univariate look at the wheat yield data.

Many of the methods associated with spatial analysis assume that the data represent a sample from a normal distribution. Wheat yields are anticipated to be normally distributed, and Figure 7.12 provides visual support for that assumption. The Kolmogorov–Smirnov (*see* Chapter 2) test did not reject the hypothesis that the data are from a normal distribution. However, if spatial correlation is present, common statistical methods, such as confidence intervals, analysis of variance, and the Kolmogorov–Smirnov test, are only approximate. Yet, at this time, we see no evidence of a need to transform these data.

Summary statistics were also computed. On the basis of 500 observations, the sample mean is 3.95, and the sample standard deviation is 0.46. The minimum, median, and maximum values are 2.39, 3.94, and 5.16, respectively.

A measure of the asymmetry or skewness of a population distribution is

$$\frac{E[(X - \mu)^3]}{\sigma^3} \tag{7.22}$$

Notice that because the deviations from the mean are cubed, the signs of the deviations are maintained, and the effects of large deviations are emphasized. Division by σ^3 removes the effect of scale so that multiplying all values by a constant does not change the skewness. Zero skewness indicates the distribution is symmetric. Positive skewness occurs when the positive deviations tend to be larger than the negative ones, indicating that the right-hand tail of the distribution

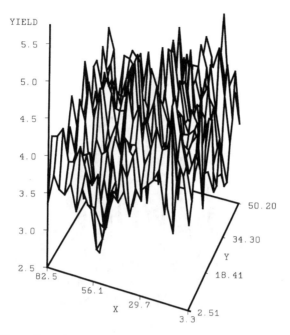

Figure 7.11. Three-dimensional plot of wheat yields from a uniformity trial (Mercer and Hall, 1911).

is heavier than the left one. Negative skewness implies that the negative deviations tend to be larger than the positive ones. Based on a sample, skewness is estimated by

$$\left(\frac{n}{(n-1)(n-2)}\right) \frac{\sum_{i=1}^{n} (x_i - \bar{x})^3}{s^3}.$$

For our example, the estimate of skewness is $-.0257$, indicating only an insignificant departure from symmetry.

The heaviness of the tails of a distribution can affect the behavior of many statistics. Kurtosis is a measure of tail heaviness. The population kurtosis may be defined as

$$\frac{E(X - \mu)^4}{\sigma^4} - 3$$

although the subtraction of 3 is sometimes omitted. An estimate of kurtosis is

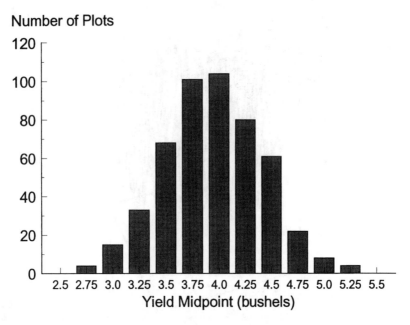

Number of Plots

Figure 7.12. Histogram of wheat yields from a uniformity trial From Mercer and Hall, (1911).

$$\frac{n(n + 1)}{(n - 1)(n - 2)(n - 3)} \frac{\sum_{i=1}^{n} (x_i - \bar{x})^4}{s^4} - \frac{3(n - 1)^2}{(n - 2)(n - 3)}.$$

If the population is normal, the kurtosis is zero. Positive values of kurtosis indicate the tails are longer than those for the normal. This is the manner in which the *t*-distribution departs from the normal. Negative kurtosis indicates that the tails are shorter, thus, values in those regions have smaller probabilities than would be observed with the normal distribution. The sample value for kurtosis is -0.1066, a value that is close to the 0 expected for the normal distribution.

Small sample estimates of skewness and kurtosis are highly variable and may consequently be misleading. They should only be trusted if the sample size is very large as it often is in studies of spatial correlation.

SAS® (1990a) was used to conduct the test for normality and to compute the summary statistics discussed thus far. Assuming that *X* and *Y* denote the *X* and *Y* coordinates of a plot and that YIELD represents the wheat yield for a plot, the following statements are needed:

PROC UNIVARIATE:

VAR YIELD;

Bivariate scatter plots are useful for revealing directional trends or discontinuities. Consider some bivariate scatter plots for the wheat yields as in Figures 7.13 to 7.15. The wheat yields are plotted as a function of their X locations (Figure 7.13), as a function of their Y locations (Figure 7.14), and as a function of their $X + Y$ location (Figure 7.15). Any linear combination of directions of the form $aX + bY$ could be plotted.

When sound biological or ecological reasons exist, a researcher may remove an observation from the data set so that its presence does not unduly affect the analysis. Alternatively, determining reasons for the existence of such extreme observations may be a valuable aid in determining the ecology of wheat yields. In spatial analysis, not only should attention be given to observations that differ significantly from other values in the data set, but also to those observations that are unusual with respect to their neighbors. This implies that local stationarity is anticipated. Because there are so many data points, these graphs have limited utility for revealing directional trends in this case.

An alternative method of summarizing directional trends is plotting row sample means and medians and column sample means and medians as in Figures 7.16

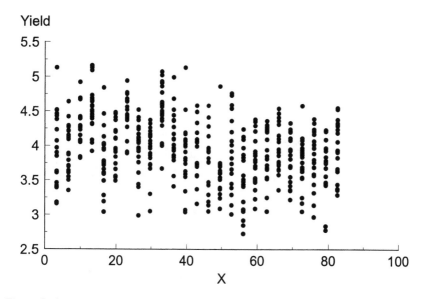

Figure 7.13. Bivariate scatter plot of wheat yields. From Mercer and Hall, (1911) against their X location.

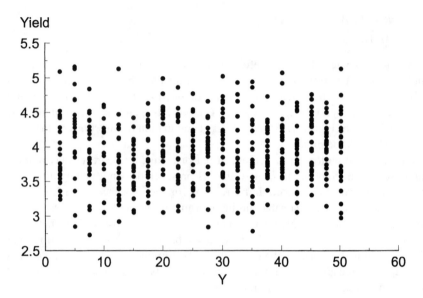

Figure 7.14. Bivariate scatter plot of wheat yields (Mercer and Hall, 1911) against their *Y* location.

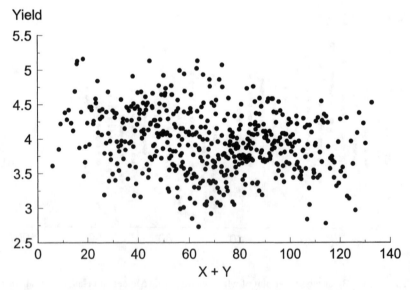

Figure 7.15. Bivariate scatter plot of wheat yields (Mercer and Hall, 1911) against their *X* + *Y* location.

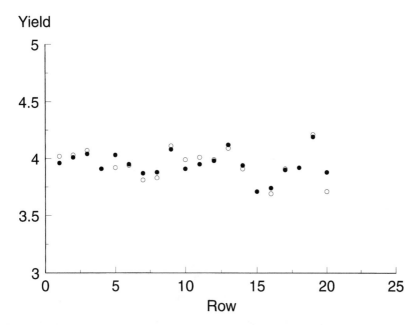

Figure 7.16. Row means (●) and medians (○) plotted against the row number. Original data from Mercer and Hall (1911).

and 7.17. There appears to be a downward trend of the column means and medians indicating a trend in the column (east–west) direction, but little or no trend in the row (north–south) direction. Information for graphs such as these can be obtained using the following SAS® code:

```
PROC SORT; BY X;
PROC UNIVARIATE; BY X;
VAR YIELD;
OUTPUT OUT = ROW   MEAN = ROWMEAN   MEDIAN = ROWMED;
PROC SORT; BY Y;
PROC UNIVARIATE; BY Y;
VAR YIELD;
OUTPUT OUT = COL   MEAN = COLMEAN   MEDIAN = COLMED;
```

Two data sets have been produced: ROW and COL. ROW has the mean and median of each row and COL those for each column.

The column and row means and medians may also be used to determine the presence of extreme observations. The median is a resistant summary statistic.

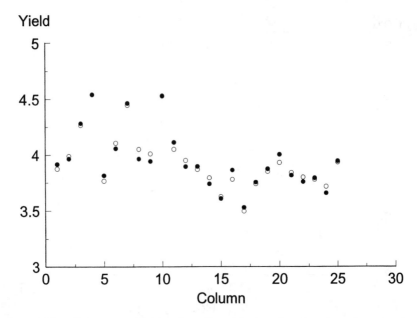

Figure 7.17. Column means (●) and medians (○) plotted against the column number. Original data from Mercer and Hall (1911).

By comparing the sample median to the sample mean, a nonresistant summary statistic, rows or columns that may have atypical observations can be identified. In general, the mean and median should be close. If they are not, then the corresponding row or column should be scanned for possible outliers.

The question then arises as to how far apart the mean and median should be before effort is expended on the search for outliers (Cressie, 1993). A test statistic for testing the null hypothesis that the row mean and row median are equal is

$$u \equiv \frac{(\bar{Z} - \tilde{Z})\sqrt{n}}{0.7555\hat{\sigma}} \tag{7.26}$$

where $\bar{Z}$ is the row mean, $\tilde{Z}$ is the row median, and $\hat{\sigma}$ is an estimate of the standard deviation. A resistant measure of the standard deviation is

$$\hat{\sigma} = \frac{\text{interquartile range}}{2(0.6745)} \tag{7.27}$$

The interquartile range is the difference in the third quartile and the first quartile of the values in the row. If the absolute value of u is around 3 or greater than 3, that row should be examined for outliers (Cressie, 1993). The same approach could be taken for ascertaining whether the column means differed significantly from the column medians.

To compute these for each row, the following SAS® code was used:

```
PROC SORT; BY X;
PROC UNIVARIATE; BY X;
VAR YIELD;
OUTPUT OUT = ROW  MEAN = MEAN  MEDIAN = MEDIAN
  QRANGE = QRANGE  N = N;
DATA WORK;
SET ROW;
SIGHAT = QRANGE/(2 * 0.6745);
U = SQRT(N) * (MEAN - MEDIAN)/(0.7555 * SIGHAT);
```

Sorting and performing PROC UNIVARIATE by Y, instead of X, results in the column values of u. Column and row values for u are given in Table 7.3. In no case is the difference in mean and median statistically significant.

It should be noted that when the covariances between observations are positive (observations closer together tend to be more alike than those farther apart). The test statistic in equation (7.26) is liberal in that it highlights more rows and columns for possible outliers than it should.

Another method of detecting outliers is based on the concept of local stationary. The idea is that a value should be similar to surrounding values. Therefore, an effort is made to detect observations that are atypical with respect to surrounding values. Figures 7.18 and 7.19 are bivariate plots of $Z(s)$ and $Z(s + he)$, where h is the constant 3.3 when e is a unit-length vector in the east–west direction (Figure

Table 7.3. Test Statistics u for the Test of the Null Hypothesis That the Mean and Median Are Equal for the Wheat Data

Row	1	2	3	4	5	6	7	8	9	10
u	−0.87	−0.52	−0.35	0.08	1.68	0.23	0.62	1.08	−0.42	−1.62
Row	11	12	13	14	15	16	17	18	19	20
u	−0.87	−0.07	0.40	0.30	−0.50	0.65	−0.15	−0.05	−0.36	2.25
Column	1	2	3	4	5	6	7	8	9	10
u	0.34	−0.35	0.41	0.02	0.95	−0.64	0.49	−1.71	−1.41	−0.04
Column	11	12	13	14	15	16	17	18	19	20
u	1.06	−0.84	0.57	−0.74	−0.42	0.78	0.34	0.23	0.34	0.96
Column	21	22	23	24	25					
u	−0.32	−0.90	0.18	−0.97	0.12					

Based on Mercer and Hall (1911).

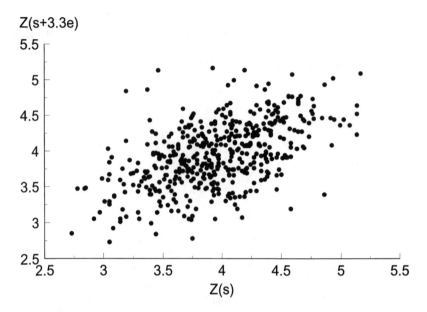

Figure 7.18.   Bivariate scatter plot of $Z(s + 3.3e)$ versus $Z(s)$, where e is in the east–west direction.

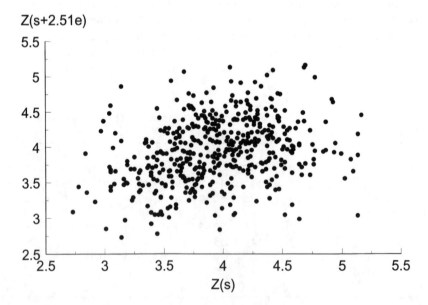

Figure 7.19.   Bivariate scatter plot of $Z(s + 2.5e)$ versus $Z(s)$ where e is in the north–south direction.

7.18) or h is the constant 2.5 when e is a unit length vector in the north–south direction (Figure 7.19). Again, no obvious outliers were detected. Other values of h can be used to construct additional graphs.

Suppose that the data set WHEAT consists of the variables I, J, and YIELD, where I is the row, J is the column, and YIELD is the wheat yield in the ith row and jth column. Then, a SAS® program that prepares the data for a plot in the east-west direction follows:

```
DATA WHEAT2;
SET WHEAT;
J = J - 1;
YIELD2 = YIELD;
DROP YIELD;
PROC SORT DATA = WHEAT; BY I J;
PROC SORT Data = WHEAT2; BY I J;
DATA XDIR;
MERGE WHEAT WHEAT2;
BY I J;
```

A plot of YIELD2 by YIELD produces the graph in Figure 7.18.

Through this exploratory data analysis, we have begun to understand the nature of the data. The normal distribution provides a good model of the data. No outliers have been detected, but a trend appears to exist in the east-west direction. We are now ready to consider the underlying assumptions of spatial models.

Intrinsic Stationarity

Suppose the yield of wheat at a location is a realization of a random process $\{Z(s): s \in D\}$ where D is the sampling region of interest. Recall from above that $Z(s)$ is the value of the random variable of interest at location s. In our example, $Z(s)$ is the wheat yield associated with a plot whose location is $s = (j(3.3), i(2.51))$, where i is the row and j the column of the plot. Z could represent the amount of nitrogen in a plot, the number of insects in a sampling unit, or another variable of interest. Spatial models generally assume that the random process Z is intrinsically stationary. Intrinsic stationarity occurs if the following two conditions are satisfied:

$$(1) \ E[Z(s + h) - Z(s)] = 0$$

$$(2) \ Var[Z(s + h) - Z(s)] = 2\gamma(h)$$

where $\mathbf{h}$ is a directional vector of length h.

The first condition implies that the mean value of Z does not change over the sampling region; that is, the region is homogeneous. Because, in Example 7.3, we have already observed a trend in the wheat yields in the east-west direction, the first condition for intrinsic stationarity has been violated. Median polishing is introduced in the next section as a means of removing the trend so that the residuals from median polishing can be used to model spatial variation.

$2\gamma(\boldsymbol{h})$ is known as the variogram, and $\gamma(\boldsymbol{h})$ is the semivariogram. The semivariogram is the crucial parameter in geostatistics. It is a quantification of the spatial correlation structure of the sample data set. We focus on modeling $\gamma(\boldsymbol{h})$ and using the resulting model to gain insight into the spatial structure present in the sampling region. Further, the process Z is said to be isotropic if $2\gamma(\boldsymbol{s} - \boldsymbol{t}) = 2\gamma(\|\boldsymbol{s} - \boldsymbol{t}\|)$ where

$$\|\boldsymbol{s} - \boldsymbol{t}\| = \left(\sum_{i=1}^{d} (s_i - t_i)^2 \right)^{1/2} \tag{7.28}$$

is the Euclidean distance (in d dimensions) between $\boldsymbol{s}$ and $\boldsymbol{t}$. Therefore, an isotropic process Z is one for which the dependence between $\boldsymbol{s}$ and $\boldsymbol{t}$ depends on the distance between the two points, but not the direction of the points from each other. A process that is not isotropic is anisotropic. After the section on median polishing, we consider methods of estimating and modeling the semivariogram.

Median Polishing

Two types of variation may be present in spatial data, large-scale and small-scale variation. These are modeled separately. Large-scale variation results when the mean is varying over the sampling region. We can represent the mean μ at the point $\boldsymbol{s}$, which is in row k and column l as the sum of the overall mean, a row effect, and a column effect as follows:

$$\mu(\boldsymbol{s}) = a + r_k + c_l \tag{7.29}$$

where a, r_k, and c_l, represent the overall mean, the kth row mean, and the lth column mean, respectively. Small scale variation at the point $\boldsymbol{s}$ may be represented by $\delta(\boldsymbol{s})$. Then the value of the process Z at $\boldsymbol{s}$ is

$$\begin{aligned} Z(\boldsymbol{s}) &= \mu(\boldsymbol{s}) + \delta(\boldsymbol{s}) \\ &= a + r_k + c_l + \delta(\boldsymbol{s}). \end{aligned} \qquad \boldsymbol{s} \in D \tag{7.30}$$

where D is the set of all points in the study region. Through median polishing, estimates of a, r_k, c_l, and δ are obtained and will be denoted by $\bar{a}$, $\tilde{r}_k$, $\tilde{c}_l$, and $\tilde{\delta}$, respectively. Estimates of a, r_k, and c_l are used to model large-scale variation. The residual δ's are the basis upon which small-scale variation is modeled. If there is no large-scale variation, $r_k = 0$ and $c_l = 0$, for all k and l. In this case, $Z(\boldsymbol{s}) = a + \delta(\boldsymbol{s})$, and the variation of $Z(\boldsymbol{s})$ and $\delta(\boldsymbol{s})$ are equal.

Tukey (1977) introduced median polishing as a means of taking the medians out. The purpose of median polishing as used here is to remove the large-scale variation so that we can model small-scale variation. This is accomplished by iteratively sweeping the medians out of the rows, then out of the columns, until the medians of all rows and columns are zero (or not significantly different from zero). Suppose the data are in a matrix of p rows and q columns. An additional row and column are added to the data matrix, and zeroes are placed initially in each entry of the $(p + 1)$th row and the $(q + 1)$th column. A median polishing algorithm given by Cressie (1993) begins by sweeping the medians from the rows using

$$Y_{kl} \equiv Y_{kl} - med(Y_{kl}: l = 1, \ldots, q), \quad k = 1, \ldots, p + 1, l = 1, \ldots, q \quad (7.31)$$

where $med(y_1, y_2, \ldots, y_n)$ is the median of $y_1, y_2, \ldots, y_n$. Thus, the first step is to find the median of the values in the first q columns (the data columns) of row k and subtract that median from each of the q values. This is done for each of the $(p + 1)$ rows. So that an estimate of the relationship in (7.30) is maintained, the median of the first q columns for each row is added to the $(q + 1)$th column in that row as expressed in the following equation:

$$Y_{k,q+1} \equiv Y_{k,q+1} + med(Y_{kl}: l = 1, \ldots, q), \quad k = 1, \ldots, p + 1. \quad (7.32)$$

After medians are swept from the rows, they are swept from the columns. The equation

$$Y_{kl} \equiv Y_{kl} - med(Y_{kl}: l = 1, \ldots, q), \quad k = 1, \ldots, p, l = 1, \ldots, q + 1 \quad (7.33)$$

computes the median of the first p rows in each column and subtracts that median from each of those p-values. This process is performed for each of the $(q + 1)$ columns. Again, so that the estimates of the values in (7.30) can be obtained, the median of each of the $(q + 1)$ columns is added to the $(p + 1)$th value in that column as given below:

$$Y_{p+1,l} \equiv Y_{p+1,l} + med(Y_{kl}: k = 1, \ldots, p), \quad l = 1, \ldots, q + 1. \quad (7.34)$$

The process of sweeping the rows and then the columns is repeated until further sweeping does not change the resulting matrix of values. This occurs when the row and column medians are all zero. A simple example is used to illustrate the process.

Example 7.4

Consider the table of values shown in Figure 7.20(a). The first three rows and three columns represent the data. The fourth row and fourth column are the aug-

1	4	7	0
2	5	8	0
3	6	9	0
0	0	0	0

-3	0	3	4
-3	0	3	5
-3	0	3	6
0	0	0	0

0	0	0	-1
0	0	0	0
0	0	0	1
-3	0	3	5

Figure 7.20. (a) Original data matrix augmented with a row and a column of zeroes. (b) Matrix after sweeping medians from rows. (c) Matrix after sweeping medians from columns.

mented row and column. Zeroes are placed in the augmented row and column. After sweeping the rows, the second table is obtained.

For example, the median of the first 3 values in row 1 (1, 4, 7) is 4. This median is subtracted from each of the first three values and added to the 4th, the $(q + 1)$th one, in that row. This process is repeated for each row. Because the last row is all zeroes, the median is zero and no change occurs in that row. The next step is to sweep out the columns.

To illustrate the process consider the first column in the table after the medians have been swept from the rows. The median of the values in the first three rows $(-3, -3, -3)$ is -3. This median is subtracted from each of the first three rows and added to the 4th, the $(p + 1)$th one. For the last column, the median of 4, 5, and 6 is 5. This value is subtracted from each of the 4, 5, and 6, and then added to the 4th row. Because each of the values in the body of the table is now zero, further sweeping of medians from rows and columns would have no effect.

The estimated row effects are in the first $p = 3$ rows of the last column, the estimated column effects are in the first $q = 3$ columns of the last row. The estimate of the overall mean is in the $(p + 1, q + 1)$ position in the table. Finally, the residuals are in the body of the table. Notice that the relationship in (7.30) is maintained. For example, in the 3rd row, 2nd column,

$$Y_{32} = \bar{a} + \bar{r}_3 + \bar{c}_2 + \bar{\delta} = 5 + 1 + 3 + 0 = 9. \qquad (7.35)$$

This example is uninteresting spatially because all the residuals are zero, indicating no small-scale variation is present.

Now that median polishing has been introduced, we will apply it to the wheat yield data to remove the large-scale variation.

Example 7.5

ECOSTAT was used to perform the median polishing for this data set. To do this, select *Geostatistics* from the *Spatial Statistics* menu. First the original data set must be entered. The process is started by pressing the *Enter Data* command button. The user is asked to identify the data set. This must be in the form of an ASCII file. The first line has the number of rows, the number of columns, and the total number of points. It is assumed that rows and columns are each numbered sequentially beginning with 1, so the total number of points should be the number of rows times the number of columns. To median polish, press the command button by that name. It is assumed that the data set already entered is the one to be median polished. If it is a new data set, the new data set must be entered first. Two output data sets are constructed and must be named by the user. One has the number of points in the data set followed by the X, Y, and residual associated with each point. The row, column, and overall effects are in the second data set.

Beginning with the initial wheat data, the rows and then the columns are swept until the medians of all rows and all columns are less than a user-specified value, 0.001 in this example. The residuals, row effects, column effects, and overall mean, rounded to the nearest 0.01, are given in Figure 7.21. Notice that we have simply decomposed the yields into these component parts. The yield in the 17th row and second column can be obtained by adding the estimated overall effect, the estimated row 17 effect, the estimated column 2 effect, and the residual for row 17, column 2; that is,

$$z_{17,2} = \bar{a} + \bar{r}_{17} + \bar{c}_2 + \tilde{\delta} = 3.88 + (-0.02)$$
$$+ 0.14 + (-0.12) = 3.88. \qquad (7.36)$$

The yield of the plot in row 17, column 2, given in Figure 7.10, is 3.87. Because all values in Figure 7.21 have been rounded to two decimal places, the yield may, as it does in this case, differ in the second decimal place from the sum of the effects given in Figure 7.10.

The estimate of trend for the plot in the *i*th row and *j*th column is $\bar{a} + \bar{r}_i + \bar{c}_j$. A three-dimensional plot of the estimated trend for the wheat yields is given in Figure 7.22. Response surface methodology is used to model this large-scale variation. However, for this example, the estimated trend for the wheat yields is too erratic for a good model of this trend. The residuals, denoted by $\tilde{\delta}_{ij}$, are shown in Figure 7.23. The small-scale variation, but not the large-scale variation, is

248 / Statistical Ecology

NORTH (columns) / **WEST** (rows)

																									Row Effects
-0.23	0.04	-0.29	0.56	-0.84	0.33	0.22	-0.11	0.02	-0.70	0.22	0.00	0.63	0.02	0.01	-0.13	-0.01	-0.36	0.34	-0.32	-0.61	-0.76	-0.98	0.39	0.54	0.10
0.18	0.06	-0.23	0.04	0.12	-0.45	-0.00	0.09	-0.12	-0.24	-0.21	0.14	0.07	0.04	-0.06	0.00	-0.17	0.02	0.29	0.29	-0.32	-0.44	-0.84	0.21	-0.08	0.13
0.60	0.12	0.00	0.07	-0.16	0.26	0.18	-0.44	-0.87	-1.08	-0.26	0.00	-0.27	-0.04	-0.04	0.23	-0.18	-0.39	0.22	0.22	0.31	-0.32	-0.41	0.08	0.34	0.15
0.16	0.64	-0.18	-0.41	-0.28	-0.13	0.10	0.48	-0.95	0.00	-0.02	-0.56	0.22	-0.60	0.23	0.87	0.51	0.02	0.48	0.18	-0.05	0.07	-0.51	-0.27	-0.24	-0.02
-0.25	0.14	0.55	0.05	-0.14	-0.07	-0.15	0.00	-0.47	0.34	-0.34	-0.41	-0.23	0.22	0.04	0.11	0.20	0.40	0.27	-0.32	-0.44	-0.28	-0.40	0.10	0.30	0.12
-0.56	-0.42	-0.13	0.30	-0.13	-0.35	0.00	-0.17	0.28	-0.20	-0.22	0.21	-0.09	0.00	-0.12	0.36	0.44	0.62	0.45	-0.32	0.37	0.15	0.43	0.05	0.44	-0.04
-0.55	-0.48	0.01	-0.05	-0.47	-0.46	0.54	0.04	0.33	0.29	-0.06	-0.15	0.51	0.00	0.21	0.93	0.53	0.64	0.01	0.15	-0.42	-0.75	-0.38	-0.93	-0.42	-0.03
-0.28	-0.60	-0.12	0.25	0.00	-0.15	0.07	-0.59	0.11	0.39	-0.06	-0.82	-0.07	0.19	0.24	1.13	0.42	0.00	-0.23	-0.22	0.14	0.07	-0.03	-0.21	0.41	-0.06
0.00	-0.62	0.21	-0.19	-0.24	-0.15	0.00	-1.27	0.14	0.20	-0.70	-0.54	-0.01	-0.05	0.01	0.83	0.56	0.23	-0.19	0.26	0.13	-0.41	0.11	-0.51	0.45	0.21
-0.38	-0.32	0.00	-0.07	0.33	0.13	0.21	-0.46	-0.05	-0.18	-0.21	0.05	-0.50	-0.35	-0.46	0.49	0.60	0.28	0.24	-0.01	0.19	0.18	0.10	-0.92	0.00	0.02
-0.42	-0.43	-0.46	-0.01	0.18	0.10	0.29	-0.15	0.10	-0.27	0.07	0.20	0.19	-0.56	-0.22	0.00	-0.12	0.04	0.47	0.55	0.32	0.19	-0.01	-0.33	-0.65	0.05
0.64	-0.34	-0.46	-0.05	-0.22	0.29	-0.01	0.00	-0.33	-0.08	0.02	-0.88	0.11	-0.69	1.30	0.64	0.01	-0.33	0.08	0.50	0.21	0.11	0.19	-0.21	-0.09	0.03
0.60	-0.00	-0.06	0.00	0.00	0.26	-0.17	-0.29	-0.32	0.01	0.78	-0.17	0.08	-0.91	-0.30	-0.25	-0.04	-0.32	0.44	0.08	0.59	0.01	0.03	0.28	0.07	0.16
0.73	0.10	0.17	0.54	0.44	-0.03	-0.58	-0.17	0.05	0.34	-0.10	-0.26	-0.48	0.13	-0.06	-0.26	0.06	-0.09	0.00	-0.50	0.30	0.57	-0.25	0.07		-0.03
-0.01	0.69	0.33	-0.10	0.47	0.19	-0.25	0.35	0.00	-0.29	0.33	0.19	0.01	-0.08	0.48	-0.01	-0.11	0.00	-0.50	0.00	0.12	-0.30	0.13	-0.03	-0.22	-0.29
1.53	0.04	0.17	0.01	0.16	-0.35	0.00	0.23	0.27	0.02	-0.48	-0.21	-0.53	-0.17	-0.08	-0.14	-0.40	-0.38	-0.42	0.08	-0.49	0.02	0.11	0.21	-0.10	-0.16
0.49	-0.12	0.13	0.00	-0.57	-0.54	-0.50	0.38	0.21	0.02	0.24	0.16	-0.08	0.13	0.01	-0.22	-0.40	-0.07	-0.29	-0.01	0.02	0.06	0.38	-0.04	0.19	-0.02
0.68	0.16	0.20	-0.49	1.12	-0.06	0.01	0.25	0.00	-0.26	0.53	0.33	0.70	0.08	0.12	-0.49	-0.69	-0.62	-0.12	-0.11	-0.34	-0.02	-0.10	0.15	-0.16	-0.06
-0.21	-0.04	0.14	0.39	0.38	-0.05	0.02	-0.17	0.01	0.33	0.91	0.04	0.31	-0.15	-0.30	-0.98	-0.92	-0.71	-0.28	-0.09	-0.24	-0.22	0.18	0.17	0.00	0.30
0.00	0.35	0.32	0.77	0.02	0.31	-0.22	0.06	0.02	0.00	0.54	0.74	-0.01	0.62	-0.15	-0.25	-0.13	-0.01	-0.43	-0.02	-0.15	-0.11	0.16	-0.39		-0.15

Column Effects:

-0.12	0.14	0.38	0.60	-0.09	0.18	0.56	0.18	0.14	0.73	0.18	0.05	-0.02	-0.07	-0.34	-0.19	-0.40	-0.10	-0.04	0.06	0.00	-0.04	-0.02	-0.13	0.02	3.88

Figure 7.21. Residuals, column effects, row effects, and overall total from median polishing wheat yield data (Mercer and Hall 1911).

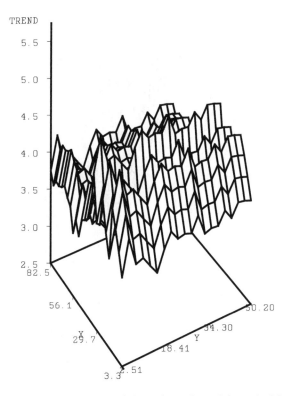

Figure 7.22. Three-dimensional plot of the estimated trend from the Mercer and Hall (1911) wheat yield data.

exhibited in the residuals. Therefore, it is the residuals from median polishing that are used in the analysis of spatial correlation. However, before beginning, another EDA should be conducted on the residuals. This is not presented in full here. For the residuals, the plot of column means and medians is shown in Figure 7.24. As can be seen by the plots, the trend in the east-west direction is not present in the residuals. The residuals now satisfy the first condition of intrinsic stationarity. The next step is to quantify the spatial correlation and how it changes as the distance between plots increases.

As in Figures 7.18 and 7.19, we construct bivariate scatter plots. For $h = 3.3$ and direction east, the scatter plot is shown in Figure 7.25 with the reference line $x = y$, a 45-degree line passing through the origin. We also developed scatter plots for values of $h = 6.6$, 9.9, and 13.2. These give us an opportunity to view the relationship between wheat yields for neighboring plots as well as those two, three, and four plots apart. If the yields from plots h units apart are similar, the points are close to the reference line. When the yields become less similar, the

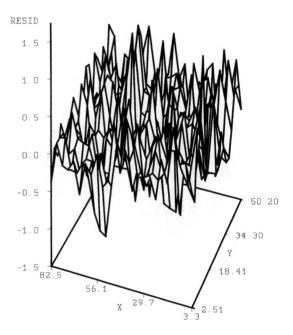

Figure 7.23. Three-dimensional plot of the residuals from the median polishing the Mercer and Hall (1911) wheat yield data.

cloud of points become broader and more diffuse. This does occur as may be seen by comparing Figures 7.25 to 7.28. It is not a surprise, as we would expect yields from plots close together to be more alike than yields from plots further apart. However, for this example, it is difficult to determine how much greater the spread in the points is just by looking at the plots. Next, we attempt to quantify this change in spatial correlation as the distance between plots increases.

The Semivariogram

Several approaches have been taken to quantifying the change in spatial correlation as a function of distance. The correlation coefficient of an h-scatterplot as a function of $h\mathbf{e}$, $\rho(\mathbf{h})$ is called the correlation function or correlogram. The correlogram is used extensively in time series analysis. Again using the points from the plots in 7.25 to 7.28, we can compute the correlation in the points $Z(s)$ and $Z(s + h\mathbf{e})$, for $h = 3.3, 6.6, 9.9,$ and 13.2 and $\mathbf{e} = (1, 0)$ which is the easterly direction. Recall that the reason h is increasing by multiples of 3.3 is that the length of each plot in the east–west direction is 3.3 m. Thus we are considering the correlation between adjacent plots and those plots 2, 3, and 4 apart. The correlations were 0.345, 0.207, 0.149, and 0.117 for $h = 3.3, 6.6, 9.9,$ and 13.2,

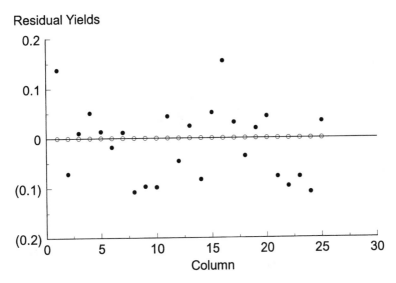

Figure 7.24. Column means (●) and medians (○) plotted against the column number for the residuals from the Mercer and Hall (1911) data after median polishing.

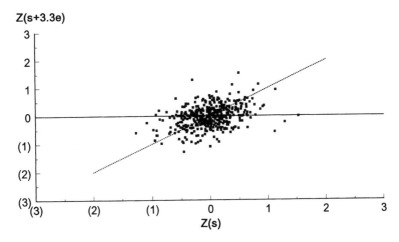

Figure 7.25. Scatter plot of $Z(s + 3.3e)$ where e is the unit vector in the east direction against $Z(s)$. The 45-degree line is shown for reference.

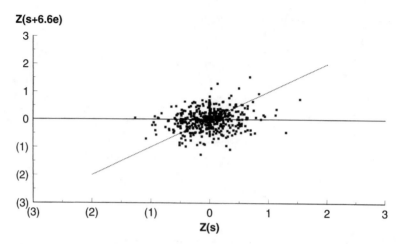

Figure 7.26. Scatter plot of $Z(s + 6.6e)$, where *e* is the unit vector in the east direction against $Z(s)$. The 45-degree line is shown for reference.

respectively. These are plotted in Figure 7.29. As expected, the correlation decreases as the distance between plots increases. Eventually, the plots are far enough apart, say *a* meters, so that the yield from one plot is independent from the yields of plots that are *a* or more meters from it. At that point, $h = a$, the correlation is zero, and remains zero for all $h > a$.

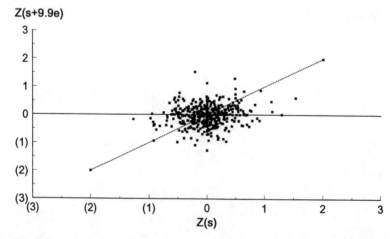

Figure 7.27. Scatter plot of $Z(s + 9.9e)$, where *e* is the unit vector in the east direction against $Z(e)$. The 45-degree line is shown for reference.

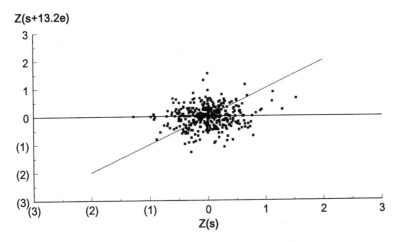

Figure 7.28. Scatter plot of $Z(s + 13.2e)$, where e is the unit vector in the east direction against $Z(e)$. The 45-degree line is shown for reference.

Instead of the correlation, we could consider the covariance as a function of he, producing the covariogram $C(h)$. Because the correlation of $Z(s)$ and $Z(s + he)$ is their covariance divided by their standard deviations, the appearance of the covariogram is similar to that of the correlogram. For the wheat yields, we computed the covariance in $Z(s)$ and $Z(s + he)$ for $h = 3.3$, 6.6, 9.9, and 13.2 and $e = (1, 0)$, which is the easterly direction. This gives us the covariance between adjacent plots and those 2, 3, and 4 plots apart in the east-west direction.

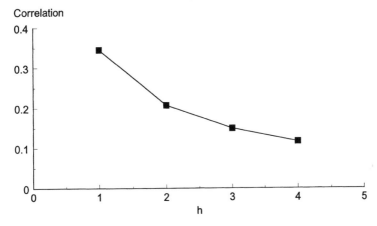

Figure 7.29. Plot of correlation for increasing values of h.

The covariances were 0.047, 0.029, 0.020, and 0.016 for $h = 3.3, 6.6, 9.9$, and 13.2, respectively. These are plotted in Figure 7.30. As we expected, the shape of the covariogram is the same as that of the correlogram, but the scale differs.

Another possible index for the spread of the points is the moment of inertia about the line $x = y$, which is calculated as follows:

$$\text{Moment of inertia} = \frac{1}{2n} \sum_{i=1}^{n} (x_i - y_i)^2. \qquad (7.37)$$

The moment of inertia is one-half the average squared difference between the x and y coordinates (between the yields of the two plots) of each pair of points (plots) on the h-scatterplot. The factor of 1/2 is present because we are interested in the perpendicular distances of the points from the line $x = y$. For general plots, the moment of inertia has little or no meaning as the distance of the points from the 45-degree line is not of interest. However, for the h-scatterplot, the increasing distance from this line indicates the increasing dissimilarity of the plot yields as the distance between plots increases. As shown in Figure 7.31, the moment of inertia increases with h, instead of decreasing as we observed for the correlogram and covariogram. The semivariogram reflects the relationship between the moment of inertia, also called the *semivariance,* and h.

To understand how the semivariogram, correlogram and covariogram are related, first recall that the semivariogram $\gamma(h)$ is the primary parameter in geostatistics and is defined by

$$2\gamma(h) = Var[Z(s + h) - Z(s)] \qquad (7.38)$$

where h is a directional vector of length h. Consider the relationship,

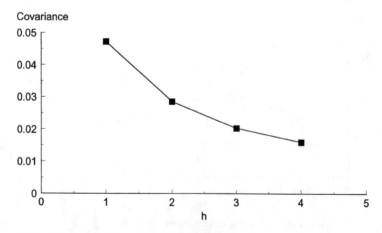

Figure 7.30. Plot of covariance for increasing values of h.

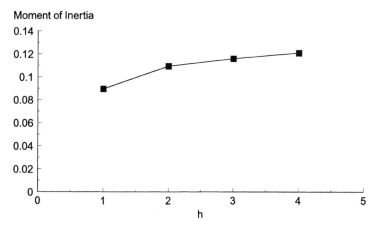

Figure 7.31. Plot of moment of inertia for increasing values of h.

$$Var[Z(s_1)] - Z[s_2] = Var[Z(s_1)] + Var[Z(s_2)] - 2Cov[Z(s_1), Z(s_2)]. \quad (7.39)$$

If $Z(\cdot)$ is intrinsically stationary, $Var[Z(s_1)] = Var[Z(s_2)] = C(\mathbf{O})$ and $Cov[Z(s_1), Z(s_2)] = C(s_1 - s_2)$, giving

$$Var[Z(s_1) - Z(s_2)] = 2[C(\mathbf{0}) - C(\mathbf{s}_1 - s_2)]. \quad (7.40)$$

Therefore,

$$2\gamma(\mathbf{h}) = 2[C(\mathbf{0}) - C(\mathbf{h})] \quad (7.41)$$

and so

$$\gamma(\mathbf{h}) = C(\mathbf{0}) - C(\mathbf{h}) \quad (7.42)$$

where $\gamma(\mathbf{h})$ is the semivariogram. In some work, the terms variogram and semivariogram are used interchangeably. Because they differ by a factor of 2, we will endeavor to keep the distinction clear.

The semivariogram is used to provide insight into the nature of the spatial variation present. Recall that $C(\mathbf{0})$ is a measure of the variability of $Z(s)$ and that $C(\mathbf{h})$ is a measure of the covariance between observations $\mathbf{h}$ units apart. When there is positive correlation between plots $\mathbf{h}$ units apart, $C(\mathbf{h})$ is positive, reducing the value of the semivariogram. Because units close together are expected to have a greater correlation than that of units farther apart, we anticipate the semivariogram will be smaller for units closer together, increasing as the units become farther apart. At some point, the units will be far enough apart to be independent.

At that point, $C(h)$ will be zero, and the semivariogram will become level. Features of the semivariogram that have particular meaning are the nugget, sill, and range (*see* Figure 7.32).

Initially, as the distance between pairs increases, the semivariogram usually increases. Eventually, the semivariogram may reach a plateau for which increasing distances between pairs no longer result in increasing values of the semivariogram. The plateau is called the *sill,* and the distance between pairs at which the semivariogram reaches the plateau is called the *range.* Pairs of points closer than the range have some correlation; pairs further apart than the range are independent.

The value of the semivariogram must be zero for $h = 0$. However, as $h \to 0$, we may have $\gamma(h) \to c_0 > 0$. Then c_0 is called the nugget effect. In this case, $Z(\cdot)$ is not continuous and is highly irregular (Cressie, 1993). The nugget effect has two possible origins. First, suppose that more than one measurement is taken at a location and that the results differ; the measurements fluctuate about the true value of Z at that point. This is measurement error and is reflected in the estimate of the semivariogram. Further, there can be microscale variation; that is, variability among pairs of points that are closer than any for which measurements were made. If this variation is continuous, then it should be 0 at $h = 0$. Unless some pairs are very close, it is not known whether the microscale variation is continuous or not. Thus, in practice, it is not easy to determine c_0 from data whose spatial separations are large.

Semivariograms must be conditionally negative-definite; that is,

$$\sum_{i=1}^{m} \sum_{i=1}^{m} a_i a_j \gamma(s_i - s_j) \leq 0,$$

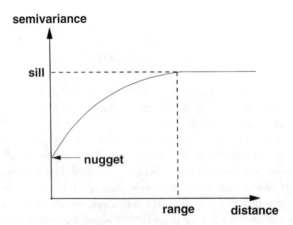

Figure 7.32. Nugget, sill, and range of a semivariogram.

for any finite number of spatial locations $\{s_i: i = 1, 2, \ldots, m\}$ and real numbers $\{a_i: i = 1, 2, \ldots, m\}$ satisfying

$$\sum_{i=1}^{m} a_i = 0.$$

This condition ensures that estimates of variance are non-negative.

Several conditionally negative-definite semivariogram models have been proposed. The four we consider are the linear, spherical, exponential, and Gaussian (*see* Figure 7.33). The linear model has semivariogram

$$
\begin{aligned}
\gamma(\boldsymbol{h};\boldsymbol{\theta}) &= 0, & \boldsymbol{h} &= \boldsymbol{0}, \\
&= c_0 + b_l\|\boldsymbol{h}\|, & \boldsymbol{h} &\neq \boldsymbol{0}, & (7.43)
\end{aligned}
$$

where $\boldsymbol{\theta} = (c_0, b_1)'$. The nugget effect is c_0, and the slope is b_1. This model has no sill or range, indicating the presence of spatial dependence for all the distances estimated by the semivariogram. Because we expect this spatial dependence to eventually disappear, a semivariogram described by this model indicates that we have not observed the full semivariogram or that large scale trends are present. By computing the semivariogram for larger values of $\boldsymbol{h}$, we expect to observe both the sill and the range.

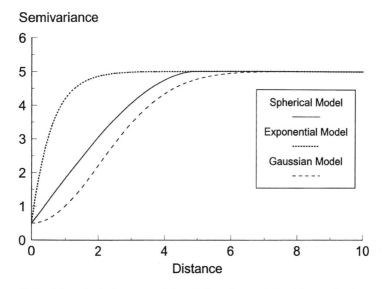

Figure 7.33. The spherical, exponential, and Gaussian models of the semivariogram.

Perhaps the most commonly used semivariogram model is the spherical model:

$$\gamma(h;\theta) = 0, \qquad h = 0,$$
$$= c_0 + c_s[1.5(\|h\|/a_s) - 0.5(\|h\|/a_s)^3], \qquad 0 < \|h\| \le a_s,$$
$$= c_0 + c_s, \qquad \|h\| > a_s, \qquad (7.44)$$

where $\theta = (c_0, c_s, a_s)'$. For this model, the nugget effect is $c_0 \ge 0$, the range is $a_s \ge 0$, and the sill is $c_0 + c_s$ where $c_s \ge 0$. In fitting this model to a sample semivariogram, it is often useful to know that the tangent at the origin reaches the sill at about two-thirds of the range.

The exponential model has the form

$$\gamma(h;\theta) = 0, \qquad h = 0,$$
$$= c_0 + c_e[1 - \exp(-\|h\|/a_s)], \qquad h \ne 0, \qquad (7.45)$$

where $\theta = (c_0, c_e, a_e)'$. Here the nugget effect is $c_0 \ge 0$ and the sill is $c_0 + c_e$ where $c_e \ne 0$. The model approaches its sill asymptotically. Therefore, the practical range is defined to be that distance at which the semivariogram value is 95% of the sill.

The Gaussian model has the form

$$\gamma(h;\theta) = 0, \qquad h = 0$$
$$= c_0 + c_g[1 - \exp(-3\|h\|^2/a_g^2), \qquad h \ne 0, \qquad (7.46)$$

where $\theta = (c_0, c_g, a_g)'$. Similar to the exponential model, the nugget effect is $c_0 \ge 0$ and the sill is $c_0 + c_g$, where $c_g \ne 0$. This model also approaches its sill asymptotically, and the practical range is again defined to be that distance at which the semivariogram value is 95% of the sill.

Because we cannot observe $Z(\cdot)$ for all points in the region but must rely on a sample, we cannot compute the semivariogram exactly. In fact, we do not have points for all possible values of h that can be used in its estimation. However, we can obtain good estimates of the semivariogram at some values of h and then model the full semivariogram based on these points. To do this, we specify the lag distance, that is, the distance between points for the estimated semivariogram. If the data are taken on a somewhat regular grid, the grid spacing usually serves as a good lag distance. If the data are random, the average distance between neighbors tends to work well (Journel and Huijbregts 1978, p. 194). Next, we should consider the lag tolerance; that is, how close to h must the distance of two points be in order to be used in the estimation process. Often $h/2$ is used because that allows all pairs of points to be used in the estimation of the semivariogram. Smaller values can be used. This excludes some points from the estimation process, but it may make the structure of the semivariogram clearer.

In ECOSTAT, a lag tolerance of $h/2$ is assumed. The user specifies the lag distance h. All points separated no more than h units are used to estimate the first point on the semivariogram. Points at least h units apart, but no more than $2h$ units apart are used to estimate the second point on the semivariogram. The kth point on the semivariogram is estimated using all points at least $(k - 1)h$ units apart, but no more than kh units apart. Because points that are various distances apart are used to estimate the semivariogram, it must be decided what distance h is most representative for that point. Traditionally, the midpoint of the interval or the average distance has been used. We report both so that the choice is left to the user. The number of lags to be estimated must also be given. Journel and Huigbregts (1978, p. 194) suggest that the fit should be only up to one-half the maximum possible lag.

The classical estimator of the semivariogram proposed by Matheron (1962) is

$$\hat{\gamma}(h) \equiv \frac{1}{2|N(h)|} \sum_{i=1}^{N(h)} [Z(s_i) - Z(s_i + h)]^2 \tag{7.47}$$

where $\hat{\gamma}(h)$ is the estimated semivariogram for lag h and $N(h)$ is the number of points separated by h. This estimator is based on the method-of-moments and is referred to as the classical estimator. Because the summation is over a squared term, this estimator is sensitive to atypical observations. By transforming the problem to estimation of location for an approximately symmetric distribution, Cressie and Hawkins (1980) developed a robust estimator of the semivariogram:

$$\bar{\gamma}(h) = \frac{\left(\dfrac{1}{|N(h)|} \sum_{N(h)} |Z(s_i) - Z(s_j)|^{1/2} \right)^4}{2 \left(0.457 + \dfrac{0.494}{|N(h)|} \right)}. \tag{7.48}$$

This estimator is referred to as the robust estimator.

Example 7.6

For the wheat data, we are now ready to estimate the semivariogram for the residuals from median polishing. To do so, we must have an input data set. The first line gives the number n of points in the data set. The following n lines have x, y, and z (the location (x, y), and the observed z at that location) with at least one space between each but no commas or other notations separating these three values. Assuming that we have entered the data after having chosen *Geostatistics* from the *Spatial Statistics* menu, we press the *Semivariogram* command button. The user is asked for the input data set. For this example, we use the one from median polishing. Further, the output data set, the distance between lags, and the number of lags must be specified. Because the distance between plots in the east-west direction is 3.3 m and in the north–south direction is 2.51 m, we want to choose an h of at least 3.3. We chose a distance of 4 between lags. The distance from the southwest corner plot to the northeast corner plot is about 97 m. We chose 14 lags, giving us a distance between pairs up to 56 m (about 58% of the largest distance, 97 m). If the estimated semivariogram does not exhibit a sill, then we will need to add more lags. Based on these input values, the results in Table 7.4 are displayed on the screen and in the output data set.

It should be emphasized here that the classical and robust estimates displayed here are estimates of the *semivariogram,* not the variogram. Notice that with the exception of the first lag, the midpoint and average distance are close. Also, the robust estimate is consistently lower than the classical one. This is not always the case. It is recommended that at least 30 pairs be used to estimate the semivariogram at each distance. This is not a problem in this example as at least 955 pairs were used for each h. The fact that we have ignored direction and only considered the distance separating pairs implies that we are assuming that the spatial process is isotropic. We later consider whether this appears to be a reasonable assumption. A plot of both the classical and robust estimates are shown in Figure 7.34.

The next step is to fit a model to the semivariogram. Several approaches have been used to estimate the parameters of the semivariogram model: (1) by eye or hand, (2) least squares, (3) weighted least squares using the number of pairs as the weight, (4) iteratively re-weighted least squares using the weight

Table 7.4. Classical and Robust Estimates of the Semivariogram for Mercer and Hall's (1911) Wheat Data

Lag	$N(h)$	Midpoint of lag interval	Average distance	Classical estimate	Robust estimate
1	955	2	2.91	0.0984	0.0894
2	3,985	6	5.93	0.1195	0.1098
3	5,662	10	9.75	0.1290	0.1188
4	8,343	14	13.89	0.1351	0.1262
5	8,411	18	17.92	0.1427	0.1290
6	10,331	22	21.85	0.1457	0.1343
7	10,611	26	26.02	0.1516	0.1376
8	9,998	30	29.99	0.1420	0.1317
9	9,947	34	33.89	0.1368	0.1259
10	10.263	38	37.95	0.1355	0.1267
11	8,625	42	41.95	0.1400	0.1284
12	7,899	46	45.91	0.1412	0.1308
13	6,446	50	49.90	0.1369	0.1236
14	5,104	54	53.76	0.1248	0.1172

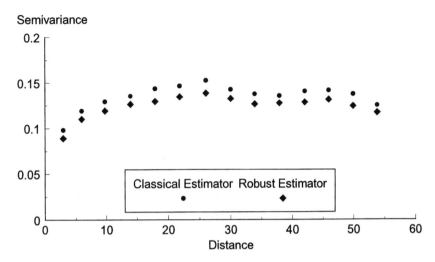

Figure 7.34. Classical and robust estimates of the semivariograms.

$$w(\boldsymbol{h};\boldsymbol{\theta}) = \frac{N(\boldsymbol{h})}{(\gamma(\boldsymbol{h};\boldsymbol{\theta}))^2},$$ (7.49)

or (5) maximum likelihood. It has been shown that when these models are fit using weighted least squares with weight given in equation (7.49), the results are never bad and are often quite good (Cressie, 1993). Gotway (1991) illustrated how to use SAS®'s (1990b) NLIN (NonLINear regression) procedure to estimate the parameters of the model using this approach. First, the data set containing semivariogram estimates must be read into SAS®. The output data set from ECO-STAT may be read in directly. Assume that the number of points, midpoint of the lag interval, average distance between points in the lag interval, classical estimate of the semivariogram, and robust estimate of the semivariogram are given SAS® variable names N, MIDPT, H, CSEMI, and RSEMI, respectively. The statements given below fit the spherical model based on the classical estimator of the semivariogram:

PROC NLIN MAXITER = 500   METHOD = DUD;
PARMS C0 = .05
       CS = .07
       AS = 10;
IF H < AS THEN DO;
*MODEL CSEMI = C0 + CS * (1.5 * H/AS − 0.5 * (H/AS)**3);*

(continued)

(continued)

```
DER.C0 = 1;
DER.CS = 1.5 * H/AS − 0.5 * (H/AS)**3;
DER.AS = − 1.5 * CS * H/(AS * AS) + 1.5 * CS * (H/AS)**3/AS;
_WEIGHT_ = N/(C0 + CS * (1.5 * H/AS − 0.5 * (H/AS)**3))**2;
END;
ELSE DO;
MODEL CSEMI = C0 + CS;
DER.C0 = 1;
DER.CS = 1;
DER.AS = 0;
_WEIGHT_ = N/(C0 + CS)**2;
END;
```

To estimate the parameters for the robust model, RSEMI is used instead of CSEMI in the statements above. In the PARMS statement, initial values are given for C0, CS, and AS. These can be chosen based on inspection of Figure 7.33 and the estimated values. It appears that the nugget effect C0 is about 0.05. The values tend to level out at about 0.12, giving the values for $CS = 0.12 − C0$. Finally, the range (distance at which the semivariogram becomes level) appears to be around 10. The statements beginning with DER give the derivative of the model with respect to the parameter following the period. These are not needed if METHOD = DUD is specified as it is here. However, if there is not convergence, other methods may be tried that do require use of these derivatives. To check for convergence, look for the successive reduction in the weighted residual sum of squares with each iteration. The message "FAILED TO CONVERGE" means just that, convergence was not reached. However, "CONVERGENCE AS-SUMED" usually also means that convergence was not achieved. If there is a convergence problem, another METHOD = option could be chosen. DUD is used here because it usually works well when others fail. Specifying initial values for the parameters closer to the true parameter values may address convergence problems so these values should also be rechecked. If the values of the empirical semivariogram are highly variable, convergence problems can occur. One way to address these is to look at a graph of the values. If the values are highly variable, the problem is usually with the values at large lags. Dropping one or two of these may also lead to convergence.

To fit the exponential model, the following code could be used:

```
PROC NLIN MAXITER = 500  METHOD = DUD;
PARMS C0 = .05
      CE = .07
```

```
            AE = 10;
MODEL CSEMI = C0 + CE * (1 - EXP(-H/AE));
DER.C0 = 1;
DER.CE = 1 - EXP(-H/AE);
DER.AE = -CE * H * EXP(-H/AE)/(AE * AE);
_WEIGHT_ = N/(C0 + CE * (1 - EXP(-H/AE))))**2;
OUTPUT OUT = EXPO P = PRED R = RESID;
```

The statements above serve the same purpose as those for the spherical model, but the last line is new. (A similar statement could have been used for the exponential model.) The Output statement is requesting SAS® to construct a new data set based on the results of the nonlinear regression. Variables in the original data set are also included in this new data set named EXPO. In addition, two new variables are included. PRED is the SAS® name that we have requested for the predicted values of the semivariogram based on the estimated model. RESID represents the residual from the estimated model.

The following code fits the Gaussian model:

```
PROC NLIN MAXITER = 500  METHOD = DUD;
PARMS C0 = .05
      CG = .07
      AG = 10;
MODEL CSEMI = C0 + CG * (1 - EXP(-H * H/(AG * AG)));
DER.C0 = 1;
DER.CE = 1 - EXP(-H * H/(AG * AG));
DER.AE = -2 * CG * H * H * EXP(-H * H)/(AG * AG))/AG**3;
_WEIGHT_ = N/(C0 + CG * (1 - EXP(-H * H/(AG * AG))))**2;
```

If least squares is used, PROC REG can be used to fit the linear model. Programming can also be used to fit the iteratively weighted least squares as in the models above. However, NLIN can also be used for this purpose using the code below:

```
PROC NLIN MAXITER = 500  METHOD = DUD;
PARMS C0 = 0.05
      B = .1;
MODEL CSEMI = C0 + B * H;
DER.C0 = 1;
DER.B = H;
_WEIGHT_ = N/(C0 + B * H)**2;
```

The models were fit to the wheat yield data as illustrated in the example below.

Example 7.7

The semivariogram output from ECOSTAT is first read into SAS®. After looking at the empirical semivariogram in Figure 7.34, we do not expect the linear model to fit well for either the classical or robust estimators so this model was not estimated.

The fit of the spherical, exponential, and Gaussian models to the classical and robust estimates of the semivariogram is shown in Figures 7.35 to 7.40. The estimates of c_0, c_s, and a for the models are given in Table 7.5.

The exponential model tends to provide the better fit. Therefore, based on the robust estimator, we estimate the nugget effect to be 0.033 bushels squared. The sill is estimated to be $0.033 + 0.096 = 0.129$. The range is estimated to be that distance for which the semivariance is 95% of the sill. The range can be found by solving equation (7.45) for h or by iterative evaluation of that equation. Using either approach, the estimate of the range is found to be 10.4 m. Therefore, we estimate that there is some spatial correlation between observations less than 10.4 m apart. If plots are more than 10.4 m apart, their wheat yields are estimated to be independent with variability of about 0.129 bushels squared. Here, the classical estimates were used because they provide more conservative estimates.

Recall that the process Z is said to be isotropic if the dependence between $Z(s)$ and $Z(s + h)$ is a function of the magnitude, but not the direction of h. If this dependency is a function of the direction as well as the magnitude of h, the process Z is anisotropic. Anisotropies result when the underlying physical process evolves

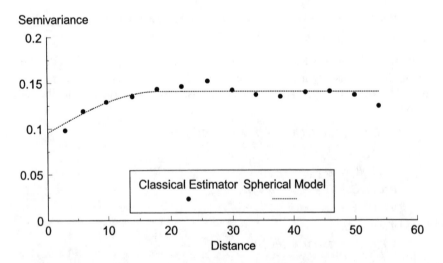

Figure 7.35. The spherical model fitted to the classical estimates of the semivariogram.

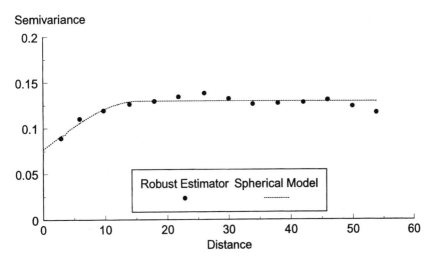

Figure 7.36.   The spherical model fitted to the robust estimates of the semivariogram.

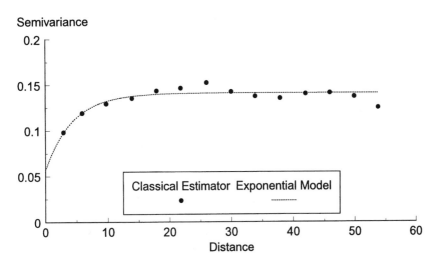

Figure 7.37.   The exponential model fitted to the classical estimates of the semivariogram.

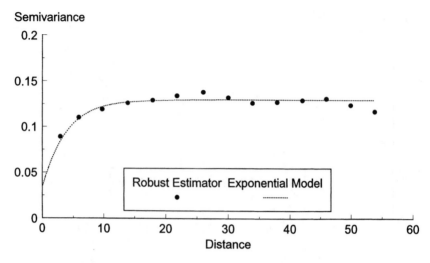

Figure 7.38. The exponential model fitted to the robust estimates of the semivariogram.

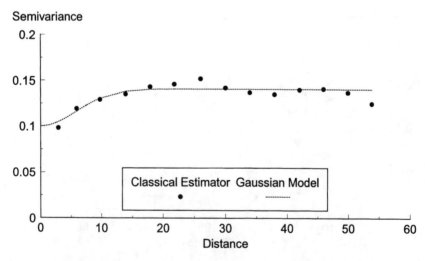

Figure 7.39. The Gaussian model fitted to the classical estimates of the semivariogram.

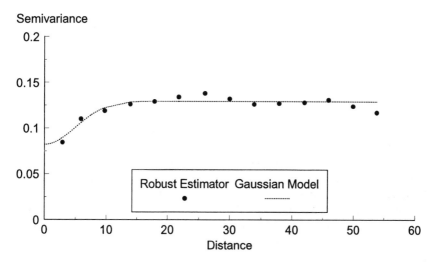

Figure 7.40. The Gaussian model fitted to the robust estimates of the semivariogram.

differentially in space. As an example, in agriculture, the dependency down a row of a crop may differ from that across rows of the crop, leading to anisotropy.

In ECOSTAT, we have provided a means to estimate the semivariogram across a row or down a column. We continue to assume that observations have been taken on a rectangular grid. The semivariogram is estimated in the x-direction and then in the y-direction. If process Z is isotropic, the semivariograms in the two directions are equivalent, and their estimates should therefore be similar. The precision of the estimated semivariogram for an isotropic process could be improved by estimating an omnidirectional semivariogram.

Example 7.8

Again look at Mercer and Hall's (1911) wheat data. Working with the residuals, we can press *Directional* on the *Geostatistics* submenu of ECOSTAT. The empirical

Table 7.5. Estimated Parameters for the Spherical, Exponential, and Gaussian Semivariogram Models

Model	Classical estimates of semivariogram			Robust estimates of semivariogram		
	$\hat{c}_0$	$\hat{c}_s$	$\hat{a}$	$\hat{c}_0$	$\hat{c}_s$	$\hat{a}$
Spherical	0.095	0.045	18.087	0.077	0.053	15.336
Exponential	0.057	0.084	4.329	0.033	0.096	3.870
Gaussian	0.100	0.041	14.451	0.082	0.047	12.247

semivariagrams down rows and down columns are computed and put in an output data set named by the user. In Figure 7.41, the robust estimates as well as the original exponential model fit when ignoring direction are shown. For small distances the semivariance computed in the east-west direction (down the rows) appears to be less than that in the north-south direction (down columns). If this difference is viewed to be important, then a model could be fit for each direction.

To this point, we have not needed to make any distributional assumptions; that is, the methods are valid whether the data are normally distributed or not. The next step may be to estimate the yield at points for which no data were taken and then to produce a map of the field based on these estimates. The process of estimating the yield at points for which no data were taken is known as *kriging*. Confidence intervals may be set on these estimated values, and this requires the assumption of normality. When working with presence–absence data, the kriged values can be interpreted as the probability of presence at that point. The interested reader is encouraged to explore these and other methods in the rapidly developing area of geostatistics.

Summary

The traditional measures of aggregation capture the relationship between mean and variance for the population distribution. However, they provide no insight into the possible spatial correlation among observed values that could be present

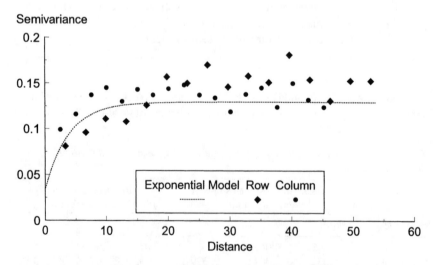

Figure 7.41. Estimated semivariograms down rows and columns with the exponential model fitted to the empirical semivariogram ignoring direction.

in the study region. To draw inference about spatial correlation, we must consider not only the value of the variable of interest, but also its location relative to other observed values.

Moran's *I* and Geary's *c* were early measures of this spatial correlation. Tests help determine whether the distribution is random, regular, or aggregated. Unfortunately, these measures, as originally introduced, consider only information from units that share a common boundary. If there is spatial correlation, the extent is unknown. Both *I* and *c* have been extended so that information from more distant neighbors can be used to gain an insight into the extent of the spatial correlation. However, we have chosen a more general approach.

General approaches to spatial correlation have been to extend ideas from time series to two or more dimensions and to adopt geostatistical methods. Although the two are directly related, we have chosen to consider geostatistics here. First, an exploratory data analysis is conducted. If there is a trend across the study region, it is removed through median polishing. Then the spatial correlation is investigated through estimation of the semivariance. The semivariogram may be modeled. The linear, spherical, exponential, and Gaussian models are four models often used for this purpose. Directional effects may also be explored. The semivariogram permits assessment of the strength and extent of the spatial correlation.

Exercises

1. Beall (1939) reports the results of a study to assess methods of making efficient estimates of insect populations. The study was conducted in a potato field planted near Chatham, Ontario. Plants were spaced a little more than a foot apart. Fifty-eight rows were planted, and the field was about as long as it was wide. However, only 96 feet of 48 rows (124 feet across) were included in the study. Strings were run perpendicular to the rows at two-foot intervals. The observational unit was the plants in a 2-foot section of row between two strings. For each unit, the number of adult Colorado potato beetles (*Leptinotarsa decemlineata* Say) was recorded. The data are given following the exercises and in the ECOSTAT file potato.dat.

    A. Using Moran's *I* and Geary's *c*, assess the presence of spatial correlation.

    B. Use geostatistics to assess whether spatial correlation is present and to model that correlation if it is.

2. Johnson (personal communication) assessed the velvetleaf (*Abutilon theophrasti* Neducys) population in corn and soybean fields in southeastern Nebraska. Observations were collected in June. The data for two soybean fields (Velvet1.dat and Velvet2.dat) and two corn fields (Velvet3.dat and Velvet4.dat) are included in the ECOSTAT data files. On each line, the first observation is the *X*-coordinate, the second is the *Y*-coordinate, and the last is

the velvetleaf count in one row-foot. Complete the following for each data set.

A. First plot the points in space. Notice that the field was irregular. Many programs for geostatistics can accept such data. However, to use ECO-STAT, you must first eliminate points until you have the same number of points in each row and the same number of points in each column.

B. Using *Moran's I* and *Geary's c*, assess the presence of spatial correlation.

C. Use geostatistics to assess whether spatial correlation is present and to model that correlation if it is.

Beall's (1939) Colorado potato beetle data

Direction of Rows

8

Spatial Point Patterns

Introduction

In the previous chapters, we have dealt primarily with organisms that have a discrete microhabitat that serves as a natural choice of sampling unit. Data may consist of the number of insects on a plant, the number of eggs in a nest, or the number of leaves on a plant. In other cases, the organisms were distributed in a continuum and a contrived sampling unit was employed. Examples include the numbers of plants in a quadrat in a prairie grass region, the number of birds in a transect, or the number of fish in a seine. These methods often work in a continuum if we are interested in estimating the population density of the organisms. However, sometimes the focus of attention is the pattern of the organisms within the continuum. Locations of nests in a breeding colony, of trees in a forest, and of weeds in a prairie are examples of biological patterns. A data set of this type is called a spatial point pattern, and the locations are referred to as events. Spatial patterns may exhibit complete spatial randomness (csr), aggregation, or regularity.

Appropriate use of geostatistics has sometimes been confused with the use of spatial point analysis. Each is useful, but in different circumstances. Geostatistics is the appropriate statistical approach when each sampling unit in the population has associated with it a response variable Z, and the interest of the scientist is in describing the spatial correlation associated with Z. In practice, the sampling unit is often a quadrat or a plant so that the number of insects on a plant, the amount of nitrogen in a core of soil, or the yield of wheat in a quadrat are examples of a response variable Z. The biological questions to be addressed concern whether observations close together are more, or less, alike than those further apart and the strength and extent of these correlations. Mathematically, the data can be thought of as a realization of a random process

$$\{Z(s): s \in D\}$$

where D is a *fixed* subset in R^d with positive d-dimensional volume. R^d is the d-dimensional set of real numbers.

Alternatively, spatial point analysis is the appropriate statistical approach when the emphasis is on the location of organisms within a sampling region and the process by which these organisms were dispersed within the region. Sampling considerations usually exclude mobile organisms from consideration in this case. The most extensive applications have tended to be in forestry, where the processes resulting in the observed spatial distribution of trees have been of interest. The biological questions to be addressed center about whether the pattern of these trees are aggregated, random, or regular. If the pattern is not random, further questions may deal with the underlying processes that gave rise to these nonrandom patterns. Mathematically, we can think of a spatial point process as a stochastic model governing the location of events $\{s_i\}$ in some set X, a subset of R^d. The result of this stochastic model is a spatial point pattern that can be defined through the spatial locations of events $s_1, s_2, \ldots$ in X.

The purpose of this chapter is to introduce spatial point patterns and to present methods that may be used to determine whether these patterns are aggregated, random, or regular. Ripley (1981) has a limited review of the analysis of spatial point patterns. Diggle (1983) provides a detailed discussion of the analysis of spatial point patterns. Cressie (1993) reviews current theoretical and applied approaches. These works have served as a foundation for developing this chapter.

Complete Spatial Randomness

Complete spatial randomness (csr) is the standard by which all patterns are judged. Because of its central place in the study of spatial point patterns, we need to first understand what is meant, both biologically and mathematically, by csr. For purposes of discussion, the points will be taken to be plants. To say that plants in a region exhibit csr is equivalent to saying that every point on the ground within the area of study is as likely as every other point to be the location of an individual plant. If a spatial point pattern exhibiting csr is sampled with randomly placed quadrats, the expected distribution of the number of plants per quadrat is Poisson with parameter $\lambda |A_i|$, where λ is the mean number, or intensity, of plants in a unit of the study area and $|A_i|$ is the area of the ith quadrat. The distance between a randomly selected plant and its nearest neighbor is exponentially distributed with mean $1/\lambda$. Further, the distance between a randomly chosen point and the nearest plant is exponentially distributed with mean $1/\lambda$.

More formally, for a spatial point pattern to be csr, the points must be distributed as a homogeneous Poisson process. Two conditions must be met for a process to be a homogeneous Poisson process: (1) the number of events in any study region A with area $|A|$ has a Poisson distribution with mean $\lambda |A|$, and (2) the N events in a region A are an independent random sample from a uniform distribution on A. Condition (1) implies that the intensity is constant throughout the study region. Processes satisfying condition (1) are said to be stationary. Condition (2) indicates that there is no interaction among events. From condition (2), we have that a csr process is isotropic. By isotropic, we mean that there are no directional effects. For example, the probability of observing an event in the northwest corner of a plot is no more likely than in any other part of the plot. However, one should note that processes that do not have csr could also be both stationary and isotropic. Formal mathematical definitions of these properties are given in Diggle (1983) and Cressie (1993). The definition of csr agrees well with intuitive ideas about randomness.

Three spatial point patterns are shown in Figures 8.1 to 8.3. Notice that the csr pattern shown in Figure 8.1 exhibits some clustering. To realize that this is to be expected, first recall that the distance between a randomly selected plant and its nearest neighbor is exponentially distributed. From Chapter 1, we know that the exponential distribution is a highly skewed distribution with much of its probability near 0. In this case, the exponential distance between a plant and its nearest neighbor indicates a high probability associated with fairly short distances. Therefore, we expect some signs of clustering in the csr patterns. This clustering due to randomness can be contrasted with true clustering, or aggregation, of a spatial point pattern as in Figure 8.2 and with a regular spatial point pattern pictured in Figure 8.3. At what point does the natural clustering of points in a pattern exhibiting csr become too much or too little to be due strictly to randomness? Statistical

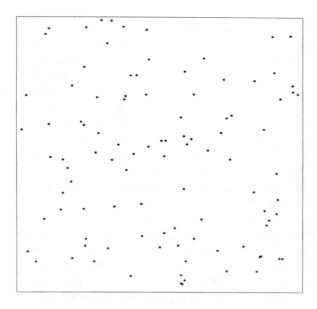

Figure 8.1. Plot of 100 plants randomly dispersed in a unit square.

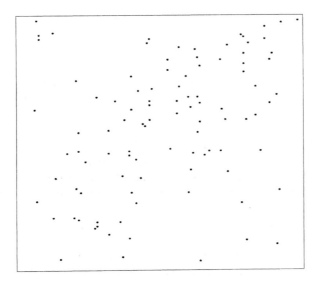

Figure 8.2. Aggregated pattern of 100 plants distributed in a unit square.

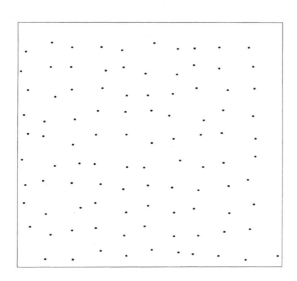

Figure 8.3. Regular pattern of 100 plants in a unit square.

methods that help us address this question are the focus of the remainder of this chapter.

Consider the data that may be obtained from the study region. An intensive map of the organisms within the region could be constructed as in Figures 8.1 to 8.3. Alternatively, quadrats could be randomly or systematically placed throughout the study region and the number of events within each quadrat observed. In this case the analysis consists of computation and testing based on the aggregation indices discussed in Chapter 7. Finally, the experimenter could observe the distance of the kth nearest event from a randomly selected event or point within the study region. Methods for detecting the departures from complete spatial randomness based on mapped data and nearest neighbor techniques are discussed in the remainder of this chapter. Because inferences are stronger from mapped data, we introduce those methods first.

The starting point for testing the hypothesis of complete spatial randomness when the data are mapped is condition (2) above; that is, the events within the region constitute an independent random sample from a uniform distribution on the region. Therefore, the inferences are conditional on the number of events N in the study region A. We consider two analysis methods: inferences based on the function $K(h)$ and Monte Carlo methods.

$K(h)$ and $L(h)$ Functions

The K and L functions (Bartlett, 1964; Ripley, 1976, 1977) are becoming the standard statistical method for quantifying the different regions of the point process. To gain an understanding of these functions, first consider a circle of radius h centered on an arbitrary event in the study region (*see* Figure 8.4). Consider first a circle contained wholly within the study region. The area of the circle is πh^2. For a process that has csr, the expected number of other events within that circle is $\lambda \pi h^2$ (λ times the area of the circle). For a given h, we could place a circle of radius h about each of the events in a mapped area and compute the average number of other events in these circles of size h. Ignoring for the moment the problems that arise by circles that do not lie wholly within the region, we could then think of this average as an estimate of $\lambda \pi h^2$ if the process has csr. We expect larger values than $\lambda \pi h^2$ if there is aggregation and smaller ones if the pattern is regular. In order to obtain a measure of spatial dependence that does not depend on the intensity λ, we divide this average by the intensity or its estimate. The subsequent quantity should be approximately equal to πh^2 if the process has csr, larger than πh^2 when there is aggregation, and less than πh^2 when the pattern is regular. This approach could be repeated for varying h, and the results plotted as a function of h. This is the conceptual foundation for the K function.

The K function is defined as follows:

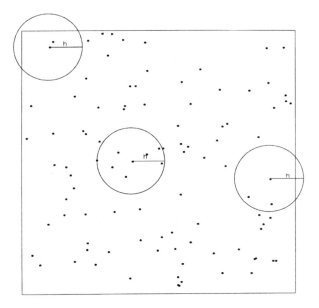

Figure 8.4. Circles of radius h centered on events in a point pattern.

$$K(h) = \lambda^{-1}E[\text{number of other events within distance } h \text{ of an arbitrary event}]$$
$$h \geq 0. \quad (8.1)$$

Intuitively, an arbitrary event is a randomly selected event from the finite population of events within the study area. Cressie (1993) argues that the K function (Bartlett, 1964; Ripley, 1976, 1977) captures the spatial dependence between different regions of the point process.

K is estimated by replacing the expectation in equation (8.1) with the empirical average. From the complete map of events, let $(x_1, x_2, \ldots, x_N)$ denote the $N = N(A)$ (where $N(A)$ represents the number of events in the study region A) locations of all events in a bounded study region A. Define

$$\hat{K}_1(h) \equiv \hat{\lambda}^{-1} \sum_{\substack{i=1 \\ i \neq j}}^{N} \sum_{j=1}^{N} I(\|x_i - x_j\| \leq h)/N, \qquad h > 0. \quad (8.2)$$

where the indicator function $I(\cdot)$ is one if the expression inside the parentheses is true and zero otherwise. Notice that it is enough to consider the distances between all pairs of events. Then the number of events within a circle centered on an event can be quickly determined for any value of h. The estimate given in equation

(8.2) is biased because it has no correction for the edge effects; that is, no correction is made for the circles that do not lie fully in the study region (*see* Figure 8.4).

In assessing the spatial point process, the impact that events close to the edge have on the statistical methods must be considered. As noted in the previous paragraph, failure to do so usually results in biased estimators. At least three approaches to correcting for edge effects have been considered (Cressie, 1993). The first approach is to construct a guard area around the perimeter of *A*. Although no circles are constructed about events in the guard area, the events in the guard area are counted when they lie within a circle centered on an event in the study region. Second, a rectangular study region may be regarded as a torus, so that events near opposite sides are considered close. An equivalent interpretation is to regard the study region as the center plot in a 3 × 3 grid of plots. Each of the nine plots in the grid is identical to the study region. Finally, finite-sample corrections may be made to the distribution theory for specific test statistics. All three approaches have been used to propose edge-corrected estimators of *K*. The method used most consistently is the latter and is due to Ripley (1976). It is the one considered here.

Suppose we are interested in estimating *K*, but that several circles of radius h, centered on events in the study region *A* do not lie fully within the region (*see* Figure 8.4). Because the full circle does not lie totally within the region, the number of events within the circle tend to be less than would have been the case if the circle was fully contained in the region. In fact, if w is the proportion of the area of the circle that lies within *A*, we expect only $w\lambda\pi h^2$ events of *A* to be within the circle. To adjust the estimate so that the expectation of the number of points within a circle is $\lambda\pi h^2$ for all circles, we need to divide the number of events in the circle by w. Based on the proportion of the circumference (and not the area) of the circle lying within *A*, Ripley (1976) proposed an edge-corrected estimator that uses information on events for which the circle does not fully lie within the study region:

$$\hat{K}_2(h) = \hat{\lambda}^{-1} \sum_{\substack{i=1 \\ i \neq j}}^{N} \sum_{j=1}^{N} w(\mathbf{x}_i, \mathbf{x}_j)^{-1} I(\|x_i - x_j\| \leq h)/N, \qquad h > 0, \qquad (8.3)$$

where $w(x_i, x_j)$ is the proportion of the circumference of a circle centered at x_i, passing through x_j, that is inside the study region *A*. Note that in general $w(x_i, x_j) \neq w(x_j, x_i)$. Also, although not shown explicitly in our notation, w is also a function of h because the proportion of the circumference of the circle lying within the study region depends on the radius h of the circle.

Diggle (1983) states that explicit formulae for $w(x_i, x_j)$ can be given for simple shapes (such as rectangular or circular) of region *A*. First, suppose that *A* is the rectangle $(0, a) \times (0, b)$. Let $d_1 = \min(x_1, a - x_1)$ and $d_2 = \min(x_2, b - x_2)$;

thus d_1 and d_2 are the distances from x_1 and x_2, respectively, to the nearest vertical and horizontal edges of A. Two cases are considered in the computation of $w(x_1, x_2)$:

(1) If $h^2 \leq d_1^2 + d_2^2$,
$$w(\mathbf{x}_1, \mathbf{x}_2) = 1 - \pi^{-1}\left[\cos^{-1}\left(\frac{\min(d_1, h)}{h}\right) + \cos^{-1}\left(\frac{\min(d_2, h)}{h}\right)\right].$$

(2) If $h^2 > d_1^2 + d_2^2$,
$$w(\mathbf{x}_1, \mathbf{x}_2) = \frac{3}{4} - (2\pi)^{-1}\left[\cos^{-1}\left(\frac{d_1}{h}\right) + \cos^{-1}\left(\frac{d_2}{h}\right)\right]. \tag{8.4}$$

Notice that if the circle is completely within the sampling region, the first inequality above is satisfied, and the weight reduces to one. The weight decreases as the proportion of the circle within the square decreases. The above formulae hold for values of h in the range of $0 \leq h \leq .5 \min(d_1, d_2)$.

Now suppose that A is a circle centered at the origin and with radius a. Let r be the Euclidean (straight-line) distance from (x_1, x_2) to the center of the circle. Again, two cases are need in the computation of $w(x_1, x_2)$:

(1) If $h \leq a - r$,
$$w(\mathbf{x}_1, \mathbf{x}_2) = 1$$

(2) If $h > a - r$,
$$w(\mathbf{x}_1, \mathbf{x}_2) = 1 - \pi^{-1} \cos^{-1}\left(\frac{a^2 - r^2 - h^2}{2rh}\right) \tag{8.5}$$

These formulae hold for values of h between 0 and a.

The edge-corrected estimator $\hat{K}_2$ is an unbiased estimator of K under the assumptions of stationarity and isotropy when λ is known. The estimate $\hat{\lambda} = N/|A|$ is used in equations (8.2) and (8.3). Use of the estimated intensity does not seem to affect unbiasedness too much (Ripley, 1981).

Under the assumption of csr, $K(h) = \pi h^2$. $K(h)$ tends to be greater than πh^2 if the events are aggregated and less than πh^2 if the events are regular. Even under csr, we expect to observe variation about πh^2 because K is not known but estimated. How much greater (or smaller) than πh^2 must $\hat{K}$ be before we conclude that the observed spatial point process does not possess csr? To address this question, we will use simulation envelopes.

A simulation envelope provides an upper and lower bound on $K(h)$, or any other function, based on an assumed underlying spatial point process. Because we want to detect departures from csr, we want to generate upper and lower bounds on $K(h)$ for a process that has csr. We begin by generating 100 point patterns, each with N events, $s_1, s_2, \ldots, s_N$, from a uniform distribution on A. $K(h)$ is estimated for each of the 100 realizations of csr. The largest and smallest values of $\hat{K}$ for each h define the simulation envelope. If the estimate of $K(h)$ based on the mapped data lies within the simulation envelope, then the estimate of $K(h)$ is consistent with that expected under csr. If $\hat{K}$ is above the simulation envelope, the data are exhibiting clustering; the data indicate regularity if $\hat{K}$ is below the simulation envelope.

When considering measures of aggregation in Chapter 7, a value of 0 or 1 was often taken as the value associated with randomness. With the same motivation of having a single value associated with csr, the function L has been defined to be

$$L(h) = \sqrt{\frac{K(h)}{\pi}} - h \qquad (8.6)$$

Under the assumption of csr, $L(h)$ is 0. Under regularity, $L(h)$ tends to be less than 0, whereas $L(h)$ tends to be greater than 0 under clustering. $L(h)$ is estimated by replacing $K(h)$ in equation (8.6) with its estimate. $\hat{L}_1$ and $\hat{L}_2$ denote the estimates of $L(h)$ when using the estimators $\hat{K}_1$ and $\hat{K}_2$, respectively. Simulation envelopes can be constructed for $L(h)$ based on csr, permitting a measure of the departure of a process from csr.

Example 8.1

Louda (1992, personal communication) is studying three species of native thistles: Wavyleaf, Platte, and Flodmans. These thistles are indigenous, perennial species in midgrass prairie. Platte thistle (*Cirsnim canescens*) occurs in midgrass prairie on sandy soils in Nebraska, Colorado, and Wyoming. Flodman's thistle (*C. flodmans*) occurs in moist areas of the upper Great Plains while Wavyleaf thistle (*C. undulatum*) extends throughout the whole Great Plains Region. Although densities are highest in the open of old disturbances, such as blowouts in sand prairie, plants often occur throughout the adjacent grassland. The life history of these thistles is typical of short-lived perennials. Seedlings establish in early spring from seed released in the previous growing season. After the seedling year, the plant grows and persists as a taprooted rosette for a variable number of years. During its flowering year, it bolts and flowers. Platte thistle dies after bolting whereas Wavyleaf and Flodman's thistles can flower again in subsequent years. Seeds tend to be dropped in the vicinity of the parent plant (Louda and Potvin, 1995; Louda et al., 1995). The thistles are usually observed in clumps, reflecting both vegetative regrowth and patterns of seed fall.

Sampling was conducted at three sites within Nebraska: Arapaho Prairie, Niobrara Valley Preserve, and an eastern farm. At each site, plots of 12 m × 12 m were marked. The location of each thistle within a plot was recorded. A plot of the wavyleaf thistles in one of these plots on the Arapaho Prairie is given in

Figure 8.5. The data are in ECOSTAT file s-wl-al.dat. The purpose of our analysis is to determine whether this spatial point pattern is due to randomness.

ECOSTAT was used to compute the estimators $\hat{K}_1$, $\hat{L}_1$, $\hat{K}_2$, and $\hat{L}_2$ and simulation envelopes for these functions based on csr (*see* Figures 8.6 to 8.9). To do this, choose *Point Pattern* from the *Spatial Statistics* menu. The plot must be designated as either circular or rectangular. If it is circular, the radius must be recorded. The length and width must be specified if the plot is rectangular. Then the *Enter Data* command button is pushed. You will then be asked to provide the name of the data file. This must be an ASCII file. The number of events in the plot n is on the first line. Each of the next n lines has the x and y coordinates of an event. The number of pairs of points and the least and greatest distances between events are displayed. To estimate the K and L functions and their simulation envelopes, press the *Estimate K* button. You must specify an output data set in which the variables will be output in the following order: h; lower simulation bound, estimate, and upper simulation bound for $\hat{K}_1$; lower simulation bound, estimate, and upper simulation bound for $\hat{L}_1$; lower simulation bound, estimate, and upper simulation bound for $\hat{K}_2$; and lower simulation bound, estimate, and upper simulation bound for $\hat{L}_2$. These are also displayed in a data grid on the screen.

Although $\hat{K}_1$ and $\hat{L}_1$ are intuitively appealing, they exhibit strong bias because no correction is made for edge effects. Note that the simulation envelopes based on a random point process do not enclose the anticipated csr values for K (Figure 8.6) or L (Figure 8.7). Because $\hat{K}_1$ and $\hat{L}_1$ are biased estimators of K and L,

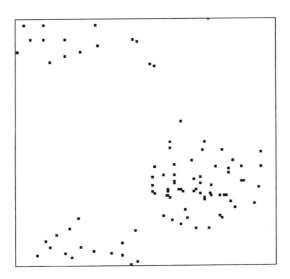

Figure 8.5. Location of the wavyleaf thistles in a 12-m $\times$ 12-m area on the Arapaho Prairie.

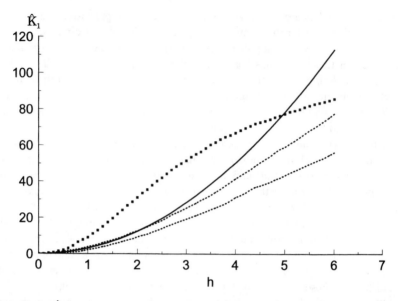

Figure 8.6. $\hat{K}_1$ ($\cdots$), csr simulation envelope (dashed lines), and K_1 under csr (solid line) for wavyleaf thistles in a 12-m $\times$ 12-m quadrat.

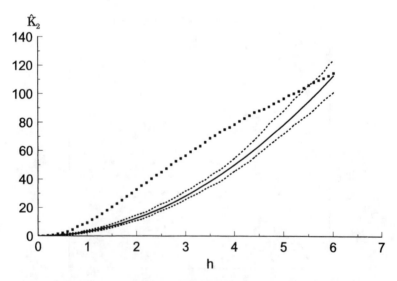

Figure 8.7. $\hat{L}_1$ ($\cdots$), csr simulation envelope (dashed lines), and L_1 under csr (solid line) for wavyleaf thistles in a 12-m $\times$ 12-m quadrat.

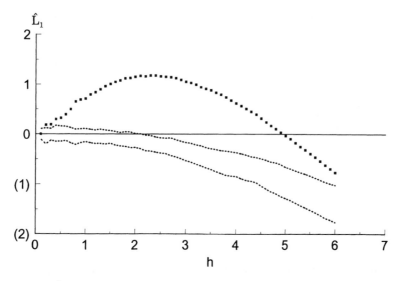

Figure 8.8. $\hat{K}_2$ ($\cdots$), csr simulation envelope (dashed lines), and K_2 under csr (solid line) for wavyleaf thistles in a 12-*m* $\times$ 12-*m* quadrat.

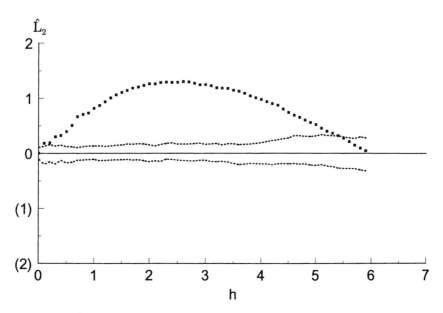

Figure 8.9. $\hat{L}_2$ ($\cdots$), csr simulation envelope (dashed lines), and L_2 under csr (solid line) for wavyleaf thistles in a 12 m $\times$ 12 m quadrat.

respectively, the simulation envelopes provide a better standard by which to judge the departure from csr than the known values of K and L under csr, πh^2 and 0, respectively. Although $\hat{K}_2$ and $\hat{L}_2$ are somewhat biased due to the estimation of the intensity λ, the simulation envelopes based on csr do cover the csr values of K and L. For $h = 0.1$ m; estimates of K and L lie within the csr simulation envelopes and are thus consistent with csr. However, for values of h greater than 0.1, the clustering that is biologically expected is statistically visible.

For the quadrat presented in this example, three clusters of plants could be observed. This was not the case in all quadrats. In Figure 8.10, the wavyleaf thistles within a 12-m $\times$ 12-m quadrat on a Nebraska farm are shown. Here there is a single cluster of plants in the center of the upper portion of the quadrat. However, the estimates of $K(h)$ and $L(h)$ are remarkably similar to those for the quadrat in Figure 8.5.

Monte Carlo Tests

Monte Carlo tests are useful for testing stochastic models for spatial point patterns because even simple models lead to intractable distribution theory. To employ Monte Carlo methods, we begin by identifying a statistic U and denote its sample value by u_1. Then $(s - 1)$ independent random samples from the distribution of U under a simple hypothesis H are drawn. For each sample, u_i, the value of statistic U for sample $i = 2, 3, \ldots, s$, is computed. Let $u_{(j)}$ denote the jth largest of the u_i, $i = 1, 2, \ldots, s$, values. Then, under H,

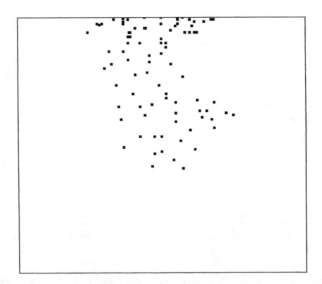

Figure 8.10. Plot of wavyleaf thistles in a 12-m $\times$ 12-m quadrat on a Nebraska farms.

$$P(u_1 = u_{(j)}) = \frac{1}{s}, \quad j = 1, 2, \ldots, s. \tag{8.7}$$

An exact, lower one-sided test of size k/s rejects H if u_1 ranks kth largest or less. This assumes that there is a unique ranking of the u_i values. If tied values are possible as when U is a discrete random variable, then a conservative rule is to choose the least rank of u_1. Two-sided tests of size k/s result if rejection of H occurs when the rank of u_1 is less than or equal to $k/2$ or greater than $(s - k/2)$. Because the loss of power resulting from a Monte Carlo test is slight (Hope 1968), $s = 100$ is adequate for a one-sided test at the 5% significance level. Marriott (1979) further concluded that if $s = 100$ is judged to be sufficient for a test at the 5% level, $s = 500$ should be used for a test at the 1% level, and proportionately larger sample sizes are needed at smaller levels.

For the n events in a study region, the $n(n - 1)/2$ associated inter-event distances can be used to establish the test statistic U (Diggle, 1983). The theoretical distribution of the distance T between two events independently and uniformly distributed in A depends on the size and shape of A. Bartlett (1964) developed a closed form expression for the distribution function of T for the most common cases of square and circular A. For a unit square, the distribution function H of T is

$$H(t) = \pi t^2 - 8t^3/3 + t^4/2, \quad 0 \le t \le 1$$

$$= \frac{1}{3} - 2t^2 - \frac{t^4}{2} + \frac{4(2t^2 + 1)\sqrt{t^2 - 1}}{3 + 2t^2 \sin^{-1}(2_t^{-2} - 1)}, \quad 1 < t \le \sqrt{2}. \tag{8.8}$$

The corresponding expression for H when A is a unit circle is

$$H(t) = 1 + \frac{2(t^2 - 1)\cos^{-1}(t/2) - t(1 + t^2/2)\sqrt{1 - t^2/4}}{\pi}, \quad 0 \le t \le 2. \tag{8.9}$$

Notice that as long as A is square or circular, the position of the events within A can be represented as if A is either a unit square or a unit circle, respectively. Assume that A has one of these shapes or another shape for which $H(t)$ is known.

The next step in constructing a Monte Carlo test of csr is to calculate the empirical distribution function (EDF), $\hat{H}_1(t)$, of inter-event distances. $\hat{H}_1(t)$ is the observed proportion of inter-event distances t_{ij} that are at most t and may be represented as follows:

$$\hat{H}_1(t) = \frac{\sum\limits_{i>j}^{n} \sum\limits_{i=1}^{n} I(t_{ij} \le t)}{n(n - 1)/2} \tag{8.10}$$

where I is the indicator function; $I(j) = 1$ if j is true and $I = 0$ if j is false. In this case, I would be 1 if $t_{ij} \leq t$ and 0 otherwise so that the numerator is the number of inter-event distances less than or equal to t.

A plot of $\hat{H}_1(t)$ against $H(t)$ should be roughly linear if the pattern is csr. Note that for the Monte Carlo test, $\hat{H}_1(t)$ is the observed value of the statistic U. To assess whether the observed departures from linearity are statistically significant, the conventional approach is to find the sampling distribution of $\hat{H}_1(t)$ under the hypothesis of csr. However, this is a complex task because of the dependence between inter-event distances with a common endpoint. To proceed, we calculate the empirical distribution functions $\hat{H}_i(t)$, $i = 2, 3, \ldots, s$, from each of $s - 1$ independent simulations of n events independently and uniformly distributed on A. Based on these EDFs, upper and lower simulation envelopes are defined as

$$U(t) = \max(\hat{H}_i(t), \quad i = 2, 3, \ldots, s) \tag{8.11}$$

and

$$L(t) = \min(\hat{H}_i(t), \quad i = 2, 3, \ldots, s), \tag{8.12}$$

respectively. Under csr,

$$P(\hat{H}_1(t) > U(t)) = P(\hat{H}_1(t) < L(t)) = \frac{1}{s}. \tag{8.13}$$

These simulation envelopes can be plotted on the graph with $H(t)$ and $\hat{H}_1(t)$.

Example 8.2

Consider again the Wavyleaf thistles in an Arapaho Prairie plot, as shown in Figure 8.5. ECOSTAT was used to conduct the analysis. The same process of data entry was followed as in Example 8.1. To do the Monte Carlo test, press the *Monte Carlo Test* button. You must specify two data sets, one for output and the other for a work space. First, the theoretical and empirical distribution functions for interpoint distances are computed. Then 100 Monte Carlo simulations of interpoint distances for 93 points (because that is the number of Wavyleaf thistles in the plot) are conducted under csr. These are used to establish upper and lower boundaries of the simulation envelope. Each line of the output data set represents a different value of t in the following order: t, $\hat{H}_1(t)$, lower simulation bound, $H(t)$, and upper simulation bound. The output can then be fed into a standard graphical package. The results are shown in Figure 8.11 and indicate that the thistles exhibits clustering. This is consistent with results based on the K and L functions in Example 8.1.

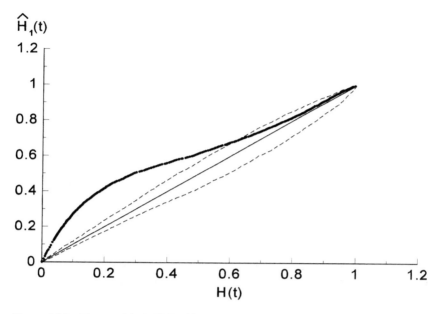

Figure 8.11. The empirical CDF of interpoint distances ($\cdots$) and the theoretical CDF (solid line) and simulation envelope (dashed line) under csr.

Diggle (1983) suggested two possible approaches to constructing an exact Monte Carlo test of csr. First, we could choose t_0 and define $u_i = \hat{H}_i(t_0)$. As described above, the rank of u_1 among the u_i provides the basis of a test because under csr all ranking of u_i are equiprobable. However, this approach seems questionable unless there is a natural choice of t_0 for the stated problem. Alternatively, u_i could be defined as a measure of the discrepancy of $\hat{H}_i(t)$ and $H(t)$ over the range of t, such as

$$u_i = \int [\hat{H}_i(t) - H(t)]^2 dt. \tag{8.14}$$

Then the test could be based on the rank of u_1. The problem with this approach is that, when working with inter-event distances, the resulting test is very weak. No single test statistic should be permitted to contradict the conclusions derived from a critical inspection of the EDF plot.

If the region A has a size and/or shape for which the theoretical distribution function $H(t)$ is unknown, a test can still be carried out if, in equation (8.14), $H(t)$ is replaced by the average of the Monte Carlo EDFs,

$$\bar{H}(t) = \frac{1}{s-1} \sum_{j \neq i} \hat{H}_j(t). \tag{8.15}$$

Although the u_i are no longer independent under csr, the required property that all rankings of u_1 are equiprobable still holds. The graphic procedure then consists of plotting $\hat{H}_1(t)$, $U(t)$, and $L(t)$ against $\bar{H}(t)$. Further, because $\hat{H}_1(t)$ involves the simulations of csr and not the data, it is an unbiased estimate of $H(t)$.

Nearest Neighbor Techniques

Mapping all events in a sampling region permits using the most powerful techniques for detecting departures from csr and testing models leading to aggregated or regular spatial point patterns. Recent technology advancements have reduced the costs of constructing point maps for many applications. However, for various reasons, it is not always possible to have a point map of the region. Yet, a test for csr may still be of interest. In these circumstances, nearest neighbor techniques are often appropriate. As illustrated in Figure 8.12, these methods are based on the distances from (1) a randomly chosen event to its nearest neighbor (Z), or (2) a randomly chosen point to the nearest event (X).

The difficulty with the methods requiring measurements from randomly selected events is field implementation. A random selection of events could be made from a map of the events. Because none of the nearest neighbor techniques are

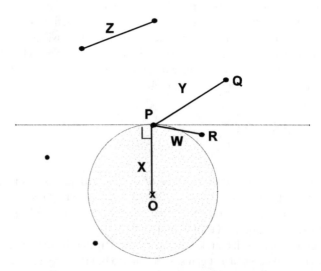

Figure 8.12. Measurements between events ($\cdot$) and random points (x) used in nearest neighbor methods. An observation for T-square sampling is illustrated.

as powerful as those based on mapped points that were discussed earlier, they cannot generally be recommended if a map exists. Thus, although we briefly mention two nearest neighbor methods requiring the random selection of events, we focus on methods based on a random selection of points as they tend to be implemented more readily in the field.

Hopkins (1954) suggested a method that involves measuring the distance X from a randomly selected point to the nearest event and the distance Z from a randomly selected event to its nearest neighbor. Because the searches for the nearest events can be thought of as concentric circles centered at the randomly chosen point or event, the areas searched are πX^2 and πZ^2. Under csr, the areas πX^2 and πZ^2 are independent. Based on a random sample of n points and n events, the test statistic

$$A = \frac{\sum_{i=1}^{n} X_i^2}{\sum_{i=1}^{n} Z_i^2}$$

has an F-distribution with $2n$ and $2n$ degrees of freedom under csr. The areas searched should not overlap, constraining the number of observations that can be taken. Numerous studies (Diggle 1975, 1977; Diggle et al., 1976; Hines and Hines, 1979; Cressie, 1993) have compared nearest neighbor methods, and the Hopkins test has been shown to be the most powerful. However, it is generally recognized as impractical because of the need to randomly select events.

Byth and Ripley (1980) suggested a modification of the Hopkins test that is easier to implement in the field. They recommended establishing a fairly regular grid of $2m$ points. From one-half of the points, referred to as the m sample points, the distance of the point to the nearest event is recorded. Around the remaining m points, a small plot is laid out. The size of the plot is set so that there is an average of about five events per plot. Each event in each plot is recorded. A random selection of m events is made from those enumerated events and the distance from each selected event to its nearest neighbor recorded. They considered both the statistic proposed by Hopkins and an alternative one,

$$B = \frac{1}{n} \sum_{i=1}^{n} \frac{X_i^2}{X_i^2 + Z_i^2}.$$

Under csr, B has an asymptotic normal distribution with mean 0.5 and variance $1/12n$. A is more powerful for clustering and B more powerful for regular patterns. Although the procedure compares favorably and is more easily implemented than that of Hopkins, it requires substantial effort and may still be impractical to implement in the field.

Pielou (1959) proposed a test based on the n distances from randomly selected points to the nearest events, X_i, $i = 1, 2, 3, \ldots$ (*see* Figure 8.12). The area of the circle with radius X_i and centered at the random point O is πX_i^2. The average area of n such circles is $\pi \Sigma X_i^2 / n$. The expected number of events in this area is then

$$C = \frac{\pi \lambda}{n} \sum_{i=1}^{n} X_i^2.$$

Because the distance from a randomly selected point to the nearest event, under the hypothesis of csr, is exponentially distributed with mean $1/\lambda$, the expected value of C is 1, and the variance is $1/n$. If the events are more regular, the expected value will be less than one, and it will be greater than one if the events are aggregated. The statistic C was presented as an index for measuring departures from csr, *not* as a test statistic. To formally test the null hypothesis of csr, Pielou suggests the statistic

$$D = 2\pi \lambda \sum_{i=1}^{n} X_i^2.$$

Under csr, D has a chi-squared distribution with $2n$ degrees of freedom. Large values of D lead to rejection of the hypothesis of csr.

A difficulty that usually arises when implementing Pielou's method in the field is that the intensity λ is unknown and must be estimated. A random sample of m randomly placed quadrats is drawn to obtain the estimate $\hat{\lambda}$. Replacing λ by its estimate in the statistic C, Mountford (1961) showed that the distribution is still asymptotically normal with mean 1 and variance $(1 + (n + 1)/(m\lambda))/n$.

T-square sampling (Besag and Gleaves, 1973) is another distance method. It can be readily applied in the field, making it a popular approach. First, a point O is randomly selected in the plot. The distance X from the point O to the nearest event P is recorded. Then a line perpendicular to the segment (OP) is constructed (*see* the dashed line in Figure 8.12). This perpendicular line divides the sampling area (plane) into two regions (half-planes). The distance Y from the event P to the nearest event Q *not* in the half-plane containing the sample point is measured. After this process is repeated for n randomly selected point, the test statistic

$$E = 2 \frac{\displaystyle\sum_{i=1}^{n} X_i^2}{\displaystyle\sum_{i=1}^{n} Y_i^2}$$

is computed. Under csr, E has an F distribution with $2n$ and $2n$ degrees of freedom. Small values of E indicate a more regular spatial pattern, and large values provide evidence of an aggregated spatial pattern. T-square sampling has been found to be powerful against a variety of alternatives (Diggle et al., 1976; Hines and Hines, 1979).

The above tests are based on the assumption that the nearest-neighbor measurements are independent. If the sampling region is small, this assumption may be violated. Cressie (1993) found that in such cases, the hypothesis of csr is rejected too often; that is, if the spatial pattern has csr, the hypothesis of csr is rejected more often than the stated size of the test indicates. Therefore, he cautioned that if test statistics are close to critical values, simulation methods are more appropriate than relying on standard significance levels.

Example 8.3

Consider again the Wavyleaf thistle plot pictured in Figure 8.5. Suppose we did not have the map. Yet we wanted to test whether the pattern of thistles in the plot was consistent with complete spatial randomness. Further, assume that we have decided to take $n = 10$ observations. The results of sampling are given in Table 8.1.

For each observation, we selected a point at random and found the nearest event. After drawing a line perpendicular to the segment from the point to the event, we found the nearest event in the half-plane that did not contain the point. The distances from the point to the nearest event and from that event to the second event were recorded. The data were put in an ASCII file. The first line of the file is the number of observations, $n = 10$. The next 10 lines each have an identification of the point, the distance X to the nearest event and the distance Y from that event to the nearest event in the half-plane that does not include the point.

Table 8.1. Ten Observations Collected Using T^2 *Sampling*

Point	Nearest event	Nearest event not in half-plane	X_i	Y_i
(3.49, 11.59)	(3.7, 11.7)	(5.4, 11.0)	0.241	1.838
(8.65, 6.64)	(8.7, 6.0)	(8.6, 5.6)	0.640	0.412
(5.38, 6.27)	(7.1, 6.0)	(7.1, 5.7)	1.737	0.300
(2.51, 10.36)	(2.3, 10.2)	(1.6, 9.9)	0.262	0.762
(5.33, 0.76)	(4.9, 0.5)	(4.4, 0.6)	0.507	0.510
(7.09, 5.58)	(7.1, 5.7)	(7.1, 6.0)	0.123	0.300
(7.78, 7.46)	(7.6, 7.0)	(7.1, 6.0)	0.496	1.118
(3.80, 9.84)	(3.8, 10.7)	(3.7, 11.7)	0.856	1.005
(2.81, 1.88)	(2.5, 1.8)	(1.9, 1.5)	0.316	0.671
(1.04, 7.67)	(1.6, 9.9)	(2.3, 10.2)	2.296	0.762

To use ECOSTAT, first select *Nearest Neighbor* from the *Spatial* menu and then choose *T-square*. After entering the data set name, the sample values, the test statistic, degrees of freedom and p-value are displayed. In our case, the test statistic is $E = 1.295$. If the spatial pattern exhibits csr (H_0 is true), E is an observation from an F-distribution with 20 and 20 degrees of freedom. The p-value is 0.28. Therefore, we do not reject the hypothesis of csr.

At this point, we have developed methods to assess departures from csr. If we do not have csr, what is the next step? Numerous models have been developed for patterns that are aggregated or regular. The fit of each model can be assessed using essentially the same approaches that we have used in evaluating csr. The interested reader is encouraged to explore these models (for examples, *see* Cressie, 1993, and Diggle, 1983).

Summary

Spatial point patterns arise through natural biological processes. The first step in studying these patterns is to assess whether or not the pattern is random; that is, whether or not the pattern results from a homogenous Poisson process. The most powerful tests result when all events in the sampling region are mapped. The K and L functions can be used to assess departures from complete spatial randomness (csr). The intuitive estimators of these functions are biased due to sampling near the edges. $\hat{K}_2$ and $\hat{L}_2$ are bias-corrected estimators. Simulation envelopes for these estimators provide a basis by which to judge whether there is significant departures from complete spatial randomness. Monte Carlo tests provide another method for testing for complete spatial randomness based on mapped data.

Sometimes resources are not available to map the events within a region. Yet the goal of the study may still be to determine whether the pattern is consistent with complete spatial randomness. Nearest-neighbor methods are appropriate in this setting. Hopkins's (A) has generally been recognized as the most powerful against a variety of alternatives (Holgate, 1965; Diggle et al., 1976; Cressie, 1993). However, the method relies on a random selection of events. This is difficult if the events are not mapped. (If they are mapped, we can use procedures for mapped data and not nearest neighbor techniques.) This random selection of events has limited the use of Hopkins's method. Byth's and Ripley's (B) was developed as a modification of Hopkins's (A) that would be easier to implement in the field. Yet the practicality of using this modification remains doubtful. Pielou's (C) has been shown to be powerful against a variety of alternatives (Payendeh, 1970), but this method requires an independent estimate of intensity. Besag's and Gleaves' T-square sampling method is easy to implement in the field and has been shown to be powerful against a variety of alternatives (Diggle et al., 1976; Hines and Hines, 1979).

Exercises

1. Consider again the setting described in Example 8.1. One data set for each of the Wavyleaf thistle (s-wl-ob.dat), Platte thistle (s-pt-ah1.dat), and the Flodmans thistle (s-fd-nz.dat) are in the ECOSTAT data sets. Each line gives first the X- and then the Y-coordinate of a thistle.

   A. Use the K and L functions to determine whether the spatial pattern of the Wavyleaf thistles is random within each plot.

   B. Use Monte Carlo methods to determine whether the spatial pattern of the Wavyleaf thistle is random within each plot.

   C. Use the K and L functions to determine whether the spatial pattern of the Platte thistles is random within each plot.

   D. Use Monte Carlo methods to determine whether the spatial pattern of the Platte thistle is random within each plot.

   E. Use the K and L functions to determine whether the spatial pattern of the Flodmans thistles is random within each plot.

   F. Use Monte Carlo methods to determine whether the spatial pattern of the Flodmans thistle is random within each plot.

2. As reported by Diggle (1983), Numata (1961) mapped the location of 65 Japanese black pine saplings within a square having each side of length 5.7 m. The data are given in the Appendix.

   A. Use the K and L functions to determine whether the spatial pattern of the Japanese black pine seedlings is random.

   B. Use Monte Carlo methods to determine whether the spatial pattern of the Japanese black pine seedlings is random.

3. Crick and Lawrence (1975) and Ripley (1977) mapped the centers of 42 biological cell within a unit square. The data are given in the appendix and in the ECOSTAT file cell.dat.

   A. Use the K and L functions to determine whether the spatial pattern of the biological cells is random.

   B. Use Monte Carlo methods to determine whether the spatial pattern of the biological cells is random.

4. Ripley (1977) presents the map of redwood saplings within a one-mile square area that represents a portion of the data in Strauss (1975). The data are given in the appendix.

   A. Use the K and L functions to determine whether the spatial pattern of the redwood saplings is random.

    B. Use Monte Carlo methods to determine whether the spatial pattern of the redwood saplings is random.

5. Choose one of the thistle data sets from exercise 1.

    A. Use Pielou's method to test for spatial randomness.

    B. Use T-square sampling to test for spatial randomness.

    C. Use Diggle's method to test for spatial randomness.

    D. Use Hines and Hines's method to test for spatial randomness.

    E. Discuss the design you would use and the factors you considered for parts *A* to *D*.

    F. Compare and contrast the results from the tests.

6. Consider again the Japanese black pine saplings in exercise 2.

    A. Use Pielou's method to test for spatial randomness.

    B. Use T-square sampling to test for spatial randomness.

    C. Use Diggle's method to test for spatial randomness.

    D. Use Hines and Hines's method to test for spatial randomness.

    E. Discuss the design you would use and the factors you considered for parts *A* to *D*.

    F. Compare and contrast the results from the tests.

7. Consider again the biological cells in exercise 3.

    A. Use Pielou's method to test for spatial randomness.

    B. Use T-square sampling to test for spatial randomness.

    C. Use Diggle's method to test for spatial randomness.

    D. Use Hines and Hines's method to test for spatial randomness.

    E. Discuss the design you would use and the factors you considered for parts *A* to *D*.

    F. Compare and contrast the results from the tests.

8. Consider again the redwood seedlings in exercise 4.

    A. Use Pielou's method to test for spatial randomness.

    B. Use T-square sampling to test for spatial randomness.

    C. Use Diggle's method to test for spatial randomness.

    D. Use Hines and Hines's method to test for spatial randomness.

    E. Discuss the design you would use and the factors you considered for parts *A* to *D*.

    F. Compare and contrast the results from the tests.

Appendix

		Japanese black pine saplings			
X	Y	X	Y	X	Y
0.51	0.51	5.59	4.50	0.68	3.76
3.36	0.11	1.20	4.50	2.96	2.96
4.90	0.74	2.96	5.30	2.68	3.82
2.39	1.25	2.05	5.47	4.16	4.16
0.11	2.34	2.17	0.17	0.68	4.79
0.46	3.36	3.82	0.74	1.82	4.73
1.77	3.02	5.59	0.11	3.93	5.30
5.36	3.31	3.53	1.20	2.45	5.47
3.36	3.82	0.40	2.39	2.74	0.17
5.36	4.45	0.68	3.59	4.50	0.17
0.97	5.42	2.39	2.79	0.63	1.77
2.22	4.50	2.11	3.88	5.07	1.31
2.05	5.53	4.33	3.76	3.65	2.45
1.65	0.11	5.53	4.90	0.97	3.31
3.70	0.91	1.65	4.79	5.19	2.96
5.07	0.46	3.31	4.73	2.96	3.82
2.74	0.74	2.22	5.47	5.07	4.22
0.17	2.51	2.22	1.03	0.63	5.36
0.46	3.59	4.16	0.74	2.00	4.90
1.82	2.96	0.11	1.03	4.39	5.30
1.94	3.88	4.16	1.31	3.63	5.53
3.76	3.88	2.96	2.39		

		Biological Cells			
X	Y	X	Y	X	Y
0.350	0.025	0.212	0.337	0.987	0.512
0.062	0.362	0.150	0.500	0.775	0.850
0.938	0.400	0.525	0.650	0.087	0.187
0.462	0.750	0.625	0.950	0.900	0.262
0.462	0.900	0.825	0.125	0.862	0.525
0.737	0.237	0.650	0.362	0.637	0.812
0.800	0.387	0.725	0.512	0.575	0.212
0.337	0.750	0.237	0.787	0.600	0.475
0.350	0.962	0.775	0.025	0.175	0.650
0.637	0.050	0.450	0.287	0.175	0.912
0.325	0.287	0.562	0.575	0.400	0.162
0.350	0.600	0.862	0.637	0.462	0.425
0.737	0.687	0.237	0.150	0.062	0.750
0.487	0.087	0.337	0.462	0.900	0.775

(*continued*)

(*continued*)

Redwood Seedlings

X	Y	X	Y	X	Y
0.364	0.082	0.381	0.836	0.703	0.279
0.898	0.082	0.483	0.770	0.627	0.344
0.864	0.180	0.203	0.426	0.559	0.639
0.966	0.541	0.102	0.574	0.263	0.852
0.864	0.902	0.186	0.557	0.263	0.697
0.686	0.328	0.441	0.098	0.441	0.754
0.500	0.598	0.839	0.082	0.237	0.475
0.483	0.672	0.780	0.164	0.136	0.574
0.339	0.836	0.898	0.779	0.483	0.148
0.483	0.820	0.678	0.246	0.898	0.189
0.186	0.402	0.610	0.344	0.949	0.525
0.203	0.525	0.585	0.574	0.966	0.959
0.186	0.541	0.220	0.836	0.729	0.262
0.483	0.082	0.407	0.852	0.644	0.361
0.898	0.098	0.508	0.754	0.525	0.656
0.780	0.123	0.220	0.459	0.288	0.852
0.898	0.754	0.119	0.574	0.441	0.820
0.746	0.902	0.500	0.098	0.186	0.377
0.644	0.279	0.898	0.164	0.203	0.500
0.525	0.574	0.763	0.148	0.119	0.623
0.220	0.795	1.000	0.836		

From Diggle, 1983, by permission of Academic Press, London.

9

Capture–Recapture: Closed Populations

Introduction

The methods for estimating the population density of organisms presented in earlier chapters are applicable when the sampling unit is well defined. Examples of these sampling units include a plant, a leaf, a fruit, a nest, and a quadrat. Brock (1954) found it impractical to use these methods when studying reef fish populations because the terrain was irregular (*see* Seber 1982, p. 28). Further, the effective use of well-defined sampling units requires the organisms of interest to remain relatively immobile. For example, if a plant is the sampling unit and an insect the species of interest, then it is assumed that all insects are observed, including those that leave the plant due to the approach of the sampler. With more mobile animals, this assumption is clearly violated, leading to severe underestimates of population density. Deer, fish, birds, rabbits, flies, and rodents are but a few of the animal species whose mobility require alternative sampling methods. Difficulties also arise if the population is sparse. Either very large sampling units or a large number of them is required to estimate population density with an acceptable level of precision.

Two broad categories of approaches that have been developed for these instances are capture models and transect sampling. We will consider capture–recapture methods and removal statistics in this and the next chapter. Transect sampling will be covered in Chapter 11.

Suppose we are interested in estimating the number N of small mammals in a region or the number of fish in a lake. Not only are they highly mobile, their size, coloration, and/or habitat make them difficult to observe even when they are still; that is, they have poor sightability. Often these animals may be captured using traps, seines, and so forth. If the method of capture is such that the organism may be returned to its habitat and subsequently recaptured, then capture-recapture methods are a possible means of sampling and estimating, the size of the population. If the capture method makes it impractical to recapture, or perhaps even to return, the animal to the population, removal statistics should be considered.

When using capture methods, a distinction is made between open and closed populations. A closed population is one in which no births, deaths, or migration into or out of the population occur between sampling periods. Models based on closed populations are generally used for studies covering relatively short periods of time, such as trapping each day for three consecutive days. Although the population may not be fully closed, this assumption may still be reasonable for a short time span. If a population is not closed, it is open; that is, births, deaths, immigration, and emigration may result in population fluctuations during the course of a study. Closed populations will be the topic of this chapter and open populations that of the next.

This chapter is designed to serve as an introduction to some of the most common capture-recapture methods. Seber (1982) is a classical general reference for both open and closed population models, and Seber (1986) updates this earlier work. Pollock (1991) provides an overview of this area of research and suggests areas for future efforts. Otis et al. (1978) and White et al. (1982) consider design and analysis issues for closed populations. We have drawn heavily on all of these works in developing this chapter.

Lincoln–Petersen Model

The Lincoln–Petersen model is an intuitive one that has been discovered and rediscovered numerous times. Laplace (1786) derived the model in the process of estimating the population of France in 1783. He began by obtaining a register of births for the whole country and also for a set of parishes for which the population size was known. He reasoned that the ratio of births (n_1) to the population size (N) for the whole country should be approximately the same as the ratio of births (m_2) to the populations size (n_2) for the parishes; that is,

$$\frac{n_1}{N} \approx \frac{m_2}{n_2}. \tag{9.1}$$

By solving for N in (9.1), Laplace obtained the intuitive estimator of population size:

$$\hat{N}_{LP} = \frac{n_1 n_2}{m_2}. \tag{9.2}$$

In 1896, the Danish fisheries biologist C.G.J. Petersen (*see* White et al. 1982) independently derived the estimator in (9.2) as he sought to estimate the size of the plaice (a marine fish) population. In 1930, Frederick Lincoln again discovered the estimator in the course of a study designed to estimate the abundance of waterfowl in North America. The work by Petersen and Lincoln made ecologists aware of the estimator's great potential in population studies; consequently, it became known as the Lincoln-Petersen estimator.

What are the features of a capture-recapture study for which the Lincoln–Petersen estimator is appropriate? The goal of the study is to estimate the population size N of a given organism within a specified region. A random sample of n_1 animals is caught, marked, and released. Later, after the marked animals have had time to be thoroughly mixed in the population, a second sample of n_2 animals is captured. Of these n_2 animals, m_2 have marks, indicating that they were also in the original sample of size n_1. The Lincoln-Peterson model is based on the assumption that the proportion of marked animals in the second sample should be the same as the proportion of marked animals in the population, leading again to equation (9.1) and subsequently to the Lincoln–Petersen estimator in (9.2). The process of capturing, marking, and recapturing animals to estimate population size has become known as the capture-recapture or mark–recapture method.

If n_1 and n_2 are determined in advance of the study (fixed), then the distribution of the number marked in the second sample (m_2) has a hypergeometric distribution. However, if as is often the case, n_1 and n_2 are random, the likelihood is composed of a series of conditional binomials. In either case, the Lincoln-Petersen estimator is the maximum likelihood estimator (MLE). The approximation in equation (9.1) reflects the fact that sampling variation will cause these to rarely be exactly equal. The estimated variance of $\hat{N}_{LP}$ quantifies this sampling variation:

$$s^2_{\hat{N}_{LP}} = \frac{n_1^2(n_2 + 1)(n_2 - m_2)}{(m_2 + 1)^2(m_2 + 2)}. \tag{9.3}$$

Although $\hat{N}_{LP}$ is a best asymptotically normal estimate of N, both $\hat{N}_{LP}$ and $s^2_{N_{LP}}$ are biased, and the bias can be large for small samples (Chapman, 1951). However, when $n_1 + n_2 \geq N$, Chapman's modified estimate,

$$N^*_{LP} = \frac{(n_1 + 1)(n_2 + 1)}{(m_2 + 1)} - 1, \tag{9.4}$$

is unbiased when n_1 and n_2 are set in advance of the study. At times, it is not possible to set n_2 in the study's design phase as the sample size may depend on the time or effort available for sampling. In these cases, N^*_{LP} is still approximately unbiased. Robson and Regier (1968) investigated the bias of N^*_{LP} and state that it is essential that

$$\frac{n_1 n_2}{N} > 4 \tag{9.5}$$

so that the bias is small. Further, they showed that if the number of recaptures m_2 is 7 or more, we can be 95% confident that $n_1 n_2 / N > 4$, so that the bias of N_{LP}^* is negligible. An approximately unbiased estimator of the variance of N_{LP}^*, given by Seber (1970c) and Wittes (1972), is

$$s_{N_{LP}^*}^2 = \frac{(n_1 + 1)(n_2 + 1)(n_1 - m_2)(n_2 - m_2)}{(m_2 + 1)^2(m_2 + 2)}. \tag{9.6}$$

Occasionally, the second sample is taken with replacement; that is, the animals may be captured more than once during the second trapping occasion. In such cases, the number of recaptures may even exceed the number of marked animals in the population. Under such conditions, the assumption of the hypergeometric distribution used to derive equations (9.4) and (9.6) is no longer appropriate. Sampling with replacement leads to the binomial distribution. Bailey (1951, 1952) worked with a binomial approximation to the hypergeometric distribution. In this case, the Lincoln–Petersen estimator is still the maximum likelihood estimator of N; however, because it is biased with respect to the binomial distribution, Bailey suggested using

$$\tilde{N}_{LP} = \frac{n_1(n_2 + 1)}{(m_2 + 1)}. \tag{9.7}$$

The variance of this estimate can be estimated using

$$s_{\tilde{N}_{LP}}^2 = \frac{n_1^2(n_2 + 1)(n_2 - m_2)}{(m_2 + 1)^2(m_2 + 2)} \tag{9.8}$$

Confidence Intervals

As $N \to \infty$, $\hat{N}_{LP}$, N_{LP}^*, and $\tilde{N}_{LP}$ are normally distributed. However, because the reciprocal of these estimators is more nearly normal and more symmetrically distributed than the estimator, confidence intervals based on the probability distribution of m_2 provide better coverage (Ricker, 1958). Let $p = n_1/N$ be the proportion of marked animals in the population. Exact confidence limits for p when N is known and n_1 unknown have been tabulated (*see* Chung and DeLury, 1950). Here N is unknown and n_1 known. No tables exist for this case so approximate methods are used. The binomial, Poisson, and normal distributions may be used to approximate the hypergeometric distribution of m_2. Unfortunately, the best choice when N is unknown has not been thoroughly investigated. With that in mind, we will follow the general guides suggested by Seber (1982).

To determine which approximation to use, begin by letting $\hat{p} = m_2/n_2$. The choice of approximation will be based on the values of $\hat{p}$ and m_2 as follows:

1. If $\hat{p} < 0.1$ and $m_2 \leq 50$, use the Poisson approximation.
2. If $\hat{p} < 0.1$ and $m_2 > 50$, use the normal approximation.
3. If $\hat{p} > 0.1$, use the binomial approximation.

First suppose $\hat{p} < 0.1$ and $m_2 \leq 50$. A confidence interval for the mean $(n_1 n_2 / N)$ of a Poisson distribution using the relationship between the gamma and Poisson distributions (*see* Chapter 1) is first developed. This gives upper and lower confidence limits on m_2. To obtain the confidence interval for N, equation (9.2), (9.4), or (9.7) is evaluated using each limit, along with the values of n_1 and n_2. (The choice should be the same as the one used to estimate N). Notice that the lower (upper) limit for m_2 gives the upper (lower) limit for N because we use the reciprocal of the lower (upper) limit for m_2 in computing the limits for N. In the accompanying software ECOSTAT, equal-tail confidence intervals are given instead of the shortest $(1-\alpha)100\%$ confidence intervals. Chapman (1948) developed a table for the shortest 95% confidence intervals.

Next assume that $\hat{p} < 0.1$ and $m_2 > 50$. Following Seber (1982), we would use a normal approximation to obtain a $(1 - \alpha)$ 100% confidence interval for p of the form

$$\hat{p} \pm \left(z_{\alpha/2} \sqrt{\frac{(1 - f)\hat{p}(1 - \hat{p})}{n_2 - 1}} + \frac{1}{2n_2} \right) \tag{9.9}$$

where $z_{\alpha/2}$ is the $(1 - \alpha/2)$th quantile of the standard normal distribution (*see* Chapter 1) and $f = n_2/N$ is the unknown sampling fraction. If the estimate of $f(\hat{f} = m_2/n_1)$ is less than 0.1, it can be neglected. The term $1/(2n_2)$ is the correction for continuity and is often negligible. From equation (9.2) and the definition of $\hat{p}$, we have

$$\hat{N} = \frac{n_1}{\hat{p}}. \tag{9.10}$$

By substituting the lower and upper confidence limits for p in for $\hat{p}$ in (9.9), we have the respective upper and lower confidence intervals for N. As with the Poisson approximation, the lower (upper) limit for p is used to compute the upper (lower) limit for N. Again, equal-tail confidence intervals, and not the shortest possible confidence intervals, are given in ECOSTAT.

Finally, suppose that $\hat{p} > 0.1$. In this case, more than 10% of the population has been marked. Generally, this occurs if the population size is small or sampling is intense. A confidence interval on p is based on the binomial distribution. The limits are computed using the incomplete beta function (*see* Chapter 1). These are then used in equation (9.8) to obtain a confidence interval for N.

Example 9.1

Ayre (1962) raised a colony of foraging ants, *Camponotus herculeanus* (Linné), in the laboratory at 23°C and 50% relative humidity. The foraging area was 2 × 4 feet. In a capture–recapture experiment, 40 ants were captured and marked on the dorsum of the thorax with a "Techpen." The ants were allowed to mix for 20 hours. Because only a few of the workers were visible at any given time, the totals of nine hourly observations were used in the recapture period. Of the 76 recaptures, 28 were marked. It is possible, even probable, that the same ant was observed more than once during the recapture period. Thus sampling was with replacement.

Using Bailey's estimator (9.7), the population size of the nest is estimated to be

$$\tilde{N}_{LP} = \frac{40(76 + 1)}{(28 + 1)} = 106.21.$$

Therefore, the population of the ant nest is estimated to be 106 ants. The estimated standard deviation of this estimate is

$$S_{\tilde{N}_{LP}} = \sqrt{\frac{40^2(76 + 1)(76 - 28)}{(28 + 1)^2(28 + 2)}} = 15.3$$

or 15 ants. The estimated proportion of marked ants in the population is

$$\hat{p} = \frac{28}{76} = 0.37.$$

Therefore, the binomial approximation is used to set the confidence interval on the population size N.

ECOSTAT may be used for this as well as for the earlier computations. One first selects the *Capture–Recapture* menu. The menu selections are then *Lincoln–Petersen,* followed by *Estimation.* After indicating whether sampling is with or without replacement, the data are to be entered from the keyboard. The *Estimate N* button will display both the original Lincoln–Petersen and either the Bailey or the Chapman population estimates, depending on whether sampling is with or without replacement. The standard errors of each are also given. By pressing the *Set CI* button, the confidence limits, based on each estimator are developed and displayed. ECOSTAT automatically uses the approximation that is recommended. Suppose a 95% confidence interval is desired. The resulting limits are 82.2 and 153.9. Therefore, we are 95% confident that the ant population is between 82 and 154. We will return to this example shortly.

Example 9.2

Migratory water fowl have been captured and banded at widely dispersed banding stations for many years. Hunters who harvest a banded bird are asked to return the band with information on date and place the bird was taken. In separate surveys, the total number of birds harvested during the season is estimated. One reason for collecting this information is to monitor the impact of hunting on the bird populations. Lincoln (1930) gathered data from the banding files of the Bureau of Biological Survey for the years 1920 to 1926. Only data from stations banding at least 100 birds were used.

In 1922, a total of 3,774 birds were banded across the United States. During the following shooting season, 572 bands were returned. This was 15.16% of the total harvest. Based on this information, a total of 3,773 ($=3572/0.1516$) ducks and geese were harvested. The population size is then estimated to be

$$N_{LP}^* = \frac{(3774 + 1)(3773 + 1)}{(572 + 1)} - 1 = 24862.6$$

or 24,863 ducks and geese. The estimated standard deviation of this estimate is

$$S_{N_{LP}^*} = \sqrt{\frac{(3774 + 1)(3773 + 1)(3774 - 572)(3773 - 572)}{(572 + 1)^2(572 + 2)}} = 880.24$$

or 880 birds. Again the proportion of captured birds in the second sample is more than 0.1 indicating the use of the binomial approximation. In this case, the sample sizes are so large that a normal approximation is also appropriate. ECOSTAT was used to compute 90% confidence limits. From these, we are 90% confident that the duck and geese population was between 23,113 and 26,841 birds in 1922.

As in all applications, the estimates are based on the assumption that the data are accurate. Because of his concern about data accuracy, Lincoln encouraged all hunters to report all banded birds and to furnish reports of their seasonal harvest.

Sample Size Considerations

Notice from equation (9.2) that if none of the n_2 animals in the second sample is marked ($m_2 = 0$), $\hat{N}$ is undefined. Further, by looking at equation (9.3), it is evident that as m_2 increases the estimate of variance decreases. Increasing m_2 can be accomplished by increasing n_1 and/or n_2. Yet, this also results in increased costs. Therefore, an important question to address before initiating the study is, "How large should n_1 and n_2 be to achieve the desired precision within the cost constraints of the study?" Robson and Regier (1968) considered the problem of determining n_1 and n_2 so that $\hat{N}$ is within a stated proportion D ($0 < D < 1$) of N with $(1 - \alpha)$ 100% confidence; that is,

$$P(|\hat{N} - N| < DN) \geq 1 - \alpha. \qquad (9.11)$$

This may be alternatively expressed as

$$1 - \alpha \leq P\left(-D < \frac{\tilde{N} - N}{N} < D\right). \qquad (9.12)$$

What level should be used for D and $(1 - \alpha)$? Robson and Regier (1968) suggested three standards depending on the application:

1. $D = 0.50$ and $(1 - \alpha) = 0.95$ for preliminary studies or management surveys where only a crude estimate of population size is needed.
2. $D = 0.25$ and $(1 - \alpha) = 0.95$ for more accurate management work.
3. $D = 0.10$ and $(1 - \alpha) = 0.95$ for research on population dynamics.

Sample size requirements can be expressed in terms of the numbers of animals captured during the two capture periods, n_1 and n_2, or in terms of the number of animals initially caught and marked and the number of marked animals in the second sample, n_1 and m_2, respectively. Chapman (1952) referred to a design specifying n_1 and n_2 as "direct" and one specifying n_1 and m_2 as "inverse." Following Robson and Regier (1968), we will consider only the direct method. Although this leads to slightly larger sample sizes, there is little, if any, practical difference unless D is relatively large (≥ 0.5) and N is small (Chapman, 1952, p. 289).

Writing $\hat{N}$ in terms of m_2, n_1, and n_2 as in equation (9.2), substituting into equation (9.12), and rearranging the terms, we have

$$1 - \alpha \leq P\left(\frac{n_1 n_2}{(1 + D)N} < m_2 < \frac{n_1 n_2}{(1 - D)N}\right) \qquad (9.13)$$

Recalling that for fixed n_1 and n_2, m_2 has the hypergeometric distribution, equation (9.13) can be evaluated for potential values of N, n_1, and n_2. ECOSTAT can be used to determine this probability. To access this section of ECOSTAT, choose *Lincoln–Petersen* from the *Capture-Recapture* menu and *Sample Size* on the submenu. The first decision is whether N is believed to be less than or in excess of a thousand. If N does not exceed 1,000, the exact probabilities based on the hypergeometric distribution are computed after values are set for n_1, n_2, D, and N.

If N exceeds 1,000, a normal approximation of the hypergeometric distribution is used. This is accomplished by setting

$$\mu = \frac{n_1 n_2}{N} \qquad (9.14)$$

and

$$\sigma^2 = \frac{n_1(N - n_1)n_2(N - n_2)}{N^2(N - 1)} \tag{9.15}$$

After substituting into equation (9.13) and rearranging the terms, we have

$$1 - \alpha = \Phi\left(\frac{p}{1 - p}\sqrt{\frac{n_1 n_2(N - 1)}{(N - n_1)(N - n_2)}}\right)$$

$$- \Phi\left(-\frac{p}{1 + p}\sqrt{\frac{n_1 n_2(N - 1)}{(N - n_1)(N - n_2)}}\right) \tag{9.16}$$

where $\Phi(\cdot)$ is the cumulative standard normal distribution. Equation (9.16) is used to determine $1 - \alpha$ when $N > 1,000$. In this case, sample sizes less than 40 should not be considered, ensuring that the normal approximation is sufficiently close. This is not unduly restrictive. For example, suppose $N = 1000$ and $n_1 = n_2 = 40$. Based on equation (9.5), the estimate of N is significantly biased. Further, using ECOSTAT, we find that the probability that the estimated population size is within 30% ($D = 0.3$) of the true population is only 0.33. Therefore, for populations of more than 1,000, we want n_1 and n_2 to be larger than 40. This will ensure that the normal approximation will be valid and that sample sizes will yield sufficiently precise estimates.

As in other circumstances, the sample size requirements depend on knowledge of the population parameter of interest, N. Although sampling is unnecessary if N is known, an ecologist should have some estimate of the sample size. This could be based on historical records or previous sampling. A plausible range of N values could be evaluated to be sure that the sample size is large enough to yield the desired precision.

Example 9.3

Suppose that we are preparing to estimate the deer population in a specified area using capture–recapture methods. Past records indicate the population has fluctuated but tends to be around 250 deer. Because the estimates are to be used for management purposes, we set $D = 0.25$. Also, it has been suggested to set $n_1 = 35$ and $n_2 = 30$. Based on these values, the confidence level is 41%, instead of the desired 95%. Obviously, more data need to be collected to obtain the desired precision in our estimate!

Assumptions

The Lincoln–Petersen model is intuitively appealing and simple to apply, leading to its extensive use today. An ecologist should give some thought to the basic biological assumptions underlying the model before implementing it. First, the population size will be estimated only for the catchable population. For many species, some portion of the population is simply not catchable. For example, consider Example 9.1 again. Because Ayre (1962) had raised the colony in the laboratory, he knew that the population was about 3,000. Our estimate of 106 was not even close! The difficulty was that only a small portion of the population ever foraged so that the vast majority of the population was uncatchable. In such circumstances, other methods of population estimation must be used.

Suppose now that we believe that most, if not all, of the population is catchable. Three critical assumptions must be made for the valid use of this model: (1) the population is closed, (2) marks are not lost or overlooked, and (3) all animals are equally likely to be captured in each sample. Each of these will be examined in more detail.

1. The Population is Closed

A population is rarely, if ever, truly closed. However, careful planning of the study will often permit this assumption to hold at least approximately. A primary consideration is the length and timing of the study. The study duration must be relatively short, such as trapping on consecutive nights. Despite the desire to keep the study period short, care must be taken that the duration is long enough to meet the needs of the model. Enough time must be spent during each capture period to obtain sufficient sample sizes. Further, adequate time must be given for the marked animals to reassimilate into the population before recapture is initiated. Releasing the animals from the point of capture (instead of from a centralized location) may reduce the time needed in this regard.

Timing is also an essential consideration. Avoiding breeding and migration periods are critical if the assumption of a closed population is to be valid. If periods in which the population is known to be changing rapidly cannot be avoided, then the open population models discussed in the next chapter should be used.

Seber (1982) and Otis et al. (1978) consider a number of common departures from the closed population assumption and evaluate their impact on the population estimate. Sometimes animals die during the process of capture and marking. If there are accidental deaths caused by the process of catching and marking the animals, the estimate of population size is valid provided that n_1 is the number of animals returned alive to the population, and not the number captured in the first sample.

Suppose that natural mortality occurs during the study period. As long as deaths are occurring at random, so that the average probability of surviving until the time of the second sample is the same for both the marked and unmarked members of the population, the Lincoln-Petersen estimator is valid.

Recruitment into the sampled population is another concern. For example, if the trapping method will only catch animals above a certain size threshold, as in the case of fish seines, young may be recruited between the sampling periods. If recruitment occurs, $\hat{N}$ overestimates the initial population size. If there is recruitment but no mortality between the two sample periods, $\hat{N}$ is a valid estimator of population size at the time the second sample is taken. If both recruitment and mortality occur between the two sample periods, $\hat{N}$ overestimates both the initial and final population size.

2. Marks Will Not Be Lost or Overlooked

For capture–recapture studies, captured animals must be marked so that they can be clearly identified as having been captured. Numerous methods of marking exist. Animals may be painted or dyed, giving a quick visual means of identifying previously captured animals. Bands or rings are commonly used to mark fish, birds, and large mammals. Mutilation, such as notching a fin, is often used for fish, amphibians, insects, and reptiles. Radioactive isotopes are frequently employed to mark animals. Details on these and other methods are covered by Southwood (1978) and Seber (1982). New marking methods are constantly being developed. Experienced field ecologists are a source of invaluable advice on the use of various marking methods.

Statistically, some important assumptions are made about the marking process. One assumption is that the marks are not lost or overlooked. Therefore, a mark that disappears with adverse weather or one that becomes difficult to observe as the hair of the animal grows may not be appropriate. A simple approach to accounting for lost marks is to use two different and distinguishable marks for each animal caught in the initial sample. In the second sample, the number of recaptures with just one mark and those with two are recorded (Beverton and Holt, 1957; Gulland, 1963; Seber, 1982). Denote the two types of marks by A and B. Then we define the following terms:

π_i: the probability that a mark of type i is lost by the time of the second sample ($i = A, B$).

π_{AB}:   the probability that both marks are lost by the time of the second sample

m_i: the number in the second sample having only mark i ($i = A, B$)

m_{AB}:   the number in the second sample having both marks A and B

m_2:   the number of recaptures, i.e., the number in sample 2 that were also in sample 1

The number of animals in the first sample is

$$n_1 = r_A + r_B + r_{AB} + \text{[number losing both marks]} \qquad (9.17)$$

If the population is estimated based on the number of animals with at least one mark, the estimate is too small because the number of animals losing both marks has been ignored. However, an adjustment can be made to compensate for these lost marks. Assuming that the marks are independent of each other, the maximum likelihood estimators of π_i, $i = A, B$, and m_2 are

$$\hat{\pi}_A = \frac{m_B}{m_B + m_{AB}}, \tag{9.18}$$

$$\hat{\pi}_B = \frac{m_A}{m_A + m_{AB}}, \tag{9.19}$$

and

$$\hat{m}_2 = c(m_A + m_B + m_{AB}), \tag{9.20}$$

respectively, where

$$c = (1 - \hat{\pi}_A\hat{\pi}_B)^{-1} \tag{9.21}$$

The population estimator

$$\hat{N}_{AB} = \frac{n_1 n_2}{\hat{m}_2} \tag{9.22}$$

is an approximately unbiased estimator of N. Given n_1 and n_2, an estimate of the variance of $\hat{N}_{AB}$ is

$$s^2_{\hat{N}_{AB}} = \frac{\hat{N}^3_{AB}}{n_1 n_2} \hat{\pi}_A \hat{\pi}_B \left[\frac{1}{(1 - \hat{\pi}_A)(1 - \hat{\pi}_B)} \right] + \frac{\hat{N}^3_{AB}}{n_1 n_2} \left[1 + \frac{2\hat{N}_{AB}}{n_1 n_2} + 6\left(\frac{\hat{N}_{AB}}{n_1 n_2}\right)^2 \right]. \tag{9.23}$$

Example 9.4

Beverton and Holt (1957 as discussed by Gulland 1963) reported the results of a double-tagging experiment. In 1947, 516 plaice had two pairs of ebonite Persen discs attached by silver wire, one on the front and one on the rear. They reported 13 fish were returned with only the front tag and 12 with the rear tag. The ratio of the number of double-tagged plaice returned with one tag to the number returned with two tags was estimated to be 0.35. Therefore, the number of double tagged fish returned was probably 71 (25/0.35 = 71.43). Using A and B to denote the front and rear tags, respectively, we can compute the following maximum likelihood estimates:

$$\hat{\pi}_A = \frac{12}{12 + 71} = 0.14,$$

and

$$\hat{\pi}_B = \frac{13}{13 + 71} = 0.15.$$

Before finding the MLE of m_2, we must first compute c:

$$c = [1 - 0.14(0.15)]^{-1} = 1.02.$$

Then we have

$$\hat{m}_2 = 1.02(13 + 12 + 71) = 97.92.$$

Therefore, the estimate of population size is

$$\hat{N}_{AB} = \frac{516 n_2}{97.92} = 5.27 n_2.$$

If the total number of fish in the second sample is 382, the population size is estimated to be $\hat{N}_{AB} = 5.27(382) \approx 2,013$ fish. The estimated variance of this estimate is

$$s^2_{\hat{N}_{AB}} = \frac{2013^3}{516(382)} (0.14)(0.15)\left(\frac{1}{(1 - 0.14)(1 - 0.15)}\right)$$

$$+ \frac{2013^3}{516(382)}\left(1 + \frac{2(2013)}{516(382)} + 6\left(\frac{2013}{516(382)}\right)^2\right) = 43443,$$

or a standard deviation of 208 fish.

Another assumption is that the capture and subsequent marking does not affect the animal's behavior or life expectancy. Thus, a mark that increases an animal's susceptibility to predation is clearly unsatisfactory. Further, marked and unmarked animals should have equal catchability, an assumption that we will now consider in more detail.

3. All Animals Are Equally Likely to Be Captured in Each Sample

In designing the study, the researcher must choose a method of capturing the animals of interest. The method will vary, depending on the species and the environment. Seines for fish and live traps for small mammals are two examples. If a capture–recapture study is planned, an important assumption is that capture does not affect the animal's behavior, its life expectancy, or its probability of being subsequently captured. This implies that a captured animal will become completely mixed in the population after being returned to its habitat and that all animals, whether marked or unmarked, have equal catchability. Equal catchability has two aspects: (1) individuals of both sexes and all age categories are sampled in proportion to their occurrence in the population, and (2) all individuals are equally available for capture regardless of their location within the habitat.

Statistical tests for equal catchability require at least three trapping occasions so we will defer that discussion until later. However, it is important to realize that careful consideration of the biology of the species can help reduce or eliminate problems arising from differential catchability. Some traps are known to attract more of one gender than another. Often traps are laid out on a grid. This ensures complete coverage of the study area. Yet, it is necessary to assume that every member of the population has an equal chance of encountering a trap. Often small mammals express territoriality or have well-defined home ranges. Either may cause some population members to have little or no probability of being captured. Otis et al. (1978) recommend placing at least four traps per home range to help avoid differential catchability.

Multiple Recapture Models

Schnabel (1938) produced a pioneering work extending the Lincoln-Petersen estimator to more than one period of marking and releasing the animals. Her work was directed toward a study to estimate the fish population of a lake. Seining stations were located at various points on the lake. Samples were taken at periodic intervals, usually about 24 hours. The fish were tagged and released. For each sample, the total number, the number of unmarked fish, and the number of re-captures were recorded. Assuming that each sample was random and that the population was closed during the sampling period, an estimate of the total number of fish in the lake was obtained.

Schnabel's work was the first of many papers that have been written to address aspects of multiple recapture studies. Initial work was largely fragmented, considering various aspects of such studies. However, in 1974, Pollock proposed a more unified approach to modeling capture–recapture data. This work was more fully developed by Otis et al. (1978), White et al. (1982), and Pollock and Otto (1983) and will be used here.

The models are for closed population models so the concerns addressed when discussing the Lincoln-Petersen estimator remain pertinent here. Each animal has a unique mark or tag so that individual capture histories may be recorded. Carothers (1973, p. 146) stated that equal catchability is an unattainable ideal. Therefore, efforts have been directed toward more accurately modeling departures from equal catchability using methods that are both biologically directed and mathematically tractable. This has led to concentrating on three sources of variation in trapping probabilities:

1. Capture probabilities vary with time or trapping occasion.
2. Capture probabilities vary due to behavioral responses to the traps.
3. Capture probabilities vary from individual to individual.

Capture probabilities may vary with time. For example, a rain on one trapping night may decrease activity and thus lower the probability of capture when compared to other trapping nights. If different capture methods or baits are used on separate trapping occasions, the capture probabilities will probably vary. The Lincoln–Petersen estimator implicitly assumes that capture probabilities vary with time so it is the maximum likelihood estimator when this is the only source of variation in capture probabilities and there are $t = 2$ trapping occasions.

Once captured the behavior of the animal may be altered. Some animals become "trap happy" while others become "trap shy". Once captured the animals respond differently, changing their probability of capture. These behavioral responses to the trap may be accounted for in the model.

Finally, some individuals may have a higher probability of capture than others. Common sources of variation are age and gender. If the species has home ranges, some animals may not have as much accessibility to a trap as others. Social dominance is another factor that could affect the individual's probability of capture. Because each individual could have a unique capture probability, models with this restriction require additional distributional assumptions.

Before considering this class of models, it is important to realize that the correct choice of model is critical. The estimate of population size may be quite biased if the wrong model is used, and the estimate of the standard deviation of this estimate is also affected. The tests designed to choose among models are based on large sample properties. They require a sufficiently large population to be valid. If the population is small ($N < 100$), work by Menkens and Anderson (1988) suggests that it may be better to divide the sampling occasions into two periods: those in the first half of the study and those in the second half of the study. The Lincoln–Petersen estimator could then be used. However, if high individual heterogeneity and trap response are present either singly or together, they found the relative bias of the Lincoln–Petersen estimator to be unacceptably high (>20%). Careful design is essential so that these factors can be kept to a minimum. Further, the work by Menkens and Anderson serves as a call for further research. Although considerable progress has been made in developing tests for choosing among models, more work is clearly needed.

We consider a set of eight models as summarized in Table 9.1. It is assumed that sampling is conducted on at least five occasions (Otis et al., 1978). The first model assumes that there is no variation in capture probabilities due to time,

Table 9.1. Capture–Recapture Models for Closed Populations Supported by ECOSTAT

| Model | Source of variation in capture probabilities | | | Estimator |
	Time	Trap response	Heterogeneity	
M_0				Maximum likelihood
M_t	X			Maximum likelihood
M_b		X		Maximum likelihood
M_h			X	Jackknife
M_{bh}		X	X	Jackknife
M_{bt}	X	X		Leslie's regression
M_{ht}	X		X	Leslie's regression
M_{bht}	X	X	X	Leslie's regression

behavior response, or individual variation. Models addressing each of these sources of variation, as well as all possible combinations are also considered. First, we must extend the notation used in developing the Lincoln–Petersen model to account for more than two sampling periods. Define the following:

t: number of samples

e_j: known relative time effect of the jth sample on capture probabilities

$E_j = \sum_{i=1}^{j-1} f_i(j = 1,3, \ldots, t; F_1 = 0)$: Units of effort expended prior to the ith sample

C: sample coverage

C_j: sample coverage of the first j samples, $j = 1, 2, \ldots, t$: $C_t = C$

Parameters

N: population size

p_{ij}:   probability that individual i will be captured during sampling period j

Statistics

n_j: the number of animals captured during sampling period i, $i = 1, 2, \ldots, t$

n: total number of captures in the study ($n = n_1 + n_2 + \ldots + n_t$)

m_j: the number of animals recaptured during sampling period j

m: sum of the m_j, $\sum_{j=1}^{t} m_j$

M_j: number of *marked* animals in the population at the time of the jth sample

M: sum of the M_j, $\sum_{j=1}^{t} M_j$

f_k: number of animals captured exactly k times during the t sampling periods

u_j: number of unmarked animals captured during sampling period j ($u_j = n_j - m_j$)

$\{X_\omega\}$:   the set of possible capture histories

$f_k^{(j)}$:   number of animals captured exactly k times that were first captured on the jth occasion

R_{jl}:   number of animals recaptured on the lth occasion that were first caught on the jth occasion.

Note that for a study with t sampling periods, M_{t+1} represents the total number of different animals trapped during the study. If no mention is made of units of effort, they are assumed to be constant.

Thought should be given to how the data will be recorded, before the experiment actually begins. The fullest analysis requires individual capture histories. This requires that each animal receives a unique tag. Further, careful records are kept showing the sampling periods that each animal is caught. A one (zero) is used to denote that the animal was (was not) captured during a sampling period. An example of some capture histories is given Table 9.2. Animal 1 was caught during the first and fourth sampling periods, but not during the others. Animal 2 was caught during the first three sampling periods but not for any later period. Animal 5 was caught only during the last sampling period. The capture histories X_ω, $\omega = 1, 2, \ldots M_{t+1}$, are vectors of the 0's and 1's for each animal. Thus $X_1 = (1\ 0\ 0\ 1\ 0\ 0)$. ECOSTAT accepts capture history data that have been saved in an ASCII file. Such files can be created easily using a word processor or a spread sheet. Spaces should separate entries on a line. The first line should give the number of individuals M_{t+1} and the number of sampling periods t. Each of the next M_{t+1} lines contains information on an individual animal. The first entry is the animal's identification (no spaces can appear within this entry). Then the t values giving the capture history follow.

Sometimes it is not possible to give each animal individual tags. In these cases, only summary data are collected. That is, for each sampling period, the number

Table 9.2.   Individual Capture Histories

Animal	Sampling period					
	1	2	3	4	5	6
1	1	0	0	1	0	0
2	1	1	1	0	0	0
3	0	1	0	0	0	0
4	0	1	0	0	1	1
5	0	0	0	0	0	1

of animals caught n_j and the number of those that are marked m_j are recorded. Some models can be fitted and tests conducted with such data, but not all models and not all tests can be considered, ECOSTAT accepts summary data, again in the form of an ASCII file. The first line has t, the number of sampling periods. The next t lines each have an identification for the sampling period, n_j, and m_j for that period.

In capture-recapture studies, the sampling effort is assumed to be the same for each sampling period. For example, the same number of traps are set on each night. Several factors can cause the sampling effort to vary for each sampling period. For example, Sylvia (1995) was studying small mammal populations during a rainy spring and summer. On each rainy night, a number of the traps were sprung by the rain, thereby reducing the catch effort. If such problems are transient and can be anticipated, waiting a night may help avoid difficulties. However, even with the best of efforts, this is not always possible and significant variation in catch effort can occur. Methods have been developed to account for differential catch effort. Again, if individual capture histories are available, along with the information on catch effort, then the fullest analysis can be conducted. Summary data may be used to estimate the parameters for some models. ECOSTAT accepts only summary data for differential catch-effort models, and it must be in the form of an ASCII file. The first line gives the number of sampling periods t. The next t lines have an id for the sampling period, followed by the number of animals caught n_j, the catch–effort e_j, and the number of animals recaptured m_j. Although we have referred to e_j as representing the number of traps, it could also be a function of an environmental variable such as temperature that is known to affect the probability of trapping an animal (*see* Lee and Chao, 1994, for a fuller discussion).

Model M_0: Constant Capture Probabilities

The simplest model assumes that there is no variation in capture probabilities. Each individual is equally at risk on each sampling occasion. Further, the probability of capture remains constant through time. This model is designated as M_0 and has only two parameters: the population size N and the probability of an individual being captured on any trapping occasion $p (p = p_{ij}$ for all i and j). The probability distribution of the set of possible capture histories is given by

$$P[\{X_\omega\}] = \frac{N!}{\left(\prod_\omega X_\omega!\right)(N - M_{t+1})!} p^n (1 - p)_{tN-n} \qquad (9.24)$$

If $t = 2$, the MLE of N is

$$\hat{N}_0 = \frac{(n_1 + n_2)^2}{4m_2}. \tag{9.25}$$

For $t > 2$, there is no closed form solution for $\hat{N}_0$. The MLE is the unique root greater than M_{t+1} of

$$\left(1 - \frac{M_{t+1}}{\hat{N}_0}\right) = \left(1 - \frac{\sum\limits_{j=1}^{t} n_j}{t\hat{N}_0}\right)^t. \tag{9.26}$$

We must solve iteratively for $\hat{N}_0$. (Recall, we encountered the same problem when determining the MLE of the negative binomial parameter k in Chapter 1.) The maximum likelihood estimator of the constant probability of capture is

$$\hat{p} = \frac{\sum\limits_{j=1}^{t} n_j}{t\hat{N}_0}. \tag{9.27}$$

The estimated asymptotic variance of $\hat{N}_0$ is (Darroch, 1959):

$$s_{\hat{N}_0}^2 = \frac{\hat{N}_0}{(1 - \hat{p})^{-1} - t(1 - \hat{p})^{-1} + t - 1}. \tag{9.28}$$

We should note that the minimal sufficient statistic is $\{n, M_{t+1}\}$. The minimal sufficient statistic is the lowest dimensional set of statistics that provides all the information that the sample has to offer about the unknown parameters. In this case, if we know the total number of animals captured and the total number of different animals captured, we have all the information that the sample can provide in estimating the population size N and the probability of capture p. For this model, individual capture histories are not needed.

Rarely is model M_0 biologically correct. The model implies that each time sampling occurs, the probability each animal is captured is constant. It does not vary with time or with animal. Animals that have been caught are as likely to be caught during the next sampling period as animals that have not been caught. Both the animals and the environment cause departures from this ideal. Although model M_0 can hold approximately, it will more likely serve as a baseline against which we judge departures from this ideal.

Model M_t: *Capture Probabilities Vary With Time*

For this model, the capture probabilities are assumed to change from one sampling period to the next. However, at any given sampling period, the probability of capture is the same for all individuals in the population. Therefore, the parameters for the model are the population size N and the probability of capture on the jth occasion $p_j = p_{ij}$ for all $i = 1, 2, \ldots, N$. The classical multiple capture–recapture study has the same set of assumptions. Here the probability of the set of possible capture histories is

$$P[\{X_\omega\}] = \frac{N!}{\left(\prod_\omega X_\omega!\right)(N - M_{t+1})!} \prod_{j=1}^{t} p_j^{n_j}(1 - p_j)^{N - n_j}. \qquad (9.29)$$

The Lincoln–Petersen estimator is the MLE when there are only two sampling periods, $t = 2$; that is, $\hat{N}_t = n_1 n_2 / m_2$. Schnabel's (1938) work was also directed toward this model and was based on the assumption that the n_j's are known. Because sampling intensity is usually constant (as when a fixed number of traps is set each night), n_j is seldom fixed. Darroch (1958) was the first to account for this additional level of mathematical complexity. ECOSTAT computes the MLE for model M_t, which is the unique root greater than M_{t+1} of

$$\left(1 - \frac{M_{t+1}}{\hat{N}_t}\right) = \prod_{j=1}^{t}\left(1 - \frac{n_j}{\hat{N}_t}\right). \qquad (9.30)$$

It should be noted here that ECOSTAT finds the root of equation (9.30) and rounds to the closest integer, as opposed to searching over the integers for the value of N that will maximize the likelihood. Otis et al. (1978) developed the software CAPTURE for the analysis of capture–recapture studies in closed populations. It is the most complete software currently available for analysis of these studies. They adopted the latter approach. This may cause our results to be different from that of CAPTURE, and slightly less than optimal. However, the difference is small (usually no more than 1), and we do not believe it to be a practical concern.

The MLE of the probability of capture on the jth sampling occasion, which is constant for all members of the population is

$$\hat{p}_j = \frac{n_j}{\hat{N}_t}. \qquad (9.31)$$

The estimated asymptotic variance of $\hat{N}_t$ is

$$s_{\hat{N}_t}^2 = \hat{N}_t\left(\frac{1}{\prod_{j=1}^{t}(1 - \hat{p}_j)} + t - 1 - \sum_{j=1}^{t}(1 - \hat{p}_j)^{-1}\right)^{-1} \qquad (9.32)$$

(Darroch, 1958).

Several estimators of N based on models that allow capture probabilities to vary through time have been developed (*see* Ricker, 1975; Seber, 1982, Chapter 4). Because of its optimality properties, we will rely on the MLE. However, the Schnaubel estimator is often used in this circumstance because it does provide a closed form estimator of N and is a fairly good approximation to the MLE. It has the form

$$\hat{N}_t^* = \frac{\sum_{j=2}^{t} n_j M_j}{\sum_{j=2}^{t} m_j}. \tag{9.33}$$

If the proportion of the population that is caught during each sampling period (n_j/N) and the proportion of the population that is marked (M_j/N) are always less than 0.1 for each j, a more precise estimate is

$$\tilde{N}_t^* = \frac{\sum_{j=2}^{t} n_j M_j}{1 + \sum_{j=2}^{t} m_j} \tag{9.34}$$

(Seber, 1982). Notice that the summation is from 2 to t. Because no animals are marked prior to the first sampling period, the term for $j = 1$ is always zero. The variance of $\tilde{N}_t^*$ is estimated by

$$s_{\tilde{N}_t}^2 = \tilde{N}_t^{*2} \left(\frac{\tilde{N}^2}{\sum\limits_{j=2}^{t} n_j M_j} + 2 \frac{\tilde{N}_t^{*2}}{\left(\sum\limits_{j=2}^{t} n_j M_j\right)^2} + 6 \frac{\tilde{N}_t^{*3}}{\left(\sum\limits_{j=2}^{t} n_j M_j\right)^3} \right) \tag{9.35}$$

(Chapman, 1952, as corrected by Seber, 1982).

Tests for the Model

How do we know when to use model M_t instead of the simpler M_0? We will look at two tests. The first tests the hypothesis $H_0: p_j = p, j = 1, 2, \ldots, t$, against H_t: not all the p_j are equal. This is explicitly testing whether the probability of capture changes with time. Rejection of the null favors the use of model M_t over model M_0. Otis et al. (1978) recommend the following test statistic for this test:

$$T_{0t} = \sum_{j=1}^{t} \frac{n_j}{1 - \hat{p}_j} - \frac{\left(\sum\limits_{j=1}^{t} \dfrac{n_j}{\hat{p}_j(1 - \hat{p}_j)}\right)^2}{\sum\limits_{j=1}^{t} \dfrac{n_j}{\hat{p}_j^2(1 - \hat{p}_j)}}. \tag{9.36}$$

Under H_0, T_{0t} has an approximate χ^2 distribution with $(t - 1)$ degrees of freedom. Large values of the test statistic lead to rejection of the null hypothesis in favor of the alternative of model M_t. Summary data are sufficient for this test.

The second test is a goodness-of-fit test of model M_t proposed by Leslie (1958) and adopted by Otis et al. (1978). The test assesses whether capture probabilities vary among sampling occasions but not among individuals (H_0) or whether capture probabilities vary with both sampling occasions and individuals (H_1). More formally, we test H_0: $p_{ij} = p_j$ versus H_1: not all $p_{ij} = p_j$, $i = 1, 2, \ldots, M_{t+1}$, and $j = 1, 2, \ldots, t$. The test statistic is

$$T_t = \sum_{j=1}^{t-3} \left(\sum_{k=1}^{t-j+1} \frac{f_k^{(j)}(k - 1 - \hat{\mu}_j)^2}{\hat{\mu}_j - \sum\limits_{l=j+1}^{t} R_{jl}^2/u_j^2} \right) I_j \tag{9.37}$$

where

$$\hat{\mu}_j = \sum_{k=1}^{t-j+1} \frac{f_k^{(j)}(k - 1)}{u_j} \tag{9.38}$$

and I_j is 1 if $u_j \geq 20$, and is 0 otherwise. The indicator function I_j is used to include only those "cohorts" of newly identified animals first captured on the ith occasion that consist of at least twenty individuals as suggested by Leslie. Further, a new cohort must be subjected to at least three subsequent sampling occasions to be used in the test. Under the null hypothesis, T_t has an approximate χ^2 distribution with

$$\sum_{j=1}^{t-3} (u_j - 1)I_j \tag{9.37}$$

degrees of freedom. Rejection of the null hypothesis indicates a lack of fit of model M_t and implies that either capture probabilities vary among individuals or some individuals are having a behavioral response to the traps. Although the model can be estimated and the first test conducted using summary data, the second test requires capture histories.

Model M_b: Behavioral Response to Capture

Model M_b is based on the assumption that the initial capture affects the probability of capture on subsequent occasions; that is, an animal exhibits a behavioral response to capture and becomes either trap happy or trap shy. However, capture probabilities do not vary with time, and all individuals have the same probability of capture. The existence of trap response has been well documented (Geis, 1955; Tanaka, 1956, 1963; Flyger, 1959; Bailey, 1968; Pucek, 1969). For model M_b, it is assumed that on each sampling occasion, all marked animals have one probability of capture, and all unmarked animals have another probability of capture. The model has three parameters: the population size N, the initial probability of capture p, and the probability of recapture c.

Pollock (1974) found the probability distribution of the set of possible capture histories $\{X_\omega\}$ for model M_b to be

$$P[X_\omega] = \frac{N!}{\left(\prod_\omega X_\omega!\right)(N - M_{t+1})!} \, p^{M_{t+1}}(1p)^{N-M_{t+1}-M.}c^{m.}(1 - c)^{M.-m.} \quad (9.40)$$

where $M. = \Sigma M_j$ is the sum over j of the number of marked animals in the population at the time of the jth sampling occasion, and $m.$ is the total number of marked animals caught.

In this case, the minimal sufficient statistic is $\{M_{t+1}, m., M.\}$. Therefore, if we know the total number of distinct animals captured, the total number of recaptures, and the sum of the number of marked animals in the population at the time of the first t samples (ΣM_j), we have all the information the sample can provide for estimating the model parameters. Because the model has three parameters and the minimal sufficient statistic is three-dimensional, each parameter is identifiable and can be estimated. However, surprisingly, estimation of the recapture probability c is independent of estimation of both the population size N and the initial capture probability p. The practical implication of this is that after its initial capture, an animal provides no additional information on the parameters N and p. Therefore, the estimates of N and p are exactly the same as they would be if an animal was removed from the population after initial capture. Consequently, the analysis is exactly the same as that for removal experiments (which we will discuss later in this chapter). These arise when the capture physically removes the animal from the population. Examples include capture through kill trapping or electrofishing. Here the removal is a consequence of marking.

Statistical analysis of removal experiments was first proposed by Moran (1951). Zippin (1956, 1958) provided a basic statistical analysis of these studies that we will use. For this model, valid MLEs for N and p are obtained from the data when the criterion

$$\sum_{j=1}^{t} (t + 1 - 2j)(n_j - m_j) > 0 \quad (9.41)$$

is satisfied (Seber and Whale, 1970). This criterion tests whether the population is being sufficiently depleted by the removal of new animals. The probability of failure to satisfy the criterion will usually be small for large population size N, and the probability will decrease as the number of sampling occasions t increases. If the condition is satisfied, the MLEs of N and p can be found by maximizing

$$\frac{1 - \hat{p}}{\hat{p}} - \frac{t(1 - \hat{p})^t}{(1 - (1 - \hat{p})^t)} = \frac{\sum_{j=1}^{t} (j - 1)n_j}{M_{t+1}} \tag{9.42}$$

and

$$\hat{N}_b = \frac{M_{t+1}}{1 - (1 - \hat{p})^t}. \tag{9.43}$$

The estimated asymptotic variance of $\hat{N}_b$ is given by

$$s_{\hat{N}_b}^2 = \frac{\hat{N}_b(1 - \hat{p})^t(1 - (1 - \hat{p})^t)}{(1 - (1 - \hat{p})^t)^2 - t^2\hat{p}^2(1 - \hat{p})^{t-1}}. \tag{9.44}$$

An estimate of the variance of $\hat{p}$ is

$$s_{\hat{p}}^2 = \frac{\hat{p}^2(1 - \hat{p})^2(1 - (1 - \hat{p})^t)}{\hat{N}_b[(1 - \hat{p})(1 - (1 - \hat{p})^t)^2 - \hat{p}^2t^2(1 - \hat{p})^t]}. \tag{9.45}$$

The MLE of the recapture probability c is

$$\hat{c} = \frac{m_.}{M_.}, \tag{9.46}$$

and the estimated variance of $\hat{c}$ is

$$s_{\hat{c}}^2 = \frac{\hat{c}(1 - \hat{c})}{M_.}. \tag{9.47}$$

Notice that summary data are sufficient to estimate this model.

Tests for the Model

As with the model M_t, we consider two tests relating to model M_b. The first tests whether model M_b is more appropriate than model M_0, and the second tests the fit of model M_b to the data. For the first test, we want to know whether capture affects the probability of capture. This leads to testing the null hypothesis H_0: p

$= c$ versus the alternative H_i: $p \neq c$. Otis et al. (1978) proposed using the test statistic

$$T_{0b} = \frac{(\hat{p} - \hat{c})^2}{s_{\hat{p}}^2 + s_{\hat{c}}^2}.$$ (9.48)

Under H_0, T_{0b} has an approximate χ^2 distribution with one degree of freedom. Rejection of the null leads to rejecting model M_0 in favor of model M_t.

An overall goodness-of-fit test for model M_b may be based on combining two independent tests (Otis et al., 1978). The first tests whether the probability of capture is constant over time; that is, H_0: $p_j = p$ versus H_1: Not all $p_j = p, j = 1, 2, \ldots, t$. This test, due to Zippin (1956), is based on the test statistic

$$T_{b1} = \sum_{j=1}^{t} \frac{[u_j - \hat{N}_b \hat{p}(1 - \hat{p})^{j-1}]^2}{\hat{N}_b \hat{p}(1 - \hat{p})^{j-1}} + \frac{[\hat{N}_b - M_{t+1} - \hat{N}_b(1 - \hat{p})^t]^2}{\hat{N}_b(1 - \hat{p})^t}$$ (9.49)

which has an approximate χ^2 distribution with $(t - 2)$ degrees of freedom when H_0 is true. However, for the approximation under the null to be good, it may be necessary to pool some of the data just as we did for the goodness-of-fit tests in Chapter 2. As suggested by Otis et al. (1978), if $\hat{N}_b \hat{p}(1 - \hat{p})^{r-1} < 2$ and $\hat{N}_b \hat{p}(1 - \hat{p})^{s-1} \geq 2$ for $s = 1, 2, \ldots, r - 1$, the terms corresponding to $r, r + 1, \ldots, t$ are combined into a single term, reducing the degrees of freedom associated with T_{b1} to $r - 2$. That is, if the expected value from a sampling period drops below two, then data for that period are combined with those from all subsequent periods in computing the test statistic.

A second test, independent of the first, evaluates whether the probability of recapture probability is constant over time. It is based on the variance test for homogeneity of binomial proportions (Snedecor and Cochran, 1989, p. 204). For the null hypothesis H_0: $c_j = c$ for $j = 2, 3, \ldots, t$, the test statistic is

$$T_{b2} = \frac{\sum_{j=2}^{t} M_j(\hat{c}_j - \hat{c})^2}{\hat{c}(1 - \hat{c})}$$ (9.50)

where

$$\hat{c}_j = \frac{m_j}{M_j},$$ (9.51)

and $\hat{c}$ is the maximum likelihood estimator of c given in equation (9.46). The test statistic T_{b2} has an approximate χ^2 distribution with $(t - 2)$ degrees of freedom when H_0 is true.

Because the two tests are independent and the test statistics have approximate χ^2 distributions, they may be combined to form an overall test statistic for the goodness-of-fit of model M_b

$$T_b = T_{b1} + T_{b2} \tag{9.52}$$

When model M_b is true, T_b will have an approximate χ^2 distribution with $(2t - 4)$ degrees of freedom ($t + r - 4$ if pooling occurs in the first test). Large values of the test statistic lead to rejection of the null hypothesis and the conclusion that model M_b does not fit the data.

Model M_h: Heterogeneity of Capture Probabilities

Model M_h has as a basis the assumption that each member of the population has its own capture probability independent of every other member of the population. It is further assumed that there is no behavioral response to capture and that the probability of capture is constant for all trapping occasions so that the individual's capture probability is unchanged during the course of the study. Cormack (1968) stated that a test for this heterogeneity among individual capture probabilities is impossible to develop unless an independent test is conducted on a population of known size that is "representative" of the population of interest. Because this is rarely, if ever, possible, this model was largely ignored for a number of years even though Eberhardt (1969) noted that several data sets indicated that the capture probabilities did often vary with individuals.

The heterogeneity model M_h was first developed by Burnham (1972) and Burnham and Overton (1978, 1979). Under this model, it is assumed that $p_{ij} = p_i$, $i = 1, 2, \ldots, N$. The probability distribution of the capture histories $\{X_\omega\}$ has an $(N + 1)$-dimensional set of minimal sufficient statistics. Because there are only N individuals in the population, there are too many nuisance parameters (p_i's) to permit estimation of N without additional restrictions. Burnham (1972) assumed that the p_i's are a random sample from some probability distribution with distribution function $F(p)$, $p \in [0, 1]$. He evaluated the properties of the estimator under the assumption that $F(p)$ was a member of the beta distribution and found that approach unacceptable. As an alternative, Burnham and Overton (1978, 1979) assumed an arbitrary distribution function and derived a jackknife estimator that is expressible as a power series in $1/t$. These investigators found that the kth-order jackknife estimator is a linear function of the capture frequencies. This estimator for orders 1 to 5 is given in Table 9.3. Notice that the kth order jackknife may be expressed in the form

$$\hat{N}_{hk} = \sum_{j=1}^{k} a_{jk} f_j \tag{9.53}$$

Table 9.3   Jackknife Estimators $\hat{N}_{hk}$ *of Population Size Under Model* M_h *for* k = 1 *to* 5

Order (k)	Estimator of population size $\hat{N}_{hk}$
1	$M_{t+1} + \left(\dfrac{t-1}{t}\right)f_1$
2	$M_{t+1} - \left(\dfrac{2t-3}{t}\right)f_1 - \dfrac{(t-2)^2}{t(t-1)}f_2$
3	$M_{t+1} + \left(\dfrac{3t-6}{t}\right)f_1 - \left(\dfrac{3t^2-15t+10}{t(t-1)}\right)f_2 + \dfrac{(t-3)^3}{t(t-1)(t-2)}f_3$
4	$M_{t+1} + \left(\dfrac{4t-10}{t}\right)f_1 - \left(\dfrac{6t^2-36t+55}{t(t-1)}\right)f_2$ $+ \left(\dfrac{4t^3-42t^2+148t-175}{t(t-1)(t-2)}\right)f_3 - \dfrac{(t-4)^4}{t(t-1)(t-2)(t-3)}f_4$
5	$M_{t+1} + \left(\dfrac{5t-15}{t}\right)f_1 - \left(\dfrac{10t^2-70t+125}{t(t-1)}\right)f_2$ $+ \left(\dfrac{10t^3-120t^2+485t-660}{t(t-1)(t-2)}\right)f_3 - \left(\dfrac{(t-4)^5-(t-5)^5}{t(t-1)(t-2)(t-3)}\right)f_4$ $+ \dfrac{(t-5)^5}{t(t-1)(t-2)(t-3)(t-4)}f_5$

From Burnham and Overton, 1978, by permission of Oxford University Press.

where f_j is the number of animals captured exactly j times during the t sampling periods. Therefore, the variance may be estimated by

$$s^2_{\hat{N}_{hk}} = \sum_{j=1}^{k} (a_{jk})^2 f_j - \hat{N}_{hk}. \tag{9.54}$$

Also, this is the first model that we have discussed for which summary data are not sufficient to estimate the parameters. In order to determine how many animals were caught once, how many twice, and so forth, we need to have capture history data.

We now have five possible estimators of the population size N (*see* Table 9.3). An objective approach to selecting one of these was outlined by Burnham and Overton (1978, 1979). With these estimators, as with other jackknife estimators, the bias of the estimator decreases as the order k increases. However, the variance increases with increasing order. The goal is to increase the order of the jackknife if the reduction in bias is significant compared to the increase in variance. Therefore, they suggested to begin by testing the hypotheses

$$H_{01}: E(\hat{N}_{h2} - \hat{N}_{h1}) = 0 \quad \text{versus} \quad H_{11}: E(\hat{N}_{h2} - \hat{N}_{h1}) \neq 0. \tag{9.55}$$

If H_{01} is not rejected, we conclude that the reduction in absolute bias due to using $\hat{N}_{h2}$ rather than $\hat{N}_{h1}$ is small relative to the increased variance of $\hat{N}_{h2}$. Consequently, we use $\hat{N}_{h1}$ as the estimator of population size. However, if H_{01} is rejected, then we conclude that the reduction in absolute bias achieved by using $\hat{N}_{h2}$ instead of $\hat{N}_{h1}$ is significant even relative to the increased variance of $\hat{N}_{h2}$. Therefore, we prefer $\hat{N}_{h2}$ to $\hat{N}_{h1}$ as the estimator. We then test whether using $\hat{N}_{h3}$ resulted in significantly less bias relative to the increased variance as opposed to using the estimator $\hat{N}_{h2}$. If so, the use of $\hat{N}_{h4}$ instead of $\hat{N}_{h3}$ is evaluated. The process is continued until further decreases in bias are not large enough to offset the increased variance. For each null hypothesis H_{0k}, the test statistic is

$$T_k = \frac{\hat{N}_{h,k+1} - \hat{N}_{hk}}{\sqrt{\widehat{Var}(\hat{N}_{h,k+1} - \hat{N}_{hk}|M_{t+1})}} \tag{9.56}$$

where

$$\widehat{Var}(\hat{N}_{h,k+1} - \hat{N}_{hk}|M_{t+1}) = \frac{M_{t+1}}{M_{t+1} - 1}\left(\sum_{j=1}^{t} b_j^2 f_j - \frac{(\hat{N}_{h,k+1} - \hat{N}_{hk})^2}{M_{t+1}}\right) \tag{9.57}$$

and

$$b_j = a_{j,k+1} - a_{jk}, \qquad j = 1, 2, \ldots, t. \tag{9.58}$$

The test statistic T_k has an approximate standard normal distribution if the null hypothesis is true. Therefore, we would reject H_{0k} if the absolute value of T_k becomes large.

The final jackknife estimate tends to overestimate the population size N in most cases (Pollock and Otto, 1983; Lee and Chao, 1994).

Tests for the Model

As with the earlier models, two tests are developed. The first tests whether there is sufficient evidence of heterogeneous capture probabilities that model M_h is preferred to model M_0. The second is an overall goodness-of-fit test for model M_h. Because of the same identifiability issues that prevented the development of MLEs to estimate the parameters of model M_h, it is not possible to develop a valid likelihood ratio test for choosing between M_0 and M_h. However, we still want to test H_0: $p_i = p$, for $i = 1, 2, \ldots, N$, versus H_1: not all p_i are equal. To do this, Otis et al. (1978) recommend comparing the observed frequencies f_i to their expected values under model M_0 using a Pearson's chi-squared test. The resulting test should be sensitive to departures from M_0 in the direction of M_h. The test statistic is

$$T_{0h} = \sum_{j=1}^{t} \frac{(f_j - \hat{f}_j)^2}{\hat{f}_j} \tag{9.59}$$

where

$$\hat{f}_i = \hat{N}_0 \binom{t}{i} \hat{p}^i (1 - \hat{p})^{t-i} \tag{9.60}$$

and $\hat{N}_0$ and $\hat{p}$ are the MLEs of N and p under model M_0. If H_0 is true and M_0 is the true model, the test statistic in equation (9.59) has an approximate χ^2 distribution with $(t - 2)$ degrees of freedom. Large values of the test statistic T_{0h} will lead to rejection of the null hypothesis in favor of model M_h.

Burnham (1972) suggested that an overall goodness-of-fit test of model M_h is equivalent to testing H_0: $p_{ij} = p_i$ versus H_1: not all $p_{ij} = p_i$, $i = 1, 2, \ldots, M_{t+1}$. Notice that the null hypothesis assumes that the capture probabilities are heterogeneous but do not change over time. Rejection of the null implies that these heterogeneous capture probabilities do change over time. However, if we conclude that the capture probabilities do change over time, the reason for the change could be due to variation in sampling periods, behavioral response to traps, or a combination of the two. With this in mind, Burnham proposed the test statistic

$$T_h = \frac{\sum_{i=1}^{t} \left(\frac{n_j - n.}{t} \right)^2}{\sum_{j=1}^{t} f_j \left(\frac{j}{t} \right) \left(1 - \frac{j}{t} \right)} \cdot \frac{t - 1}{t}. \tag{9.61}$$

If the null hypothesis is true, T_h has an approximate χ^2 distribution with $(t - 1)$ degrees of freedom. The test is conditional on the frequency of capture statistics $f_1, f_2, \ldots, f_t$. A large value of the test statistic leads to rejection of the null and the conclusion that model M_h fits. Burnham also suggested another test statistic for this test when f_j is large enough. Otis et al. (1978) use this alternative test statistic when f_j is greater than t for $j = 1, 2, \ldots, t$. We do not consider this test here.

Model M_{bh}: Heterogeneity of Capture Probabilities and Trap Response

Model M_{bh} is conceptually appealing. It permits the capture probability to differ from individual to individual and for the individuals to respond to capture differentially. Therefore, each member of the population has a pair of capture probabilities: the probability p_i that the ith animal will be captured given that it has not previously been captured assuming the probability is constant for all trapping occasions and the probability c_i that the ith animal will be recaptured given that it has been captured at least once previously. Pollock (1974) introduced this model and proposed a generalized removal estimator building on some of the ideas that were used in developing an estimator for model M_b. A maximum likelihood estimator exists and is used in the program CAPTURE (Otis et al. 1978). Pollock and Otto (1983, equation 17) developed a jackknife estimator of population size

$$\hat{N}_{bh} = \sum_{j=1}^{t-1} u_j + tu_t \tag{9.62}$$

The variance of $\hat{N}_{bh}$ is estimated using

$$s_{\hat{N}_{bh}}^2 = \sum_{j=1}^{t-1} u_j + t^2 u_t - \hat{N}_{bh}. \tag{9.63}$$

Simulation studies by Pollock and Otto (1983) and Lee and Chao (1994) suggest that this is the best overall estimator currently available for this model. However, Lee and Chao noted that (1) if the number of captured animals is relatively large, the jackknife estimator tends to severely overestimate N; and (2) the variability of the estimator sometimes increases with the number of sampling occasions t. Summary data are sufficient to estimate the population size under model M_{bh}. However, capture history data are needed to perform the test (given below) for this model.

Tests for the Model

Pollock (1974) developed a test for H_0: model M_b fits versus the alternative H_1: model M_{bh} should be used. The test statistic depends on both the number of times animals were caught exactly k times f_k, $k = 1, 2, \ldots, t$, and the number of animals caught exactly k times that were first caught in the jth sampling period, $f_k^{(j)}$, j, $k = 1, 2, \ldots, t$. A goodness-of-fit test is formed by pooling $(t - 1)$ independent χ^2 tests. The kth test is conditional on the value of f_k and has $(t - k)$ degrees of freedom. The overall test statistic is

$$T_{bh} = \sum_{k=1}^{t-1} \sum_{j=1}^{t-k+1} \frac{\left(f_k^{(j)} - \left[\dfrac{\left(\dfrac{t-j}{t-k-j+1} \right)}{\left(\dfrac{t}{k} \right)} \right] f_k \right)^2}{\left[\dfrac{\left(\dfrac{t-j}{t-k-j+1} \right)}{\left(\dfrac{t}{k} \right)} \right] f_k}. \tag{9.64}$$

Under H_0, T_{bh} has an approximate χ^2 distribution with

$$\sum_{k=1}^{t-1} (t - k) = \frac{t(t - 1)}{2} \tag{9.65}$$

degrees of freedom. As with test statistic T_{b1}, terms may be combined, reducing the number of degrees of freedom. Here the quantity

$$\left[\frac{\binom{t - i}{t - k - i + 1}}{\binom{t}{k}} \right] f_k \tag{9.66}$$

is checked to be sure it exceeds 2. If not, terms are combined and degrees of freedom are correspondingly reduced.

Models M_{bt}, M_{ht}, M_{bht}

The three remaining models look at the other possible combinations of relaxing the three primary assumptions underlying capture-recapture models. Model M_{bt} allows for a behavioral response resulting from initial capture and for the probability of capture varying over time, but each individual has the same probability of initial capture or recapture at a given trapping occasion. Model M_{ht} assumes that each individual has a possibly different probability of capture and that this probability can change with trapping occasion; however, there is no behavioral response to trapping. Model M_{bht} is the most general model. It permits each individual to have an initial capture probability and a behavioral response to trapping that individually impacts that animal's recapture probability. Further, these probabilities can change with trapping occasion. These three models were discussed by Pollock (1974) and Otis et al. (1978), but at that time, no methods of estimating the parameters were available. Under maximum likelihood estimation, not all parameters are identifiable. Therefore, estimation of this model requires additional constraints on the parameters.

Lee and Chao (1994) used the regression methods developed by Leslie and Davis (1939), DeLury (1947), and Ricker (1958) to provide population estimators for these models. Because DeLury's estimator severely underestimated the population size for most cases and Ricker's estimator is not independent of the choice of scale of the effort, we present Leslie's approach here.

Before presenting the method, we should first note that Leslie and Davis (1939) developed this model for cases in which the catch effort was not constant over time. For example, the number of traps varied over the course of an experiment. However, Lee and Chao (1994) suggested that variation in capture probabilities through time is equivalent to variation in catch effort. If a variable, possibly an environmental one such as temperature, is strongly correlated with these capture probabilities, then that variable could be used to represent catch effort. That is, it is enough to know the relative time effects and to use those as the catch effort.

Assuming that sampling is a Poisson process with regard to effort and that all individuals have the same probability p_j of being caught in the jth sample, Leslie and Davis concluded that the number of unmarked individuals caught in sample j should be linearly related to the total number of animals already marked; that is,

$$E(n_j|M_j) = p_j(N - M_j)$$
$$= ke_j(N - M_j) \tag{9.67}$$

where k is the average probability that an animal is captured with one unit of effort and e_j is the number of units of effort expended on the jth sample. Then defining

$$Y_j = \frac{n_j}{e_j} \tag{9.68}$$

the catch per unit effort, we have

$$E(Y_i|M_i) = k(N - M_i) \tag{9.69}$$

The least-squares estimators of k and N are

$$\hat{k} = -\frac{\sum_{j=1}^{t} Y_j(M_j - \bar{M})}{\sum_{j=1}^{t} (M_j - \bar{M})^2} \tag{9.70}$$

and

$$\hat{N} = \bar{M} + \frac{\bar{Y}}{\hat{k}} \tag{9.71}$$

where $\bar{M}$ is the average number of M_i, $i = 1, 2, \ldots, t$. Notice that if e_j is the same for all j and if n_j is used instead of Y_j, the estimate of k increases by a factor of e_j, but the estimate of N does not change. (This confirms that it is enough to know the relative time effects.) The estimate of N is approximately unbiased and has an estimated variance of

$$s_{\hat{N}}^2 = \frac{\hat{\sigma}^2}{\hat{k}^2} \left(\frac{1}{t} + \frac{(N - \bar{M})^2}{\sum\limits_{j=1}^{t} (M_j - \bar{M})^2} \right). \tag{9.72}$$

where the sampling variance $\hat{\sigma}^2$ is computed using

$$\hat{\sigma}^2 = \frac{\sum\limits_{j=1}^{t} (Y_j - \hat{k}(\hat{N} - M_j))^2}{t - 2}. \tag{9.73}$$

Although this method has been used mainly in fishery research (Seber, 1982, p. 299), it provided estimates with relatively good properties for this set of models (Lee and Chao, 1994) and perhaps should be considered more broadly. The fit of the model is generally assessed by visual inspection of the graph of observed and predicted values of Y_j versus M_j.

Recent work by Lee and Chao (1994) and Evans, et al. (1994) has now provided methods for estimating the parameters for models M_{bt}, M_{ht}, and M_{bht}. Lee and Chao use the concept of sample coverage to obtain estimates. Sample coverage is defined as the total probabilities of the observed species. This is generalized to capture–recapture studies by defining sample coverage of a capture–recapture experiment to be the proportion of the total individual probabilities associated with the captured animals; that is, the sample coverage C is defined to be

$$C = \frac{\sum\limits_{j=1}^{t} \sum\limits_{i=1}^{N} p_i I[\text{the } i\text{th animal is captured}]}{\sum\limits_{i=1}^{N} p_i}. \tag{9.74}$$

Here I is the indicator function. I[the ith animal is captured] is 1 if the animal is captured on at least one sampling occasion and 0 if it is never captured. If all p_i's are equal, $C = M_{t+1}/N$, which is the proportion of distinct animals observed. A natural estimator for the case of $p_{ij} = p$ for all i and j is

$$\hat{N}_0 = \frac{M_{t+1}}{\hat{C}} \tag{9.75}$$

where $\hat{C}$ is an estimator of C. To obtain population estimates for more general models, Lee and Chao explored the relationship between population size N and the coefficient of variation (CV) of the capture probabilities using the sample coverage. This was building on the work by Cormack (1966) and Carothers (1973, 1979) that indicated the CV of the capture probabilities,

$$\text{CV} = \gamma = \frac{\sqrt{\sum_{i=1}^{N} \frac{(p_i - \bar{p})^2}{N}}}{\bar{p}}, \tag{9.76}$$

provides insight into the effect of heterogeneity under various models. The effects in the models need only be defined in a relative sense, that is, up to a multiplicative constant. Although the mean $\bar{p} = \Sigma\Sigma p_{ij}/N$ is not uniquely defined, the CVγ and the sample coverage C are. Further, to obtain unique estimates when the probabilities of capture vary with time, it is assumed that the relative time effects are known. This assumption is satisfied if the probability for capture on the jth occasion is proportional to a known covariate. Lee and Chao (1994) developed closed-form estimators of N for the set of models under consideration. These estimators seem to have good potential especially for models M_h, M_{ht}, and M_{bht} when the CV is at least 0.4 and the data are sufficient to provide a stable estimator of the CV.

Fienberg (1972) represented the complete capture history for a t-sample multiple-recapture study as an incomplete 2^t contingency table with one unobservable cell. These models do not correspond to the models of Otis et al. (1978). Cormack (1989) presented loglinear models equivalent to M_0 and M_t, along with approximations to M_b and M_{tb}. Alho (1990) and Huggins (1989, 1991) used logistic regression models with auxiliary variables to estimate population size N when the individual capture probabilities are heterogeneous. Evans et al. (1994) developed a unified approach to estimating the models using loglinear models. Constrained maximum likelihood was used to estimate the parameters for all eight models. This is exciting work, and loglinear models will probably play an increasing role in the analysis of recapture data. However, currently computational issues do not permit ready implementation. Further, the behavior of the methods is not fully understood over a broad range of population sizes or sampling intensities.

As we have briefly outlined, capture-recapture models for closed populations have been an area of active research for many years. Several approaches have been suggested, and new ideas are constantly being developed. It is unclear what the best approach is for a given circumstance. A comprehensive study comparing these approaches under various conditions would be valuable.

Model Selection

We have taken a limited approach to model selection by considering one or two tests for most models. Sometimes these tests give conflicting results. In general, we will concentrate on how to estimate the model parameters assuming we know the model. This is an idealized situation that is often assumed in statistics, especially at the elementary level. However, it is evident that the larger problem is in discerning which model is correct. Biological complexity requires models with an extremely large number of parameters if the circumstances giving rise to data are precisely reflected. Yet much of the variation in the data is often attributable to a few primary factors. Discerning these and putting them in the model without adding factors that contribute little to the biological understanding is the challenge that we face. CAPTURE (Otis et al. 1978) uses a series of seven tests and a discriminant function to help decide among models. The interested reader should get this more complete program.

Confidence Intervals

The asymptotic 95% confidence intervals based on the normal distribution have the general form

$$\hat{N} \pm 1.96 s_{\hat{N}}. \tag{9.77}$$

The distribution of $\hat{N}$ tends to be positively skewed. Therefore, even if the coverage of the confidence interval approaches that of the nominal level, the population size N is above the upper limit more often than it is below the lower limit. Also, sometimes the lower limit of intervals based on equation (9.77) lies below M_{t+1} even though the population must be at least as large as the number of different animals captured. Therefore, Burnham et al. (1987) and Rexstad and Burnham (1991) suggest using intervals based on the assumption that the number of individuals in the population that are *not* captured ($N - M_{t+1}$) is lognormally distributed. A $(1 - \alpha)$ 100% confidence interval is

$$\left(M_{t+1} + \frac{\hat{N} - M_{t+1}}{C}, \quad M_{t+1} + (\hat{N} - M_{t+1})C \right) \tag{9.78}$$

where

$$C = e^{z_{\alpha/2} \sqrt{\ln(1 + s_{\hat{N}}^2/(\hat{N} - M_{t+1})^2)}}. \tag{9.79}$$

Example 9.5

Conner and Labisky (1985) describe a study of raccoons conducted on a 918-hectare area on the east bank of the St. Johns River in southeastern Putnam County, Florida. No hunting or commercial trapping has been allowed in the area since the 1930s. For 20 nights from January 9 to 18 and from January 22 to the 31, 1981, 48 live traps were set. Fish heads were used to bait the traps. Upon capture, the raccoons were tagged in each ear using numbered metal ear tags. Table 9.4 presents the summary data from the study.

To use ECOSTAT to fit and test models for these data, we must first construct an ASCII file for the data. The first line has the number of sampling occasions, 20. The next 20 lines each have an identification for the sampling period, the

Table 9.4   Summary Data for Raccoon Capture–Recapture Study

Day	No. captured, n_i	No. marked (m_i)
1	8	0
2	5	1
3	2	0
4	2	0
5	5	2
6	5	3
7	6	2
8	5	3
9	2	0
10	5	4
11	7	4
12	6	3
13	6	5
14	8	6
15	5	3
16	4	3
17	5	2
18	2	1
19	5	4
20	3	2

From Conner and Labisky, 1985, by permission of The Wildlife Society.

number captured during that period n_i, and the number of marked in the sample m_i. It is assumed that the sampling occasions are listed in chronological order. Then *Capture–Recapture* is chosen from the *Closed Population* menu. After specifying the name of the data set, the data may be displayed. Before estimating the population size, a model must be chosen from the menu. Different models can be fit. Appropriate tests for each model are conducted. Confidence intervals on the population size can also be set.

First, model M_0 is fit to the data. Based on this model, the raccoon population is estimated to be 59, and the estimated standard deviation of this estimate is 5. By setting a 90% confidence interval, we can be 90% confident that the population consists of 53 to 69 raccoons (assuming the model is correct). Model M_t produces an estimate of the raccoon population of 58. However, when testing whether model M_t would be preferred to model M_0, the value of the test statistic T_{0t} is 18.71, with 19 degrees of freedom and a p-value of 0.4756. Therefore, we would not choose M_t over M_0. (We should also note that we cannot test the fit of model M_t without information on individual capture histories.) Next, model M_b was fit. The raccoon population under this model is estimated to be 63, and the estimated standard deviation of this estimate is 11. The test statistic T_{0b} is used to determine whether there is enough evidence to conclude that model M_b would be preferred to model M_0. In this case, $T_{0b} = 0.32$ with 1 degree of freedom and a p-value of 0.57. Hence, we do not choose M_b over M_0.

Model M_h cannot be fit with the summary data here. However, Conner and Labisky report that model M_h did fit better than model M_0 ($T_h = 15.49$, df $= 10$, and $p = 0.69$). Under model M_h, the estimated population of raccoons was estimated to be 66. Confidence intervals on the population size under model M_0 and model M_h overlapped. Further, the selection criterion used by CAPTURE indicated that either model M_b (1.00) or model M_h (0.99) were most representative of the data.

We also fit model M_{bh}. The population size was estimated to be 67, and the estimated standard deviation of this estimate is 20. The test for this model could not be conducted because it requires capture history data.

From this example, we see that we lose flexibility if individual capture histories are not available. We could not fit one of the two most likely models! The estimates of population size ranged from 58 to 67, depending on the model. Although the absolute range is not large, 67 is 15% larger than 58. In this situation, we recognize the possible range in estimates. Depending on the application, we could consider the impact each extreme would have on our conclusions. For example, if we want to know whether the number of raccoon is sufficient to sustain the population, we could address this question using both 58 and 67 as the population estimates. If the answer is the same for both estimates, we have some confidence in our conclusions. If the answer differs, we may need to repeat the study.

Example 9.6

Edwards and Eberhardt (1967) considered issues in using live traps to estimate cottontail (*Sylvilagus floridanus*) populations. In a classical study, a 40-acre, rabbit-proof enclosure was established on the Olentangy Wildlife Experiment Station in Delaware County, Ohio. Wild cottontail rabbits were collected from the State Game Farm in Urbana, Ohio, on October 19 and 20, 1961. The rabbits were gathered by drive-netting to avoid capturing a disproportionate number of trap-prone animals. Because 3 to 4 rabbits per acre are common in the wild, an effort was made to get 120 to 160 rabbits for release in the enclosure. From the drive-netting, 135 were collected, marked, and released. After being placed in the enclosure, the cottontails were given four days to adjust to the new environment. Then trapping began. Livetrapping was conducted on 18 consecutive nights from October 24 to November 10, 1961. Of the 135 rabbits, 76 were captured at least once. The capture-history data for each animal was gathered and is in file rabbit.dat. The first line of this ASCII file has the number of animals captured at least once ($M_{19} = 76$) and the number of sampling occasions ($t = 18$). Each of the next 76 lines contains an identification for the captured rabbit and the capture history for that animal. Summary data are given in Table 9.5.

Table 9.5   Summary Data for a Capture–Recapture Study of Cottontail Rabbits, Conducted by Edwards and Eberhardt (1967)

Date	No. captured (n_i)	No. marked (m_i)
October 24	9	0
October 25	8	2
October 26	9	6
October 27	14	3
October 28	8	4
October 29	5	4
October 30	16	8
October 31	7	4
November 1	9	3
November 2	33	2
November 3	8	7
November 4	14	5
November 5	2	1
November 6	5	0
November 7	11	5
November 8	0	0
Nobember 9	5	5
November 10	9	7

Models M_0, M_b, M_h, M_t, and M_{bh} were fit using ECOSTAT. The results are presented in Table 9.6.

Notice that the estimates of population size range from 96 to 159. Tests indicated that models M_h and M_t, but not M_b, were preferred to model M_0. However, models M_b and M_h did not fit the data well, and the fit of model M_t could not be assessed because the u_i values were not large enough. Model M_{bh} was preferred

Table 9.6   Results of Fitting Closed Population Capture–Recapture Models to Edwards and Eberhardt's (1967) Cottontail Rabbit Data

Models	$\hat{N}$	$s_{\hat{N}}$	Tests
M_0	97	7	
M_b	97	12	$T_{0b} < 0.01$ (df = 1, $p = 0.96$)
			$T_b = 103.12$ (df = 16, $p = .001$)
M_h	159	22	$T_{0h} = 60.22$ (df = 16, $p < 0.0001$)
			$T_h = 141.73$ (df = 17, $p < 0.0001$)
M_t	96	7	$T_{0t} = 45.31$ (df = 16, $p = 0.0001$)
			u_i is too small
M_{bh}	110	25	$T_{bh} = 4039.3$ (df = 17, $p < 0.0001$)

to model M_b, based on the test statistic T_{bh}. Therefore, based on the limited analysis available through ECOSTAT, we would consider models M_t and M_{bh}. Using CAPTURE, Rexstad and Burnham (1991) found that model M_t was the most appropriate.

Removal and Catch Effort Models

The models discussed thus far have assumed that animals can be captured, marked, and recaptured. Sometimes, it is more practical to remove animals from a population. Often live traps may not be effective when capturing some animals such as small mammals, and kill trapping may be used. When sampling streams, electrofishing is often used to obtain fish counts. The fish may be removed during the process and returned once sampling is complete. Commercial or sports catches may be used to estimate fish or wildlife populations. In each of these, marking and recapturing the individuals is not reasonable. Further, the intensity of sampling may vary with time. We need methods that permit estimation of population size N based on the number of animals removed from the population on two or more sampling occasions that may represent different levels of sampling intensity. Approaches to this problem differ for open and closed populations. In this section, we will explore these models for closed populations.

First, suppose that we have only two sampling occasions and that animals are removed from the population through the capture process. We assume the following:

1. The population is closed except for removals.
2. The sampling effort is constant across sampling occasion.
3. The probability of capture is constant across sampling occasions.

Then estimates of the capture probability are $n_1/\hat{N}$ and $n_2/(\hat{N} - n_1)$ based on the first and second sampling occasions, respectively. Equating these two estimates of probability, we can obtain an estimate of population size

$$\hat{N} = \frac{n_1^2}{n_1 - n_2}. \tag{9.80}$$

Even if the assumptions of the model hold, the probability that $n_2 > n_1$ is positive in which case the method fails to produce a valid estimate. The positive probability of a method failing is a problem faced by other removal models as well.

Suppose now that there are more than two sampling occasions. Based on the same assumptions, we can obtain an estimate of the population size N using methods first developed by Moran (1951) and Zippin (1956, 1958). Otis et al. (1978) and White et al. (1982) note that this model is statistically equivalent to the trap response model M_b. The key to recognizing the equivalence is to note that in the presence of trap response, only first capture information is used in the estimation of N. Therefore, whether an animal is marked or permanently removed, the useful information is the same. If individuals have heterogeneous responses to trapping, model M_{bh} is appropriate. We have already discussed estimation of these models in the sections on models M_b and M_{bh}. In the past, generalized removal models have been used for model M_{bh} (Otis et al., 1978); however, based on the results by Pollock and Otto (1983) and Lee and Chao (1994), we will use the jackknife estimator given in equation (9.62).

Example 9.7

Leslie and Davis (1939) discuss the analysis of an experiment designed to estimate the number of rats in a 22.5-acre area in Freetown, Sierra Leone. The experiment consisted of 18 trapping occasions over a period of 6 weeks. For each trapping occasion, 210 traps were set in houses in an effectively random manner. Trapping removed the rats from the population. Several species of rats were captured, with *Rattus rattus* the most common. The data are presented in Table 9.7.

Based on these data, we want to estimate the size of the *Rattus rattus* population. First, an ASCII file is constructed. The first line has the number of sampling occasions. The next 18 lines has an identification of the sampling occasion and the number of animals removed from the population on that occasion. It is assumed that the sampling occasions are listed in chronological order. First, model M_b is fit to the data, giving an estimated population size of 522 with an estimated standard deviation of 34. Because heterogeneity among the rats is expected, model M_{bh} is also fit, yielding an estimated population size of 513 with an estimated standard deviation of 46. However, we are unable to conduct the test that compares the fit of model M_b to that of model M_{bh}. Based on our understanding of the biology and the goals of the study, we use the estimate that seems most appropriate. Fortunately, in this case, the difference in the two is relatively small.

Data from some sources do not permit assumption (2) of equal sampling effort on each sampling occasion to hold. For example, information from commercial fishing may provide the number of fish caught and the effort associated with that catch for each sampling occasion; however, the effort may vary. Adjustments need to be made for the variation in catch effort so that the sampling occasions are comparable. This was the motivation leading to the development of catch effort models by Leslie and Davis (1939), DeLury (1947), and Ricker (1958). Leslie's model was considered in the section on the models M_{bt}, M_{ht}, and M_{bht}.

Seber (1982) recommended maximum likelihood estimation for the catch effort model, assuming that sampling is a Poisson process with regard to effort. (Recall this assumption was also made in the development of Leslie's regression model.) Again, using k as the catchability coefficient, the maximum likelihood estimators of N and k are solutions of

Table 9.7 Results of Leslies and Davis's (1939) Capture–Recapture Study of Rats in Freetown, Sierra Leone

Trapping occasion	Rattus rattus	Rattus norvegicus	Mus musculus	Crocidura spp.	Mastomys spp.	Total
		No. of rats trapped				
				Other species		
1	49	7	3	.	.	10
2	32	4	3	.	.	7
3	31	5	7	.	.	12
4	34	5	4	.	.	9
5	16	8	1	.	.	9
6	33	7	2	.	.	9
7	22	5	1	.	.	6
8	27	4	5	.	.	9
9	17	5	11	.	.	16
10	19	1	3	1	.	5
11	18	10	3	.	1	14
12	16	3	3	1	.	7
13	18	2	2	.	.	4
14	12	1	2	.	.	3
15	14	1	5	.	.	6
16	12	3	2	1	.	6
17	17	3	2	.	.	5
18	7	2	5	.	.	7

$$\hat{k}E_{t+1} = -log\left(1 - \frac{M_{t+1}}{\hat{N}}\right) \tag{9.81}$$

and

$$\hat{N}E_{t+1} = \sum_{j=1}^{t} f_j(M_j + 0.5n_j) + \frac{M_{t+1}}{\hat{k}}. \tag{9.82}$$

These two equations are solved iteratively, starting with the regression estimates. The estimated variance of $\hat{N}$ is

$$s_{\hat{N}}^2 = \frac{M_{t+1}}{\hat{N}(\hat{N} - M_{t+1})} + \frac{1}{2\hat{N}^2} - \frac{1}{2(\hat{N} - M_{t+1})^2}. \tag{9.83}$$

Little is known about the relative performance of the MLEs and the regression estimators. Further, models allowing for heterogeneity among individuals in the presence of a variable catch effort have not been developed. An alternative approach to modeling when catch effort varies was discussed by Pollock et al. (1984), who modeled capture probability as a linear logistic function of effort. This appears to be an area in which additional research would be useful.

Example 9.8

DeLury (1947) presents a day-by-day record of lobster catches. The data were collected in 1944 in the Tignish area of Prince Edward Island. The number of traps varied from day to day. The size of the lobsters were fairly constant throughout the sampling period so the use of pounds of lobster instead of number of lobster would not seem to be a problem. Migration and natural mortality were thought to be minimal during the sampling period. Also, molting times were avoided so recruitment through growth to legal size should not be a concern. Therefore, the assumption of a closed population seems reasonable. The date, pounds of lobsters caught, and number of traps fished for each day of the study as well as the catch per unit effort and cumulative catch are presented in Table 9.8.

Figure 9.1 illustrates the plot of catch per unit effort by cumulative catch. Clearly, this is not a plot of a straight line. However, if we ignore those values prior to May 23, the values do appear to be linear. Therefore, we will disregard those values prior to May 23. Also, we can avoid some of the extremely large numbers if we consider the catch in 1,000s of pounds and the effort in 1,000s of traps. The numbers that will be used for the analysis are presented in Table 9.9.

ECOSTAT may be used to estimate the lobster population. First, the data must be put in an ASCII file. The first line of the file has the number of sampling occasions (17). The next 17 lines each represent a sampling occasion and has an identification for the sampling occasion followed by the number of lobster removed (measured in 1,000s of pounds) and the catch effort (measured in 1,000s of traps). Estimation can be based on either Leslie's regression or maximum likelihood with the maximum likelihood methods being preferred.

First, consider using Leslie's regression. The lobster population is estimated to be 121,000 pounds, and the estimated standard deviation of this estimate is 8,920 pounds. Further, we are 95% confident that the lobster population is between 106,000 and 141,000 pounds. The fit of the model to the data can be viewed in Figure 9.2.

Now suppose maximum likelihood estimates are obtained. The lobster population is estimated to be 122,000 pounds, and the estimated standard deviation of this estimate is 54,000. The estimates of population size agree well. However, the measure of precision associated with the two estimates are quite different.

Change-in-Ratio or Selective Removal Models

Suppose that a population can be divided into two type of individuals, such as male and female or juvenile and adult. We will call them x-types and y-types.

Table 9.8. Record of Daily Lobster Catches in 1944 in the Tignish Area of Prince Edward Island

Date	Pounds of lobster	No. of traps	Catch per unit effort (n_i/e_i)	Cumulative catch
May 2	147	200	0.739	0
May 3	2,796	3,780	0.740	147
May 4	6,888	7,174	0.960	2,943
May 5	7,723	8,850	0.873	9,831
May 8	5,330	5,793	0.920	17,554
May 9	8,839	9,504	0.930	22,893
May 10	6,324	6,655	0.950	31,732
May 11	3,569	3,685	0.969	38,056
May 12	8,120	8,202	0.990	41,625
May 13	8,084	8,585	0.942	49,745
May 15	8,252	9,105	0.906	57,829
May 16	8,411	9,069	0.927	66,081
May 17	6,757	7,920	0.853	74,492
May 18	1,152	1,215	0.948	81,249
May 20	1,500	1,471	1.020	81,401
May 22	11,945	11,597	1.030	82,901
May 23	6,995	8,470	0.826	94,846
May 24	5,851	7,770	0.753	101,841
May 25	3,221	3,430	0.939	107,692
May 26	6,345	7,970	0.796	110,913
May 27	3,035	4,740	0.640	117,258
May 29	6,271	8,144	0.770	120,293
May 30	5,567	7,965	0.699	126,564
May 31	3,017	5,198	0.580	132,131
June 1	4,559	7,115	0.641	135,148
June 2	4,721	8,585	0.550	139,707
June 5	3,613	6,935	0.521	144,428
June 6	473	1,060	0.446	148,041
June 7	928	2,070	0.448	148,514
June 8	2,784	5,725	0.486	149,442
June 9	2,375	5,235	0.454	152,226
June 10	2,640	5,480	0.482	154,601
June 12	3,569	8,300	0.430	157,241

From DeLury (1947).

Further assume that a differential change in the numbers of the two types of organisms occurs during the observation period, i.e., from time t_1 to time t_2. For example, suppose our goal is to estimate the deer population of an area, and the observation period is during the hunting season. Hunting permits result in more males being harvested than females. Thus, we have some aspects of a removal

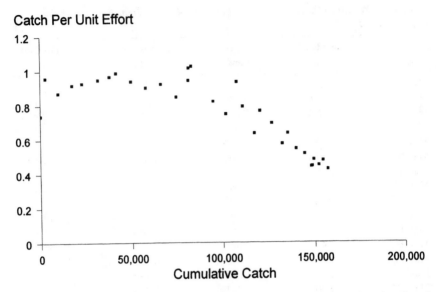

Figure 9.1. The catch per unit effort of DeLury's (1947) lobster full data set plotted against cumulative catch.

Table 9.9. Set of 1944 Daily Lobster Catches Used in the Analysis

Date	Pounds of lobster (1,000s)	No. of traps (1,000s)	Catch per unit effort (n_i/e_i)	Cumulative catch (1,000s)
May 23	6.995	8,470	0.826	0
May 24	5.851	7,770	0.753	6.995
May 25	3.221	3,430	0.939	12.846
May 26	6.345	7,970	0.796	16.067
May 27	3.035	4,740	0.640	22.412
May 29	6.271	8,144	0.770	25.447
May 30	5.567	7,965	0.699	31.718
May 31	3.017	5,198	0.580	37.285
June 1	4.559	7,115	0.641	40.302
June 2	4.721	8,585	0.550	44.861
June 5	3.613	6,935	0.521	49.582
June 6	.473	1,060	0.446	53.195
June 7	.928	2,070	0.448	53.668
June 8	2.784	5,725	0.486	54.596
June 9	2.375	5,235	0.454	57.380
June 10	2.640	5,480	0.482	59.755
June 12	3.569	8,300	0.430	62.395

Based on DeLury (1947).

Catch Per Unit Effort

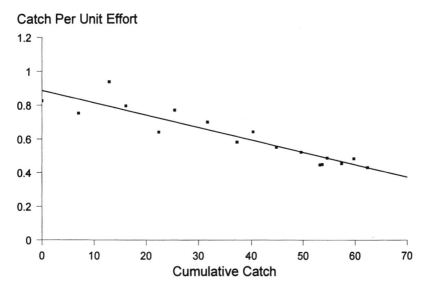

Cumulative Catch

Figure 9.2. The catch per unit effort of DeLury's (1947) lobster data collected from May 23, to June 12, 1944, and the fitted model plotted against cumulative catch.

experiment with the extra feature that the removals from the two groups are at different rates. Our goal is to take advantage of this additional information to obtain an estimate of the population size. Change-in-ratio models and selective removal models are two terms that have been used for models of these experiments.

Kelker (1940, 1944) first developed the traditional model that applies to a population divided into two types with one removal of known size from each type. Paulik and Robson (1969) provided a statistical synthesis of these models. The assumptions are as follows:

1. The population is closed except for removals.
2. The population can be divided into two or more types.
3. Pre- and post-removal type ratios are known or can be estimated.
4. The number of removals of each type is known.

Following Seber (1982), define

X_i = number of x-type animals in the population at time i

Y_i = number of y-type animals in the population at time i

$N_i = X_i + Y_i$ = total population size at time i

$P_i = X_i/N_i$ = proportion of the population of the x-type at time i

$R_x = X_1 - X_2 =$ number of x-type animals removed from the population

$R_y = Y_1 - Y_2 =$ number of y-type animals removed from the population

$R = R_x + R_y = N_1 - N_2 =$ total number removed from the population

Again, R_x and R_y are assumed to be known. Although we are referring to these as removals, Rupp (1966) has noted that the theory continues to be valid if R_x and R_y are negative and therefore represent additions.

The proportion of x-types at time 2, P_2, may be written as

$$P_2 = \frac{X_1 - R_x}{N_1 - R} = \frac{P_1 N_1 - R_x}{N_1 - R}. \tag{9.84}$$

Solving for N_1, we have

$$N_1 = \frac{R_x - R P_2}{P_1 - P_2}. \tag{9.85}$$

If $\hat{P}_1$ and $\hat{P}_2$ are estimates of P_1 and P_2, respectively, N_1, X_1, and N_2 can be estimated by

$$\hat{N}_1 = \frac{R_x - R \hat{P}_2}{\hat{P}_1 - \hat{P}_2}, \tag{9.86}$$

$$\hat{X}_1 = \hat{P}_1 \hat{N}_1, \tag{9.87}$$

and

$$\hat{N}_2 = \hat{N}_1 - R = \frac{R_x - R \hat{P}_t}{\hat{P}_1 - \hat{P}_2}. \tag{9.88}$$

The estimated variances of these estimates are

$$s_{\hat{N}_i}^2 = \frac{\hat{N}_1^2 s_{\hat{P}_1}^2 + \hat{N}_2^2 s_{\hat{P}_2}^2}{(\hat{P}_1 - \hat{P}_2)^2}, \quad i = 1, 2 \tag{9.89}$$

and

$$s_{\hat{X}_i}^2 = \frac{N_1^2 \hat{P}_2^2 s_{\hat{P}_1}^2 + N_2^2 \hat{P}_1^2 s_{\hat{P}_2}^2}{(\hat{P}_1 - \hat{P}_2)^2}, \quad i = 1, 2. \tag{9.90}$$

Note that the variance of $\hat{N}$ increases rapidly as the difference in $\hat{P}_1 - \hat{P}_2$ decreases. Therefore, to be effective, a significant proportion of the x-types must be removed from the population. If only a small proportion is removed, this is not a practical approach to population estimation.

We have one more thing to consider before this method is completely defined: the estimation of $\hat{P}_1$. Suppose the proportion of x-types is estimated through binomial sampling at times t_i, $i = 1, 2$. That is, at time t_i, n_i animals were sampled with replacement and x_i are found to be of x-type. Then $\hat{P}_i = x_i/n_i$ is an unbiased estimator of P_i with estimated variance

$$s^2_{\hat{P}_i} = \frac{\hat{P}_i(1 - \hat{P}_i)}{n_i}. \tag{9.91}$$

Chapman (1954) showed that in this case, equations (9.86) and (9.89) may be written as

$$\hat{N}_1 = \frac{N_1(R_x(n - x_2) - R_y x_2)}{x_1(n - x_2) - x_2 y_1} \tag{9.92}$$

and

$$s^2_{\hat{N}_1} = \frac{\dfrac{x_1 y_1}{n_1} + \dfrac{x_2 y_2}{n_2}}{(\hat{P}_1 - \hat{P}_2)^2}, \tag{9.93}$$

respectively. If sampling is without replacement, then the hypergeometric is the true model. The MLEs of N_1 and X_1 are still $\hat{N}_1$ and $\hat{X}_1$. The asymptotic variances are given by (9.89) and (9.90), respectively, with

$$s^2_{\hat{P}_i} = \frac{\hat{P}_i(1 - \hat{P}_i)}{n_i} \cdot \frac{(N_i - n_i)}{(N_i - 1)}. \tag{9.94}$$

Thus far we have assumed that n_1 and n_2 are predetermined. More often a fixed amount of effort is used in sampling, and n_1 and n_2 are random. However, the variances of $\hat{N}_1$ and $\hat{X}_1$ are virtually unchanged.

Paulik and Robson (1969) presented three methods of constructing confidence intervals. Because it is generally better to avoid using estimators with high variability in the denominator of a fraction ($\hat{P}_1 - \hat{P}_2$ here), we use the interval based on inverting a confidence interval on $1/\hat{N}_1$. From the delta method,

$$s^2_{1/\hat{N}_1} = \frac{(R_x - \hat{P}_1 R)^2}{(R_x - \hat{P}_2 R)^4} s^2_{\hat{P}_2} + \frac{1}{(R_x - \hat{P}_2 R)^2} s^2_{\hat{P}_1}. \tag{9.95}$$

The approximate $100(1 - \alpha)\%$ confidence interval for $1/N_1$ is

$$\frac{1}{\hat{N}_1} \pm z_{\alpha/2} s_{1/\hat{N}_1}. \tag{9.96}$$

This is inverted and reversed to give the confidence interval for N_1.

In planning change-in-ratio experiments, we need to determine n_1 and n_2, so that the estimate of N_1 has a prescribed accuracy. Suppose we want to estimate N_1 within a given proportion p of N_1 with $(1 - \alpha)$ 100%; that is,

$$Pr(|\hat{N}_1 - N_1| < DN_1) \geq 1 - \alpha. \tag{9.97}$$

Seber (1982) noted that choosing $n_1 = n_2 = n$ represents a near optimal allocation of effort and is assumed here. Let the x-type be defined so that R_x/R is greater than P_1, which is greater than P_2. Paulik and Robson (1969) showed that

$$Pr(|\hat{N}_1 - N_1| \leq DN_1) = \Phi\left(\frac{D(P_1 - P_2)\sqrt{n}}{\sqrt{(1 + D - u)^2 P_2(1 - P_2) + (1 + D)^2 P_1(1 - P_1)}}\right)$$
$$- \Phi\left(\frac{-D(P_1 - P_2)\sqrt{n}}{\sqrt{1 - D - u)^2 P_2(1 - P_2) + (1 - D)^2 P_1(1 - P_1)}}\right)$$
$$= 1 - \alpha \tag{9.98}$$

where Φ is the cumulative standard normal distribution and $u = R/N_1$ represents the rate of exploitation.

At times, the removals R_x and R_y are not known exactly and must be estimated. Suppose $\hat{R}_x$ and $\hat{R}_y$ are independent unbiased estimates of R_x and R_y, and $\hat{R} = \hat{R}_x + \hat{R}_y$. Then N_1 may be estimated by

$$\hat{N}_1 = \frac{\hat{R}_x - \hat{P}_2\hat{R}}{\hat{P}_1 - \hat{P}_2}. \tag{9.99}$$

Using the delta method, the estimated variance of this estimator is

$$s_{\hat{N}_1}^2 = \frac{\hat{N}_1^2 s_{\hat{P}_1}^2 + \hat{N}_2^2 s_{\hat{P}_2}^2 + (1 - \hat{P}_2)^2 s_{\hat{R}_x}^2 + \hat{P}_2^2 s_{\hat{R}_y}^2}{(\hat{P}_1 - \hat{P}_2)^2}. \tag{9.100}$$

This traditional model has been extended to three or more types of animals (Otis, 1980; Udevitz, 1989), to more than one removal time (Chapman, 1955; Pollock et al., 1985; Udevitz, 1989), and to several regions (Heimbuch and Hoenig, 1989). The interested reader would be encouraged to refer to these sources and to Pollock (1991), as further study of these methods would be valuable.

Example 9.9

Conner et al. (1986) report on a study of the deer (*Odocoileus* spp.) population in and around Remington Farms near Chestertown, Maryland. During October and November, 1980, 54 road counts were conducted both in the morning and evening. Of the 1,347 deer observed, 120 had antlers and 1,126 were antlerless. The remaining 101 deer could not be seen well enough to determine with certainty whether they had antlers. During the 1-week hunting season, 110 deer were harvested, 56 with antlers and 54 without antlers. A post-season survey was conducted during December 1980, and January 1981. After 52 road counts, again taken both in the morning and evening, 1,346 deer were observed, 43 with antlers, 1,086 without antlers, and 217 unknown. We want to estimate the deer population size.

The change-in-ratio method assumes that the number of removals of each type is known. It is likely that the number of removals exceeds the number reported. For example, a deer may be killed but not recovered, perhaps due to movement after being shot. Other deer may be poached, causing them to be removed but not reported. Either factor results in the population size being underestimated (Paulik and Robson, 1969).

Another assumption that is difficult to meet is that antlered and antlerless deer are equally sightable; that is, the probability of seeing a given deer is the same, whether it has antlers or not. However, P_i, $i = 1, 2$, tends to be underestimated (Conner et al., 1986), probably because antlered deer have a higher probability of being observed than is true of antlerless deer. This results in the population size being overestimated (Seber, 1982). Seber developed a correction to the change-in-ratio estimator that assumes the difference in sightability of the two groups is known. Although the estimate of population size is going to be biased, the estimate of the population of antlered deer will be unbiased (Seber, 1982)! Further, Conner et al. (1986) note that if the differential sightability remains constant from year to year, the change-in-ratio method provides a precise and accurate index of population size.

Given the assumptions and keeping in mind their possible violations, we use the change-in-ratio method. The estimates of the proportion of antlered deer pre- and post-season are

$$\hat{P}_1 = \frac{120}{1,246} = 0.096$$

and

$$\hat{P}_2 = \frac{43}{1,129} = 0.038,$$

respectively. The estimated standard deviations of these estimates are

$$s_{P_1} = \sqrt{\frac{0.096(1 - 0.096)}{1246}} = 0.008$$

and

$$s_{P_2} = \sqrt{\frac{0.038(1 - 0.038)}{1129}} = 0.006.$$

The pre- and post-season deer populations are estimated to be

$$\hat{N}_1 = \frac{56 - 110(0.038)}{0.096 - 0.038} = 890$$

and

$$\hat{N}_2 = 890 - 110 = 780,$$

respectively. Because the number of removals is assumed to be known, subtracting the number of removals from the pre-season estimate of deer population has no effect on the variability of the estimate. Therefore, the estimated standard deviation of the population size estimate is

$$s_{\hat{N}} = \sqrt{\frac{890^2(0.008^2) + 780^2(0.006^2)}{(0.096 - 0.038)^2}} = 149$$

for both pre- and post-season. The size of the antlered deer population prior to the season is estimated to be

$$\hat{X}_1 = 0.096(890) = 86$$

with an estimated standard deviation of

$$s_{\hat{X}_1} = \sqrt{\frac{890^2(0.038^2)(0.008^2) + 780^2(0.096^2)(0.006^2)}{(0.096 - 0.038)^2}} = 5.7.$$

ECOSTAT can be used to perform these calculations. Choose *Removal* from the *Closed Population* Menu and *Change-in-Ratio* from the submenu. Then enter the values observed in the study. Press the *Estimate* button to see the estimates of parameters. After setting the level of confidence and pressing the *Set CI* button, the confidence limits are computed for a $(1 - \alpha)$ 100% confidence interval on

N_1. In this example, based on 90% confidence, the limits are 698 and 1,228. Therefore, we are 90% confident that the pre-season deer population was between 698 and 1,228.

Density Estimation

Although the emphasis in this chapter has been on estimation of population size, often these estimates are converted to population density estimates. Conceptually, the population density is the population size divided by the area of the sampled region. Difficulties arise in determining the area of the sampled region. If the area enclosed by the trapping grid is used, the population density will tend to be overestimated. This results because some animals will be captured even though only a portion of their home range lies within the trapping grid. Although it is recognized that the effectively trapped area is larger than that of the trapping grid, how much larger is seldom clear. The most common approach is to define a boundary strip about the grid so that the effective trapping area is the trapping grid plus the boundary strip. Dice (1938) assumed that the boundary strip should have a width equal to a half of the average diameter of the animal's home range. The problem then becomes one of estimating this width W. Otis et al. (1978) recommend designing the study so that both the density D and the strip width W can be jointly estimated. Should there be a possibility that density estimates will be desired, then this resource should be consulted before conducting the study.

Summary

In ecological studies, capture-recapture experiments are used extensively to estimate population size. The Lincoln–Petersen estimator is intuitively appealing and has nice optimality properties. Recent work has concentrated on modeling capture-recapture studies in which marking occurs in more than one sampling period. A series of eight models have been developed. The first assumes that the probability of capture remains constant through time and for each individual. There are three common departures from this ideal: (1) variation in capture probabilities across sampling periods, (2) heterogeneous capture probabilities among individuals, and (3) trap response. Models allowing these departures from constant capture probabilities singly and in all combinations account for the other seven models. Progress has been made in estimating the parameters of these models and in developing alternative approaches, but this remains an active area of research. Removal models are appropriate when the individual cannot be returned to the population after capture. Catch effort models allow for differential sampling effort over time. A catch effort model that accounts for variation in capture probabilities through time would be useful, but is not yet available. Change-in-ratio models are appropriate when there are two or more types in the population and, during a single removal period, more individuals of one type are captured than of the other type.

Exercises

1. Consider again the study by Lincoln (1930) discussed in Example 9.2. The percentages of returns from banding stations banding at least 100 ducks and geese in a season for 1920 to 1926 are given in the table below:

Year	Banded	Return	Percentage	Year	Banded	Return	Percentage
1920	238	31	13.03	1924	2,256	332	14.65
1921	382	52	13.61	1925	1,795	214	11.92
1922	3,774	572	15.16	1926	4,891	444	9.08
1923	4,103	438	10.68	—	—	—	—

   Estimate the population for each year.

2. Roe-deer are common in most European countries, including Denmark. Anderson (1962) reported the results of a study designed to estimate the deer population in two forests on a large experimental game farm. Deer were trapped in fenced enclosures with a gate at one end from January 3, to February 10, 1958. A few deer died and some were moved to enclosures for other experiments. The remaining were given a collar: blue for fawns born during the summer of 1957 and green-red for older deer. Some deer that had been born in 1955 and given white collars as fawns in 1956 had their collar replaced with a white collar. During March, April, and possibly May, an individual was given the job of combing the woods and their immediate surroundings with binoculars. Each deer observed was recorded, and it was noted whether it had a collar and, if so, its color was noted. The data can be summarized as follows: 122 without collar, 154 with blue collar, and 186 with either a white or a green-red collar.

   A. Estimate the population of 1,956 fawns and the standard deviation of the estimate. Set a 95% confidence interval on the population size.

   B. Estimate the population of older roe-deer and the standard deviation of the estimate. Set a 95% confidence interval on the population size.

   C. Based on these results, suggest a design for 1959.

3. Green and Evans (1940) reported the results of a study of snowshoe hares (*Lepus americanus*) in the Lake Alexander area of Minnesota from 1932 to 1939. The population size was estimated each year as part of a study of the fluctuations in abundance of the hares. Trapping began in either October or November and extended to April or May of the following year when hares

stopped entering traps in appreciable numbers probably because of the abundance of food. Most of the trapping season was used to trap and tag the hares. A small metal ear band with a unique number was clipped to a captured hare's ear using special pliers. Although the bands made only a small perforation in the ear, they did appear to stay on throughout the animal's life without causing irritation. During the final two and one-half weeks of the trapping period (usually beginning early in April), the entire area was retrapped. For sampling, the area was divided into stations. During retrapping, each station was operated for six-trap days, usually by having three traps at a station for 2 days or occasionally having six traps at a station for 1 day. Some of the data are given below:

Year	n_1	n_2	m_2
1932–1933	948	421	167
1933–1934	876	329	154
1934–1935	659	272	105

A. Estimate the population of snowshoe hares in the Lake Alexandra area for each of the three years given in the table. Estimate the standard deviation of the estimate, and set a confidence interval on the population size. Be sure to present your results using complete sentences in the context of the experiment.

B. Mortality was estimated to be high (43 and 50%) during the winters of 1935–1936 and 1936–1937. Assume that mortality occurred at similar rates in the years given above. What impact will this have on the population estimates?

C. Snowshoe hares lived in a restricted home range, seldom moving more than 0.125 miles from the area in which they were originally trapped. Under these circumstances, what are some aspects of design that you would want to consider.

D. In later years, the area trapped during the period of banding the hares was restricted to a portion of the total area called Mile Area (because its area was slightly larger than a square mile). However, during the retrap period, the entire area originally trapped (which included an area in addition to the Mile Area) was covered. What statistical issues would you anticipate when working with the data in these years?

4. Hayne (1949b) reported the results from a live-trapping experiment conducted at East Lansing, Michigan, between July 19 and August 4, 1942. Traps for the meadow vole (*Microtus pennsylvanicus pennsylvanicus* Ord) were set in a rectangular grid at 60-foot intervals covering 8.6 acres. Captures of only adult females were reported. Traps were tended each morning

(M) and evening (E). Below are the results from the evening of July 19 to the morning of July 23. Hayne computed estimates of the adult female meadow vole population by using consecutive trapping times (each line in the table) and the Lincoln–Petersen estimator. He could get no estimate in three cases.

Trapping date and time of day in July		No. of captures		
First interval	Second interval	First interval	Second interval	Common to both
19E	20M	8	19	0
20M	20E	19	10[a]	2
20E	21M	10[a]	23	0
21M	21E	23	9	0
21E	22M	9	14	0
22M	22E	14	9[a]	2
22E	23M	9[a]	21	2

[a]One dead animal in the group.

A. Using the Lincoln–Petersen estimator, estimate the vole population based on the information in each line of the table. (You should have seven estimates.) Did you encounter the same problem as Hayne in finding estimates for each entry?

B. For the cases that permitted estimates in A, set 90% confidence intervals on the size of the female adult vole population.

C. Another study to estimate the adult female meadow vole population in this area is to be conducted. Suggest a design for the study. Explain the process you went through in developing this design.

5. Wood (1963) conducted a study of climbing cutworms in a small field (0.25 to 0.5 acre) with an average stand of blueberries from 1956 to 1962. Although a wide variety of cutworms might be present in any given year, five species constituted more than 90% of the population: *Actebia fnnica* Tausch., *Graphiphora smithi* Snell., *Graphiphora collaris* G. & R., *Syngrapha epigaca* Grt., and *Eucirrhoedia pampina* Gn. In each year, the field was swept with an insect net at night. The captured insects were taken back to the laboratory, examined, and marked with Fluorescent 2266, using either a camel hair brush or an atomizer. Larvae injured during the collecting or handling were discarded. The marked larvae were systematically redistributed in the field the same night of capture. The field was sampled again on the following two or three evenings. Records were kept of the total capture and the number marked. The data are presented in the table below:

Year	No. of larvae marked and released in first sample	No. of larvae in second sample	No. of recaptures
1956	125	107	8
1958	75	143	6
1959	93	114	3
1961	200	122	14
1962	1,000	1,755	41

A. Estimate the population of climbing cutworms in the field for each period for which data are available. Estimate the standard deviation of each of these estimates.

B. Set a 90% confidence interval on the size of the climbing cutworm population for each year.

C. Suppose you are to design the study for the next year. What would you propose? Justify your answer.

6. Gulland (1963) worked with information from a double-tagging experiment. In 1959, 603 plaice had front and rear tags attached with stainless steel wire. Seven fish were returned within the first two years of release. Assume that 4 of these had the front tag and three had the rear tag. The ratio of single tagged fish returned (7) to the number of fish returned with both tags was 0.05. Suppose that the information on these tags came from a second sample of plaice of size 482 (This is not the case, but will serve to illustrate the process.) Estimate the population of plaice in the lake. Estimate the standard deviation of the estimated population.

7. Consider again the capture-recapture study conducted by Hayne (1949b) and discussed in exercise 4 above. In Table 2 of his paper, Hayne presented the following capture–recapture information from 8 sampling periods:

	No. of captures	
Date and time of capture	Unmarked	Marked
July 19, p.m.	8	0
July 20, a.m.	19	0
July 20, p.m.	8	2[a]
July 21, a.m.	15	8
July 21, p.m.	9	0
July 22, a.m.	5	9
July 22, p.m.	2	7[a]
July 23, a.m.	8	13

[a]One dead adult female meadow vole was found in sample.

Analyze these data using closed population models. Describe the models considered and the tests conducted. What conclusions do you draw from these data?

8. Hayne (1949b) reported the result of a study of the population size of meadow voles (*Microtus pennsylvanicus pennsylvanicus* Ord). The study was conducted during June, 1948, in a field near East Lansing, Michigan. Twenty sets of three traps each were spaced 25 feet apart along a 475-foot line. The voles were removed from the population after capture. The data from the study are presented in the table below:

Date	No. captured			
	Adult males	Adult females	All adults	All juveniles
June 13	5	8	13	3
June 14	1	2	3	1
June 15	1	1	2	1

A. Estimate the size of the adult vole population in the field. What is the standard deviation of this estimate?

B. Set a 95% confidence interval on the size of the adult vole population in the field.

C. Based on the limited description provided you about the design, what additional information would you need to assess the appropriateness of the design? Making the assumptions you need, discuss the adequacy of this study's design.

9. Carothers (1973) presents the results of a capture–recapture study of Edinburgh taxi cabs. The taxis registered within Edinburgh were the study population. These were of the "London" diesel type. Because the nearest city (Glasgow) with similar taxis is 40 miles away, it was considered highly improbable that a taxi observed within Edinburgh would be registered in some other city. It was determined that taxis are in almost continual use during working hours except for brief periods of servicing. Because there are about twice as many drivers as cabs, a cab is in use even if a driver is sick or on vacation. Therefore, immigration and emigration for the population was believed to be limited to normal vehicle turnover. The working life of a cab is about 3 years, so it was estimated that the turnover during the 2-week period of the study was less than 2%. There were 435 taxi cabs registered with the city. Allowing for those that might be out of service during the sampling period, 420 taxi cabs was the assumed population size. Sampling took place on 10 consecutive weekdays (Monday through Friday of two consecutive weeks).

Sampling was conducted according to two schemes. In scheme A, sampling points were scattered across the city. The immediate vicinity of cab ranks was avoided, and different points were chosen each day. Time of sampling was varied. The data are in data set taxia.dat. In scheme B, sampling occurred at the same points and times on each day. These data are in the set taxib.dat. The registration number of all observed cabs was recorded. The first time a cab was observed was defined to be its capture and all subsequent times as recaptures. A total of 283 and 241 cabs were observed under schemes A and B, respectively. (*See* Rexstad and Burham 1991 and Burnham et al., 1980.)

A. Estimate the number of Edinburgh taxi cabs from data collected under scheme A. Attach a measure of precision to this estimate.

B. Estimate the number of Edinburgh taxi cabs from data collected under scheme B. Attach a measure of precision to this estimate.

C. Discuss the relative merits of the two sampling schemes. What are the strengths (weaknesses) of each.

D. What conclusions would you draw based on this study?

10. Fischler (1965) reported on a study of commercial-size, male blue crabs (*Callinectes sapidus*). Three different kinds of gear were used to trap the crab; trot-lines, crab trawls, and shrimp trawls. The three gear types were not equally effective in trapping lobster. Based on the ratios of total catch over total effort, the relative efficiency of the three gears were found to be 1.00, 1.42, and 1.06 for trot-lines, crab trawls, and shrimp trawls, respectively. That is, one crab trawl per day was equivalent to 1.42 trot-lines, and one shrimp trawl per day was equivalent to 1.06 trot-lines. These factors were used to convert the weekly fishing efforts of the crab and shrimp trawls into standard units of trot-lines per day. The catches for each week were adjusted for recruitment from precommercial to commercial size so that the population under study was the population of commercial-size male crabs at the beginning of the study. Also, females were not included in the study because tagging studies indicated that the commercial-size females were emigrating out of the area during the study period. Natural mortality was negligible. On the basis of these adjustments, the data are presented in the table below:

Week	Pounds of commercial-size male blue crab	Weekly trapping effort
1	33,541	194
2	47,326	248
3	36,460	243
4	33,157	301
5	29,207	357

(*continued*)

(*continued*)

Week	Pounds of commercial-size male blue crab	Weekly trapping effort
6	33,125	352
7	14,191	269
8	9,503	244
9	13,115	256
10	13,663	248
11	10,865	234
12	9,887	227

Assuming the goal is to estimate the population of commercial-size male blue crabs in the study area, perform a complete analysis of these data.

11. Gerking (1967) reports on a study of fish in Gordy Lake, which is in the lake district of northern Indiana. Cylindrical wire traps about 4 feet long and 2.5 feet in diameter with a funnel at only one end were fished from June 2 to July 15, 1950. Traps were spread around the periphery of the lake. They were usually moved each day and at least every other day. Each day the unmarked fish in the traps were marked and released in the general vicinity of the trap. Fish above the legal length were marked by removal of the left pectoral fin, and those below the legal length were marked by removal of the right pectoral fin. Somewhat arbitrary boundaries were used to establish age groups based on lengths. These boundaries were increased 5 mm every 2 weeks to adjust for fish growth, thereby reducing the effect of recruitment. In addition to questions of recruitment, some mortality undoubtedly occurred during the sampling period. The data for age group III redear sunfish (*Lepomis microlophus*) are presented in the table below:

Date	No. captured (n_i)	No. marked (m_i)	Date	No. captured (n_i)	No. marked (m_i)
June 2	10	0	June 24	5	2
June 3	27	0	June 25	3	2
June 4	17	0	June 26	25	9
June 5	7	0	June 27	4	2
June 6	1	0	June 28	4	2
June 7	5	0	June 29	7	3
June 8	6	2	June 30	8	6
June 9	15	1	July 1	6	6
June 10	9	5	July 2	4	3
June 11	18	5	July 3	9	4
June 12	16	4	July 4	1	1
June 13	5	2	July 5	7	5
June 14	7	2	July 6	0	0
June 15	19	3	July 7	14	7

Date	No. captured (n_i)	No. marked (m_i)	Date	No. captured (n_i)	No. marked (m_i)
June 16	15	6	July 8	3	2
June 17	20	6	July 9	10	4
June 18	5	0	July 10	9	4
June 19	15	2	July 11	8	6
June 20	15	3	July 12	5	4
June 21	8	4	July 13	7	4
June 22	23	11	July 14	4	2
June 23	11	6	July 15	7	3

A. Estimate the population size of age group III redear sunfish. Attach a measure of precision to this estimate.

B. Set a 95% confidence interval on the number of age group III redear sunfish in the lake.

C. Discuss the impact of recruitment and mortality of the estimates of population size. Justify your answer.

12. Originally, Leslie and Davis (1939) used regression to obtain an estimate of the rat population described in Example 9.7. During the study, some traps would be sprung and not capture a rat. This could happen if inhabitants of the house where the trap was set had accidentally sprung the trap or if a rat had sprung it but avoided capture. However, it was assumed that there was an equal effort on each sampling occasion.

A. Estimate the rat population using Leslie's regression and assuming equal effort on each trapping occasion. Attach a measure of precision to this estimate.

B. Compare this estimate to that in Example 9.7.

C. What impact would you expect the sprung traps to have on your estimate.

13. Paulik and Robson (1969) describe a hypothetical study of a population of pheasants. In a pre-season survey, 600 of 1,400 mature birds were cocks. Following the season, a survey found only 200 of the observed 2,000 mature birds to be cocks. Based on a road check, it was estimated the 8,000 cocks and 500 hens were harvested during the season.

A. Estimate the size of the pheasant population before and after the season. Attach a measure of precision to these estimates.

B. Estimate the proportion of cocks before and after the season. Attach a measure of precision to these estimates.

C. Estimate the cock population before and after the season. Again, attach a measure of precision to these estimates.

    D.   Do you anticipate any difficulties meeting the assumptions needed for the estimator that you used? If so, which ones?

14.   Paulik and Robson (1969) describe a hypothetical study at a lake containing brook trout [*Salvelinus fontinalis* (Mitchill)] and cisco (*Coregonus* sp.). Before fishing season, gill nets were set, and 150 brook trout and 50 cisco were captured. During the first 3 weeks of the fishing season, 700 brook trout were caught and removed from the lake. After these first 3 weeks, gill nets were set again. This time 30 brook trout and 30 cisco were captured.

    A.   What are gill nets? Describe their design and use.

    B.   Estimate the brook trout population prior to the fishing season and three weeks into the season. Attach a measure of precision to these estimates.

    C.   Estimate the population of fish in the lake before the season and three weeks into the season. Attach a measure of precision to these estimates.

    D.   Estimate the proportion of brook trout in the lake before the season and 3 weeks into the season. Attach a measure of precision to these estimates.

    E.   For this study, what are the most likely causes of departures from the model assumptions? What effect will these have on the estimates you have made?

10

Capture–Recapture: Open Populations

Introduction

Capture–recapture studies were introduced in the last chapter. Historically, the goal of such studies has been to estimate population size. They are appropriate when the animals can be captured, marked, and recaptured without adversely affecting their behavior. Extensions have permitted estimation when the animals are removed from the experiment after initial capture. In the design phase of a study, the researcher must first determine whether the population is closed or open. A closed population has a constant membership; there are no births, deaths, or migration into or out of the population during the sampling period. If a population is not closed, it is open. Often, through careful design, the assumption of a closed population can hold at least approximately. If so, the closed population models covered in Chapter 9 are appropriate. An advantage of closed population models is that the parameters can be estimated with less data. However, it is not always possible to design a study so that the closed population assumption holds even approximately. In these cases, the open population models are needed. The seminal work on open population models was conducted independently by Jolly (1965) and Seber (1965). The model these investigators proposed, and numerous extensions of that model, are used extensively by animal ecologists to estimate population size and survival probabilities.

Sometimes insufficient information is collected for population size estimation, but survival rates can still be estimated. A series of models appropriate for band recovery studies have been developed for this purpose. In these studies, animals are either captured, marked, and released, or an independent release of marked animals is made. Bands are returned after the animals are retrieved by hunters. The number of banded animals released and the band return information is used to estimate survival and band recovery rates.

Seber (1982) provides a comprehensive discussion of the Jolly–Seber model. Pollock et al. (1990) and Pollock (1991) review methods for both closed and open models and suggest areas in need of further research. Lebreton et al. (1992) discuss open population models. Brownie et al. (1985) discuss models for band recovery data. We have drawn heavily on these works in developing this chapter.

Jolly–Seber Model

In 1938, Schnabel described a sample fish census that may or may not be confined to a single species. Because the process she discussed is widely used in capture–recapture studies, it is known as the Schnabel census. In a general context, a Schnabel census is conducted in the following manner. At periodic intervals, samples are collected from the population. Each animal in the sample is examined for marks (except for the first sample), given a mark, and then returned to the population. The capture–recapture history of the animals caught during the experiment is known if different marks or tags are used for different samples. Giving each animal a unique mark or tag permits individual capture histories to be recorded. In her work, Schnabel worked with fish. Seining stations were established at various points on the lake, and samples were usually taken at about 24-hour intervals. In the last chapter, we discussed some models that could be used when the population is assumed closed. Here we assume it is open. Further assumptions are needed to provide a foundation for modeling. The Jolly–Seber model requires the following assumptions:

1. Every animal present in the population at a given sampling occasion has an equal probability of capture.

2. Every marked animal present in the population immediately after a given sampling occasion has an equal probability of survival until the next sampling occasion.

3. Marks are not lost or overlooked.

4. All samples are instantaneous, and each release is made immediately after the sample.

Notice that assumptions 1, 3, and 4 were required under the Lincoln–Petersen model in Chapter 9. Because only marked animals are used to estimate survival probabilities, the assumption that marked and unmarked animals have the same survival probabilities is not needed. However, in practice, biologists will often want to use the survival estimates to refer to the whole population, leading to this additional assumption. The Jolly–Seber model allows some animals to be lost during capture and not returned to the population, adding some complexity to the notation.

In this chapter, we use the following notation:

t: number of sampling occasions

Parameters

N_i: number in the population at the time of the ith sample

M_i: number of marked animals in the population at the time of the ith sample

$U_i = N_i - M_i$:   number of unmarked animals at the time of the ith sample

B_i: number of new animals joining the population between the ith and $(I + 1)$th samples

ϕ_i: survival probability for all animals between the ith and $(i + 1)$th sample

p_i: capture probability for all animals at the time of the ith sample

Statistics

n_i: number caught in the ith sample

m_i: number of marked animals caught in the ith sample

m_{hi}:   number of marked animals in the ith sample last caught in the hth sample

R_i: number of marked animals released after the ith sample

r_i: number of marked animals from the release of R_i animals that are subsequently recaptured

z_i: number of different animals caught before the ith sample that are not caught in the ith sample, but are subsequently caught

$u_i = n_i - m_i$:   number of unmarked animals caught in the ith sample

The data can be organized in a table as illustrated in Example 10.1.

Example 10.1

In his seminal paper, Jolly (1965) presented data collected from an apple orchard population of female black-kneed capsids (*Blepharidopterus angulatus*). Thirteen samples were collected alternately on 3- and 4-day intervals. Because of both fresh emergence and immigration, the closed population methods of the last chapter would be inappropriate. A model reflecting the open nature of the population must be considered. Following the format of Seber (1982), the data are given in Table 10.1.

The m_{hi} value in Table 10.1 represent the number caught in the ith sample that were last captured in the hth sample. For example, when $h = 3$ and $i = 6$, $m_{3,6}$

Table 10.1   Values of m_{hi} Number of Black-Kneed Capsids Caught in the ith Sample Last Captured in the hth Sample

h	i:	1	2	3	4	5	6	7	8	9	10	11	12	13	
	n_i:	54	146	169	209	220	209	250	176	172	127	123	120	142	Total
	R_i:	54	143	164	202	214	207	243	175	169	126	120	120	—	r_h
1				10	3	5	2	2	1	0	0	0	1	0	0	24
2					34	18	8	4	6	4	2	0	2	1	1	80
3						33	13	8	5	0	4	1	3	3	0	70
4							30	20	10	3	2	2	1	1	2	71
5								32	34	14	11	3	0	1	3	109
6									56	19	12	5	4	2	3	101
7										46	28	17	8	7	2	108
8											51	22	12	4	10	99
9												34	16	11	9	70
10													30	16	12	58
11														26	18	44
12															35	35
Total	m_i:	0	10	37	56	53	77	112	86	110	84	77	72	95	

After Jolly (1965).

= 8; that is, 8 capsids caught in the 6th sample were last caught in the 3rd sample. The remaining $n_6 = 209$ capsids in the 6th sample were either unmarked (132) or were last caught in some sample other than the third one (2 from the first sample, 4 from the second sample, 20 from the 4th sample, and 32 from the 5th sample). For convenience, notice that the number caught during each sampling period n_i and the number of those caught that were released R_i are also recorded. By totaling each column, we obtain m_i, the number of marked capsids that were caught in each of the samples. Totaling each row gives r_h, the number of capsids released in sampling period h that were subsequently recaptured. Table 10.1 is referred to as a Method B table (Leslie and Chitty, 1951). Based on the information there, we want to estimate the population size at various points in time and to estimate the probability of survival through time. Because the Method B table is a summarization of the data, we have lost some information that is available from the individual capture histories (*see* Chapter 9). As a consequence, some estimation and testing procedures cannot be conducted based on Method B data.

Notice from Table 10.1 that the time of each marked animal's last capture is retained. Suppose an animal is caught in the second, third, and seventh samples. When captured in the seventh sample, the most important information to record is that it was last caught in the third sample (Leslie, 1952). The fact that it was also caught in the second sample is not as valuable from the perspective of the seventh sample. That information was recorded at the time of the third sample when the time of last capture was the second sample. This permits the critical information in the study to be summarized as in Table 10.1.

We begin by developing intuitive estimators for the Jolly–Seber model in the manner of Pollack et al. (1990). Initially suppose that M_i, the number of marked animals in the population at time i, is known for all $i = 2, 3, \ldots, t$. (We must later return to the estimation of the M_i values because, in an open population, they are never known.)

Based on the model assumptions, we expect the proportion of marked animals in the sample to be about the same as the proportion of marked animals in the population; that is,

$$\frac{m_i}{n_i} \approx \frac{M_i}{N_i}. \tag{10.1}$$

Solving for N_i in (10.1), we have the following estimator of N_i:

$$\hat{N}_i = \frac{n_i \hat{M}_i}{m_i}, \qquad i = 2, 3, \ldots, t - 1. \tag{10.2}$$

Notice this has the basic form of the Lincoln–Petersen estimator described in Chapter 9.

To develop the estimator of survival probability from the time of sample i to the time of sample $(i + 1)$, first note that the number of marked animals in the population immediately after sample i is $(M_i - m_i + R_i)$. To see this, recall that the number of marked animals not caught in sample i is $(M_i - m_i)$ and that R_i represents the number of animals caught in sample i and released back into the population with marks. Now M_{i+1} is the number of marked animals in the population at the time of the $(i + 1)$th sample. Therefore, a natural survival probability estimator is the ratio of these two quantities:

$$\hat{\phi}_i = \frac{\hat{M}_{i+1}}{\hat{M}_i - m_i + R_i}, \qquad i = 1, 2, \ldots, t - 2. \tag{10.3}$$

To estimate recruitment between the time of the ith and $(i + 1)$th sample, we first consider the expected number of survivors from the time of the ith to the $(i + 1)$th sample. The population size immediately after the ith sample is $(N_i - n_i + R_i)$, where $(N_i - n_i)$ represents the number in the population that are not caught in the ith sample, and R_i is the number of animals caught, marked, and released back into the population. If ϕ_i is the probability of survival from the time of the ith to the $(i + 1)$th sample, then $\phi_i(N_i - n_i + R_i)$ is the expected number of survivors from the time of the ith to the $(i + 1)$th sample. If more animals are in the population at the time of the $(i + 1)$th sample than could be accounted for by survivors, these additional members must be due to recruitment. An intuitive estimator of recruitment between the ith and $(i + 1)$th sample is therefore

$$\hat{B}_i = \hat{N}_{i+1} - \hat{\phi}_i(\hat{N}_i - n_i + R_i), \qquad i = 2, 3, \ldots, t - 2. \qquad (10.4)$$

Capture probability p_i can be estimated as the proportion of marked or total animals alive at i that are captured at i; that is,

$$\hat{p}_i = \frac{m_i}{\hat{M}_i} = \frac{n_i}{\hat{N}_i}, \qquad i = 2, 3, \ldots, t - 1. \qquad (10.5)$$

Because we know the population is open, we still need an estimator of M_i. This estimator is developed by realizing that at i, we have two distinct groups of marked animals. The first group is comprised of the $(M_i - m_i)$ marked animals not caught at time i. The other group consists of the R_i marked animals caught at time i, marked and subsequently released back into the population. Notice that, as in earlier cases, R_i may be less than n_i as a result of deaths that occur during the capture and marking process. Assuming that the probability of future capture is the same for all marked animals, the future recovery rate for the two groups should be about the same; that is,

$$\frac{z_i}{M_i - m_i} \approx \frac{r_i}{R_i}. \qquad (10.6)$$

Therefore, the estimator of M_i is given by

$$\hat{M}_i = m_i + \frac{R_i z_i}{r_i}, \qquad i = 2, 3, \ldots, t - 1. \qquad (10.7)$$

This estimator is defined only for $i = 2, 3, \ldots, t - 1$, because we need animals seen before and after each i. Notice this also impacts the values of i for which the estimators of N_i, ϕ_i, B_i, and p_i are well defined.

The estimators of ϕ_i and p_i given in equations (10.3) and (10.5), respectively, are maximum likelihood estimators. Unfortunately, all five estimators are biased. The following approximately unbiased estimators (denoted by $\tilde{\ }$) were recommended by Seber (1982, p. 204, for ϕ_i, N_i, B_i, M_i) and Jolly (1982, p. 304, for p_i):

$$\tilde{M}_i = m_i + \frac{(R_i + 1)z_i}{r_i + 1}, \tag{10.8}$$

$$\tilde{N}_i = \frac{(n_i + 1)\tilde{M}_i}{m_i + 1}, \tag{10.9}$$

$$\tilde{\phi}_i = \frac{\tilde{M}_{i+1}}{\hat{M}_i - m_i + R_i}, \tag{10.10}$$

$$\tilde{B}_i = \tilde{N}_{i+1} - \hat{\phi}_i(\tilde{N}_i - n_i + R_i), \tag{10.11}$$

and

$$\tilde{p}_i = \frac{m_i}{\tilde{M}_i}. \tag{10.12}$$

Further, Seber recommended that m_i and r_i should be greater than 10 for satisfactory performance of these bias-adjusted estimators.

We now need to quantify the variability associated with estimating these population parameters. Variability is often divided into two sources: sampling variation and nonsampling variation. Sampling variation occurs because capture probabilities are not equal to 1. Therefore, a parameter cannot be determined with certainty, and the lack of certainty is quantified in the variability of the estimate. Nonsampling variation is associated with the stochasticity of the birth and death processes; that is, parameters, such as population size at time i (N_i), are not the same for all i, but instead vary. Pollock et al. (1990) give the asymptotic variances and covariances based on sampling variation alone. Replacing the parameters and unobservable variables by their estimates and all expected values of the other random variables by their observed values, we obtain the following estimators of variance:

$$s^2_{\hat{M}_i|M_i} = (\hat{M}_i - m_i)(\hat{M}_i - m_i + R_i)\left(\frac{1}{r_i} - \frac{1}{R_i}\right), \tag{10.13}$$

$$s^2_{\hat{N}_i|N_i} = \hat{N}_i(\hat{N}_i - n_i)\left[\frac{\hat{M}_i - m_i + R_i}{\hat{M}_i}\left(\frac{1}{r_i} - \frac{1}{R_i}\right) + \frac{\hat{N}_i - \hat{M}_i}{\hat{N}_i m_i}\right], \tag{10.14}$$

$$s^2_{\hat{\phi}_i|\phi_i} = \hat{\phi}_i^2\left[\left(\frac{(\hat{M}_{i+1} - m_{i+1})(\hat{M}_{i+1} - m_{i+1} + R_{i+1})}{\hat{M}_{i+1}^2}\right)\left(\frac{1}{r_{i+1}} - \frac{1}{R_{i+1}}\right)\right.$$
$$\left. + \frac{\hat{M}_i - m_i}{\hat{M}_i - m_i + R_i}\left(\frac{1}{r_i} - \frac{1}{R_i}\right)\right], \tag{10.15}$$

$$s^2_{\hat{B}_i|B_i} = \frac{\hat{B}_i^2(\hat{M}_{i+1} - m_{i+1})(\hat{M}_{i+1} - m_{i+1} + R_{i+1})}{\hat{M}_{i+1}^2}\left(\frac{1}{r_{i+1}} - \frac{1}{R_{i+1}}\right)$$
$$+ \frac{(\hat{M}_i - m_i)(\hat{\phi}_i R_i(\hat{N}_i - \hat{M}_i))^2}{(\hat{M}_i - m_i + R_i)\hat{M}_i^2}\left(\frac{1}{r_i} - \frac{1}{R_i}\right)$$
$$+ \frac{(\hat{N}_i - n_i)(\hat{N}_{i+1} - \hat{B}_i)(\hat{N}_i - \hat{M}_i)(1 - \hat{\phi}_i)}{\hat{N}_i(\hat{M}_i - m_i + R_i)}$$
$$+ \frac{\hat{N}_{i+1}(\hat{N}_{i+1} - n_{i+1})(\hat{N}_{i+1} - \hat{M}_{i+1})}{\hat{N}_{i+1}m_{i+1}} + \frac{\hat{\phi}_i^2\hat{N}_i(\hat{N}_i - n_i)(\hat{N}_i - \hat{M}_i)}{\hat{N}_i m_i}, \quad (10.16)$$

$$s^2_{\hat{p}_i|p_i} = \hat{p}_i^2(1 - \hat{p}_i)^2\left(\frac{1}{r_i} - \frac{1}{R_i} + \frac{1}{m_i} + \frac{1}{z_i}\right), \quad (10.17)$$

$$s_{\hat{\phi}_i,\hat{\phi}_{i+1}} = -\hat{\phi}_i\hat{\phi}_{i+1}\frac{\hat{M}_{i+1} - m_{i+1}}{\hat{M}_{i+1}}\left(\frac{1}{r_{i+1}} - \frac{1}{R_{i+1}}\right), \quad (10.18)$$

and

$$s_{\hat{B}_i,\hat{B}_{i+1}} = -\hat{\phi}_{i+1}\frac{(\hat{N}_{i+1} - n_{i+1})(\hat{N}_{i+1} - \hat{M}_{i+1})}{\hat{N}_{i+1}}\left[\frac{\hat{B}_i\hat{R}_{i+1}}{\hat{M}_{i+1}}\left(\frac{1}{r_{i+1}} - \frac{1}{R_{i+1}}\right) + \frac{\hat{N}_{i+1}}{m_{i+1}}\right]. \quad (10.19)$$

All other covariance terms are zero. Seber (1982) noted that equations (10.13) to (10.19) are only valid for large expectations of n_i, m_i, R_i, r_i, and z_i and that more investigation of their small sample properties was needed. The asymptotic variances are the same whether we use the estimators that are corrected for bias or the more intuitive estimators. However, if the population is large or sampling is not intensive, estimates based on these formulae may not be well-defined. For example, if there are no marked animals in a sample, or if none of the marked animals from one sampling occasion is ever recaptured, we have division by zero in one or more of the equations (all of them, if both events occur). Manly (1977c) suggested adding ones as was done in obtaining the bias-adjusted estimators of the parameters, and this is the approach we will adopt in the accompanying software ECOSTAT. Because these are asymptotic results, one can and should question if the analysis is appropriate when m_i or r_i is zero for any i. Careful design can usually help avoid this problem. However, our experience is that even if things do not go well in an experiment, the researcher still wants to glean all the sample has to offer about the biological system. If this is the case, the analysis should be done with caution.

Pollock et al. (1990) suggest that biologists will generally be interested only in the sampling variation when estimating N_i, M_i, B_i, and p_i. However, they anticipate interest in the underlying survival probability (ϕ_i), making the sampling variance containing both sampling and nonsampling variation appropriate. This is estimated as follows (*see* Seber, 1982):

$$s^2_{\hat{\phi}_i} = s^2_{\hat{\phi}_i | \hat{\phi}_{i+1}} + \frac{\hat{\phi}_i(1 - \hat{\phi}_i)}{\hat{M}_i - m_i + R_i}. \tag{10.20}$$

Notice that accounting for both sampling and nonsampling variation results in another term when compared to accounting for sampling variation alone. We tend to think that there would be interest in the underlying population size N_i. Accounting for both sampling and nonsampling variation results in an additional term in the computation of the estimated variance of $\hat{N}_i$ just as it did for the survival probability. Seber (1982, p. 202) presents this additional term and discusses its awkward computation. Further, he concludes that except possibly when p_i is large, the contribution of nonsampling variation will be much smaller than sampling variation and can be ignored. We assume that we can ignore it here.

Manly (1971) and Roff (1973) evaluated the performance of the variance estimators. Many concluded that for small sample sizes, there is a positive correlation between the estimates of variance and the estimates of the corresponding parameters. That is, $s^2_{\hat{N}_i | N_i}$ is positively correlated with $\hat{N}_i$. As a consequence, if N_i is underestimated, the estimate appears to be more precise than it really is. Also, if N_i is overestimated, the estimate appears to be less precise than it really is.

The behavior of the variance estimators for small sample sizes and their subsequent impact on confidence intervals have led several to suggest alternative approaches to estimating confidence intervals and variances. These include use of transformations (Seber, 1982), the generalized jackknife method (Manly, 1977c), and Monte Carlo simulation (Buckland, 1980). Manly (1984) proposed transformations based on equations (10.2) and (10.3) for population size and survival probabilities. Although we agree with Pollock et al. (1990) that the arbitrariness in the transformations give them a "somewhat ad hoc appearance," they do perform well for small sample sizes. While recognizing the need for additional research in this area, we will use Manly's approach as described below.

First, transform N_i to

$$T_1(\tilde{N}_i) = \ln(\tilde{N}_i) + \ln(0.5(1 - 0.5\tilde{p}_i^* + \sqrt{1 - \tilde{p}_i^*})) \tag{10.21}$$

where $\tilde{p}_i^* = n_i / \tilde{N}$. Then the estimated variance of $T_1(\tilde{N})$ is

$$s^2_{T_1(\tilde{N}_i)} = \left(\frac{\tilde{M}_i - m_i + R_i + 1}{\tilde{M}_i + 1} \right) \left(\frac{1}{r_i + 1} - \frac{1}{R_i + 1} \right) + \frac{1}{m_i + 1} - \frac{1}{n_i + 1}. \tag{10.22}$$

Based on a simulation study, Manly proposed an approximate 95% confidence interval for a population size N_i of the form

$$\frac{(4e^{T_{1l}} + n_i)^2}{16e^{T_{1l}}} < N_i < \frac{(4e^{T_{1u}} + n_i)^2}{16e^{T_{1u}}} \tag{10.23}$$

where

$$T_{1i} = T_1(\tilde{N}_i) - 1.6s_{T_1(\tilde{N}_i)} \tag{10.24}$$

and

$$T_{1u} = T_1(\tilde{N}_i) + 2.4s_{T_1(\tilde{N}_1)}. \tag{10.25}$$

Replacing -1.6 by -2.2 and 2.4 by 3.0 gives 99% confidence limits.

Similarly, to set a confidence interval of ϕ, we begin by transforming $\tilde{\phi}_i$ to

$$T_2(\tilde{\phi}_i) = \ln\left(\frac{1 - \sqrt{1 - \tilde{\gamma}_i\tilde{\phi}_i}}{1 + \sqrt{1 - \tilde{\gamma}_i\tilde{\phi}_i}}\right) \tag{10.26}$$

where

$$\tilde{\gamma}_i = \frac{D_i}{C_i + D_i}. \tag{10.27}$$

Then $T_2(\tilde{\phi}_i)$ has an estimated variance of

$$s^2_{T_2(\tilde{\phi}_i)} = C_i + D_i \tag{10.28}$$

where

$$\begin{aligned}C_i = &\frac{(\tilde{M}_{i+1} - m_{i+1} + 1)(\tilde{M}_{i+1} - m_{i+1} + R_{i+1} + 1)}{(\tilde{M}_{i+1} + 1)^2}\left(\frac{1}{r_{i+1} + 1} - \frac{1}{R_{i+1} + 1}\right)\\ &+ \frac{\tilde{M}_i - m_i + 1}{\tilde{M}_i - m_i + R_i + 1}\left(\frac{1}{r_i + 1} - \frac{1}{R_i + 1}\right)\end{aligned} \tag{10.29}$$

and

$$D_i = \frac{1}{\tilde{M}_{i+1} + 1} \tag{10.30}$$

The approximate 95% confidence interval for a survival probability ϕ_i is then

$$\frac{1}{\tilde{\gamma}_i}\left(1 - \frac{(1 - e^{T_{2i}})^2}{(1 + e^{T_{2i}})^2}\right) < \phi_i < \frac{1}{\tilde{\gamma}_i}\left(1 - \frac{(1 - e^{T_{2u}})^2}{(1 + e^{T_{2u}})^2}\right) \qquad (10.31)$$

where

$$T_{2i} = T_2(\tilde{\phi}_i) - 1.9 s_{T_2(\tilde{\phi}_i)} \qquad (10.32)$$

and

$$T_{2u} = T_2(\tilde{\phi}_i) + 2.1 s_{T_2(\tilde{\phi}_i)}. \qquad (10.33)$$

By replacing -1.9 by -2.5 and 2.1 by 2.7, we have 99% confidence limits.

Manly suggested these asymmetric limits because symmetric limits about $T_i(\cdot)$, $i = 1, 2$, resulted in approximately the stated level of confidence. However, when the intervals failed to cover the population parameter, the parameter tended to be above the upper limit. With the asymmetric limits, there tended to be about as many misses below as above the interval. Manly suggested that these limits should be effective for populations up to at least 1,000.

Example 10.2

Consider the data for the black-kneed capsids presented in Table 10.1. To proceed manually, construct Table 10.2 that contains the values of c_{hi}, the number

Table 10.2. Values of c_{hi}, *the Number of Black-Kneed Capsid caught in the ith Sample Last Captured in or Before the hth Sample*

h	1	2	3	4	5	6	7	8	9	10	11	12	13	Total	
1	10	3	5	2	2	1	0	0	0	1	0	0		14	z_2
2		37	23	10	6	7	4	2	0	3	1	1		57	z_3
3			56	23	14	12	4	6	1	6	4	1		71	z_4
4				53	34	22	7	8	3	7	5	3		89	z_5
5					77	56	21	19	6	7	6	6		121	z_6
6						112	40	31	11	11	8	9		110	z_7
7							86	59	28	19	15	11		132	z_8
8								110	50	31	19	21		121	z_9
9									84	47	30	30		107	z_{10}
10										77	46	42		88	z_{11}
11											72	60		60	z_{12}
12												95			

Based on Jolly (1965)

caught in the ith sample that were last caught in or before the hth sample, (as in Table 3 of Jolly, 1965, and Table 5.2 of Seber, 1982). Each entry in the body of the table is obtained in the following manner. First, think about what we want to do. We want to determine the number caught in the ith sample that were last caught on or before the hth sample. Where is that information in Table 10.1? The number of capsids caught in the ith sample that were last caught in the hth sample is listed in the hth row and ith column. What about the capsids in the ith sample last caught before the hth sample? These are still in the ith column, but now we are looking at all the rows above the hth row. We add the value in the hth row and ith column of Table 10.1 and all values above the hth row but still in the ith column together and put the total in the hth row and ith column of Table 10.2. For example, suppose we want to find the entry for $h = 5$ and $i = 8$; that is, we want to know the number of capsids that are in the 8th sample that were last caught in the 5th or an earlier sample. None of them was last caught in the first or third sample, 4 were last caught in the second sample, 3 were last caught in the 4th sample, and 14 were last caught in the 5th sample (see Table 10.1). The entry for $h = 5$ and $i = 8$ in Table 10.2 is then $0 + 4 + 0 + 3 + 14 = 21$. Mechanically, to obtain the value for row h, column i in Table 10.2, we take the corresponding value in Table 10.1 and add to it the values above it in the same column.

We also use Table 10.2 to compute the z_i values. Now z_i is the number of different animals that were caught before the ith sample, not caught in the ith sample, but caught in a later sample. For example, suppose we want to determine z_8. We need to add together the number of animals that were caught in samples 9, 10, 11, 12, and 13 that were last caught before sample 8. Look again at Table 10.2. The number of capsids last caught before the 8th sample is given in row $h = 7$. We would not want to include the 86 that were caught in sample 8, but all other values in the row would be added together to give $z_8 = 59 + 28 + 19 + 15 + 11 = 132$. In each case, we exclude the number farthest to the left in the row and then sum the remaining row values. To emphasize which values are to be omitted, we have made them bold in the table.

The process of constructing the table is tedious, but not hard. If computations are being done manually, great care should be taken to avoid errors. However, once the process is complete, all the information needed to estimate the parameters for the sampling periods is now available. The bias-corrected estimates of parameters can be obtained using equations (10.8) to (10.12). The variances and covariances of the estimates are computed using equations (10.13) to (10.19). Finally, confidence intervals can be set using either the normal approximation or the more precise ones attributable to Manly.

ECOSTAT can be used to obtain the same information. Before beginning, an ASCII file needs to be constructed with the data. The first line has t, the number of samples collected . Each of the next t lines has an identification, n_i, and R_i for the ith sampling occasion 1 are on line 2, for sampling occasion 2 are on line 3, etc. The last ($t - 1$) lines contains the body of Table 10.1 for the particular study. Capsid.dat is a data set containing the information for this example.

Once the data set is prepared, we can begin ECOSTAT. From the *Capture–Recapture* menu, select *Capture–Recapture* and then *Open*. Press *Enter Data* and supply the name of your data set (capsid.dat for this example). Also, remember if the data set is not in the current directory, path information must be provided such as c:\ecostat\capsid.dat. If an error occurs, it is most likely due to incorrectly constructing the data set. Next press *Estimate N*. After ECOSTAT asks for the name of an output data set, it will compute estimates of M_i, N_i, ϕ_i, B_i, and p_i and estimates of the standard deviation of these estimates for all possible i. This same information is stored in the output data set you constructed. Similarly, by pressing *Set CI*, 95% and 99% confidence intervals for N_i and φ_i are set using Manly's approach. The information is also stored in a data set that you name. Collecting this information, we can construct Tables 10.3 and 10.4 for these data.

We want to assess the fit of the model to the data. Several different approaches have been suggested. Seber (1982, p. 223) suggests a traditional χ^2 goodness-of-fit test based on comparing observed and expected values for numbers of animals captured with each possible capture history. He emphasized that extensive pooling may be needed because of small expected cell values. Seber (1982) also suggested an indirect test of fit based on work by Leslie et al. (1953). This test considers only the marked animals. It compares the increase in the marked population at i with an estimate of this increase based on the animals previously marked using a χ^2 goodness-of-fit test. Jolly (1982) proposed an overall test. He suggested

Table 10.3. Estimated Parameters and Standard Errors for the Black-Kneed Capsids Discussed in Examples 10.1 and 10.2

ID	$\tilde{M}_i$	$\tilde{N}_i$	$\tilde{\phi}_i$	$\tilde{B}_i$	$\tilde{P}_i$	$s_{\tilde{M}_i}$	$s_{\tilde{N}_i}$	$s_{\tilde{\phi}_i}$	$s_{\tilde{B}_i}$	$s_{\tilde{P}_i}$
1	—	—	0.646	—	—	—	—	0.088	—	—
2	35	466	1.009	291	0.287	5	128	0.110	159	0.082
3	169	758	0.864	293	0.218	18	124	0.104	133	0.044
4	256	944	0.564	400	0.219	27	136	0.057	117	0.042
5	227	929	0.834	102	0.234	17	119	0.071	109	0.038
6	324	872	0.790	109	0.238	24	93	0.068	74	0.035
7	358	796	0.651	134	0.313	25	72	0.053	54	0.039
8	318	648	0.981	− 12	0.270	20	59	0.095	52	0.031
9	400	623	0.685	49	0.275	33	60	0.077	34	0.038
10	314	473	0.880	83	0.267	27	48	0.118	39	0.038
11	314	499	0.767	73	0.246	34	62	0.124	39	0.042
12	274	454	—	—	0.263	36	65	—	—	0.043
13	—	—	—	—	—	—	—	—	—	—

After Jolly (1965).

Table 10.4   95% and 99% Confidence Limits on N_i *and* ϕ_i *for the Black-Kneed Capsids Example*

ID	95% confidence interval on N_i		99% confidence interval on N_i		95% confidence interval on ϕ_i		99% confidence interval on ϕ_i	
	Lower limit	Upper limit	Lower limit	Upper limit	Lower limit	Upper limit	Lower limit	Upper limit
2	308	937	268	1,126	0.818	1	0.763	1
3	587	1,134	535	1,257	0.682	1	0.632	1
4	752	1,345	692	1,473	0.457	0.71	0.427	0.757
5	759	1,272	705	1,379	0.704	1	0.666	1
6	736	1,132	692	1,210	0.664	0.951	0.629	1
7	691	992	656	1,050	0.550	0.78	0.521	0.821
8	561	809	532	856	0.815	1	0.768	1
9	536	787	507	836	0.547	0.874	0.51	0.936
10	403	608	380	648	0.679	1	0.626	1
11	410	677	382	732	0.560	1	0.507	1
12	362	648	334	710	—	—	—	—

After Jolly (1965).

comparing the observed and expected frequencies of the triangular array of numbers in the body of the Method B table. Thus, a comparison is made between the observed and expected values for the number of animals caught in sample i that were last caught in sample h using a χ^2 goodness-of-fit test. Pollock et al. (1985) noted that Jolly's test may eliminate too much capture–history information by overpooling. As an alternative, they developed two contingency table χ^2 tests based on minimal sufficient statistics. First, we consider Jolly's test, which is the one used in ECOSTAT. Because ECOSTAT accepts data for the Jolly–Seber model in the form given in the Method B table, the information on the individual capture histories has been lost, and we are not able to perform the test proposed by Pollock et al. (1985a). However, we describe it.

First, consider the test proposed by Jolly (1982). The key to conducting this test is the computation of the expected values in the body of the Method B table. To determine, the expected value of the number captured at time i that were last caught at time h, it is important to realize that the animal must be alive and not caught between times h and i. Thus,

$$E(m_{hi}) = \frac{r_h m_i}{z_h + r_h} \prod_{k=h+1}^{i-1} \frac{z_k}{z_k + r_k}.$$

Then using Pearson's chi-squared test, the test statistic has the form

$$X^2 = \sum_{i=3}^{t} \sum_{h=1}^{i-1} \frac{(m_{hi} - E(m_{hi}))^2}{E(m_{hi})}.$$

If the model is correct, the test statistic has an asymptotic χ^2 distribution with $0.5(s - 2)(s - 3)$ degrees of freedom. As when applying this test in other circumstances, the expected values should be no less than five in no more than 20% of cases and never less than one for the asymptotic approximation to be sufficient. If this is not the case, the data may be grouped, causing a reduction in the degrees of freedom.

ECOSTAT pools across rows when necessary. This may not be the most efficient method of pooling. Further, pooling down the columns or in some other fashion based on inspection will lead to slightly different test results.

Example 10.3

When using ECOSTAT, Jolly's test of fit is automatically performed when the model parameters are estimated. Three different levels of grouping are considered. Groups are enlarged until the expectation exceeds 1 or 3 or 5, depending on the level of grouping. Results for all three are displayed. Consider again the data in Examples 10.1 and 10.2. The results from the test of fit are given in Table 10.5. For each grouping, we do not reject the null hypothesis that the model fits. Therefore, we conclude that the Jolly–Seber model fits the data within error.

The test introduced by Pollock et al. (1985) and described by Pollock et al. (1990) has two components. The first is based on $(k - 2)$ contingency tables. Following Pollock et al. (1985a, 1990), we collapse these into 2×2 tables of the form shown in Table 10.6.

Observed values are shown in Table 10.6, constructed for $i = 2, 3, \ldots, k - 1$. For example, suppose $i = 3$. O_{11} is the number of animals captured for the first time in one of the first two sampling occasions, recaptured in the third sampling occasion and recaptured again in a later sampling period. O_{12} is the number of animals captured for the first time in the third sampling occasion and subsequently recaptured. O_{21} is the number of animals captured in at least one of the first two

Table 10.5 Results of Goodness-of-Fit Test for the Joly–Seber Model Fit to the Black-Kneed Capsids

Minimum expectation	X^2	df	p
1	55.061	43	.1027
3	28.071	31	.6175
5	24.708	26	.5355

Data from Jolly (1965).

Table 10.6 Form of 2 × 2 Table Used in First Component of Goodness-of-Fit Test

	First captured before i	First captured in i	Row totals
Captured in i and recaptured	O_{11}	O_{12}	$n_{1.}$
Captured in i and not recaptured	O_{21}	O_{22}	$n_{2.}$
Column totals	$n_{.1}$	$n_{.2}$	$n_{..}$

After Pollock et al. (1990).

sampling occasions, recaptured in the third sampling occasion, but not recaptured again. O_{22} is the number of animals caught for the first time in the third sampling occasion and never recaptured. We denote row totals by n_i, $i = 1, 2$; column totals by n_j, $j = 1, 2$; and the overall total by n. A χ^2 test for independence is constructed for each i. To conduct this test, we first compute the expected values as follows:

$$E_{ij} = \frac{n_{i.} n_{.j}}{n_{..}}. \tag{10.34}$$

The test statistics have the same form as that used in Pearson's χ^2 goodness-of-fit test:

$$X_{1i}^2 = \sum_{i,j} \frac{(O_{ij} - E_{ij})^2}{E_{ij}}. \tag{10.35}$$

If any of the expected values in the table is less than 2, the table is dropped from the test. The first component is the sum over the remaining i of the individual components given in equation (10.35).

The second component of the test is based on $(k - 3)$ contingency tables. For $i = 2, 3, \ldots, k - 2$, we construct a contingency table (see Table 10.7).

The notation is similar to that in the first component. However, for $i = 3$, O_{11} is now the number of animals caught in the first sample (before the second sample) and recaptured in the third sample. O_{12} is the number of animals first captured in the second sample and recaptured in the third sample. O_{21} is the number of animals caught in the first sample, not recaptured in the third sample, but recaptured in a later sample. O_{22} is the number of animals first caught in the second sample, recaptured in a later sample, but not in the third sample. Using the same notation for row, column, and overall totals, the expected values are computed as in equation (10.34). After dropping any table with an expected frequency below two, the test statistic is computed as in equation (10.35), but denoted by X_{2i}^2.

The overall test statistic is the sum of the individual components:

Table 10.7. Form of 2 × 2 Tables Constructed When Computing Second Component of Goodness-of-Fit Test

	First captured before $i - 1$	First captured in $i - 1$	Row totals
Recaptured in i	O_{11}	0_{12}	$n_1.$
Recaptured after i but not in i	O_{21}	O_{22}	$N_2.$
Column totals	$n_{.1}$	$n_{.2}$	$n..$

After Pollock et al. (1990).

$$X^2 = \sum_{i=2}^{k-1} X_{1i}^2 + \sum_{i=3}^{k-2} X_{2i}^2 \qquad (10.36)$$

If the null hypothesis that the Jolly–Seber model is correct is true, X^2 has an approximate χ^2 distribution with ($2k - 5 -$ number of tables deleted) degrees of freedom. Large values of X^2 result in the rejection of the null leading us to conclude that the Jolly–Seber model does not fit the data.

JOLLY (Pollock et al., 1990) is an extensive computer program written for the analysis of data using the Jolly–Seber model. Arnason and Schwarz (1986) also have developed POPAN–3 for use with open models. The interested reader should investigate these programs as they are more fully developed in this area than ECOSTAT.

The Jolly–Seber model has been criticized both for being too general and too specific. The criticism for being too general arises because it is assumed that survival and birth processes are time specific and thus must be estimated for each sampling period. If there are deaths but no births, births but no death, or constant survival or birth rates, the model has more parameters than are needed and is thus too general. The idea of fitting the model with the fewest number of necessary parameters has been called the principle of parsimony. The Jolly–Seber model, as presented, suffers from the disadvantage that the parameter estimates have poor precision unless the number of recaptures is very large. For example, look again at Table 10.3. At the time of the second sample, the estimate of population size is 466, and the estimated standard deviation of the estimate is 133. We actually have very little precision in the estimate! If some parameters are unnecessary, we can often estimate the remaining parameters with greater precision.

Sometimes, especially in certain seasons of the year, evidence may indicate that births and immigration are negligible but that deaths are occurring during the course of the study. Darroch (1959) first considered the deaths only model assuming that there were no losses on capture. Jolly (1965) considered it as a special case of the Jolly-Seber model, and Seber (1982, p 217) also discusses this model.

During spring or other seasons of the year, we may be sampling a population for which deaths and emigration are negligible, but the birth rate cannot be ignored during the study period. Darroch (1959) also considered this model, again assuming that no losses occurred during capture. Jolly (1965) viewed it as a special case of the Jolly–Seber model, and Seber (1982, p. 219) also discusses the births-only model.

Jolly (1982), prompted from practical considerations, considered three generalizations of the original Jolly–Seber model. He recognized that it might be reasonable to assume that survival rates are constant, that birth rates are constant, or that both survival and birth rates are constant during the study period. Further, Jolly showed that if these more restrictive models are appropriate, the remaining parameters may be estimated with greater precision.

Not only may the Jolly–Seber model be too general, it may be too specific. As with the closed population models discussed in the last chapter, animals may have a response to traps after capture becoming either more or less likely to be recaptured. Robson (1969), Pollock (1975), and Brownie and Robson (1983) have generalized the Jolly–Seber model to account for the short-term (usually one time period) effect of marking on survival and capture probabilities.

Another possible generalization of the Jolly–Seber model accounts for different survival and capture rates associated with identifiable age groups. Pollock et al. (1990) refer to this as the cohort model. Buckland (1982) obtained similar estimators, and Loery et al. (1987) present a detailed discussion of the cohort model. Further, Pollock (1981), Stokes (1984), and Pollock et al. (1990) consider a model that allows for age classes that may have different survival rates and possibly different capture rates. Pollock et al. (1990) have developed the program JOLLY-AGE for analysis of data from these studies.

Cormack (1964) anticipated that the survival part of the Jolly-Seber model would be of interest, without consideration of population size or birth rates. Since the mid-1980's a great deal of research has been directed in this direction, and the model is generally referred to as the Cormack–Jolly–Seber model. Burnham et al. (1987) consider comparison of survival rates for different populations of animals. As an example, they compared the survival of fish that had passed through a dam's turbine to the survival of those that had not experienced that stress. They developed the program RELEASE for comparing survival rates. Lebreton et al. (1992) synthesizes research on the Cormack–Jolly–Seber model with emphasis on multiple data sets. Our discussion of the survival models will be limited.

Band recovery studies are also useful in estimating survival. The difference in these and the Cormack–Jolly–Seber (CJS) model is that band recovery studies are based on the returns of bands after the animal dies and the CJS model relies on recapture of live animals. The band recovery studies will be the focus of the next section.

Adult Band and Tag Recovery Models

Consider a fish or wildlife study conducted in the following manner. Uniquely marked animals are released. These animals have been captured from the study region before marking, or they could be new to the environment as in the case of the release of hatchery fish. Hunters or anglers subsequently capture the animals and return the band or tag to the appropriate agency. In this situation, it is not usually possible to follow unmarked animals. Therefore, only estimation of survival probabilities is possible and such estimation is the purpose of the study. Here we assume that only adults are banded. Brownie et al. (1985) present a comprehensive discussion of the models developed for data arising from such studies, including the more general case where both adults and juveniles are banded. Their work will be the foundation of our discussion.

To make the discussion more concrete, suppose that a large sample of birds is banded in August of each year for k consecutive years in a given area. A group banded in the same period is referred to as a cohort. Records are kept on the number of bands reported from dead birds in the jth year after banding. Our goal is to estimate survival probabilities from this type of data. As with other models, some basic assumptions must be made:

1. All banded individuals have the same annual survival and recovery rates, and these may vary from year to year. (This implies that survival is not affected by banding.)

2. The fate of each banded animal is independent of that of every other banded animal.

3. Bands or tags are not lost.

Consideration should be given to the timing of tagging. For birds, banding studies will generally involve catching, banding, and releasing a sample from some population at regular intervals (August of each year, in our example). Banding is usually avoided during the breeding season so that the birds are not disturbed during this critical period. Banding during migration periods or the hunting season is also avoided because of the resulting complexities in data interpretation and analysis. Therefore, most banding is done during the winter or late summer. The same concepts apply to fish populations. However, we note again that the tagged animals could be independent releases. For example, hatchery fish could be tagged and released into the population. If it is believed that survival of the hatchery fish is the same as the resident population, then inferences about the population can be made. Otherwise, the conclusions apply to survival of the hatchery fish.

A banded adult alive at the beginning of the year has several possible fates. The animal could survive the year with probability S_i. The animal could be killed by a hunter with probability K_i. If it is killed, the animal could be retrieved with probability c, or it may not be retrieved (probability $(1 - c)$). Once the banded animal is retrieved, the band may be reported, with probability λ, or not reported, with probability $(1 - \lambda)$. The animal could also die of natural causes with probability $(1 - S_i - K_i)$. These outcomes are depicted in Figure 10.1.

As before, we need to develop notation for these studies. Although the terminology will be that relating to birds, it should be realized that it also applies to fish and other animals. Define the following:

k: number of years of banding

t: number of years of recovery $(t \geq k)$

Parameters

S_i: probability an animal alive when a given cohort is banded will survive for calendar year i.

K_i: probability an animal is killed during year i

$1 - S_i - K_i$:   probability of dying from natural causes during year i

f_i: probability that an animal alive when a given cohort is banded will be retrieved by a hunter and its band reported during hunting season i

λ: probability that a hunter will report the band, given that he has killed and retrieved a banded bird

Statistics

N_i: number banded and released back into population in year i

R_{ij}:   number of band recoveries in hunting season j from birds originally banded in year i

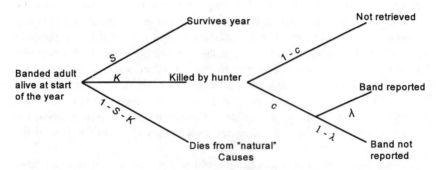

Figure 10.1. Possible yearly outcomes for adults banded at the start of a year (after Brownie et al. 1985).

T_i: number of animals banded in the *i*th or preceding years for which bands are recovered in the *i*th or subsequent years.

The data collected could be displayed in a table. For example, suppose banding is done in 3 years and band recovery over five years, the years of banding and two additional years. The data may be presented as in Table 10.8.

One other set of quantities needs to be computed from the above table. Define T_i the number of animals banded in the *i*th or preceding years for which bands are recovered in the *i*th or subsequent years. These may be computed as follows:

$$T_1 = R_1,\qquad(10.37)$$

$$T_i = R_i + T_{i-1} - C_{i-1}, \qquad i = 2, 3, , \ldots, k, \qquad(10.38)$$

and

$$T_{k+j} = T_{k+j-1} - C_{k+j-1}, \qquad j = 1, ,2, \ldots, t - k. \qquad(10.39)$$

The choice of model depends on whether the survival rate (S_i) and band recovery rate (f_i) are time or age dependent. If survival is completely time and age dependent, the model has too many parameters, leading to identifiability problems. This requires constraints to be put on the model. However, by banding only adults, we are willing to assume that survival is time dependent, but not age dependent. Further, it seems reasonable to assume that the band recovery rate is time but not age dependent. We now consider the most commonly used model within this structure.

Suppose we assume that survival, hunting, and reporting rates are year-specific but independent of the year of banding; then we have the model developed independently by Seber (1970a) and by Robson and Youngs (1971). This leads to the subscripts *i* in the above notation, indicating dependence on a specific year.

Consider a study in which banding is conducted for 3 years and band recovery is recorded for 5 years, the three years of banding and the two subsequent years.

Table 10.8. Data Collected From a Band Recovery Study

Year banded	No. banded	Year of recovery					Row totals
		1	2	3	4	5	
1	N_1	R_{11}	R_{12}	R_{13}	R_{14}	R_{15}	R_1
2	N_2	—	R_{22}	R_{23}	R_{24}	R_{25}	R_2
3	N_3	—	—	R_{33}	R_{34}	R_{35}	R_3
Column totals		C_1	C_2	C_3	C_4	C_5	—

After Brownie et al. 1985

The expected or average numbers of band recoveries expressed as functions of N_i, f_i, and S_i are given in Table 10.9.

Consider $N_1 S_1 S_2 f_3$ that appears in the third year of recovery for birds marked in the first year. For a band to be recovered in the third year, the bird had to survive the first 2 years. The probability of surviving is S_1 for the first year and S_2 for the second year. Assuming that survival is independent from year to year, the probability of surviving both years is $S_1 S_2$. The probability of band recovery in the third year, given that the bird is alive at the start of the year, is f_3. Therefore, the probability that the bird survives the first 2 years and the band is recovered in the third year is $S_1 S_2 f_3$. If we multiply by the number of birds N_1 that were banded in the first year, we have the expected value in the table. Similar arguments could be made for each entry in the body of the table.

Another thing that we need to note in this table is that S_3 and f_4 always occur together as the product $S_3 f_4$. As a result, the product $S_3 f_4$ is estimable, but S_3 and f_4 are not separately estimable. In general, under this model, the parameters f_1, $f_2, \ldots, f_k$, $S_1, S_2, \ldots, S_{k-1}$ are separately estimable. However, if $t > k$, the products $S_k f_{k+1}$, $S_k S_{k+1} f_{k+2}, \ldots$, $S_k S_{k+1} \ldots S_{t-1} f_t$ are estimable, but the individual parameters S_{k+j-1} and f_{k+j}, $j = 1, 2, \ldots, t - k$, are not. Although the estimates of the products are not of biological interest, they must be estimated to perform goodness-of-fit tests for the model.

Seber (1970a) and Robson and Youngs (1971) found the maximum likelihood estimators for f_i and S_i to be

$$\hat{f}_i = \frac{R_i C_i}{N_i T_i} \qquad i = 1, 2, \ldots, k \qquad (10.40)$$

and

$$\hat{S}_i = \frac{R_i N_{i+1}}{N_i R_{i+1}} \left(\frac{T_i - C_i}{T_i} \right), \qquad (10.41)$$

Table 10.9 Expected Numbers of Band Recoveries as Functions of N_i, f_j, and S_i

Year banded	No. banded	Year of recovery				
		1	2	3	4	5
1	N_1	$N_1 f_1$	$N_1 S_1 f_2$	$N_1 S_1 S_2 f_3$	$N_1 S_1 S_2 S_3 f_4$	$N_1 S_1 S_2 S_3 S_4 f_5$
2	N_2	—	$N_2 f_2$	$N_2 S_2 f_3$	$N_2 S_2 S_3 f_4$	$N_2 S_2 S_3 S_4 f_5$
3	N_3	—	—	$N_3 f_3$	$N_3 S_3 f_4$	$N_3 S_3 S_4 f_5$

After Brownie et al. (1985).

respectively. The MLE of S_i is biased. Adjusting for this bias, we have the estimator of survival rate S_i that we use,

$$\tilde{S}_i = \frac{R_i}{N_i} \left(\frac{T_i - C_i}{T_i} \right) \frac{N_{i+1} + 1}{R_{i+1} + 1} \qquad (10.42)$$

The bias-adjusted estimator of survival is always slightly smaller than the MLE. Estimated average recovery and survival rates are computed as

$$\hat{\bar{f}} = \frac{1}{k - 1} \sum_{i=1}^{k-1} \hat{f}_i \qquad (10.43)$$

and

$$\hat{\bar{S}} = \frac{1}{k - 1} \sum_{i=1}^{k-1} \tilde{S}_i, \qquad (10.44)$$

respectively. Notice that $\hat{f}_k$ is not included in $\bar{f}$. We want to be able to compare the variabilities of recovery and survival rates. Because S_k cannot be estimated, comparability is enhanced by not using $\hat{f}_k$.

If $t > k$, $S_k f_{k+1}$, $S_k S_{k+1} f_{k+2}$, $\ldots$, $S_k S_{k+1} \ldots S_{t-1} f_t$ are estimated by

$$\widehat{S_k S_{k+1} \ldots S_{k+j-1} f_{k+j}} = \frac{R_k C_{k+j}}{N_k T_k}, \qquad j = 1, 2, \ldots, t - k. \qquad (10.45)$$

The estimated variances of $\hat{\bar{f}}$ and $\tilde{S}$ have the form

$$s_{\hat{f}_i}^2 = \hat{f}_i^2 \left(\frac{1}{R_i} - \frac{1}{N_i} + \frac{1}{C_i} - \frac{1}{T_i} \right), \qquad i = 1, 2, \ldots, k \qquad (10.46)$$

and

$$s_{\tilde{S}_i}^2 = \tilde{S}_i^2 \left(\frac{1}{R_i} - \frac{1}{N_i} + \frac{1}{R_{i+1}} - \frac{1}{N_{i+1}} + \frac{1}{T_{i+1} - R_{i+1}} - \frac{1}{T_i} \right),$$
$$i = 1, 2, \ldots, k - 1, \qquad (10.47)$$

respectively. Approximate confidence intervals for each are based on the normal distribution.

Because the estimators of recovery and survival probabilities are based on the same sample data, some are correlated. Understanding these correlations is valuable because we might otherwise attach biological significance to what is actually a mathematical relationship between the estimators. The covariance estimators for this model are

$$s_{\tilde{S}_i \hat{f}_i} = \hat{f}_i \tilde{S}_i \left(\frac{1}{R_i} - \frac{1}{N_i} - \frac{1}{T_i} \right), \qquad i = 1, 2, \ldots, k - 1, \tag{10.48}$$

$$s_{\hat{f}_i \hat{f}_{i+1}} = 0, \qquad i > 1, \tag{10.49}$$

$$s_{\hat{f}_{i+j} \tilde{S}_i} = 0, \qquad\qquad\qquad\qquad j > 1,$$
$$= -\tilde{S}_i \hat{f}_{i+1} \left(\frac{1}{R_{i+1}} - \frac{1}{N_{i+1}} \right), \qquad j = 1, \tag{10.50}$$

and

$$s_{\tilde{S}_i \tilde{S}_{i+1}} = 0, \qquad\qquad\qquad\qquad j > 1, \tag{10.51}$$
$$= -\tilde{S}_i \tilde{S}_{i+1} \left(\frac{1}{R_{i+1}} - \frac{1}{N_{i+1}} \right), \qquad j = 1.$$

Because significant covariances may be quite large or small, we often compute the correlation so that the value is always between -1 and 1. The correlation of two estimators $\hat{X}$ and $\hat{Y}$ may be estimated using

$$\hat{\rho}_{\hat{X}\hat{Y}} = \frac{s_{\hat{X}\hat{Y}}}{s_{\hat{X}} s_{\hat{Y}}}. \tag{10.52}$$

A correlation of -1 means that as one estimator increases the other decreases in a linear way. A correlation of 1 means that both estimators increase linearly with each other. A correlation of 0 indicates no linear relationship in the two estimators. Other values indicate varying strengths of the linear relationship in the two estimators.

The estimated variance and covariances of the estimates of average recovery and survival rates are given below:

$$s_{\bar{f}}^2 = \frac{1}{(k-1)^2} \sum_{i=1}^{k-1} s_{\hat{f}_i}^2, \tag{10.53}$$

$$s_{\bar{S}}^2 = \frac{1}{(k-1)^2} \left(\sum_{i=1}^{k-1} s_{\tilde{S}_i}^2 + 2 \sum_{i=1}^{k-2} s_{\tilde{S}_i \tilde{S}_{i+1}} \right), \tag{10.54}$$

and

$$s_{\hat{S}f}^{\hat{}} = \frac{1}{(k-1)^2} \left(\sum_{i=1}^{k-1} s_{\hat{f}_i \bar{S}_i} + 2 \sum_{i=1}^{k-2} s_{\hat{f}_{i+1} \bar{S}_i} \right). \tag{10.55}$$

Before accepting the model, we must test to be sure that it fits the data adequately. We will use the Pearson's χ^2 goodness-of-fit test, just as we did with the Jolly–Seber model. Here the observed values, O_{ij}, are equal to the R_{ij} values, the number of bands returned in the jth year of recovery from animals tagged in the ith year (*see* Table 10.8). The expected values, E_{ij}, are estimated by replacing the model parameters (as in Table 10.9) with the estimates of those parameters. Then the test statistic has the form

$$X^2 = \sum_i \sum_j \frac{(O_{ij} - E_{ij})^2}{E_{ij}}.$$

These will be illustrated in the next example.

Example 10.4

Brownie et al. (1985) present data from a midwestern state where adult male wood ducks (*Aix sponsa*) were banded in August each year in 1964–1966. The data are presented in Table 10.10.

In this study, there are three cohorts, one for each year 1964–1966. Table 10.10 should be examined so that each number is fully understood. For example, in 1964, 1,603 adult birds were banded. Of these, 127 had bands returned during the 1964 hunting season, and 17 had bands returned during the 1968 hunting season. As another example, of the 1,157 adult birds banded during 1966, 61 bands were returned during the 1967 hunting season. We want to estimate the probability of wood duck survival for each year i.

ECOSTAT was used to help fit the model. First, an ASCII file containing the data must be constructed. The first line of the data file has the number of banding

Table 10.10. Data from a Banding Study of Adult Male Wood Ducks in a Midwestern State

Year banded	No. banded	Recoveries by hunting season					Row totals
		1964	1965	1966	1967	1968	
1964	1,603	127	44	37	40	17	265
1965	1,595	—	62	76	44	28	210
1966	1,157	—	—	82	61	24	167
Column totals		127	106	195	145	69	—
T_j		265	348	409	214	69	—

After Brownie et al. (1985).

periods k and the number of recovery periods t, 3 and 5, respectively, for our example. The next k lines have an identification for the banding period and the number banded during that period. The final k lines of the data set has the information in the body of Table 10.8. For our example, the first of the three lines follows: 127 44 37 40 17. Notice that as with other data sets, entries are separated by spaces. Also, we should note that ECOSTAT assumes that the banding information and the recovery information are entered in the chronological order with respect to banding.

To obtain the estimates, select *Banding* on the *Capture–Recapture* menu. Then press the *Enter Data* button. After supplying the name of the data set (duck.dat for this example), press the *Estimate Survival* button. The estimates of model parameters and the test of fit for the model will be displayed as well as placed in an output data set. The results for this example are in Table 10.11.

Therefore, the probability that a banded wood duck alive when banded will be retrieved by a hunter and its band returned during the 1964 hunting season is estimated to be 0.079 with an estimated standard deviation of 0.007. Of the wood ducks either banded alive in 1964 and surviving to 1965 or those banded alive in 1965, an estimated 4.0% were shot and had their bands returned during the 1965 hunting season. The standard deviation of this estimate is estimated to be 0.4% The probability that a bird banded in 1964 survived to 1965 is estimated to be 0.65 with an estimated standard deviation of 0.10. Further, the birds banded in 1964 and surviving to 1965 as well as the birds first banded in 1965 had an estimated 63% chance of surviving to 1966, and the standard deviation of this estimate is estimated to be 10%. Further, the average return rate from 1964 to 1966 is estimated to be 6.0% with a standard error of 0.4%. The average survival rate for this same period is 64% with a standard error of about 7%.

Although these were computed using ECOSTAT, we could have easily fit this model using a calculator. To begin, for the table above, we compute the row totals, the column totals, and the T_i. Recall the T_i represent the number of wood ducks banded in the ith or preceding years for which bands are recovered in the ith or subsequent years. These are computed using equations (10.37) to (10.39). For example, $T_i = R_i = 265$; $T_3 = R_3 + T_2 - C_2 = 167 + 348 - 106 = 409$;

Table 10.11. *Estimated Parameters for the Banding Model Fit to Data from a Wood Duck Study in a Midwestern State*

Identification	$\hat{f}_i$ (%)	$s_{\hat{f}_i}$ (%)	$\tilde{S}_i$ (%)	$s_{\tilde{S}_i}$ (%)
1964	7.923	0.675	65.117	10.372
1965	4.010	0.415	63.109	10.255
1966	6.882	0.608	—	—
Estimate of mean	5.967	0.396	64.113	6.685

After Brownie et al. (1985).

and $T_5 = T_4 - C_4 = 214 - 145 = 69$. Notice, we could not compute T_3 without first computing T_2; that is, we must compute T_1, then T_2, then T_3, etc.

Now we are ready to estimate the band recovery and survival rates. As examples, we compute

$$\hat{f}_2 = \frac{R_2 C_2}{N_2 T_2} = \frac{210(106)}{1595(348)} = 0.04$$

and

$$\tilde{S}_2 = \frac{R_2}{N_2}\left(\frac{T_2 - C_2}{T_2}\right)\frac{N_3 + 1}{R_3 + 1} = \frac{210}{1,595}\left(\frac{348 - 106}{348}\right)\left(\frac{1157 + 1}{167 + 1}\right) = 0.63,$$

$$(10.42)$$

from equations (10.40) and (10.41), respectively. Although not difficult, the computations are time consuming if done by hand or with a calculator, and care must be taken to avoid errors. However, try checking some of the computations until you feel comfortable that the correct values are being computed through ECOSTAT.

The goodness-of-fit test was conducted through ECOSTAT. The sample sizes were large enough that no pooling was required. Therefore, the test statistic was 5.921 with 8 degrees of freedom and a p-value of 0.656. This leads us to conclude that the model adequately describes the data. In a fuller analysis, we would next ask whether a model with fewer parameters also adequately describes the data. Akaike's Information Criterion (1973) is often recommended to help discern which of several models that adequately describe the data should be used (*see,* for example, Lebreton et al., 1992).

Brownie et al. (1985) present three other models of adult banding. Two are more restrictive in that one assumes that survival rates are constant from year to year and the other assumes that both survival and recovery rates are constant from year to year. The third is a generalization of the one we discussed. It permits the recovery rate for the year an animal is banded to be different from the recovery of animals previously banded. This is particularly useful when the recovery rates are higher near banding sites. Brownie et al. (1985) also have a program called ESTIMATE that will fit each of these models and test for their fit.

Sometimes both adult and juvenile animals are tagged. Brownie et al. (1985) provide an excellent overview of the models that have been developed for this purpose. Their program BROWNIE can be used to fit the models to the data.

Summary

The Jolly–Seber model is based on the independent seminal works by Jolly (1965) and Seber (1965). When the population is subject to births, deaths, immigration, and/or emigration, the closed population models of Chapter 9 are no longer suitable, and the Jolly–Seber model is usually the first open population model considered. In addition to population size, survival rates, birth rates, and the probability of capture are estimated for each sampling period. Because the model has numerous parameters, large amounts of data must be collected in order to estimate the parameters with reasonable precision. Sometimes, a more parsimonious model may describe the data well. Models allowing for deaths only, births only, or constant survival and/or capture probabilities have been developed. Extensions have also allowed for temporary trap response, cohorts, and age classes.

In this and the previous chapter, we have emphasized that determining whether a population is open or closed is the first step in developing a model for the capture–recapture data. Pollock et al. (1990) suggest that the distinction is too rigid and that a combination of the two may be needed. For example, in a multiple year study, the population is open as viewed across years. However, within a year, the population may be approximately closed. A closed population model could be used to model the data taken within each year, and then an open population model could provide the foundation for combining the information across years.

The approach presented in this chapter does not reflect the most recent developments in the study of open populations. In the past 10 years, the move has been to a generalized linear models framework. The emphasis is on estimation of the survival and recruitment parameters instead of population density. Lebreton et al. (1990) summarizes early work in this direction, but great strides have been made since then. We have chosen not to include the generalized linear model approach in this text because it requires a significantly higher level of mathematical and statistical competency. However, the interested reader is encouraged to investigate this exciting area of current research activity.

Band recovery models have found wide application, particularly in the studies of bird and fish populations. The model considered here assumes that there is only one age class so that survival and recovery rates are independent of age but may vary from year to year. More parsimonious models allow survival and/or recovery rates to be constant from year to year. Extensions permit reduced survival in the first year after banding and more than one age class.

Exercises

1. Krebs (1989, p. 38) presents capture–recapture data from a series of 11 samples of a field vole (*Microtus pennsylvanicus*) population in the southwestern Yukon. Each sample was collected over a 2-day period, and sampling periods were approximately 2 weeks apart. The Method B table for these data is given below:

i:		1	2	3	4	5	6	7	8	9	10	11	
n_i:		22	41	48	82	89	101	107	91	19	27	22	Total
h	R_i:	21	41	46	82	88	99	106	90	19	26	22	r_h
1			15	1	0	0	0	0	0	0	0	0	16
2				15	0	1	0	0	0	0	0	0	16
3					37	2	0	0	0	0	0	0	39
4						61	4	1	1	0	0	0	67
5							75	3	2	0	0	0	80
6								77	4	0	0	0	81
7									69	0	0	0	69
8										8	1	0	9
9											14	19	33
10												30	30
Total	m_i	0	15	16	37	64	79	81	76	8	15	19	

From ECOLOGICAL METHODOLOGY by Charles J. Krebs, Copyright© 1988 by Charles J. Krebs. Reprinted by permission of Addison-Wesley Educational Publishers.

Perform a complete analysis of these data. Discuss your conclusions in the context of the problem. Discuss the design issues that you might anticipate in establishing this study.

2. Jolly (1982) presents a Method B table originating from a behavioral study of a local population of butterflies (*Hypaurotis crysalus*) in a small grove of oak in Connecticut. Only data on the males are provided in the table below:

i:		1	2	3	4	5	6	7	8	9	10	11	12	13	14	15	16	17	18	19
n_i:		44	43	17	34	37	25	19	34	36	19	28	15	16	15	13	8	2	3	2
h	R_i:	44	43	17	34	37	25	19	34	36	19	28	15	16	15	13	8	2	1	—
1			4	4	4	1	0	0	2	0	0	0	0	0	0	0	0	0	0	0
2				3	5	6	3	2	0	0	1	0	0	1	0	0	0	0	0	0
3					3	2	1	0	0	1	0	0	0	0	0	0	0	0	0	0
4						7	5	2	1	0	1	0	0	0	0	0	0	0	0	0
5							7	4	2	7	0	0	0	0	0	0	0	0	0	0
6								6	5	0	2	1	1	0	0	1	0	0	0	0
7									5	6	0	0	0	0	0	0	0	0	0	0
8										14	2	4	2	0	0	0	0	0	0	0
9											10	10	4	0	0	0	0	0	0	0
10												6	1	1	0	0	1	0	0	0
11													6	2	0	1	0	0	0	0
12														4	2	1	0	0	0	0
13															6	0	0	0	0	0
14																4	4	0	0	0
15																	3	2	1	0
16																		0	2	1
17																			0	0
18																				1

A. Perform a complete analysis of these data. Discuss the results in the context of the problem.

B. Jolly (1982) found that the hypotheses of constant survival and capture probabilities could not be rejected. How does that knowledge affect your conclusions in A? How would you proceed?

3. Leslie et al. (1953) studied a population of field voles in Wales. The Method B table is presented below. No breeding occurred in October 1948, and April 1949.

h	i:	June	July	Sept	Oct	Nov	Mar	Apr	May	June	July	Sept	Nov	Apr
	n_i:	107	45	85	69	67	106	125	99	117	98	127	190	26
	R_i:	96	41	82	64	64	104	121	89	92	95	127	188	25
1948														
June			12	7	4	0	1	1	0	0	0	0	0	0
July				10	0	1	0	0	0	0	0	0	0	0
Sept					19	8	4	3	0	0	0	0	0	0
Oct						11	9	1	2	0	0	0	0	0
Nov							14	6	2	1	1	0	0	0
1949														
Mar								46	11	4	0	0	1	0
Apr									34	18	0	0	0	0
May										34	3	1	0	0
June											40	5	1	0
July												56	11	0
Sept													69	0
Nov														19

Perform a full analysis of these data. Discuss your results in the context of the experiment.

4. Anderson and Sterling (1974) present the results of a band study on adult male pintail drakes (*Anas acuta* L.). The pintails were trapped, banded, and released on Pel and Kutawagan marshes in south-central Saskatchewan in July from 1955 to 1958. The primary source of band recovery was birds shot

or found dead during the legal hunting seasons from 1955 to 1970. A secondary source was from the recaptures of banded birds in years subsequent to banding. We will consider only the primary source in our analysis.

Year banded	No. banded	1955	1956	1957	1958	1959	1960	1961	1962	1963	1964	1965	1966	1967	1968	1969	'70
1955	4417	119	113	50	49	23	16	7	6	5	5	4	3	0	0	0	0
1956	3058		117	49	34	16	11	6	8	3	3	0	1	0	1	1	0
1957	6505			179	124	59	34	26	23	22	12	6	7	2	5	2	1
1958	4840				102	54	33	32	11	22	14	6	5	4	3	3	0

Year of recovery spans columns 1955–'70.

From Anderson and Sterling, 1974, by permission of the Wildlife Society.

A. Analyze these data completely. Discuss your conclusions within the context of the study, emphasizing the biological implications of the study.

B. Show how to estimate the band recovery rates for the following years:
   1. 1955
   2. 1959
   3. 1965

C. Show how to estimate the standard deviation of each estimate in (*B*).

D. Show how to estimate the survival rates for the following years:
   1. 1955
   2. 1956

E. Show how to estimate the standard deviation of each estimate in *D*.

F. Based on the estimates of band recovery and survival rates obtained from ECOSTAT, show how to estimate the average survival and recovery rates as well as their standard errors.

G. Read the paper by Anderson and Sterling and discuss how they used the secondary information to increase the precision of their estimates.

5. Conroy and Williams (1984) discuss a banding study on ring-necked duck [*Aythya collaris* (Donovan)]. The ducks were trapped and banded from December through February in Maryland, Virginia, and North Carolina. The data are given in the table below:

Banding year	No. banded	1953	1955	1956	1958	1960	1961	1963	1964
1953	220	9	8	1	0	0	0	0	0
1955	386		36	16	4	1	0	1	0
1956	223			16	2	1	0	1	1

Year of recovery spans columns 1953–1964.

(*continued*)

(continued)

Banding year	No. banded	Year of recovery							
		1953	1955	1956	1958	1960	1961	1963	1964
1958	102				6	1	0	2	0
1960	212					11	2	3	2
1961	433						16	7	7
1963	140							4	1
1964	313								12

From Conroy and Williams, 1984, by permission of the International Biometrics Society.

A. Perform a complete analysis of these data. Discuss the analysis and the biological implications in the context of the problem.

B. How does the fact years are missing affect the fitting and interpretation of the model?

C. Conroy and Williams tested for constant survival probabilities and could not reject the hypothesis that they were constant. What factors did they need to consider in conducting the test? In what way, if any, does this affect your analysis?

6. Schwarz et al. (1988) discuss a banding study conducted from 1975 to 1982. Adult male and female mallards were banded in southwest Alberta. The birds were recovered either in Canada, in the Pacific flyway, or in the Eastern (Central, Mississippi, and Atlantic) flyways. Recovery data for male mallards in the Pacific and Eastern Flyways are presented in the table below.

Flyway	Year banded	No. of Recoveries							
		1975	1976	1977	1978	1979	1980	1981	1982
Pacific	1975	6	3	3	1	1	0	0	2
	1976		25	12	9	6	3	4	1
	1977			22	18	12	7	5	2
	1978				17	13	9	8	4
	1979					29	25	18	13
	1980						12	10	5
	1981							28	23
	1982								19
Eastern	1975	9	18	4	3	4	1	1	1
	1976		26	31	16	8	7	5	4
	1977			35	24	18	13	10	2
	1978				17	12	16	9	8
	1979					37	33	18	19
	1980						19	14	9
	1981							50	21
	1982								23

From Schwarz Burnham and Anderson, 1988 by permission of the International Biometrics Society

The number of male mallards banded from 1975 to 1982 was 453, 1,337, 1,380, 1,079, 2,253, 888, 1,924, and 1,107, respectively.

A. Model these data, ignoring the flyway and assuming that these represent all recoveries. Discuss your conclusions within the context of the problem.

B. Review the paper by Swhwarz et al. (1988). Discuss their analysis and compare the results to those you found in A.

11

Transect Sampling

Introduction

Many studies of biological populations require estimates of population size or density. Consequently, the focus of several previous chapters has been either estimation of population size or population density. In this chapter, we are again concentrating on estimating population density. Not all populations can be sampled effectively using the techniques presented thus far. For example, suppose we want to estimate the density of a plant population. We know that the plants are sparsely scattered in the region of interest. Quadrat sampling would not be effective because we would need either large quadrats or an excessive number of small quadrats to obtain a precise estimate of density. Capture–recapture methods are not applicable to this problem. An intuitive approach is to walk along a line through the region and count the number of plants that are seen on either side of the line. The line and the area on either side of the line that is searched for plants has been called a transect. If we assume that we can see all plants within w units of the line, we have strip transects. More commonly, we can observe all plants close to the line, but we count only a portion of those farther from the line. These are called line transects, and we can allow for this differential sightability in estimating the population density.

Animal population density may also be estimated using strip and line transects. Sometimes a slightly different, but related approach to sampling is taken, especially when sampling bird populations. The researcher will sit at a point and observe birds within a given radius of the point. These are known as circular plot surveys if we assume that all the birds are observed within a distance w of the point and point transects (or variable circular plots), if we assume that all birds close to the point are counted but some farther from the point may go unobserved. Because birds are often conspicuous by their bright coloration or distinctive song or call, detection may be possible even in dense habitats. Line transects are often used in open habitats, while point transects are commonly used in more closed habitats with high canopies.

Transect sampling has been used to estimate the population density of terrestrial mammals such as fruit bats, pronghorn, mice, deer, rabbits, primates, and African ungulates. Several species of dolphin, porpoise, seal, and whale have been some of the marine animals for which surveys have been reported in the literature. Amphibians, reptiles, beetles, wolf spiders, coral reef fish, and red crab densities have been estimated using transect methods. Gamebirds, passerines, raptors, and shorebirds are some of the many bird species for which transect sampling has been an effective tool for estimating population density.

Often the subject of interest will be referred to as an object of interest instead of an organism because many inanimate objects have also been surveyed. These include bird nests, mammal burrows, dead pigs, and dead deer. We use the term *object of interest.*

Transect sampling may be used regardless of the spatial distribution of the objects. It is not necessary for the objects to be distributed randomly (according to a Poisson process) in the study region. However, it is essential that the lines or points be randomly distributed in the study area.

The objects of interest may occur naturally aggregated into clusters such as herds of mammals, flocks of birds, pods of whales, prides of lions, and schools of fish. As stated by Buckland et al. (1993, p. 12), a "cluster is a relatively tight aggregation of objects of interest as opposed to a loosely clumped spatial distribution of objects." Data collection for clustered populations differs in an important way between strip transects or circular plots and line or point transects. For strip transects or circular plots, we ignore the fact that the objects occur in clusters; objects within the strip or circle are counted and those outside are not. Consequently, some objects in a cluster may be included while others are not. In line (point) transect sampling of fixed width (or diameter), the entire cluster is included (excluded) if the center of the cluster is within (beyond) the strip (plot). The number of clusters is observed. If the number within each cluster is also recorded, the average cluster size can be estimated, providing an estimate for the population density as the density of clusters times the average size per cluster.

Gates (1979) reviewed the status of line transect sampling. Seber (1982) is a traditional reference for transect studies. Burnham et al. (1980) provides an overview of both design and analysis issues. Buckland et al. (1993) synthesized much of the recent research while giving an historical perspective. We have drawn heavily on all of these in developing this chapter. Buckland et al. refer to these methods as distance sampling. We choose to use transect sampling instead to avoid confusion with the nearest neighbor techniques of Chapter 7.

Strip Transects and Circular Plots

Suppose we want to estimate the density of an object, say a plant species, within a well-defined study region. We decide to walk along k randomly located lines, each of length l_i, across the study region of area A. These plants are easy to observe so we are confident that we can observe all that are within w units of the line. Each line is walked, and the number of plants, n_i, within w units of that line is recorded (*see* Figure 11.1). Notice that the width of the strip is $2w$ because we observe for w units out on either side of the line. The total length of the lines is denoted by $L = \Sigma l_i$ and the total number of plants observed by $n = \Sigma n_i$. The area of the strip transects is $2wL$. Two assumptions are needed for estimation (Gates, 1979):

1. All objects are sighted once, and only once, within the specified distance w.

2. The strip transects randomly sample the study region.

On the basis of these assumptions, the estimated density of the plants in the study region is

$$\hat{D}_{ST} = \frac{n}{2wL} \tag{11.1}$$

where the density D is the mean number of plants per unit area in the study region and $\hat{D}_{LT}$ is the estimator of D. Assuming that the length L and the width w of the line transect are determined in advance of the study, the variance of $\hat{D}_{ST}$ is estimated by

$$s^2_{\hat{D}_{ST}} = \frac{s^2_n}{(2wL)^2}. \tag{11.2}$$

We now need to estimate s^2_n. Often the study region is sampled by walking several

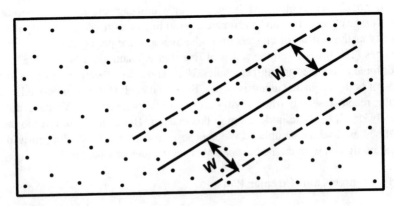

Figure 11.1. Strip transect with half-width w.

transects. A common approach is to select a random starting point and to walk parallel transects throughout the study region. Let n_i denote the number of objects observed while walking the ith line of length l_i, $i = 1, 2, \ldots, k$. Then the variance of n can be estimated by

$$s_n^2 = \frac{L}{k - 1} \sum_{i=1}^{k} l_i \left(\frac{n_i}{l_i} - \frac{n}{L} \right)^2 \tag{11.3}$$

where $n = \Sigma n_i$ is the total number of objects observed and $L = \Sigma l_i$ is the total length of the transects walked. It should be noted here that dividing a line into k segments does not represent replication, and this approach for estimating s_n^2 should not be used.

If no replicate lines are walked or if information from replicate lines is not recorded, a further assumption about the distribution of the objects in the study region is needed to obtain a valid estimate of variance. For example, suppose we make the following assumption:

3. The objects are distributed according to a Poisson process throughout the study region.

Under the three assumptions, the number of objects observed n is Poisson, and $s_n^2 = n$. Therefore, the variance of $\hat{D}_{ST}$ can be estimated by

$$s_{D_{ST}}^2 = \frac{n}{(2wL)^2}. \tag{11.4}$$

Assumption 3 may be replaced by another assumption that could lead to alternative distributions such as the negative binomial, Neyman type A, or some other distribution providing the basis for estimating the variance. Burnham et al. (1980) note that usually the objects are aggregated to some extent and suggest using $s_n^2 = 2n$ if no additional information is available.

In many bird studies, circular plots are used instead of strip transects, especially in dense habitats (*see* Figure 11.2). The area of a circular plot with radius w is πw^2. If observations are taken in k such plots, the area surveyed is $k\pi w^2$. Let n_i be the number of birds observed in a circular plot and $n = \Sigma n_i$ the total number of birds observed. Then the estimate of population density is

$$\hat{D}_{CP} = \frac{n}{k\pi w^2}. \tag{11.5}$$

Assuming that the radius of observation w is fixed, the variance of $\hat{D}_{CP}$ is estimated by

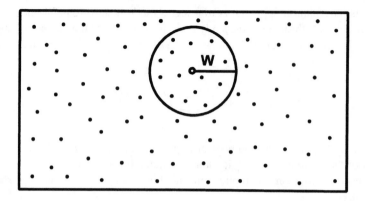

Figure 11.2. Circular plot with radius w.

$$s^2_{\hat{D}_{CP}} = \frac{s^2_n}{(k\pi w^2)^2}. \tag{11.6}$$

Usually, data are recorded for more than one, say k, plots. Following the same reasoning as we did with replicate lines, information from these replicate plots can be used to estimate the variance of n. Let n_i now denote the number of objects observed in circular plot i, $i = 1, 2, \ldots, k$. Further, assume that the radius of each circular plot is w. An estimate of the variance of n can now be written as

$$s^2_n = \frac{k}{k-1} \sum_{i=1}^{k} \left(n_i - \frac{n}{k}\right)^2. \tag{11.7}$$

If assumption 3 is again made, we would have $s^2_n = n$. Substituting this estimate into Equation (11.6) could provide the estimate of the variance of $\hat{D}_{CP}$. Another distribution could be assumed, or the suggestion by Burnham et al. (1980) of using $s^2_n = 2n$ could be followed.

Confidence intervals may be set on the density of the population using the normal approximation given by

$$\hat{D} \pm z_{\alpha/2} s_{\hat{D}} \tag{11.8}$$

where $z_{\alpha/2}$ is the $(1 - \alpha/2)$ quantile of the standard normal. However, the distribution of $\hat{D}$ is positively skewed, causing the true confidence level to be below the stated one. Further, when the interval fails to cover the true density, the density lies above the upper limit more frequently than it lies below the lower limit. Intervals based on the log-normal distribution provide better coverage (Buckland et al., 1993). Following Burnham et al. (1987, p. 212) and Buckland et al. (1993), a $100(1 - \alpha)\%$ confidence interval is given by

$$\left(\frac{\hat{D}}{C}, \hat{D}C\right) \tag{11.9}$$

where

$$C = e^{z_{\alpha/2}\, s_{\ln(\hat{D})}} \tag{11.10}$$

and

$$s_{\ln(\hat{D})} = \sqrt{\ln\left(1 + \frac{s_{\hat{D}}^2}{\hat{D}^2}\right)}. \tag{11.11}$$

Example 11.1

Robinette et al. (1974) report the results of a strip census used to estimate the population of elephants in Kenya's Travo National Park (West) in August of 1967. There are 28,857 acres in the park. Seven parallel transects 2 km apart were walked by teams of three students. The total length of the transects was 29.89 miles. It was assumed that all elements within 1,312 feet of the transect line could be seen. In all, 137 elephants were observed. The goal is to estimate the elephant population density in the park. First, we need to adopt a consistent unit of measurement as we have acres, kilometer (km), feet, and miles. We will use feet. Based on 43,560 square feet per acre, the park has 28,857(43,560) = 1,257,010,920 square feet. Recalling that there are 5,280 feet in a mile, we find that the total length of the transects was 29.89(5,280) = 157,819.2 feet. Therefore, we would estimate the density of the elephant population to be

$$\hat{D}_{ST} = \frac{137}{2(1312)(157819.2)} = 0.000000331 \text{ elephants per square foot.}$$

Reporting the elephant population on a square foot basis is not reasonable (one elephant cannot fit within a square foot area). One approach is to report the estimated density per acre. Therefore, we would estimate a density of 0.000000331(43,560) = 0.0144 elephants per acre.

Because the information on replicate lines is not available to us, we assume the elephants are distributed according to a Poisson process. Therefore the standard error of our estimate is estimated to be

$$s_{\hat{D}_{ST}}^2 = \frac{2(137)}{(2(1312)(157,819.2))^2} = 1.60 \times 10^{-15} \text{ (elephants/square foot)}^2.$$

To convert to acres, we multiply by 43,560², giving 0.00000303. Therefore, the estimated standard error of the estimated density is 0.00174 elephants/acre.

Further assume we want to set a 90% confidence interval on the density per acre. Using equations (11.9) to (11.11), we first compute

$$s^2_{\ln(\hat{D}_{ST})} = \ln\left(1 + \frac{0.00000303}{0.0144^2}\right) = 0.01449.$$

and

$$C = e^{1.645\sqrt{0.01449}} = 1.219.$$

The upper confidence limit is $\hat{D}C = 0.0144(1.219) = 0.0176$, and the lower limit is $\hat{D}/C = 0.0144/1.219 = 0.0118$ elephants per acre. Therefore, we are 90% confident that the true density of elephants in the park in 1968 was between 0.0118 and 0.0176 elephants per acre, or between 1.18 and 1.76 elephants per 100 acres).

The accompanying software ECOSTAT can be used to complete these computations. From the *Transect* menu, select *Strip Transects*. Indicate that no information is available on replicate lines. Then, after converting to consistent measurement units, enter the data from the keyboard. If this is done using feet, the estimated density is zero. The problem is that the density on a per square foot basis is less than the number of decimals displayed by ECOSTAT. One way to avoid this is to enter the w and L in 10,000s of feet; that is, $w = 0.1312$ and $L = 15.78192$. To convert back to feet, divide the displayed estimates by 10,000². Because our experience is that the variance is generally greater than that associated with a Poisson process, we have followed the suggestion made by Burnham et al. (1980) and used $s^2_n = 2n$ when no information is available on replicate lines. This is incorporated into ECOSTAT.

If information is available on replicate lines, then an ASCII file must be constructed before entering ECOSTAT. The first line should have the number of lines k and the value of w. The following k data lines should have the identification of the transect line, the length, and the number of observed objects for each of the k lines.

An important consideration in strip transects and circular plots is determination of the distance w from the line or point for which all objects closer than w are observed with certainty. During the 1930s, R.T. King suggested using the average sighting distance $\bar{r}$ as the *effective width* surveyed (Leopold, 1933; Gates, 1979). Although an intuitive understanding of effective width was present, it remained for Gates (1979) some 40 years later to provide a formal definition. He defined the effective width to be the distance ρ for which the number of observed objects closer to the line than ρ is equal to the observed number of objects farther from the line than ρ. Notice here that ρ is taken as the estimate of w, one-half of the strip width. Kelker (1945) took an alternative approach. He estimated w by looking at a histogram of the perpendicular distances of the observed objects from the line and noting the distance at which counts began to decrease. Once w was estimated, only objects closer than w units from the line were used to estimate density.

In an effort to avoid determining w, Hayne (1949a) developed an estimator based on sighting distances r_i. However, the method is poor if the average sighting angle is not about 32.7° and may not perform well even if the average sighting angle is close to this value (Buckland et al., 1993). It was obvious that a new approach was needed. In 1968, Eberhardt and Gates et al. each published papers suggesting that the sightability of objects as a function of distance from the transect be modeled. These classical papers were the beginning of a rigorous statistical treatment of line transect data.

Line and Point Transects

Suppose we are going to walk along a line and observe all objects within w units. Especially for live animals, we may spot an animal some distance in front of us. In these cases, it is often convenient to measure the sighting distance r_i and the sighting angle θ instead of the perpendicular distance x_i. Then $x_i = r_i \sin(\theta_i)$ as illustrated in Figure 11.3. Although analysis can be based on r_i and θ_i (*see* Hayes and Buckland, 1983), the methods based on sighting distance and angle perform poorly relative to methods based on perpendicular distances. Therefore, if sighting distance and angle are recorded, they are used to determine perpendicular distances which are the foundation of our analysis (*see* Figure 11.4).

In many cases, we are able to observe all objects on the line. However, even the most careful observer encounters difficulty seeing all objects within w units. One way to determine whether this is a concern is to construct a histogram of the number of objects within set distances of the line. For example, suppose $w = 100$ meters (m). We could construct a histogram of the number of objects within 10

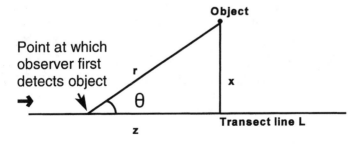

Figure 11.3. Sighting distance r and sighting angle θ of an object from a line transect.

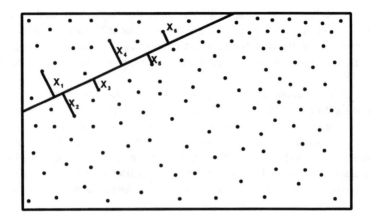

Figure 11.4. Perpendicular sighting distances from a line transect.

m of the line, from 10 to 20 m, 20 to 30 *m*, . . . , 90 to 100 m. If we see a histogram as in Figure 11.5, then we conclude that we are effectively observing all objects within *w* of the line and the strip transect methods above are appropriate.

More likely, we will see a histogram similar to that in Figure 11.6. In this case, the sightability of the objects decreases as the distance from the line increases. If

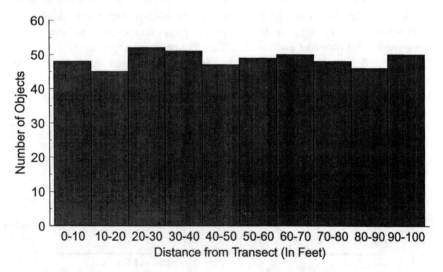

Figure 11.5. Number of objects observed at various distances from the line transect when all objects are observed.

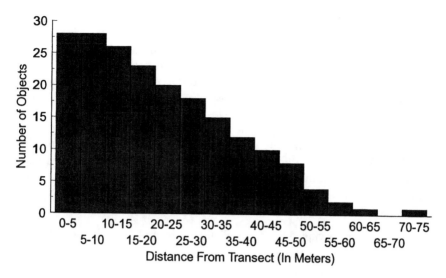

Figure 11.6. Number of objects observed at various distances from the line transect when sightability decreases as the distance from the line increases.

we use the strip transect methods given above, we will be seriously underestimating the population density because we have not accounted for the reduced probability of detection away from the line. To adjust for differential sightability, we could model the detection probability as a function of distance from the line and, based on the model, adjust the density estimate to remove the effect of this differential sightability. This is the basic concept behind line and point transects, and it will be the focus of our efforts in this section.

The modeling of line and point transect data is based on three primary assumptions (Buckland et al., 1993):

1. All objects on the line or point are detected with certainty.

2. Objects are detected at their initial location.

3. Measurements are exact.

Although theoretically the first assumption can be relaxed if the detection probability is known for some distance, practically it is of little use. Every effort should be made to ensure that *all* objects on the line or point are observed.

If the objects are plants or birds' nests, then satisfying the second assumption is not a problem. However, many animals have a tendency to move away, or more rarely toward, the observer. If this movement goes undetected, movement away from the observer results in an underestimate of the population density. We will tend to overestimate the population density if movement is toward the observer. Often histograms of the observed distance data can provide insight into movement before detection. The most obvious cases occur when there is a heaping of the distance data away from the line or point as illustrated in Figure 11.7.

Finally, there is no substitution for careful data collection. If the data are not grouped, then measuring angles near zero or distances close to the line or point are especially important. Sometimes, distances are paced, taken with a range finder, or estimated using binocular reticles. These approximate methods reduce the quality of the data, but may be necessary from a practical standpoint. If the errors are random and relatively small, then the density may still be estimated reliably, especially if the sample size is large (Gates et al., 1985). Biased measurements seriously compromise the data. Every effort should be made to adopt field methods that will keep any bias as small as possible.

Throughout this section, we use the following notation:

A: area occupied by the population of interest

k: number of lines or points surveyed

l_i: length of the ith transect line, $i = 1, 2, \ldots, k$

$L = \Sigma l_i$:   total line length

w: width of area searched on each side of the line transect or the radial distance searched around a point transect or the truncation point beyond which data are not used in an analysis

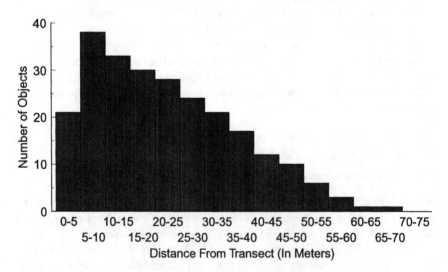

Figure 11.7.   Number of objects observed at various distances from the line transect when objects move away from the line before detection causing heaping of the data.

Parameters

D: population density (number per unit area)

N: population size in the study area

$E(s)$:   mean population cluster size

P_a: proportion of the objects detected within the area a surveyed

$g(y)$:   detection probability as a function of the perpendicular (radial) distance from the line (point)

$g(0)$:   detection probability on the line or point

$f(y)$:   probability density function of distances from the line

$f(0)$:   probability density function of distance from the line evaluated at zero distance

$h(y)$:   slope of the probability density function of distance from the point

$h(0)$:   slope of the probability density function of distance from the point, evaluated at zero distance

Statistics

$\hat{D}$:   estimated population density

n_i: number of objects observed within w units of a line or point i, $i = 1, 2, \ldots, k$

$n = \Sigma n_i$:   number of objects observed within w units of all lines or points in study region

r_i: sighting distance from the observer to the ith object

θ_i: sighting angle between a line transect and the ith object

Estimates of the parameters not explicitly stated are denoted by $\char`^$, thus $\hat{g}(y)$ is an estimate of the detection probability y units from the line or point.

Remember, in line transect sampling, the area surveyed has size $a = 2wL$, but only a proportion of the objects in the area surveyed is detected. This unknown proportion is denoted by P_a. If P_a can be estimated from the distance data using $\hat{P}_a$, the population density can be estimated using

$$\hat{D} = \frac{n}{2wL\hat{P}_a}. \qquad (11.12)$$

Notice that this equation makes sense. If we only see one-half of the objects, $\hat{P}_a$ would be about 0.5. Dividing by $\hat{P}_a$ would result in doubling the estimate of density based on only the observed objects, thereby correcting for incomplete detection.

Suppose that the detection probability changes with distance from the line as in Figure 11.8. Failure to consider the reduced probability of sighting objects as distance from the transect line increases results in an underestimate of population density as indicated by the unobserved objects in the shaded area of Figure 11.8. To model this change in detection probability, we define the detection function to be the probability of detecting an object, given that it is a distance y from the random line or point. The distance y is the *perpendicular* distance x for line transects or the radial distance r for point transects. Because $g(y)$ is a detection probability at distance y, $0 \le g(y) \le 1$ for all y. Generally, $g(y)$ will decrease with increasing distance y. We will assume $g(0) = 1$; that is, objects on the line or point are detected with certainty (probability 1). Buckland et al. (1993) describe the more general theory that is needed when this assumption is not made.

The unconditional probability of detecting an object within a strip of area $a = 2wL$ is

$$P_a = \frac{\int_0^w g(x)dx}{w}. \tag{11.13}$$

In equation (11.13), we are finding the average detection probability from 0 to w units from the line or point. Substituting (11.13) into (11.12), we have

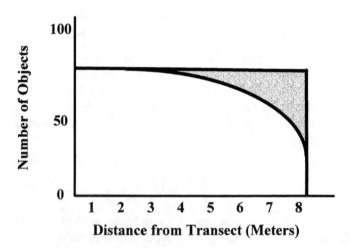

Figure 11.8. Decreasing sightability of objects as distance from the transect increases. The shaded area represents unobserved objects.

$$\hat{D} = \frac{n}{2L \int_0^w \hat{g}(x)dx}. \qquad (11.14)$$

From (11.14), we see that the integral $\int_0^w g(x)dx$ is a critical quantity, and we denote it as τ for simplicity. (Here we are departing from the notation used by Buckland et al. (1993). They use μ instead of τ because μ is the average detection probability. We have chosen τ because μ has been used extensively in this text to represent the population density.) Therefore,

$$\hat{D} = \frac{n}{2L\hat{\tau}}. \qquad (11.15)$$

We now need to determine an estimator for $1/\tau$. First note that the probability density function (pdf) of the perpendicular distance data, given that the object is detected is

$$f(x) = \frac{g(x)}{\int_0^w g(x)dx}. \qquad (11.16)$$

By assumption, $g(0) = 1$, giving us

$$f(0) = \frac{1}{\int_0^w g(x)dx}$$

$$= \frac{1}{\tau}. \qquad (11.17)$$

Therefore, the general estimator of density for line transect sampling may be written as

$$\hat{D} = \frac{n\hat{f}(0)}{2L}$$

$$= \frac{n}{2L\hat{\tau}}. \qquad (11.18)$$

For point transects, the basic ideas are the same. Let the proportion of detected objects in each sampled area be denoted by P_a. Then the estimator of density is

$$\hat{D} = \frac{n}{k\pi w^2 \hat{P}_a}. \tag{11.19}$$

The unconditional probability of detecting an object within one of the k circular plots is

$$\hat{P}_a = \int_0^w \frac{2\pi r g(r)dr}{\pi w^2}$$

$$= \frac{2}{w^2} \int_0^w r g(r)dr. \tag{11.20}$$

Substituting (11.20) into (11.19) and simplifying, we have

$$\hat{D} = \frac{n}{2k\pi \int_0^w r\hat{g}(r)dr}. \tag{11.21}$$

Defining

$$v = 2\pi \int_0^w r g(r)dr, \tag{11.22}$$

we have

$$\hat{D} = \frac{n}{k\hat{v}}. \tag{11.23}$$

From the above discussion, we can see that the critical quantity to be estimated in a point transect survey is v, just as τ is the critical quantity to be estimated in a line transect survey.

To emphasize, the statistical problem in the estimation of population density of objects is estimation of τ for line transects or v for point transects. The fit close to the line or point is especially important. This requires careful modeling and estimation of $g(y)$. Not only is $g(y)$ unknown, it varies as a result of numerous factors, including cue production, observer effectiveness, and environment. The object of interest may provide cues that lead to its detections such as a call by a bird or a splash by a marine mammal. These cues are frequently species-specific and may vary by age or sex of the animal, time of day, or season of the year causing the total count n to vary for reasons unrelated to density (Mayfield, 1981; Richards, 1981). Observer variability may be due to numerous factors, including interest in the survey, training, experience, height, fatigue, and both visual and auditory acuity (Buckland et al., 1993). Habitat type and its phenology are important environmental variables (Bibby and Buckland, 1987). Wind, precipitation, darkness, and sun angle are some of the physical conditions that can affect detection. Because so many factors can impact detection, it is important to have a flexible or "robust" model for $g(y)$.

Buckland et al. (1993) recommend selecting a class of models that are robust, satisfy the shape criterion, and have efficient estimators. Model robustness is the most important property for the detection function. This means the model can exhibit a variety of shapes that are common for the detection function. In addition, if the data can be pooled over factors that affect detection probability (e.g., different observers and habitat types) and still yield a reliable estimate of density, the model is said to be pooling robust (Burnham et al., 1980). The shape criterion requires the derivative of $g(y)$ at 0, $g'(0)$, to be zero. Practically, this means that the detection probability remains nearly certain at small distances from the line or point. This implies that the detection function have a "shoulder" near zero and excludes models that are spiked near zero. If the model is robust and meets the shape criterion, efficient estimation of model parameters, preferably through maximum likelihood, should be sought.

The individual distances may be used, or the data may be grouped in distance intervals. Further, all observed objects should be recorded and not just those within some preselected w units of the line or point. After data collection, histograms of the distance data should be constructed. These can provide insight into the presence of heaping, evasive movement, or outliers. Heaping may occur when the observer tends to round distances. This may result in more distances ending in 0 or 5 than those ending in 1, 2, 3, 4, 6, 7, 8, or 9. Angles may be more likely to be reported as 15, 30, 45, 60, 75 or 90°. Judicious grouping can often ameliorate these effects. Although the data may provide insight into the tendency of animals to move away from or toward the observer, the observer must be very alert to such movement. Outliers, a few very large distances, do not provide much information for estimating population density and are usually difficult to model. Often 5% to 10% of the largest distances are dropped before fitting a model.

The modeling process can be conceptualized in two steps (Buckland et al., 1993). First, a key function, $k(y)$ is selected as a starting point. Initially, consideration should be given to the uniform, which has no parameter, or to the half-normal, which has one parameter (*see* Figure 11.9). The hazard-rate model is another possibility for a key function (Figure 11.10), but it requires estimation of two parameters.

The second step involves selecting a series expansion, if needed, to improve the fit of the model to the distance data. Therefore, the detection function is modeled using the following general form:

$$g(y) = \text{key}(y)[1 + \text{series}(y)]$$

Possible series expansions include the cosine series, simple polynomials, and

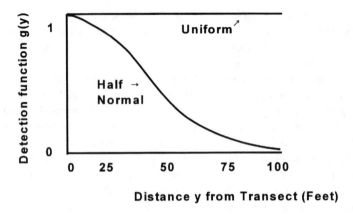

Figure 11.9. Half-normal and uniform key functions.

Hermite polynomials (Stuart and Ord, 1987, pp. 220–227). Based on these, some useful models are given in Table 11.1. Generally, it is assumed that the detection function is symmetric about the line. Therefore, even functions are considered, leading to the use of only the even powers of the polynomials and the cosine, rather than the sine function.

The combination of a uniform key function and a cosine series is the Fourier series model of Crain et al. (1979) and Burnham et al. (1980). This model has

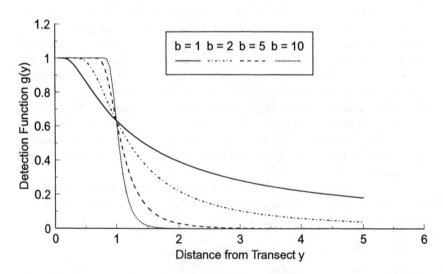

Figure 11.10. Hazard rate key function with $\sigma = 1$ and $b = 1, 2, 5,$ and 10.

Table 11.1.   Common Combinations of Key Functions and Series Expansions With the Resulting Detection Function

Key function	Series expansion	Detection function $g(y)$
Uniform	Cosine	$\dfrac{1}{w}\left[1 + \displaystyle\sum_{j=1}^{m} a_j \cos\left(\dfrac{j\pi y}{w}\right)\right]$
Uniform	Simple polynomial	$\dfrac{1}{w}\left[1 + \displaystyle\sum_{j=1}^{m} a_j\left(\dfrac{y}{w}\right)^{2j}\right]$
Half-normal	Cosine	$e^{-y^2/2\sigma^2}\left[1 + \displaystyle\sum_{j=2}^{m} a_j \cos\left(\dfrac{j\pi y}{w}\right)\right]$
Half-normal	Hermite polynomial	$e^{-y^2/2\sigma^2}\left[1 + \displaystyle\sum_{j=2}^{m} a_j H_{2j}(y_s)\right]$ where $y_s = y/\sigma$
Hazard rate	Cosine	$(1 - e^{-(y/\sigma)^{-b}})\left[1 + \displaystyle\sum_{j=2}^{m} a_j \cos\left(\dfrac{j\pi y}{w}\right)\right]$
Hazard rate	Simple polynomial	$(1 - e^{-(y/\sigma)^{-b}})\left[1 + \displaystyle\sum_{j=2}^{m} a_j\left(\dfrac{y}{w}\right)^{2j}\right]$

After Buckland et al. (1993).

been shown to perform well in a variety of situations. The uniform key function and simple polynomial form the models of Anderson and Pospahala (1970), Anderson et al. (1980), and Gates and Smith (1980).

The half-normal may be a good key function because histograms of distance data often decline rapidly with distance from the line or point (Hemingway, 1971). The use of Hermite instead of simple polynomials is a theoretical consideration based on the orthogonality of the half-normal and Hermite polynomials (Buckland et al., 1993).

Often the hazard-rate model with its two parameters provides an adequate fit to the detection function, but other terms can be added if needed. The hazard rate model was initially derived from a set of assumptions (Buckland, 1985), whereas the other two key functions were selected because of their shape.

ECOSTAT may be used to fit the uniform with one or more cosine terms (the Fourier series model). The program DISTANCE (Buckland et al., 1993) is a fully developed program for use with transect sampling. If extensive work with line transects is needed, DISTANCE will be invaluable.

Ungrouped Data

Crain et al. (1978) and Burnham et al. (1980) provide a full discussion of fitting a Fourier series to transect data. We sketch the basic steps so that some understanding of the process is gained. First, suppose the data are not grouped. Because we are assuming that the probability of detection is symmetric about the line, the detection function is even. First, an even function is expanded in a Fourier series over $[-w, w]$. After noting that the sine function is odd and the cosine function is even, this expansion can be reduced to

$$f(x) = \frac{1}{w} + \sum_{j=1}^{\infty} a_j \cos\left(\frac{j\pi x}{w}\right), \qquad 0 \leq x \leq w. \qquad (11.25)$$

Our interest is in estimating $f(0) = 1/\tau$. With that focus and noting that $\cos(0) = 1$, we have

$$f(0) = \frac{1}{w} + \sum_{j=1}^{\infty} a_j \approx \frac{1}{w} + \sum_{j=1}^{m} a_j \qquad (11.26)$$

where m is a truncation point that must be determined.

The Fourier series model is not a true probability density function because it is possible for $f(x)$ to be negative, especially as distances from the line approach w. Therefore, it is not possible to obtain maximum likelihood estimates. Instead, it is noted that

$$a_j = \frac{2}{w} E\left[\cos\left(\frac{j\pi x}{w}\right)\right]. \qquad (11.27)$$

Therefore, an unbiased estimator for a_j is

$$\hat{a}_j = \frac{2}{w} \frac{1}{n} \sum_{i=1}^{n} \cos\left(\frac{j\pi x_i}{w}\right), \qquad j = 1, 2, 3, \ldots \qquad (11.28)$$

Tolstov (1962) showed that this estimator is optimal in the sense of giving the minimum mean integrated squared error (MISE) estimator of $f(x)$ where

$$\text{MISE} = E\left[\int_0^w (\hat{f}(x) - f(x))^2 \, dx\right]. \qquad (11.29)$$

Following Kronmal and Tarter (1968) and Burnham et al. (1980), additional terms are added in (11.22) until the first m such that $\text{MISE}_{m+1} - \text{MISE}_m \geq 0$; that is, until

$$\frac{1}{w} \sqrt{\frac{2}{n+1}} \geq |\hat{a}_{m+1}|. \qquad (11.30)$$

Defining $a_0 = 2/w$, the covariance of a_j and a_h is then estimated by

$$s_{\hat{a}_j \, \hat{a}_h} = \frac{1}{n-1}\left[\frac{1}{w}(\hat{a}_{j+h} + \hat{a}_{j-h}) - \hat{a}_j \, \hat{a}_h\right], \qquad j \geq h. \tag{11.31}$$

Finally, the variance of $\hat{f}(0)$ is estimated by

$$s_{\hat{f}(0)}^2 = \sum_{i=1}^{m}\sum_{j=1}^{m} s_{\hat{a}_j \, \hat{a}_j}. \tag{11.32}$$

Obtaining the estimator of $\hat{f}(0)$ permits us to estimate the population density based on the fundamental equation (11.18). The estimated variance of $\hat{D}$ is

$$s_{\hat{D}}^2 = \hat{D}^2\left[\frac{s_n^2}{n^2} + \frac{s_{\hat{f}(0)}^2}{(\hat{f}(0))^2}\right] \tag{11.33}$$

This estimator assumes that the objects do not occur in clusters and that the probability of sighting points on the line is 1.

Akaike's (1973) Information Criterion (AIC) provides a basis for model selection, whether or not models are hierarchical (derived from adding or deleting terms). AIC is defined as

$$\text{AIC} = -2 \ln(\mathcal{L}) + 2q \tag{11.34}$$

where $\ln(\mathcal{L})$ is the log-likelihood function evaluated at the maximum likelihood estimates of the model parameters and q is the number of parameters in the model. The first term is a measure of the fit of the model. The second term is a penalty for additional terms in the model, emphasizing the desire for parsimony. The model with the smallest AIC is the preferred model. This has been illustrated by Burnham and Anderson (1992).

A Pearson's χ^2 goodness-of-fit test can be used to assess the fit of the model to the data and may aid in model selection. Suppose that n observed distances from line or point transects are divided into g groups. Let $n_1, n_2, \ldots, n_k$ be the number of observed distances within each group. Let $c_0, c_1, \ldots, c_k$ be the cut-off distances between groups. Generally, $c_0 = 0$. Further, suppose that a model with q parameters is fitted to the data and that the area under the curve between the cutoff points c_{i-1} and c_i is $\hat{p}_i$. Then

$$X^2 = \sum_{i=1}^{g} \frac{(n_i - n\hat{p}_i)^2}{n\hat{p}_i} \tag{11.35}$$

has an approximate χ^2 distribution with $g - q - 1$ degrees of freedom if the fitted model is the true model. As with other goodness-of-fit tests of this type, we would determine the cutoff points so that $n\hat{p}_i$ is 5 or more for at least 80% of the cells and never less than 1. Although a poor fit may not be of great concern, it does indicate that the model should be reviewed and perhaps alternatives sought. A poor fit could also indicate problems with the data.

Example 11.2

Burnham et al. (1980) reported a study by Laake, in which a known number of wooden stakes were placed in a sagebrush meadow east of Logan, Utah. Transect lines were walked by graduate students in the Department of Wildlife Science at Utah State University. On one survey, the observers found 68 stakes within 20 m on a single line transect 1,000 m long. The true population density was known to be $0.00375/m^2$ or 37.5 stakes/ha. The observers recorded the perpendicular distances in meters given in Table 11.2.

Table 11.2. Perpendicular Distances from the Line Transect to Stakes in a Sagebrush Meadow East of Logan, Utah

Stake	Distance (m)	Stake	Distance (m)	Stake	Distance (m)
1	2.02	24	18.60	47	9.04
2	0.45	25	0.41	48	7.68
3	10.40	26	0.40	49	4.89
4	3.61	27	0.20	50	9.10
5	0.92	28	11.59	51	3.25
6	1.00	29	3.17	52	8.49
7	3.40	30	7.10	53	6.08
8	2.90	31	10.71	54	0.40
9	8.16	32	3.86	55	9.33
10	6.47	33	6.05	56	0.53
11	5.66	34	6.42	57	1.23
12	2.95	35	3.79	58	1.67
13	3.96	36	15.24	59	4.53
14	0.09	37	3.47	60	3.12
15	11.82	38	3.05	61	3.05
16	14.23	39	7.93	62	6.60
17	2.44	40	18.15	63	4.40
18	1.61	41	10.03	64	4.97
19	31.31	42	4.41	65	3.17
20	6.50	43	1.27	66	7.67
21	8.27	44	13.72	67	18.16
22	4.85	45	6.25	68	4.08
23	1.47	46	3.59		

From Burnham et al. (1980) by permission of The Wildlife Society.

The data for stake 19 must be mistyped because the transect width is only 20 m. This stake was removed from the analysis. First, an ASCII data file needs to be developed. The first line has the number of observations, the length of the transect, and the distance maximum distance w from the line. There are as many additional lines as observations. Each line has an identification for the observation and its distance from the line.

ECOSTAT can now be used to fit the Fourier series to these data. *Line transect* is chosen from the *Transect* menu. The option for ungrouped data is selected, and the data set name is entered. To estimate the density, press the *Estimate D* button. Two cosine terms were found to be significant. The observed numbers of stakes and the estimated detection probability are shown in Figure 11.11. $f(0)$ was estimated to be 0.116. The estimated density is then

$$\hat{D} = \frac{68(0.116)}{2} (1,000) = 0.0039$$

or 39.0 stakes per hectare. Again estimating the variance of n to be twice that observed in a Poisson process, the estimated standard error is 0.000705 or 7.05 stakes per hectare. A 95% confidence interval (obtained by pressing the *Set CI* button) has lower and upper limits 0.0027 and 0.0055, respectively. Therefore, we are 95% confident that there are between 27 and 55 stakes per hectare. It

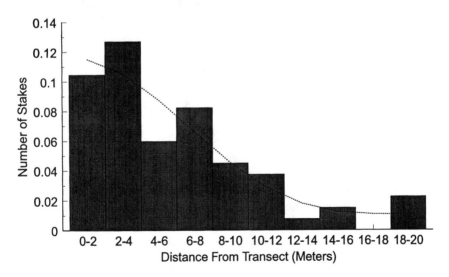

Figure 11.11. Observed and modeled number of stakes at various distances from the line transect. After Burnham et al., (1980).

is reassuring that this interval covers the true population density of 37.5 stakes per hectare!

Grouped Data

It is often difficult to obtain exact perpendicular distance data. Even when it is possible, recording accurate distances may be excessively time consuming. Recording objects as to distance classes may be more manageable. For example, instead of recording the exact perpendicular distance, the sampler determines whether an observed object is within 10 m of the line, 10 to 20 m from the line, 20 to 30 m from the line, etc. Measurements may only be needed when objects are near a class boundary to accurately place the object in the correct distance category. Judicious grouping may help us model the detection function, especially when the data exhibit excessive rounding.

Suppose we establish cut points for distance categories at $0 = c_0, c_1, \ldots, c_g = w$, where the c_i values represent perpendicular distances from the line. Further, let n_i represent the number of objects observed in distance category i, which comprises distances in the interval (c_{i-1}, c_i). Note that it is not necessary for the intervals to be of equal length. As before, $n = \Sigma n_i$.

Consider fitting a continuous detection function to grouped data. First, notice that, conditional on knowing n, the probability distribution of grouped data is multinomial with parameters n and p_i, $i = 1, 2, \ldots, g$. Given that an object is detected, p_i is the probability of its being in the ith distance category. Assuming that the underlying detection function has a cumulative distribution function $F(x)$, the probability an observation being in the ith distance category may be written as

$$p_i = F(c_i) - F(c_{i-1}). \qquad (11.36)$$

Under the Fourier series model, equation (11.36) becomes

$$p_i = \frac{c_i - c_{i-1}}{w} + \sum_{j=1}^{m} a_j \left(\frac{w}{j\pi}\right) \left[\sin\left(\frac{j\pi c_i}{w}\right) - \sin\left(\frac{j\pi c_{i-1}}{w}\right)\right]. \qquad (11.37)$$

Remember, our goal is to estimate the a_j values so that we can subsequently estimate $f(0)$. To obtain the maximum likelihood estimators, we maximize the log-likelihood function

$$\ln(\mathcal{L}(a_1, a_2, \ldots, a_g | n_1, n_2, \ldots, n_g) = \ln(A) + \sum_{i=1}^{g} n_i \ln(p_i) \qquad (11.38)$$

where A is a function of the known sample values:

$$A = \frac{n!}{n_1! n_2! \ldots n_g!} \qquad (11.39)$$

Unfortunately, there is no closed form solution for the estimated a_i values as there was with the ungrouped data. Solutions are obtained using iterative numerical methods such as the Newton–Raphson method. Estimates of the covariances are obtained by evaluating the inverse of the information matrix at the $\hat{a}_i$ values and dividing those by n. The interested reader should consult Burnham et al. (1980) and Buckland et al. (1993) for a more detailed discussion. Burnham et al. also have detailed examples. This approach is used to develop estimates in ECOSTAT.

As with ungrouped data, m must be determined. Likelihood ratio tests are one approach. We have chosen instead to use Akaike's Information Criterion given in equation (11.30). Once the final Fourier series is fit to the model, the chi-squared goodness-of-fit test described for ungrouped data is used to test the fit for grouped data as well.

Example 11.3

Anderson and Pospahala (1970, *see* also Burnham et al. 1980) reported on line transect data on duck nests from a waterfowl production study. Over the course of two summers, 539 perpendicular distances were obtained along 1,600 miles of transect line. The data were grouped into eight intervals, each 1 foot long. The data are presented in the Table 11.3.

ECOSTAT will again be used to fit a Fourier series to these data. First, an ASCII data file must be constructed. For grouped data, the first line of the data file must have the number of groups g, the length of the transect L, and the maximum distance from the line w. Each of the next g lines has the right distance cut-point for the group and the number of observed objects in that distance group. It is

Table 11.3. Number of Duck Nests in Certain Intervals of Perpendicular Distances From a Line Transect

Cell	Lower cut point	Upper cut-point	No. of nests
1	0	1	74
2	1	2	73
3	2	3	79
4	3	4	66
5	4	5	78
6	5	6	58
7	6	7	57
8	7	8	54

After Anderson and Pospahala (1970) by permission of The Wildlife Society.

assumed that the left distance cut-point for the first group is 0. Also, it is assumed that the groups are listed in order with the one closest to the line listed first and the one most distant from the line listed last. Here we had to decide whether to use miles (the transect length was 1,600 miles) or feet (the maximum distance from the line was 8 feet). We converted the distance from the line to miles from the line. (Yes, the numbers were very small.) As a result, the final estimate will be in terms of the number of nests per square mile.

To use ECOSTAT to estimate the parameters, select *Line Transect* from the *Transect* menu and indicate that the data are grouped. After entering the data set name, the model can be fit by pressing the *Estimate D* button. In this case, one cosine term is sufficient. The observed number of nests and the estimated detection probability are shown in Figure 11.12. The estimated nest density is 127.8 nests per square mile. The estimated standard deviation of this estimate is 10.4 nests. (Burnham et al. 1980 report an estimated standard deviation of 8.7 under the Poisson assumption as opposed to our assuming the variance-to-mean ratio is 2.) A 95% confidence interval was also set. Therefore, we are 95% confident that the nest density is between 109.0 and 149.8 nests per mile.

Clustered Populations

For clustered populations, n is the number of clusters observed and $E(s)$ is the expected cluster size. Let $\hat{D}_s$ be the estimated cluster density. If $\bar{s}$ is the estimated average cluster size, then $\hat{D} = \hat{D}_s\bar{s}$; that is, the estimated population density is the

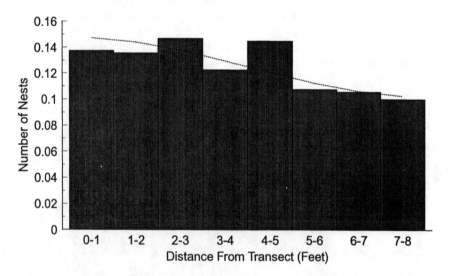

Figure 11.12. Observed and modeled number of duck nests at various distances from the line transect. After Anderson and Pospahala (1970).

estimated cluster density times the estimated size of the average cluster. This with equation (11.18) leads to the population density estimator for clustered populations

$$\hat{D} = \frac{n\hat{f}(0)\bar{s}}{2L} \qquad (11.40)$$

For point transects, we have a similar result using equation (11.23)

$$\hat{D} = \frac{n\bar{s}}{k\hat{v}} \qquad (11.41)$$

Usually, it is easier to observe large clusters, than small clusters, especially as the distance from the transect line or point increases. This is known as size-biased detection of objects. Several approaches to distinguishing between the distribution of cluster sizes in the population and the distribution of cluster sizes in the sample have been suggested (Quinn, 1979; Burnham et al., 1980; Drummer and McDonald, 1987; Drummer, 1990; Quang, 1991). We take the following approach which is also the simplest solution. Only clusters close to the line will be used to estimate cluster size. The truncation distance used to estimate cluster size need not be the same as the one used to estimate the detection function. If size bias may be severe, then truncation should be greater. The estimated detection probability $g(v)$ in the range of 0.6 to 0.8 generally identifies a truncation distance v that ensures that bias in the estimated average cluster size is small.

The variance of $\bar{s}$ is estimated by

$$s_{\bar{s}}^2 = \frac{\sum_{i=1}^{n} (s_i - \bar{s})^2}{n(n-1)} \qquad (11.42)$$

where s_i is the size of the ith cluster. This estimator is unbiased even when the individual s_i have different variances. Weighted averages, stratification, and regression estimators are other methods that may be used if there are not enough clusters to permit a good estimate of the average cluster size using this simple approach.

The estimated variance of the estimate of population density for line transect sampling is approximately

$$s_{\hat{D}}^2 = \hat{D}^2 \left(\frac{s_n^2}{n^2} + \frac{s_{\hat{f}(0)}^2}{[\hat{f}(0)]^2} + \frac{s_{\bar{s}}^2}{\bar{s}^2} \right). \qquad (11.43)$$

For point transect sampling, the equivalent expression is

$$s_{\hat{D}}^2 = \hat{D}^2 \left(\frac{s_n^2}{n^2} + \frac{s_{\hat{h}(0)}^2}{[\hat{h}(0)]^2} + \frac{s_s^2}{\bar{s}^2} \right) \tag{11.44}$$

where

$$\hat{h}(0) = \frac{1}{\displaystyle\int_0^w rg(r)dr}. \tag{11.45}$$

The estimated variability of $\hat{f}(0)$ for line transects or $\hat{h}(0)$ for point transects follows from estimation of the detection function. The estimated variability in the estimated average cluster size is given in equation (11.42). Now, recall that the number of detections from the ith line or point is n_i, $i = 1, 2, \ldots, k$. For point transects or for line transects for which all the replicate lines are the same length, the estimated variability of n is

$$s_n^2 = k \sum_{i=1}^{k} \frac{\left(n_i - \dfrac{n}{k} \right)^2}{k - 1} \tag{11.46}$$

If line i is of length l_i, the estimated variability of n from line transects is

$$s_n^2 = L \sum_{i=1}^{k} \frac{l_i \left(\dfrac{n_i}{l_i} - \dfrac{n}{L} \right)^2}{k - 1} \tag{11.47}$$

Because the distribution of $\hat{D}$ is positively skewed, the normal based confidence intervals do not have the desired level of coverage. Therefore, we again use a lognormal transformation to establish confidence limits as outlined in equations (11.9) to (11.11) (*see* Burnham et al., 1987).

Design

Transect sampling is particularly suited to estimating the population density of immobile populations (e.g., nests, dead deer, and yucca plants), flushing populations (e.g., quail, ruffed grouse, and pheasants), and slow moving populations (such as elephants, gila monsters, and desert tortoise) (Anderson et al., 1979). As with any other study, proper design and field procedures are essential if accurate estimates are to be obtained. The assumptions underlying transect sampling that we discussed earlier suggest some important field considerations. However, following Anderson and colleagues, we emphasize two of them here.

First, the transect line must be defined and maintained. Often it is not physically possible to walk a straight line in the field due to obstacles. However, every effort to walk one must be made. Several approaches have been used to help the observer keep on course. Markers may be put up along the line, or one of the field crew members can watch and direct the observer back to the line when he strays. Without defining the transect line, it is impossible to obtain accurate estimates of sighting distances and/or angles. Without a well-defined line, an observer may tend to walk toward the object once it is sighted. This tends to cause the sighting distance to be underestimated, leading to a positive bias in the estimate of population density.

Second, accurate measurements of distances and/or angles must be made. Exact measurements using a steel tape are best. Pacing or the use of a range finder leads to poorer quality in the data. However, they may be required where the terrain is too hilly or rough for the use of a steel tape. Rounding to 5 or 10 m should be avoided. Certainly, "eye-balling" the distance is not acceptable.

Travel along the line can occur by means other than walking. Horses, off-road vehicles, ships, and airplanes are some examples. Any method that does not violate the basic assumptions of transect sampling can be used. Precision of the estimates can be improved by increasing the total length of the transect lines, L. Finally, the samplers must be interested, well-trained, and competent.

Summary

Transect sampling is often an effective means of estimating population density, especially for immotile, flushing, and slow-moving populations. Strip transects and circular plots assume that all objects within w units of the line or point are observed with certainly. Early work in this area dealt with how to determine w. A histogram of the number of objects observed as the distance from the point or line increases usually indicates decreasing sightability as distance increases. If the objects on the point or line are observed with certainty and at their initial location and if measurements to these objects can be made exactly, line or point transects can be an effective means of estimating the population density. Several methods of estimating the sighting function have been introduced. The Fourier series is one of the most useful.

Exercises

1. In addition to the 1967 strip transect study described in Example 11.1, Robinette et al. (1974) give the results of a 1968 strip transect study in Travo National Park (West) in August 1968. Recall the park covers 28,857 acres. Seven parallel transects, 2 km apart, were walked a total of 29.89 miles. It was believed that all elephants within 1,640 feet of the transect were ob-

served. A total of 229 elephants were observed. Estimate the density of the elephants in the park. Set a 90% confidence interval on the estimate.

2. Edwards (1954) reported the results of a strip transect study designed to estimate the size of the moose population in Wells Gray Park, British Columbia, during March 1952. The park covers 40 square miles. Because accurate maps were not available, strips were chosen between points of known location based on previous land surveys. Consequently, most strips were located on the valley bottomlands. It is thought that slightly more moose would be seen in the bottomlands than in the rest of the range. A ground crew of two men walked transects on snowshoes and counted moose within 200 feet of the transect line. The crew traversed 17 miles of line. They observed 73 moose. Eight of the 73 were seen surrounding a haystack at the edge of a strip.

    A. Estimate the moose population density and the standard deviation of this estimate (1) with and (2) without the 8 moose that were around the haystack.

    B. Do you think the 8 moose should be included in the estimate of population size? Why or why not?

    C. Based on your decision in (B), set a 95% confidence interval on population density.

    D. Suppose you were responsible for management of the moose population in the park. Design a study to estimate the population density the year following this study.

3. As reported by Gates (1979), Evans (1975) estimated white-tailed deer populations in three different environments at the Welder Wildlife Refuge at Sinton, Texas, from 1969 to 1971. The greatest perpendicular distance from the transect line to a deer was 450, 450, and 225 yards in sandy soils, mesquite, and chaparral, respectively. The total lengths of the transects were 63.7 miles in sandy soils, 46.7 miles in mesquite, and 59.5 miles in chaparral. The data are presented in the tables below.

Sandy soils		Mesquite		Chaparral	
Distance group (yd)	No. of deer	Distance group (yd)	No. of deer	Distance group (yd)	No. of deer
0–25	308	0–25	171	0–12.5	40
25–75	616	25–75	233	12.5–37.5	65
75–125	396	75–125	243	37.5–62.5	56
125–175	231	125–175	183	62.5–82.5	42
175–225	99	175–225	35	82.5–112.5	34
225–275	62	225–275	18	112.5–137.5	25
>275	41	>275	33	>137.5	19

A. Estimate the white-tailed deer population in each of the three environments. Attach a measure of precision to your estimate.

B. Set a 90% confidence interval on the white-tail deer population size in each of the three environments.

C. Discuss design issues that you would deem important in a study such as this.

4. In Gates et al. (1968) and Gates (1979), data are given from a study of ruffed grouse that was conducted near Cloquet Minnesota from 1950 to 1958. The total length of the line transect was 382.6 miles. A total of 224 grouse were observed up to 50 yards from the transect line. The data are presented in the table below:

Distance group (yd)	No. of ruffed grouse
0–5	103
5–10	62
10–15	28.5
15–20	11.5
20–25	9
25–30	3.5
>30	6.5

A. Estimate the population density of the ruffed grouse and attach a measure of precision to the estimate.

B. Set a confidence interval on the population size of ruffed grouse.

C. Obviously, it is not possible to observe half of a grouse in a transect. Why do you think there is 0.5 grouse reported in several distance groups. Based on your reasoning, discuss the biological implication this may have on your analysis.

5. Krebs (1989, p. 124) presents a data set collected from a line transect. The sighting distance and angle of pheasants within 65 m of the line were observed. The transect was 160 m in length. The data are presented in the table below:

Observation	Sighting distance (m)	Sighting angle
1	27.6	46°
2	25.2	27
3	16.2	8
4	24.7	31
5	44.4	42
6	48.0	28

(continued)

(*continued*)

Observation	Sighting distance (m)	Sighting angle
7	13.1	2
8	6.9	18
9	23.5	48
10	5.3	29
11	14.9	36
12	23.7	31
13	36.7	56
14	10.9	68
15	24.1	0
16	61.8	46
17	27.6	22
18	8.3	18
19	16.2	27
20	25.2	34

From ECOLOGICAL METHODOLOGY by Charles J. Krebs, Copyright (©) 1988 by Charles J. Krebs. Reprinted by permission of Addison-Wesley Educational Publishers

A. Estimate the population density of pheasants in the area. Attach a measure of precision to this estimate.

B. Set a 95% confidence interval on the pheasant population density.

12

Degree-Day Models

Introduction

The development of plants and poikilothermic (cold-blooded) animals is highly dependent on temperature. It has been said that "the hotter it is the faster they grow." This is true, up to a point, where higher temperatures retard or even stop development. Temperature-driven models are available for a large number of plants such as corn, cotton, alfalfa, and several vegetables as well as many pests of these crops. These have had several practical impacts. For example, when adverse weather conditions delay planting, a farmer may reference historical data to assess the probability that a long-season variety will have ample time to mature before a killing freeze. If it is unlikely that the crop will have time to mature, he may consider planting a shorter-season variety or an alternative crop.

Relating temperature to development has (broadly) followed two approaches. The first is to develop mathematical relationships relating temperature to growth (Wagner et al. 1984; *see* Figure 12.1). These curvilinear approaches can be necessary when most development occurs at temperatures near the upper or lower limits for growth, but curvilinear descriptions come at the expense of considerable complexity. A related approach, the biophysical model of Sharpe and DeMichele (1977), has been advocated as providing a mathematical description of the biochemical basis for temperature and development relationships, but many workers have argued the biological basis of the Sharpe and DeMichele model is unfounded (Lamb, 1992; Lowry and Ratkowsky, 1983; Lamb et al., 1984; Higley and Peterson 1994).

The other approach is to assess the area under a temperature curve that is above some minimum threshold, with this area being called degree-days. A degree-day (DD) represents 24 hours during which the temperature is one degree above the minimum temperature at which an organism is known to start growth. This is extended in a natural way so that if the temperature is 5 degrees above this

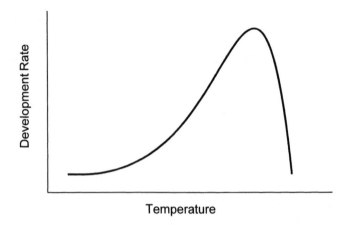

Figure 12.1. Thermal development curve showing the generalized relationship between temperature and rate of development. Reprinted with permission from Higley and Peterson, 1994. Initializing sampling programs. In L.P. Pedigo and G.D. Buntin (eds.): *Handbook of Sampling Methods for Arthropods* in Agriculture pp. 119–136. Copyright CRC Press, Boca Raton, FL.

initiating growth temperature for 24 hours, then 5 DD are accumulated. A degree-hour (HD) is one hour during which the temperature is one degree above the initiating growth temperature of the organism. Thus a degree-day is 24 degree-hours. Because temperature is seldom constant for an hour, much less a day, several approaches to estimating DD from temperature data have been presented. Because a constant relationship between temperature and development is assumed in calculating degree-days, this method is sometimes referred to as a linear estimate of development. Although development rates are not constant (linear) near minimum or maximum temperatures for development, degree-days work because much development is at temperatures in the linear portion of the development curve. Also, degree-days are easily calculated, which is an important attribute.

Using temperature and time to model plant and poikilotherm development is not recent. In 1735, Reaumur discussed the principle (Higley et al., 1986). Wang (1960) and Pruess (1983) summarize various methods of computing degree-days and critique the degree-day approach. Wagner et al. (1984) review approaches for curvilinear descriptions of the temperature and development. More recently, Higley and Peterson (1994) reviewed issues in estimating thermal development. In addition to discussing computation of degree-days and providing a program for that purpose, Higley et al. (1986) provide a discussion of the underlying assumptions to the degree-day approach. Their work has heavily influenced this chapter.

Assumptions

Emphasis has often been on the computation of degree-days. Yet, numerous assumptions are made that can have a biological impact on the utility of degree-days in any given application. Following Higley et al. (1986), we discuss eight of these:

1. All required substrates are present in sufficient quantities for optimal growth.

2. Essential enzymes are available in sufficient quantities for optimal growth.

3. Laboratory estimates of development are sufficient for field use.

4. Minimum threshold temperatures are adequately estimated.

5. Maximum threshold temperatures are adequately estimated.

6. Single minimum and maximum threshold temperatures are adequate over the life span of the organism.

7. The organism cannot regulate its own temperature.

8. The recorded temperatures represent the temperatures of the organism's microhabitat.

All degree-day models are based on the essential assumption that poikilotherm and plant development is directly related to ambient temperature and time. The need for sufficient time for the growth of any organism is obvious. The influence temperature has on the time required for development is not as direct. Through biochemical processes, an organism develops. The rate with which these processes occur depends on the availability of substrates, the availability of enzymes, and temperature. For plants, substrates include water, nutrients, and photosynthates. For animals, food and water are substrates. If a substrate is not present in sufficient quantities, the organism will not develop at an optimal rate even though temperature and all other environmental factors are optimal.

Growth rates are also negatively impacted if the availability of enzymes is not sufficient for optimal growth. For some organisms, enzyme concentrations are regulated by hormones through factors such as day length or genetics. In these cases, the organism's growth may not follow day-degree accumulations well. Different enzymes may need different temperatures for optimal reactions. Consequently, fluctuating temperatures often result in a different rate of development when compared to constant temperatures (Howe, 1967). The average temperatures and the amplitude and frequency of fluctuations determine the magnitude of difference between development time under constant and fluctuating temperatures (Campbell et al., 1974).

Another key observation is that development is faster for fluctuating temperatures around a low temperature mean. This reflects what has been termed the rate summation effect, which recognizes that fluctuations at low temperatures increase development, because the temperature-development curve is concave, and decrease development at high temperatures because the curve is convex (Figure 12.2). Worner (1992) and Higley and Peterson (1994) further discuss this issue.

For most field crops, day-degree models are calibrated in the field. However, for insects and some plants, the development rates used in degree-day models are often established in the laboratory using growth chambers. In growth chambers, the relative humidity, light and temperature are all determined by the researcher. Diet is generally optimal, and temperatures are constant. These idealized conditions do not always directly relate to the field. Higley et al. (1986) suggest that development times under these conditions should be considered minimal. Further, other factors, such as photoperiod, may be critical in development. If these have not been investigated in the laboratory and they are impacting development, the value of degree-day models in the field will be limited.

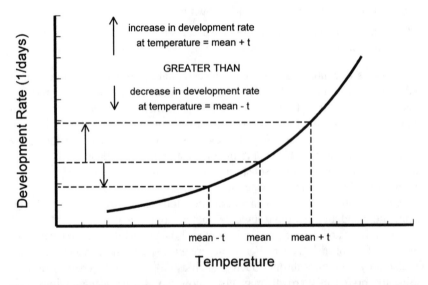

Figure 12.2. Rate summation effect at low temperatures. Because fluctuations above the mean increase development rates proportionally more than fluctuations below the mean reduce development rates, temperature fluctuations (*t*) above and below a specified mean produce a development rate greater than that predicted by the mean temperature. Reprinted with permission from Higley and Peterson, 1994. Initializing sampling programs. In L.P. Pedigo and G.D. Buntin (eds.): *Handbook of Sampling Methods for Arthropods* in Agriculture pp. 119–136. Copyright CRC Press, Boca Raton, FL.

The minimum and maximum threshold temperatures needed for growth are an integral part of any degree-day model. Both are estimated from either field or growth chamber work, leading to imprecision in degree-day computation. The minimum (maximum) threshold temperature is the lowest (highest) temperature at which development can occur. No growth occurs outside the range of temperature from minimum to maximum. Extreme temperatures outside this range can result in growth retardation or death.

Generally, the minimum threshold temperature is estimated in the following manner. The development rate of the organism (the reciprocal of the number of days required for an organism to reach maturity) is determined over a range of temperatures. A line is fit to the development rates. The line is extended to zero, and the x-intercept (the temperature for which the estimated development rate is zero) is taken as the minimum threshold temperature. At temperatures near the threshold, development may begin but not in a linear fashion. For example, in 1944, P.A. Lehenbauer (*see* Newman, 1971) determined that although corn growth begins at 41°F (5°C), growth is extremely slow below 50°F (10°C.). Extending the line to the *x*-axis ignores the slow growth near the threshold and results in the minimum threshold temperature being over estimated. Similarly, Lamb (1992) demonstrated with pea aphid, *Acyrthosiphon pisuin* (Harris), clones that development rates at low temperatures are intrinsically curvilinear and not a reflection of genetic variation among individuals. Other methods for determining the minimum threshold are available (Arnold, 1959, Kirk and Aliniazee, 1981), but the minimum remains an estimate.

Sometimes, the maximum threshold temperature is not estimated. Ignoring the maximum may or may not negatively impact our ability to model development depending on environmental factors. Returning to corn, growth does not occur above 95°F (35°C) (Newman, 1971). In some regions, temperatures rarely exceed 95°F (35°C), and degree-day models ignoring the maximum threshold work well. However, in areas where temperatures in excess of 100°F (38°C) are common, failure to recognize the maximum threshold might indicate more degree-day accumulations and hence more development than actually occurred.

The maximum threshold is difficult to estimate because there tends to be more variability in development rates at the upper level and mortality is usually high. Generally, it is estimated to be the point at which development either plateaus or declines. This point may be estimated (Funderburk et al., 1984) or calculated iteratively (Nowierski et al., 1983). No matter which method is used, the maximum threshold is estimated adding further to the imprecision in the degree-day computation.

Although single values are determined for the maximum and minimum threshold temperatures, these values vary depending on the age of the individual (Wang, 1960). Sanborn et al. (1982) showed that, for insects, each of the different life stages (e.g., egg, larvae, pupae, adult) has a different set of threshold temperatures. To reduce model complexity, single values are used for these thresholds.

Another assumption of degree-day models is that the organism's body temperature is the same as ambient temperature. Many poikilothermic organisms violate this assumption. For example, behavioral and physiological mechanisms are used by insects for thermoregulation (May 1979). An organism may seek a microhabitat that is warmer (cooler) as temperatures decrease (rise). Such behavioral mechanisms are clearly related to the last assumption of degree-day models that we will discuss.

The temperature recorded for calculations is the same as that experienced by the organism. If an insect is using the behavioral mechanism described above for thermoregulation, this is clearly violated. This assumption is usually violated to some extent. If weather station data are being used and that station is not on location, the temperature data are only an approximation of the temperature at another location.

We have discussed eight basic assumptions underlying degree-day models. With all these cautions, we may lose sight of the most important fact. Degree-day models do work extremely well. They are the foundation of numerous crop models that are impressively accurate. For example, with the corn models of today, a researcher can look at a field and state within two or three days when it was planted. Some vegetable models are even more precise. Therefore, these assumptions need to be considered when developing a model based on degree-days. Yet, if the development rate of an organism is largely temperature driven, then degree-days will undoubtedly serve as a major factor in developing growth models for that organism.

Calculating Degree-Days

Based on the discussion above, we know that computation of degree-days is always an estimate. Numerous methods of estimating degree-days have been proposed. We focus on two: the rectangular method and the sine method.

The most common approach to computing degree-days is the rectangular (or historical) method (Arnold, 1959). Originally, the method was to take the average of the daily maximum and minimum temperatures and subtract the lower threshold temperature to determine the degree-day accumulation for 24 hours; that is,

$$DD = \frac{\text{Max} + \text{Min}}{2} - L_t \qquad (12.1)$$

where Max is the daily maximum temperature, Min is the daily minimum temperature, and L_t is the lower threshold for development. Baskerville and Emin (1969) modified the approach to account for a maximum. They took the approach that if the daily maximum temperature is above the developmental maximum, then the developmental maximum is used in the average instead of the daily maximum. These temperatures are accumulated for each day.

Example 12.1

The maximum and minimum temperature thresholds for cotton are 92°F (33°C) and 60°F (16°C), respectively. The minimum and maximum temperatures for three consecutive days are 59° and 90° for day 1, 66° and 93° for day 2, and 55° and 64° for day 3. For the first day, the degree-day accumulation is computed to be

$$\frac{59 + 90}{2} - 60 = 74.5 - 60 = 14.5 \text{ degree-days.}$$

The second day, the maximum daily temperature exceeded the upper threshold for development. To account for this, the maximum development threshold is used instead of the observed maximum:

$$\frac{66 + 92}{2} - 60 = 79 - 60 = 19 \text{ degree-days.}$$

On the third day, the maximum was below the upper threshold and so no adjustments are needed:

$$\frac{55 + 64}{2} = 59.5 < 60 \rightarrow \text{no degree-days.}$$

Notice that because the mean temperature is below the minimum needed for development, no degree-days are recorded for day 3.

The accompanying software ECOSTAT can be used to compute degree-days in one of two ways. Temperature information for a particular day can be entered from the keyboard, or a data file with numerous days can be used with the cumulative degree-days being stored in a file and displayed on the screen. The data file must be an ASCII file. The first line should have the number of dates n of temperature data. Each of the next n lines has an identification for the day, the minimum temperature, and the maximum temperature separated by spaces. Press *Compute DDs* to complete the computations.

Look again at the third day in the example. Because some development usually occurs when the temperature is above the minimum threshold needed for development, even if the average temperature for the day is not, another modification of the historical method has been considered. If the minimum daily temperature is below the lower threshold needed for development, then the minimum is set to this limit. Thus there are corrections when temperatures exceed either the maximum or minimum development thresholds. Using thresholds of 86° and 50°F (30° and 10°C) and computing degree-days in this manner is the standard adopted in the Weekly Weather and Crop Bulletin (Pruess, 1983).

Example 12.2

Consider the temperature data in Example 12.1 again. Using the approach of the Weather Bureau, but the thresholds established for cotton, determine the degree-days for the three days. That is, we keep the thresholds of 60°F (16°C) and 92°F (33°C), but adjust the maximum or minimum if either falls outside these limits. Degree-days for the first day change slightly because the minimum was below the lower threshold:

$$\frac{60 + 90}{2} - 60 = 75 - 60 = 15 \text{ degree-days.}$$

Because the minimum for the second day was above the lower threshold, and we adjusted the maximum in the first example, no changes are made in the degree-day calculations for the second day. However, on the third day, we had no degree-days using the historical method, even though the temperature surpassed the lower threshold. Adjusting the minimum, we now have the following for the third day:

$$\frac{60 + 64}{2} - 60 = 62 - 60 = 2 \text{ degree-days.}$$

When daily temperatures span the developmental minimum, the historical method under estimates the degree-days, and the Weather Bureau approach over estimates degree-days.

Models that give a closer approximation to the heat available for the growth of organisms are degree-day models constructed with the use of sine or cosine waves. For early applications of this approach, *see* Hartstack, (1973) and Hartstack et al., (1976). The premise here is that sine or cosine functions better describe the daily temperature curve and better account for instances when daily temperatures span the developmental minimum or maximum. A basic sine wave is plotted in Figure 12.3. The maximum occurs at 1 and the minimum at −1. Further, 360° or 2π radians are used for a complete cycle. We want to adjust this curve so that it covers a 24-hour period with a maximum and a minimum at 1 and −1, respectively.

Most models assume that the maximum and minimum temperatures occur at the same time each day, although on any day this might not be the case. In fact, the time from the early morning minimum to the afternoon maximum tends to

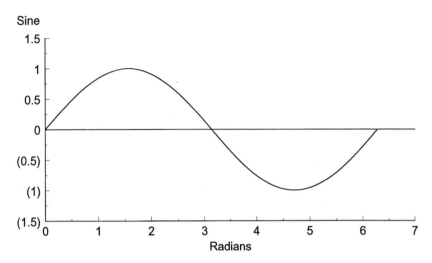

Figure 12.3. Sine wave to be used as foundation for temperature model.

be closer to 9 or 10 hours and that from the afternoon maximum to the early morning minimum is usually 14 or 15 hours. However, using 12-hour spacings tends to work well in practice. Also, to compute the degree-days for a 24-hour period, we must have that day's maximum and minimum and the next day's minimum. The temperature at j hours past the time of the day's minimum for the sine-wave model is

$$\left(\frac{\text{Max} + \text{Min}}{2}\right) + \left(\frac{\text{Max} - \text{Min}}{2}\right) \sin(-1.5708 + .2618j) \qquad (12.2)$$

where Max is the maximum daily temperature, Min is the minimum daily temperature, and j is the number of hours past the minimum for that day.

Notice that for the first 12 hours ($j = 1, 2, \ldots, 12$), the minimum for the equation is the minimum immediately preceding that day's maximum. However, for the last 12 hours ($j = 13, 14, \ldots, 24$), the minimum is the one immediately following the day's maximum and is taken from the following day. Of course the modeled temperatures for non-integer hours, such as 12.23 hours past the day's minimum could also be computed. The angle is assumed to be in radians. The $-1.5708 = -\pi/2$ is an adjustment so that the minimum occurs at hour zero. The $0.2618 = 2\pi/24$ is the proportion of the cycle from minimum to maximum and back to minimum completed in an hour. The computations will be illustrated in the example below.

Example 12.3

From Example 12.1, we have the minimum and maximum for the first day are 59°F and 90°F (15°C and 32°C), respectively. The minimum for the second day is 66°F (19°C). The average of the minimum and maximum is 74.5, and one-half of their difference is 15.5. To estimate the temperature at one hour past the recorded minimum, we use equation (12.2) with Max = 90°, Min = 59°; that is,

$$74.5 + 15.5 \sin(-1.5708 + 0.2618(1)) = 59.5.$$

At $j = 12$, the maximum of 90 is reached. Then the minimum of 66° for the following day is used with the maximum of 90°. In this case, the average is 78 and half of their difference is 12. For $j = 15$, we have

$$78 + 12 \sin(-1.5708 + 0.2618(15)) = 86.5.$$

The estimated sine wave is displayed in Figure 12.4. Table 12.1 gives the estimated temperatures for each hour, as well as the HDs (degree hours), based on the threshold temperatures of 92°F and 60°F (33°C and 16°C) for cotton.

For the 24-hour period, there was a total of 189 + 204 = 393 degree-hours, or 393/24 = 16.4 degree-days. Notice that there were no degree hour accumulations for the first hour because the modeled temperature at that time was below the minimum threshold of 60°F (16°C).

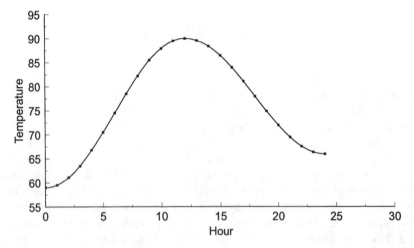

Figure 12.4. Sine wave fitted to a minimum of 59°F (15°C), followed by a maximum of 90°F (32°C), and a minimum of 66°F (19°C).

Table 12.1. Estimated Temperatures and Degree-Hours Based on Sine Model[a]

j	Temperature	Degree-hours	*j*	Temperature	Degree-hours
1	59.5	0.0	13	89.6	29.6
2	61.1	1.1	14	88.4	28.4
3	63.5	2.5	15	86.5	26.5
4	66.8	6.8	16	84.0	24.0
5	70.5	10.5	17	81.1	21.1
6	74.5	14.5	18	78.0	18.0
7	78.5	18.5	19	74.9	14.9
8	82.2	22.2	20	72.0	12.0
9	85.5	25.5	21	69.5	9.5
10	87.9	27.9	22	67.6	7.6
11	89.5	29.5	23	66.4	6.4
12	90.0	30.0	24	66.0	6.0
Totals		189.0	Totals		204.0

[a]Consecutive minimum, maximum, and the following day's minimum temperatures are 59°, 90°, and 66°F, respectively. Upper and lower threshold values are 92°F and 60°F, respectively.

In the preceding example, we refined our estimation of degree-days. However, only the points for each hour were estimated. By taking advantage of the features of the sine function, the degree-day accumulation for this model can be computed exactly. To do so, care must be taken to consider the relationship between the maximum and minimum temperatures and the maximum and minimum threshold temperatures. This results in six cases:

1. Both the minimum and maximum daily temperatures are above the upper and lower thresholds.

2. Both the minimum and maximum daily temperatures are below the upper and lower thresholds.

3. The minimum temperature is above the minimum threshold and the maximum temperature is below the maximum threshold.

4. The minimum temperature is below the lower threshold and the maximum temperature is between the two threshold temperatures.

5. The minimum temperature is between the threshold temperatures and the maximum is above the upper threshold temperature.

6. The minimum temperature is below the lower threshold temperature and the maximum is above the upper threshold.

Allen (1976) presented the formulae associated with computing the degree-days for each. Several investigators have developed computer programs to calculate the degree-days (e.g., Allen, 1976; Higley et al., 1986). We have included such a program in ECOSTAT. In this case, it is assumed that the data are in an ASCII file. The first line has the number of days *n* of temperature data. Each of the next *n* lines has the minimum and maximum for a day. It is assumed that the days are entered sequentially. Although both the minimum and maximum must be given for the last day, only the minimum is used in the degree-day computations with the maximum temperature from the preceding day.

Example 12.4

The daily maximum and minimum temperatures for Grand Island, Nebraska, during the 31 days of July, 1995, are recorded in the file temp.dat. Temp.dat is an ASCII file. The first line of the file has 31, the number of days for which temperature data are given. Each of the next 31 lines has a day identification that can have no spaces, the minimum temperature, and the maximum temperature with one or more spaces separating the three. The minimums and maximums are plotted in Figure 12.5, and their values are given in the Table 12.2.

To use ECOSTAT to compute the degree-days for this time period, we start with the *Temperature* menu and choose the estimation approach, either *Rectangular* or *Sine*. The maximum and minimum threshold values must be entered in the appropriate boxes. Then indicate the method by which the temperature data will be entered, from the keyboard or from a data file. Press *Data File* on the menu to indicate that these data are in a data file. Then enter the name of the data file.

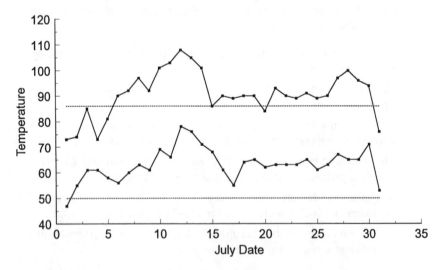

Figure 12.5. Maximum and minimum temperatures in degrees Fahrenheit (°F) at Grand Island, Nebraska, during July 1995. The dashed lines mark the upper and lower temperature thresholds for development of 86°F and 50°F (30°C and 10°C) respectively.

Table 12.2. Daily Maximum and Minimum Temperatures for Grand Island, Nebraska, During July 1995

Date in July	Minimum	Maximum	Date in July	Minimum	Maximum
1	47	73	17	55	89
2	54	74	18	64	90
3	61	85	19	65	90
4	61	73	20	62	84
5	58	81	21	63	93
6	56	90	22	63	90
7	60	92	23	63	89
8	63	97	24	65	90
9	61	92	25	61	89
10	69	101	26	63	90
11	66	103	27	67	97
12	78	108	28	66	99
13	76	105	29	66	96
14	71	101	30	71	94
15	68	86	31	53	76
16	61	90			

Press the *Compute DD* to calculate the degree-days. The degree-days accumulated for each day are shown on the screen and stored in a user-specified data set. The total number of degree-days is displayed on the screen. In this case, the thresholds for corn (86°F = 30°C and 50°F = 10°C) are used. The total degree-day accumulation for July, 1995, in Grand Island, Nebraska, is estimated to be 834 using the rectangular method and 816 using the sine-wave method.

Summary

Degree-day models are simple temperature models. They are based on a number of assumptions, eight of which we discussed. Various other methods have been developed for quantifying the heat available for plant and poikilothermic animal development as well as how development rates are impacted by temperature (*see* Wagner et al., 1984, 1991; Pedigo and Zeiss, 1996). In spite of all the cautions, one fact is inescapable. Temperature is the primary environmental factor driving the development of many organisms, and temperature models have proven to be extremely useful in numerous cases.

Exercises

1. Daily maximum and minimum temperatures at Grand Island, Nebraska, during July 1994, are presented in the table below.

Date in July	Minimum	Maximum	Date in July	Minimum	Maximum
1	59	91	17	63	82
2	59	78	18	68	88
3	65	80	19	67	88
4	62	88	20	60	80
5	65	89	21	57	81
6	65	87	22	56	83
7	58	78	23	59	89
8	53	77	24	65	76
9	53	80	25	62	85
10	64	88	26	56	78
11	66	87	27	52	80
12	60	84	28	50	78
13	59	79	29	54	81
14	59	79	30	62	87
15	61	80	31	63	93
16	58	80			

A. Estimate the degree-days for these data based on the rectangular method.

B. Estimate the degree-days for these data based on the sine wave method.

C. Compare and contrast the degree-days accumulated in July 1994, to those in July 1995, given in Example 12.4. Given that corn is the dominant crop around Grand Island, what questions or concerns might arise?

2. Daily maximum and minimum temperatures at Brownsville, Texas, during July 1995, are presented in the table below.

Date in July	Minimum	Maximum	Date in July	Minimum	Maximum
1	73	90	17	77	92
2	75	93	18	78	94
3	79	93	19	77	94
4	81	94	20	75	95
5	82	93	21	80	95
6	78	90	22	79	94
7	75	91	23	79	95
8	75	91	24	77	93
9	73	90	25	78	95
10	74	92	26	77	94
11	72	94	27	76	95
12	74	92	28	75	97

(continued)

(*continued*)

Date in July	Minimum	Maximum	Date in July	Minimum	Maximum
13	74	93	29	75	98
14	75	92	30	76	95
15	75	93	31	79	98
16	75	93			

    A. Estimate the degree-days for these data based on the rectangular method.

    B. Estimate the degree-days for these data based on the sine-wave method.

3. Daily maximum and minimum temperatures at Brownsville, Texas, during July 1994, are presented in the table below.

Date in July	Minimum	Maximum	Date in July	Minimum	Maximum
1	78	93	17	74	96
2	76	93	18	74	95
3	77	93	19	73	95
4	76	93	20	75	95
5	80	93	21	77	95
6	80	93	22	76	97
7	78	94	23	77	96
8	79	94	24	77	96
9	76	94	25	76	96
10	74	96	26	78	96
11	76	94	27	75	97
12	80	94	28	75	95
13	80	94	29	75	95
14	78	96	30	75	95
15	77	94	31	79	93
16	77	95			

    A. Estimate the degree-days for these data based on the rectangular method.

    B. Estimate the degree-days for these data based on the sine-wave method.

4. Daily maximum and minimum temperatures at Tampa, Florida, during July 1995, are presented in the table below.

Date in July	Minimum	Maximum	Date in July	Minimum	Maximum
1	74	91	17	73	85
2	74	89	18	74	88
3	76	90	19	78	92
4	74	93	20	75	92
5	78	96	21	78	89
6	76	96	22	79	90
7	75	97	23	77	93
8	73	93	24	76	92
9	78	92	25	74	91
10	75	91	26	74	85
11	80	91	27	75	87
12	72	88	28	75	90
13	72	92	29	73	87
14	72	92	30	73	89
15	74	93	31	76	92
16	74	92			

A. Estimate the degree-days for these data based on the rectangular method.

B. Estimate the degree-days for these data based on the sine-wave method.

5. Daily maximum and minimum temperatures at Tampa, Florida, during July 1994, are presented in the table below.

Date in July	Minimum	Maximum	Date in July	Minimum	Maximum
1	72	92	17	72	93
2	73	85	18	72	93
3	73	85	19	73	92
4	73	88	20	74	91
5	73	91	21	73	89
6	71	89	22	71	90
7	72	91	23	74	89
8	74	91	24	75	89
9	75	93	25	75	87
10	75	88	26	74	90
11	71	93	27	76	90
12	72	91	28	75	91
13	74	92	29	71	88
14	74	92	30	71	89
15	74	92	31	72	92
16	73	93			

A. Estimate the degree-days for these data based on the rectangular method.

B. Estimate the degree-days for these data based on the sine-wave method.

6. Daily maximum and minimum temperatures at Bakersfield, California, during July 1995, are presented in the table below.

Date in July	Minimum	Maximum	Date in July	Minimum	Maximum
1	63	94	17	98	85
2	64	96	18	101	88
3	62	95	19	100	92
4	63	93	20	96	92
5	69	99	21	94	89
6	67	97	22	92	90
7	65	97	23	91	93
8	72	100	24	93	92
9	69	96	25	93	91
10	62	92	26	97	85
11	59	87	27	104	87
12	54	81	28	104	90
13	57	87	29	94	87
14	67	96	30	98	89
15	70	103	31	103	92
16	75	97			

A. Estimate the degree-days for these data based on the rectangular method.

B. Estimate the degree-days for these data based on the sine-wave method.

7. Daily maximum and minimum temperatures at Bakersfield, California, during July 1994, are presented in the table below.

Date in July	Minimum	Maximum	Date in July	Minimum	Maximum
1	69	103	17	72	102
2	68	101	18	73	101
3	66	99	19	69	98
4	68	98	20	71	97
5	65	98	21	73	94
6	65	96	22	67	97
7	71	100	23	64	94
8	73	102	24	62	90
9	73	104	25	65	96

(continued)

(*continued*)

Date in July	Minimum	Maximum	Date in July	Minimum	Maximum
10	74	106	26	69	100
11	68	106	27	70	101
12	68	104	28	72	102
13	71	104	29	72	104
14	72	105	30	73	100
15	69	105	31	68	99
16	71	104			

    A. Estimate the degree-days for these data based on the rectangular mehod.

    B. Estimate the degree-days for these data based on the sine-wave method.

8. Obtain the July 1995, maximum and minimum temperatures for Seattle, Washington.

    A. Estimate the degree-days for these data based on the rectangular method.

    B. Estimate the degree-days for these data based on the sine-wave method.

9. Obtain the July 1994, maximum and minimum temperatures for Seattle, Washington.

    A. Estimate the degree-days for these data based on the rectangular method.

    B. Estimate the degree-days for these data based on the sine-wave method.

10. Obtain the July, 1995, maximum and minimum temperatures for Albany, New York.

    A. Estimate the degree-days for these data based on the rectangular method.

    B. Estimate the degree-days for these data based on the sine-wave method.

11. Obtain the July, 1994, maximum and minimum temperatures for Albany, New York.

    A. Estimate the degree-days for these data based on the rectangular method.

    B. Estimate the degree-days for these data based on the sine-wave method.

12. Consider the six locations for which July temperature data have been collected: Grand Island, Nebraska; Brownsville, Texas; Tampa, Florida; Bakersfield, California; Seattle, Washington; and Albany, New York.

    A. Determine the principal commercial crop grown in the region surrounding each of these cities.

B. Based on research articles, find the upper and lower temperature thresholds and the degree-day requirements for each crop.

C. From the limited temperature data gathered here, discuss the impact of temperature on the commercial crops in each region.

D. Discuss other environmental factors that have an impact on the choice of cropping system.

13

Life-Stage Analysis

Introduction

Many animals have well-defined stages of development. For insects, this is especially evident. Typically, an adult lays eggs. After a period of time, larvae emerge from the eggs. Larvae may go through several growth stages, 4 to 20 is common. These are called instars and are characterized by increasing larval size. The larvae pupate, and the adults emerge from the pupae. In the study of such populations, counts or estimates of the number of individuals in the various development stages are recorded at a series of points in time. Based on these data, researchers may want to estimate the total number of individuals that enter each stage, the average time spent in each stage, the survival probability for each stage, the mean entry time to each stage, or the unit time survival rate. A researcher may be interested in determining not only the mortality for each stage, but whether the mortality is density dependent. Interest may be focused on the stability of the population and factors affecting that stability. We refer to the statistical approach to answering these questions as life-stage analysis. Caughley (1977a), Southwood (1978) and Pedigo and Zeiss (1996) discuss life-table construction and the use of life tables in understanding population dynamics. Manly (1990) summarizes the statistical work in this area. These works serve as the foundation for much of this chapter.

For some populations, all individuals enter at the beginning of the sampling period. For example, bollworm moths *Heliothis zea* (Boddie) may lay eggs across a cotton field on a given night. The eggs in the field constitute the individuals within the population. They are described as a single cohort because all entered the population within a short enough time frame for us to assume that it was at the same time. For many mammals, young are born in the spring of the year. Although the time span during which the young enter the population is longer than the preceding bollworm example, their life span is also longer, being measured in years instead of days. Therefore, the young born in a given season could also serve as a cohort. In these examples, the analysis of the stage-frequency data consists of modeling the development and survival of this single cohort through time.

More often individuals may enter the population over an extended period of time covering much of the sampling period, leading to multi-cohort data. Because some individuals are developing and leaving the early life stage while others are still entering the population, the distribution of the durations of stages is confounded with the distribution of entry times. This makes estimation of survival probabilities for each stage more difficult.

Whether one is following a single cohort through time or working with multiple cohorts, life-stage analysis is based on estimates of the numbers of population members in each life stage at several points in time. Suppose that we are working with a population that has q life stages and that samples are taken at n points in time. The data may be displayed as in Table 13.1. Although some insight may be gained from laboratory studies, these results do not translate directly to the field where populations are impacted by variation in temperature, weather, predators, prey, and so forth. Therefore, the methods we discuss have been developed for field use.

When working with field instead of laboratory populations, sampling is generally used to provide estimates of the number in the population in each life stage at each sampling point. Quadrat sampling based on fixed or random sample sizes, capture-recapture methods, and line and point transects may be used to provide these estimates. These methods have been covered in earlier chapters. However, there are some special concerns when collecting life-stage data. First, we must take care to sample a constant fraction of the population during each sampling period or to adjust the population estimates for differences. One way to do this is to divide the sampling region into N equal-sized plots and select n of these at random and without replacement for the sample at time t_i. The number of individuals within each life stage totaled over the n sample plots provides one row of data as in Table 13.1. The process is repeated at each sampling occasion, keeping n fixed.

Table 13.1. Structure of Life-Stage Data

Sample time	Sample frequency in stage				
	1	2	3	. . .	q
t_1	f_{11}	f_{12}	f_{13}	. . .	f_{1q}
t_2	f_{21}	f_{22}	f_{23}	. . .	f_{2q}
$\vdots$					
t_n	f_{n1}	f_{n2}	f_{n3}	. . .	f_{nq}

Random sampling permits the assumption of a Poisson distribution as a basis for modeling. As noted in earlier chapters, the Poisson assumption of the variance equaling the mean is frequently violated by the estimated variance being significantly greater than the mean. This will result in large, goodness-of-fit statistics. By having counts from the n plots, the mean and variance for each life stage can be estimated. If the variance is significantly greater than the mean by a constant factor, a heterogeneity factor H, which is the estimated variance-to-mean ratio (*see* Chapters 1 and 7), can be computed. If this ratio is constant for all stage frequencies, then the heterogeneity factor can be used to adjust the variances of population parameters.

Sometimes there is a need to change the sampling fraction during the course of the experiment; that is, the sample size n is changed. For example, suppose that after the first two sampling periods, it becomes clear that the population is smaller than anticipated and more plots must be included in the sample for the stage-frequency counts to be large enough for analysis. Then either the counts must be adjusted to a common sampling fraction or allowances must be made in the estimation process, the latter approach providing more accuracy.

Another problem arises when there is differential sightability among the life stages. Experiments may be needed to estimate the sightability of each stage (McDonald and Manly, 1989; Schneider, 1989). As an example, suppose, based on experimentation, it is found that 80% of the eggs, 60% of the larvae, and 90% of the adults are counted in samples. Then the number of observed eggs, larvae and adults would be divided by the correction factors 0.8, 0.6, and 0.9, respectively, to provide estimates of the numbers in these life stages. The standard errors of these adjusted totals are the standard errors of the original counts divided by the correction factor.

Finally, we should note that taking a random sample is not necessarily the best approach for obtaining good estimates of population parameters. In a study of the yellow birch lace bug *Corythucha pallipes,* Munholland (1988) found that data from random sampling was too erratic to be analyzed alone. Data were also collected by repeatedly counting the stages in a part of the population of interest. These latter data were less variable and could be analyzed. Therefore, repeatedly sampling a fixed proportion of the population is another possible approach to obtaining population estimates. However, if this is done, the assumption of Poisson sampling errors is no longer reasonable, and simulation is probably needed to determine the accuracy of the estimates.

Environmental conditions, especially ambient temperature, may have a major impact on development time, survival rates, and other population parameters. If there are temperature fluctuations, basing time on physiological time instead of days may be more appropriate. Physiological time is generally based on degree-days (*see* Chapter 12). This can have a major effect on the analysis. For example, van Straalen (1985) concluded that the time from November to March in 1979 represented only 3 weeks of physiological time in his study of Collembola *Or-*

chesella cincta (Linné). Just because physiological time is the appropriate time scale for development does not necessarily mean that it is the right one for survival. Survival rates may be more constant in terms of calendar time than physiological time. In such cases, the model needs to allow for both time scales.

Stage duration and survival probabilities may also be confounded (Braner and Hairston 1989). For example, suppose that the duration of a given life stage is not the same for all individuals; that is, some develop more rapidly than others. Further assume that mortality occurs at a constant rate through time. Whereas individuals who develop rapidly spend less time in the stage and are more likely to survive, slower developing individuals spend more time in the stage and are more likely to die. The survivors of a stage then represent a biased sample for the estimation of the mortality-free distribution of stage duration. The difference in the mortality-free distribution of stage duration and the distribution of stage duration for survivors should be relatively small compared to estimation errors unless stage mortality is high.

One of the goals of life-stage analysis is usually the construction of a life table, and this is where we will begin.

Life Tables

Life tables were originally developed by human demographers. They are one of the principal tools of the insurance industry. We are interested in them because they are also one of the primary methods of studying animal population dynamics. Life tables display population survival and mortality provide in a systematic manner. They may be either time specific or age specific (Southwood, 1978). An age-specific life table is based on a single, real cohort. The original size of the cohort and the number surviving each stage are either known or, more likely, estimated. A time-specific life table is developed by sampling a stationary population that has considerable overlap of generations. The age structure of the sample is determined and used to develop the life table. First, consider the following example of an age-specific life table.

Example 13.1

Caughley (1966) presented data on a cohort of female Himalayan thar. The age and number surviving in the cohort over time is recorded in the table below:

Because the animals were not born on the same day, they do not form a cohort in the strictest sense of the word. Yet, they were all born during the same season and for practical purposes do serve the role of a cohort.

What information can be gained from such data? Perhaps interest is in the probability of survival to a specified age or the probability of death before reproductive age. The probability of surviving a particular life stage, such as survival through the first year, may be of interest. Alternatively, attention could be focused on mortality instead of survival probabilities. The frequencies in Table 13.2 can be used to derive estimates for each of these, and the results are generally presented as part of the life table. Because each sampling time corresponds to an age for this cohort, we use X to denote the age interval at which the sample was taken at time t_i and f_X to denote the number of individuals in age interval X at the time of the sample. In addition, the following symbols are used:

Table 13.2. Life Table for Female Himalayan Thar[a]

i	t_i	f_i
1	0	205
2	1	96
3	2	94
4	3	89
5	4	79
6	5	68
7	6	55
8	7	43
9	8	42
10	9	22
11	10	15
12	11	10
13	12	6
14	>12	11

[a]The number of female thar (f_i) alive during the birth season ($t_i = 0$) and each subsequent year.

Table 13.3. Life Table for Himalayan Thar.

Age X	f_x	l_x	d_x	q_x
0	205	1.000	109	0.532
1	96	0.467	2	0.021
2	94	0.461	5	0.053
3	89	0.433	10	0.112
4	79	0.387	11	0.139
5	68	0.331	13	0.191
6	55	0.269	12	0.218
7	43	0.209	11	0.256
8	32	0.155	10	0.312
9	22	0.109	7	0.318
10	15	0.073	5	0.333
11	10	0.047	4	0.400
12	6	0.029		
>12	11			

l_x: proportion of cohort surviving to age interval X (f_x/f_0)

d_x: stage-specific mortality, number from cohort dying during age interval X

q_x: stage-specific mortality rate, the number dying in the age interval divided by the number of survivors at the beginning of the age interval (d_x/f_x)

No set of life table symbols has become standard across all branches of ecology. Symbols other than l_x, d_x, and q_x may be used to denote the stage-specific survival rate, stage-specific mortality, and stage-specific mortality rate, respectively. Care should be taken to determine the meaning of each life stage symbol for the reference being consulted.

Example 13.2

Consider again the data in Example 13.1. Based on the frequency of female thar alive at each age, we can develop more columns for the life table (see Table 13.3).

Notice that mortality is high (0.532) during the first year of life. During the second year of life, mortality is quite low (0.021), but it increases steadily each subsequent year. Only about 3% of the population survives to age 12.

Caughley (1977a) considers six methods of collecting data for life tables. In Method 1, a cohort of animals is followed through life and the number dying in successive intervals of time are recorded. Instead of observing deaths, the number of live animals within a cohort is observed at regular time intervals for Method 2. The ages of death of animals marked at birth is Method 3. In this method, the marked animals are treated as a cohort. Method 4 is based on estimating q_x by comparing the number of animals aged x in a population with the number of these that subsequently die before age $x + 1$. This is the method most frequently used for human populations. These methods require no assumptions on the rate of population increase or the stability of the age distribution.

Methods 5 and 6 do require both an assumption of a stable age distribution and a known rate of increase. Define the finite rate of increase to be $e^r = N_{t+1}/N_t$, and let the population at time 0 be N_0. Then the population at time t is $N_t = N_0 e^{rt}$. In Method 5, ages at death are recorded. Assuming that r is zero (e^{rt} is one), the observed frequencies of death within each age class are used in the d_x column to develop the remainder of the life table. For Method 6, the age distribution is estimated, and the frequency of the zero-age class is determined from fecundity rates. Each age frequency is multiplied by e^{rx} and divided by the number in the zero-age class to obtain the l_x column values. Caughley (1977) discusses the difficulties associated with these latter two methods as they have been used incorrectly often. If possible, one of the first four methods should form the foundation for the life table. If either Method 5 or 6 is used, care should be taken to ensure the validity of the results (*see* Caughly (1977) for further discussion).

Key Factor Analysis

Suppose a population has distinct generations as in Examples 13.1 and 13.2. Then an analysis of the stage frequencies permits us to estimate stage survival rates from generation to generation. We could then study the variation in these survival rates to help us determine which sources of variation are particularly important in population dynamics. A key point to notice is that we are focusing on the *variation* in survival rates, and not the survival rates. Morris (1957) was the first to emphasize the importance of studying the variation. He noted that if the survival rate is constant, it has little or no effect on the population dynamics. However, changes in the survival rate can have a major impact on the population. For example, for each generation, if the mortality of Himalayan thar is about 50% due to predation, weak young, etc., then it has little impact on population changes. We could simply take the number of thar born in a year, divide by 2, and begin the life table with the first year of life. The same could be said if this mortality rate was 80% or even 90%. As long as the mortality rate does not vary, it can impact the potential rate of increase or decrease of the species but not population changes. For this reason, a life table for one generation provides little insight into the population dynamics of the species. A better understanding can be gained by comparing the variation in survival rates for different stages and determining how this variation can impact numbers in subsequent stages. We will use a set of methods known as factor analysis for this purpose.

Before reviewing methods of factor analysis, we need to discuss some terminology and some ideas. Varley and Gradwell (1970) reviewed the terms density-dependent mortality, inverse density-dependent mortality, delayed density-dependent mortality, and density-independent mortality. Although there have been many arguments as to the correct usage of these terms, they are important ideas for key factor analysis so we define them. A population with density-dependent mortality is one for which the death rate increases as density increases. In this case, a large population has a high mortality rate, and a small population has a low mortality rate. Inverse density-dependent mortality occurs when the death rate decreases as the population density increases. A population with delayed density-dependent mortality is one for which the effects of population density are not immediate, but instead delayed for a period of time. This could occur if an increase in size of the study population leads to growth in the population of a predator or prey, resulting in increased mortality for the study population the next generation. Density-independent mortality does not relate in any way to population density. Common causes are food supplies or weather. Notice that a population may be subjected to more than one type of mortality. Density-independent mortality is present to some extent in most populations. Further, a population may have inverse density-dependent mortality when the density is below a certain threshold but density-dependent mortality above that threshold.

Two other ideas need to be at least mentioned. First, the fecundity rate of a species may have little relationship to population size. For some species, almost all individuals die before reaching maturity. Therefore, survival rate is often much more important than fecundity rate when studying population dynamics. Second, some populations are closed in the limited sense that the number of individuals entering stage 1 may be calculated directly from the number in the last stage in the previous generation. The survival rate for stage 1 then reflects lowered fecundity as well as mortality. In other cases, the relationship between the two generations is not as clear, and more thought must be given to obtaining a valid estimate of the maximum in stage 1.

Example 13.3

Southern (1970) presented life tables for the tawny owl *Strix aluco* L. in Wytham Wood, near Oxford. Six life stages were considered. Stage one was the maximum number of eggs possible for the adult pairs present. The number of pairs consisted of adults in the area surviving from the previous year, fledgling from the previous year that survived and remained in the area, and adults that immigrated from outside the area. Southern estimated the maximum mean clutch size to be 3.0 eggs. To estimate the maximum number of eggs possible from the adult pairs present, he took the number of pairs plus the number of pairs that laid again after losing the first clutch and multiplied that total by 3. Stage two is the maximum number of eggs possible from the *breeding* pairs, and stage 3 consists of the eggs actually produced. Eggs that hatch are in stage 4. Stage 5 consists of fledgling young. The young remaining the next spring are stage 6. The data for years 1949 to 1959 are presented in Table 13.4.

Notice that no pairs bred in 1958. Also, sampling errors resulted in the observed frequencies in stage 5 being larger than those in stage 4 in four different years. What stage(s) is (are) most critical to the population dynamics of the tawny owl? This is the question we try to address as we consider methods of key factor analysis. Our analysis follows that of Manly (1990).

Varley and Gradwell's Method

The ideas of a key factor analysis were developed in classical papers by Morris (1957, 1959) and Varley and Gradwell (1960). The graphic approach of Varley and Gradwell is intuitively appealing. Let $N_1, N_2, \ldots, N_q$, be the numbers entering development stages $1, 2, \ldots, q$, respectively. The stage specific survival rates are then

$$w_j = \frac{N_{j+1}}{N_j}. \tag{13.1}$$

Table 13.4. Life Table for the Tawny Owl from 1949 to 1959

Year	Stages					
	1	2	3	4	5	6
1949	60	54	50	34	26	7
1950	75	66	57	31	25	9
1951	63	36	25	6	6	6
1952	72	51	43	16	21	9
1953	75	48	32	20	20	13
1954	84	60	50	15	17	17
1955	90	12	8	4	4	4
1956	99	66	48	23	24	12
1957	102	66	66	26	20	10
1958	93	0	0	0	0	0
1959	110	81	65	25	28	12

From Southern (1970).

Therefore, assuming survival in one stage is independent of that in another stage, the probability of surviving to adult stage is

$$w_1 w_2 \ldots w_{q-1} = \frac{N_2}{N_1} \frac{N_3}{N_2} \ldots \frac{N_q}{N_{q-1}} = \frac{N_q}{N_1}. \tag{13.2}$$

Now, taking the base 10 logarithm of both sides (and reversing the equation), we have

$$\log\left(\frac{N_q}{N_1}\right) = \log(w_1) + \log(w_2) + \ldots + \log(w_{q-1}), \tag{13.3}$$

or

$$K = k_1 + k_2 + \ldots + k_{q-1}, \tag{13.4}$$

where $K = -\log(N_q/N_1)$ and $k_j = -\log(w_j)$. Because we are taking logarithms of values between 0 and 1, the logarithms are always negative. By taking the negative of the logarithm, all terms are positive. Now, K and the k_j values are plotted against time. The key factor is the stage for which the variation in the k value most closely follows variation in K.

Although inspection of the graph is commonly used to determine the key factor, more objective measures have been suggested. One approach is to estimate the correlations between K and the k for each stage. The one with the highest correlation is the key factor.

Example 13.4

Consider again the tawny owl data in Example 13.3. Because no pairs bred in 1958, all but the first stage had zero individuals. This indicates that there was a severe overwinter loss, but no additional information about the key factors is available. Because the logarithm of zero is undefined, the k_j values are not defined so we dropped this year from the analysis. The accompanying software ECOSTAT can be used to compute the K and k_j values. To do so, press *Key Factor* on the *Life Stage* menu. Then press *Enter Data* and give the name of the data set for analysis. It is assumed that the data set is an ASCII file. The first line of the data set has the number of generations G and the number of stages q that are to be analyzed. Each of the next G lines has an identification for the generation, followed by the number in each stage, beginning with the first stage. The data for this example are in owl.dat. Because 1958 was not included in the analysis, only $G = 10$ generations are analyzed. ECOSTAT computes the K and k_j, $j = 1, 2, 3,$ 4, 5, for the remaining years. These are presented in Table 13.5.

K is plotted across time, and the k_j values are plotted across time in Figure 13.1. Looking at the graphs, it appears that k_i is most closely correlated with K. The correlation between K and each k_j is given in Table 13.6.

Both the plots in Figure 13.1 and the correlations indicate a close relationship in K and k_1, leading us to identify k_1 as a key factor. Notice that k_1 represents loss due to birds failing to breed. Although the relationship is not as clear, notice that k_5 and K appear to have an inverse relationship; that is, high values of k_5 occur with low values of K and vice versa. This is also reflected in the negative correlation of the two. Therefore, the loss of young owlets after they leave the nest

Table 13.5. Computed K and k_j Values for the Tawny Owl Data

Time	K	k_1	k_2	k_3	k_4	k_5
1949	0.933	0.046	0.033	0.167	0.117	0.570
1950	0.921	0.056	0.064	0.265	0.093	0.444
1951	1.021	0.243	0.158	0.620	0.000	0.000
1952	0.093	0.150	0.074	0.429	−0.118	0.368
1953	0.761	0.194	0.176	0.204	0.000	0.187
1954	0.694	0.146	0.079	0.523	−0.054	0.000
1955	1.352	0.875	0.176	0.301	0.000	0.000
1956	0.916	0.176	0.138	0.320	−0.018	0.301
1957	1.009	0.189	0.000	0.405	0.114	0.301
1959	0.962	0.133	0.096	0.415	−0.049	0.368
Mean	0.947	0.221	0.099	0.365	0.008	0.254
SD	0.175	0.238	0.061	0.141	0.078	0.201

After Manly (1990).

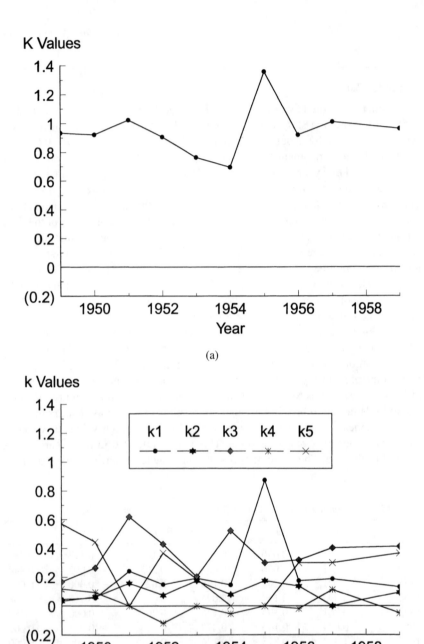

Figure 13.1. (a) *K* as a function of time used in the key factor analysis of Southern's (1970) tawny owl data. (b) $k_j, j = 1, 2, 3, 4, 5$, as a function of time used in the key factor analysis of Southern's tawny owl data. After Manly (1990).

Table 13.6. Correlations Between K *and the* k_i *Values for the Tawny Owl Data* *(Southern 1970)*

k_j	k_1	k_2	k_3	k_4	k_5
Correlation between k_j and K	0.797	0.228	−0.045	0.178	−0.177

due to death and emigration appears to be a density-dependent mortality factor. The other stages seem to be relatively unimportant.

Regression of K *against* k_j *values*

Another key factor approach is to perform a regression of K against the k_j values (Smith, 1973; Podler and Rogers, 1975). Let k_{ij} be the k value for the jth stage in the ith generation, $\bar{k}_j$ be the mean k value for stage j, K_i the K value for generation i, $\bar{K}$ be the mean of K for all generations, and G the number of generations of data. The slope from the regression has the form

$$b_j = \sum_{i=1}^{G} \frac{(k_{ij} - \bar{k}_j)(K_i - \bar{K})}{\sum\limits_{i=1}^{G} (K_i - \bar{K})^2}. \tag{13.5}$$

Notice that we then have

$$\sum_{j=1}^{q-1} b_j = 1, \tag{13.6}$$

leading the b_j values to often be interpreted as the fraction of the total variation of K, as measured by $\Sigma(K_j - \bar{K})^2$, that is accounted for by k_j (Manly 1990). However, because some b_j values may be negative and others greater than 1, this interpretation can be confusing. None the less, the key factor is the k value with the largest associated b value.

Smith (1973) suggested that once the key factor is identified as, say, k_p, the residual "killing power" in stages other than p is $K_i' = K_i - k_{ij}$ in the ith generation. The covariance between k_j and K_i' is a measure of the contribution of stage j to this residual. Thus, the second most important k-value could be determined. The process could be repeated to obtain an ordering of the k-values as to their relative importance. Alternatively, regressions of the residual killing powers on the k-values could be used to obtain this ranking.

After determining key factors, plots of k values against the logarithms of the corresponding population densities are generally constructed. If there is a tendency for a positive relationship between k_j and the logarithm of the density, we have an indication of density-dependent survival. A negative relationship represents inverse density-dependent survival. The correlation between k_j and the logarithm of density could be used to give an objective assessment of this relationship. Regression could also be used. However, standard regression methods are not appropriate except in special circumstances, and we do not use that approach here (Kuno 1971, Ito 1972, Royama 1977, Slade 1977).

Example 13.5

Consider the tawny owl data from Examples 13.3 and 13.4 again. To determine the first key factor, K was regressed on the k_j's, $j = 1, 2, 3, 4, 5$, using the following SAS® (1990) code:

```
PROC REG;
MODEL K = k1 k2 k3 k4 k5;
```

The results are presented on the first line in Table 13.7.

As anticipated, k_1 is identified as the first key factor because its estimated regression coefficient of 1.082 is greater than any other. This confirms that survival during the first stage is a clear key factor; that is, from the available pairs of tawny owls, the number that breed is a critical factor in the population dynamics of the species. Next $K_i' = K_i - k_{i1}$ was computed for each generation i and regressed on the k_j values, $j = 1, 2, 3, 4,$ and 5. The estimated regression coefficient of 1.089 identifies k_5 as the second key factor. Notice again that k_5 appears to have a reciprocal relationship with K (*see* Figure 13.1). Although the strength of the relationship is not as strong as that in k_1, it is present. Therefore, from the number of fledgling young, those that remain in the area the next spring tend to be inversely related to the change in population, pointing again to the presence of density-dependent mortality.

Notice all k_j values are included in each regression. The periods in the table following the identification of a key factor simply indicate that we do not need to consider that factor in assessing subsequent key factors. Therefore, k_3 and k_4

Table 13.7. *Regression Coefficients from Regressions of* K *and Residual Killing Powers on the* k_j *Values for the Tawny Owl Data*

Step	k_1	k_2	k_3	k_4	k_5	Key factor
1	1.082	0.056	−0.036	0.079	−0.204	1
2		−0.267	−0.050	0.232	1.084	5
3		0.125	0.945	−0.070		3
4		0.291		0.709		4

From Southern (1970) after Manly (1990).

are the third and fourth key factors, respectively. Yet, their role appears to be relatively unimportant.

The plots of the k_j values by the logarithm of the number in stage j for $j = 1$, 2, 3, 4, and 5 (called R_j) are shown in Figure 13.2. Notice that in Figure 13.2(e), k_5 appears to increase as the logarithm of N_5 increases, indicating a density-dependent effect.

Manly Method

A third approach to key factor analysis was suggested by Manly (1977b, 1979, 1990). This method was developed because the others that we have considered do not make use of the order in which mortality operates through stages. For example, high-density-dependent mortality in one stage may remove almost all the variation introduced by variable mortality in the preceding stages (Manly, 1990). Although the density-dependent k-value may have little or no relationship to the K-values, the stage is important for population dynamics. An approach to identifying this type of key factor is to estimate the parameters of a population model that allows for density-dependent mortality. From these parameters, it is possible to estimate the effect of each life stage on the variation in the numbers entering the final life stage. In this case, a key factor is defined as a life stage that substantially either increases or decreases this variation.

As above, N_j is the number entering stage j in one generation, and $k_j = -\log(N_{j+1}/N_j)$, using base 10 logarithms. Now let

$$R_j = \log(N_j), \tag{13.7}$$

so that

$$k_j = R_j - R_{j+1}. \tag{13.8}$$

Further assume that the k_j values are linearly related to the logarithms of N_j; that is,

$$k_j = \tau_j + \delta_j R_j + \epsilon_j, \tag{13.9}$$

where τ_j and δ_j are constants, and the ϵ_j values are random errors with mean zero and variance $Var(\epsilon_j)$.

From (13.8) and (13.9), we can write

$$R_j = R_{j-1} - k_{j-1} = (1 - \delta_{j-1})R_{j-1} - \tau_{j-1} - \epsilon_{j-1}. \tag{13.10}$$

Repeated use of this result allows the logarithm of the number entering the final stage to be written as

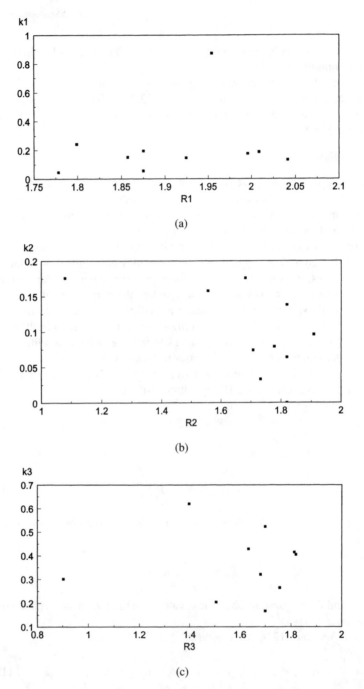

Figure 13.2. Plots of (a) k_1, (b) k_2, (c) k_3, (d) k_4, and (e) k_5 values against R values (R_j = $\log(N_j)$, where N_j is the number of individuals in stage j, for Southern's (1970) tawny owl data to search for evidence of density-dependent survival. After Manly (1990).

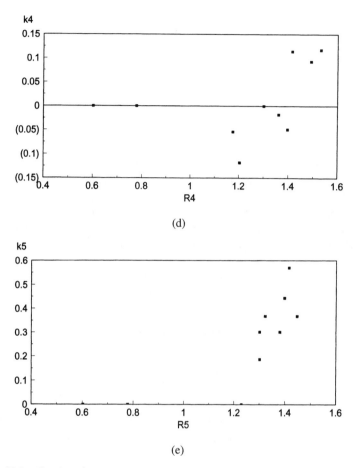

Figure 13.2. Continued.

$$R_q = \theta_0 R_1 - \sum_{j=1}^{q-1} \theta_j \tau_j - \sum_{j=1}^{q-1} \theta_j \epsilon_j, \tag{13.11}$$

where

$$\theta_j = (1 - \delta_{j+1})(1 - \delta_{j+2}) \ldots (1 - \delta_{q-1}), \qquad j = 0, 1, \ldots, q - 2, \tag{13.12}$$
$$= 1, \qquad j = q - 1.$$

From the above, we can see that the expected value of R_q is

$$E(R_q) = \theta_0 E(R_1) - \sum_{j=1}^{q-1} \theta_j \tau_j, \tag{13.13}$$

and the variance is

$$Var(R_q) = \theta_0^2 \, Var(R_1) + \sum_{j=1}^{q-1} \theta_j^2 \, Var(\epsilon_j). \tag{13.14}$$

Equations (13.13) and (13.14) are used differently depending on whether or not the results for different generations are related. We consider the case where they are unrelated first.

When the results for different generations are unrelated, the number entering stage 1 in a given generation, N_1, is a random variable that is independent of the number in the last stage during the previous generation. Now define

$$
\begin{aligned}
A_u &= \theta_0^2 \, Var(R_1), & u &= 0, \\
&= \theta_u^2 \, Var(\epsilon_u), & u &= 1, 2, \ldots, q - 1.
\end{aligned}
\tag{13.15}
$$

Notice that we can now rewrite equation (13.14) as

$$Var(R_q) = \sum_{j=0}^{q-1} A_j. \tag{13.16}$$

Therefore, an A_j value represents the relative importance of random variation in the jth stage in determining the random variation in the last stage. The A_j values should be reported as percentages of $Var(R_q)$.

Sometimes the A_j values alone do not identify an important stage. If much of the variation in the stage survival rate is attributable to a density-dependent response to the number entering the stage, the A value may be quite small because the k value is actually removing variation present from earlier stages. Yet, this stage is important because the final stage would be more variable if it was not present. In this case, the role of the stage is clarified if we determine how $Var(R_q)$ would be expected to change if the stage were made constant. This is accomplished by setting the δ and ϵ values for that stage to zero. If this is done for stage u, then the value of $Var(R_q)$ is expected to change to

$$B_u = \sum_{j=0}^{u-1} \frac{A_j}{(1 - \delta_u)^2} + \sum_{j=u+1}^{q-1} A_j. \tag{13.17}$$

If B_u is very different from the observed variance of R_q, the variation in stage u

survival would seem to be important to the population dynamics. As with A_j, it is recommended to express the B_j values as percentages of $Var(R_q)$.

Suppose now that the results are related from one generation to the next. Although this relationship could occur in numerous ways, we consider only one. Specifically, assume that the number in stage 1 in generation i is obtained by multiplying the number entering the last stage in generation $(i - 1)$ by a constant C. Typically, this constant is the average or maximum number of eggs produced per adult. Thus, if N_{ij} is the number entering stage j in generation i, $N_{i1} = CN_{i-1,q}$ so that

$$R_{i1} = \log(N_{i1}) = \log(N_{i-1,q}) + \log(C) = R_{i-1,q} + \log(C). \quad (13.18)$$

Substituting this result into equation (13.13), we have that the relationship between the mean values of R_q in generations $(i - 1)$ and i is

$$E(R_{iq}) = \theta_0[E(R_{i-1,q}) + \log(C)] - \sum_{j=1}^{q-1} \theta_j \tau_j. \quad (13.19)$$

Therefore, if the population is stable, with $E(R_{iq}) = E(R_{i-1,q})$, this mean equals

$$E(R_q) = \frac{\theta_0 \log(C) - \sum_{j=1}^{q-1} \theta_j \tau_j}{1 - \theta_0}, \quad (13.20)$$

as long as θ_0 does not equal 1.

To compute the variance of R_q, first notice that $Var(R_{i1}) = Var(R_{i-1,q})$ because R_{i1} and $R_{i-1,q}$ only differ by a constant, $\log(C)$. Then, from equation (13.14), the variation of R_q from one generation to the next may be expressed as

$$Var(R_{iq}) = \theta_0^2 Var(R_{i-1,q}) + \sum_{j=1}^{q-1} A_j, \quad (13.21)$$

where A_j is as defined by equation (13.15). Now if the variance is stable so that $Var(R_{iq}) = Var(R_{i-1,q}) = Var(R_q)$,

$$Var(R_{iq})(1 - \theta_0^2) = \sum_{j=1}^{q-1} A_j. \quad (13.22)$$

Because both sides of the above equation must be positive, this requires $\theta_0^2 < 1$. If this occurs, the stable variance is

$$Var(R_q) = \sum_{j=1}^{q-1} \frac{A_j}{1 - \theta_0^2}. \tag{13.23}$$

If $\theta_0^2 > 1$, the population is undergoing an explosion, and the concept of key factor has no meaning.

If $\theta_0^2 < 1$, analysis continues in a manner similar to the case in which the generations were unrelated. The relative importance of the random variation in stage u in determining the long-term variance of R_q is represented by A_u as a percentage of ΣA_j. To determine whether stage u has a strong influence in the variation in later stages by reducing the variation produced in earlier stages, the survival rate in stage u is first made constant by changing δ_u and $Var(\epsilon)$ to zero. The resulting expected change in ΣA_j is given by equation (13.17) with $A_0 = 0$. Again, this change is represented as a percentage of ΣA_j.

Whether the generations are directly related or not, we must first estimate the δ and $Var(\epsilon)$ values to detect the key factors. Look again at equation (13.9). Notice that for a given j, there are G k_j values and G R_j values, one for each generation. Using linear regression and assuming G generations, the estimate of δ_j is

$$\hat{\delta}_j = \sum_{i=1}^{G} \frac{(k_{ij} - \bar{k}_j)(R_{ij} - \bar{R}_j)}{\sum_{i=1}^{G} (R_{ij} - \bar{R}_j)^2}, \tag{13.24}$$

where k_{ij} and R_{ij} are values for stage j in generation i, and $\bar{k}_j$ and $\bar{R}_j$ are mean values for the same stage. The estimate of τ_j is

$$\hat{\tau}_j = \bar{k}_j - \hat{\delta}_j \bar{R}_j. \tag{13.25}$$

For equation (13.25) to hold for data estimates, the estimates of $Var(\epsilon_j)$ and $Var(R_j)$ must be

$$s_{\epsilon_j}^2 = \frac{1}{G - 1} \sum_{i=1}^{G} (k_{ij} - \hat{\tau}_j - \hat{\delta}_j R_{ij})^2 \tag{13.26}$$

and

$$s_{R_j}^2 = \frac{1}{G - 1} \sum_{i=1}^{G} (R_{ij} - \bar{R}_j)^2, \tag{13.27}$$

respectively. Using these estimates, equations (13.11) to (13.17) are evaluated to provide estimates when generations are not directly related. Equations (13.20) and (13.23) are used instead of equations (13.13) and (13.14) when the generations are directly related; otherwise, the analysis proceeds as in the earlier case.

Example 13.6

Once again, look at the tawny owl data from the two previous examples. Because of the emigration and immigration that occurs between breeding season, the results for different years are not directly related. Therefore, we use the first approach as outlined above. This is done by pressing *Manly Key Factor* in the *Key Factor* section as in the earlier example. The option of *No Direct Relationship Between Generations* should be selected. The user is asked for the name to be given to the output data set. The results displayed in Table 13.8 are presented on the screen and saved in an ASCII file in the output data set specified by the user. On each line of the output data set, the estimated variables are listed in the following order: stage, $\hat{\tau}$, $\hat{\delta}$, s_ϵ^2, $\hat{\theta}$, A, $A\%$, B, $B\%$, and s_R^2. Periods are used when a particular variable cannot be computed for a given stage.

The key factor(s) is(are) identified by observing the $A\%$ and $B\%$ columns in Table 13.8. Remember that A_0 is the contribution of the variance of R_1, and A_j ($j > 0$) quantifies the contribution of the variance of ϵ_j to the variance of R_q. Also, B_j is the amount the variability of the last stage $[Var(R_q)]$ would be expected to change if stage j was not present. By considering the A_j and B_j values as percentages of $Var(R_q)$, we are quantifying the relative importance of the random variation in stage j in determining the variation in the last stage and in reducing the variation in the last stage, respectively. Based on this approach, k_5 is observed to be the first key factor. If it were not present, the variation in R_q (the logarithm of the number in the last stage) would be increased an estimated 248%. It is also responsible for an estimated 48% of the variation in the last stage. Because B_5 is such a large percentage, we have strong evidence of the density-dependent response that we have seen hints of in the earlier examples. However, k_1 is the second key factor and shows a density-dependent reaction that we have not observed until now. Therefore, the owlets that emigrate or die after fledgling and

Table 13.8. Estimates of Parameters from Manly's Key Factor Analysis of Southern's (1970) Tawny Owl Data

Stage	$\hat{\tau}$	$\hat{\delta}$	s_ϵ^2	$\hat{\theta}$	A	$A\%$	B	$B\%$	s_R^2
0				0.205	0.000	1.0	0.033	99.0	
1	−0.863	0.567	0.054	0.474	0.012	36.2	0.023	68.2	0.008
2	0.349	−0.148	0.002	0.413	0.000	1.3	0.030	89.8	0.055
3	0.350	0.010	0.020	0.417	0.003	10.3	0.030	90.4	0.076
4	−0.102	0.090	0.005	0.458	0.001	3.3	0.036	106.8	0.094
5	−0.406	0.542	0.016	1.000	0.016	48.0	0.083	248.2	0.083
6									0.033

After Manly (1990).

the loss from a failure of pairs to breed are the key factors in the population dynamics of the tawny owl; both are density dependent.

For the Himalayan thar in Examples 13.1 and 13.2, constructing the life table for one generation involved counting the number of thar born in a given season in that season and each subsequent year. We have not considered how the tawny owl life table was constructed for each generation. However, once presented with the life tables of several generations, key factor analysis is an effective means of determining which stage(s) is(are) most important to the population dynamics of the species. How can sample data be used to estimate the vital life table information? This question will now be the focus of our attention.

Multi–cohort Stage-Frequency Data

Because methods used to model multi–cohort data can also be used to model single cohort data, we cover the multi–cohort models first. Although three methods of analysis (the Kiritani–Nakasuji–Manly method, the Kempton method, and the Bellows and Birley model) are considered, only the first is reviewed in detail. The following example provides a basis for our discussion.

Example 13.7

Qasrawi (1966) reported on a study of the grasshopper *Chorthippus parallelus* Zett. carried out on East Budleigh Common, Devon, during 1964 and 1965. The study region was a 3,500-square-meter area over which the soil, vegetation, and climatic conditions are reasonably homogeneous. The entire region was sampled in 1964 and 71% of the region in 1965. In late summer, the adult grasshoppers lay eggs which do not hatch until the following spring. The immature grasshopper passes through four instars (stages 1 to 4) before reaching the adult stage (stage 5). The adult female lays eggs, and both males and females die before the winter season. Qasrawi focused on estimating the energy flow through the grasshopper population. One of the requirements for this was knowledge of the population dynamics over the season. To gain this understanding, he conducted the study presented here.

A rectangular grid was pegged out on the ground. On each sampling date, a number of small areas totaling A_i were chosen at random from this grid. All live grasshoppers in the sample units were removed to the laboratory. There they were counted and assigned to instar and gender. On the first two sampling occasions (May 20 and 25), the sampling fraction was 0.00143; otherwise, it was 0.002. Therefore, the counts of 5 and 6 first instar grasshoppers found on May 20 and 25, respectively, were multiplied by 0.002/0.00143 ($= 1.4$) to adjust the counts for a consistent sampling fraction of 0.002. The adjusted data are presented in Table 13.9.

Table 13.9. Qasrawi's (1996) Grasshopper Data Collected During 1964 on East Budleigh Common, Devon

Date	Day	Instar 1	Instar 2	Instar 3	Instar 4	Adult	Total
May 20	4	7.0					7.0
25	9	8.4					8.4
29	13	14					14
June 3	18	10	1				11
10	25	7	5	1			13
15	30	1	10				11
18	33	1	8	1	1		11
22	37	3	8	4	2		17
25	40	7	12	6			25
29	44		7	6	6		19
July 2	47	1	1	6	4	1	13
9	54		1	3	2	1	7
13	58		4	4	4	5	17
16	61			1	3	2	6
20	65		1	1	5	6	13
24	69		1	1	2	5	9
27	72					6	6
31	75					6	6
August 5	81				1	6	7
11	87				1	1	2
14	90					3	3
19	95					3	3
24	100					5	5
28	104					3	3
Sept. 2	109					4	4
8	115					2	2
11	118					2	2
17	124					2	2
23	130					1	1

Notice that the data do not immediately give us the nice, neat life tables that we had for the Himalayan thar and the tawny owl. Grasshoppers entered the population over a period of time. Further, we cannot simply add the columns to estimate the total number within each life stage during the season because a grasshopper may be in the same life stage on more than one sampling occasion. In spite of these concerns, it is obvious that the population is moving through the various life stages during the sampling period. To get the values in the day column, calendar days were numbered sequentially with May 17 being Day 1. The reason for this will become evident in the next example.

In describing the methods below, frequent reference will be made to the stage frequency curve. Viewing the number in stage *j* or a higher stage as a function of time produces this curve. Because we have data collected at specific points in time, we are not able to produce the full curve. However, by plotting the observed frequencies as a function of time, we can get an idea of its appearance. This was done in Figure 13.3 for the last stage. The area under the stage frequency curve is an important quantity to estimate because it is related to the number entering the stage, the survival parameter, and the duration of the stage as we will see in describing the Kiritani–Nakasuji–Manly method.

Kiritani–Nakasuji–Manly Method

Originally, Kiritani and Nakasuji (1967) developed a method that required samples to be taken at regular time intervals throughout the generation. Further, unless an independent estimate of the number entering stage 1 was obtained, only survival probabilities could be estimated. Manly (1976) modified the model to remove these two restrictions. The resulting Kiritani–Nakasuji–Manly method is fairly simple. For this method to be used appropriately, three conditions must be satisfied:

1. The survival rate *per unit time* is constant for all stages during the entire sampling period.

2. Sampling begins before, at, or shortly after the time individuals begin entering the first stage.

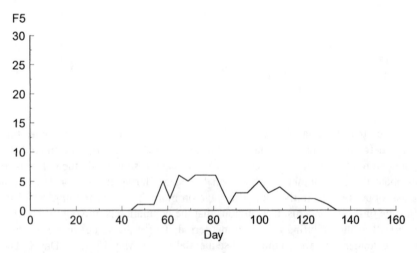

Figure 13.3. Stage–frequency curve for the last stage of Qasawri's (1966) 1964 grass-hopper data.

3. Population losses through migration are negligible.

The following notation is adopted:

q: number of stages

Parameters

$f_j(t)$:   number in the population in stage j for the fraction of the population sampled at time t

$F_j(t)$:   number in the population in stages $j, j + 1, \ldots, q$, for the fraction of the population sampled at time t

M_j: number entering stage j in the sampled fraction of the population

$g_j(x)$:   probability density function of entry to stage j

μ: mean time of entry to stage j

$e^{-\theta}$:   probability of surviving one time unit

a_j: duration of stage j

w_j: stage-specific survival rate for stage j

A_j: area under the stage-frequency curve for stage j

A_j^*:   area under the stage-frequency curve for stages $j, j + 1, \ldots, q$

D_j: area under the $tf_j(t)$ curve

D_j^*:   area under the $tF_j(t)$ curve

$p_j(t)$:   probability that an individual is in stage j at time t

$w(t)$:   probability of surviving to time t

$g_j(t)$:   the probability density function of the entry time to stage j

P_{ij}:   probability of an observation occurring in stage j at time t_i

$H_j(t)$:   probability that an individual spends t or less time units in stage j

$h_j(t)$:   probability that the duration of stage j is t time units

Statistics

f_{ij}:   sample stage frequencies

Notice that the phrase "for the fraction of the population sampled at time t" in the definition of $f_j(t)$ and $F_j(t)$. This accounts for the fact that this value is based on a sample estimate. Yet, we will be treating it as a parameter during the modeling process. We will return to this issue when assessing the fit of the model.

The individuals in stage j at time t are those that enter between time $t - a_j$ and t and survive until time t. It is enough to consider only the time since $t - a_j$ because all individuals in the stage prior to that time will have either died or moved to the next stage. Assuming the probability of survival in one time period is independent of survival in subsequent time periods, the probability of surviving $(t - x)$ time units is $e^{-\theta(t-x)}$. Further, the number entering stage j during the small time interval from x to $(x + dx)$ would be $M_j g(x)dx$. Therefore, we have in equation form

$$f_j(t) = M_j \int_{t-a_j}^{t} g_j(x)e^{-\theta(t-x)}dx. \tag{13.28}$$

Now the stage frequency curve is developed by plotting $F_j(t)$ against t for all values of time t.

The area under the stage frequency curve A_j is obtained by integrating the stage frequency curve from minus to plus infinity; that is,

$$A_j = \int_{-\infty}^{+\infty} f_j(t)dt = M_j\left(\frac{1 - e^{-\theta a_j}}{\theta}\right). \tag{13.29}$$

Therefore, the area under a stage frequency curve is related to the number entering stage j, the survival parameter θ, and the duration of the stage.

D_j, the area under the $tF_j(t)$ curve, is

$$D_j = \int_{-\infty}^{+\infty} tf_j(t)dt = A_j\left(\mu_j + \frac{1}{\theta} - \frac{a_j e^{-\theta a_j}}{1 - e^{-\theta a_j}}\right). \tag{13.30}$$

This indicates that D_j is related to the mean time of entry to stage j, the survival parameter θ, and the area under the stage frequency curve.

Equations (13.29) and (13.30) can be applied to accumulated stage frequency data. Let $F_j(t)$ be the stage-frequency curve for stages $j, j + 1, \ldots, q$. The duration of the last stage is infinite because all losses are through death. Therefore, putting $a_j = \infty$ in equations (13.29) and (13.30), we find A_j^*, the area under $F_j(t)$ and D_j^*, the area under the corresponding time-stage-frequency curve $tF_j(t)$, to be

$$A_j^* = M_j\theta \tag{13.31}$$

and

$$D_j^* = \left(\mu_j + \frac{1}{\theta}\right)A_j^*, \tag{13.32}$$

respectively.

From the two previous equations, we find the stage-specific survival rate for stage j to be

$$w_j = \frac{A^*_{j+1}}{A^*_j} = \frac{M_{j+1}}{M_j}, \qquad j = 1, 2, \ldots, q - 1; \tag{13.33}$$

the survival parameter to be

$$\theta = -\ln\left(\frac{\dfrac{A^*_q}{A^*_1}}{\dfrac{D^*_q}{A^*_q} - \dfrac{D^*_1}{A^*_1}}\right); \tag{13.34}$$

the duration of stage j to be

$$a_j = -\frac{1}{\theta}\ln(w_j), \qquad j = 1, 2, \ldots, g - 1; \tag{13.35}$$

and the number entering stage j to be

$$M_j = A^*_j \, \theta, \qquad 1, 2, \ldots, q. \tag{13.36}$$

From the above, it is evident that estimation of A^*_j and D^*_j is key to obtaining estimates of the other population parameters. This is accomplished using the trapezoidal rule. Suppose that all stage frequencies are zero at times t_1 and t_n. Let F_{ij} be the sample frequency for the number in stage j or a higher stage at time t_i. Then F_{ij} is the sample estimate of $F_j(t_i)$. By the trapezoidal rule, the estimate of A^*_j is

$$\hat{A}^*_j = \frac{1}{2}\sum_{i=1}^{n-1}(F_{ij} + F_{i+1,j})(t_{i+1} - t_i). \tag{13.37}$$

Similarly, the estimate of D^*_j is

$$\hat{D}^*_j = \frac{1}{2}\sum_{i=1}^{n-1}(t_i F_{ij} + t_{i+1}F_{i+1,j})(t_{i+1} - t_i). \tag{13.38}$$

For equally spaced data, b_i, the estimated number entering stage I between the times of samples i and $i + 1$ and surviving until the time of sample $i + 1$, is estimated to be

$$b_i = F_{i+1,1} - e^{-\hat{\theta}(t_{i+1} - t_i)}F_{i1}, \tag{13.39}$$

where F_{i1} is the number of individuals in all stages in the ith sample and

$\exp[-\hat{\theta}(t_{i+1} - t_i)]$ is the estimated probability of surviving the time t_i to t_{i+1}. Some entries will die before a sample is taken so the number entering the population between the ith and $(i + 1)$th sample times should be higher than b_i. To correct for this, we use

$$b_i^* = b_i \frac{\theta(t_{i+1} - t_i)}{1 - e^{-\theta(t_{i+1} - t_i)}}, \qquad (13.40)$$

Here b_i is multiplied by the reciprocal of the probability that an individual survives until a sample is taken assuming that entry times to stage 1 are uniformly distributed between sample times. The estimates from (13.39) and (13.40) may be negative requiring further adjustments.

Now that estimators for the parameters have been obtained, our next task is to develop a measure of variability for these estimators. No equations have been developed for this purpose. However, estimates can be obtained through simulation. A simple way of doing this is to assume that these are counts arising from a Poisson distribution. A simulated set of data is generated by replacing each observed stage-frequency with a random variate from the Poisson distribution with mean equal to the observed value. Parameter estimates are computed for the simulated data. The process is repeated, usually 100 times. The observed variation in the estimates from these 100 data sets is used as the estimate of variability of the estimator.

The estimates of variability obtained through the simulation above incorporate only sampling variation. They do not address whether the model is appropriate. To test the validity of the model, another type of simulation must be conducted. Here the model itself is used to generate data. The numbers entering stage 1 at different times are approximated. Setting these values begins by computing the estimates b_i or b_i^* for each time period. Then values are grouped and entry times are spread over the periods using the estimates as a guide. Once this is done development and survival are projected forward through time. A plot of the resulting observed and expected values allows one to assess visually how well the model fits the data.

Although graphically comparing observed and expected values gives us a feel for the fit of the model, we need a more objective measure. To obtain one, we begin be simulating 100 data sets. However, now we use the expected values from the model as the means of the Poisson random variates. For each data set, we compute

$$G = \sum_{j=1}^{g} (f_{ij} - f_j(t_i))^2, \qquad (13.41)$$

where f_{ij} is the observed stage j frequency for time t_i and $f_j(t_i)$ is the expected stage *j* frequency at time t_i. We conclude that the model provides an adequate fit if G from the sample data set is within the range of G values calculated for each of the 100 simulated data sets. If G from the sample data set is larger than the values calculated through simulation, we conclude that the model does not fit. Notice that we have not used Pearson's chi-squared test statistic here as we have in earlier chapters. For Pearson's test statistic, each term is divided by the expected value. Here the expected value may be small or even zero. Division by these values could improperly inflate the value of the test statistic. G is called the deviance and is another common measure of fit.

Example 13.8

Consider again the grasshopper data in Example 13.7. For equations (13.37) and (13.38) to be valid, no grasshoppers should have been observed during the first and last sampling dates. A few were observed on these dates, but so few that one may feel comfortable adding zeros at a date preceding the first sampling date and after the last one. This was the approach adopted by Manly (1990) and the one we will use. Zeroes were added for May 16 and September 28. Considering May 16 as Day 0 of sampling, May 20 becomes day 4 as was given in the last example.

We should note here that Manly (1990) also used the KNM method to estimate population parameters based on these data. However, his analysis was based on observations equally spaced through time. Interpolation was used to obtain this equal spacing. Our results will differ slightly from his because our analysis is based on the original sampling dates.

The cumulative stage frequencies that are so important to modeling stage frequency data are given in Table 13.10. Therefore, $F_{58,1} = 17$ implies that 17 grasshoppers were in the first stage or a higher stage in the sample taken on day 58. The stage frequency curve $F_1(t)$ is shown in Figure 13.4. We can now estimate the density under the stage frequency curve using equation (13.37). ECOSTAT will do this for us. First, an ASCII file must be formed. The first line gives the number of stages q and the number of sampling periods G separated by one or more spaces. Each of the next G lines has the time of sampling and the number in each of the q stages again separated by spaces. It is assumed that the times are given sequentially and that for each time the stages are listed from 1 to q.

From the *Life Stage* menu, choose *Multi-Cohort*. Then press *Enter Data* and identify the ASCII file containing the data. Press *KNM* and choose the option *No Iterative Calculations* to estimate the parameters of the model. The results appear on the screen and in an output file that the user identifies. Each of the q lines of output represents a stage with estimated values presented in the following order: $\hat{A}*$, $\hat{D}*$, $\hat{B}*$, $e^{-\hat{\theta}}$, $\hat{w}_j$, $\hat{a}_j$, and $\hat{M}_j$. Notice that $e^{-\hat{\theta}}$ is an estimate of the daily survival rate. Because this is assumed to be constant for all stages, it is only given for

Table 13.10.   Stage Frequencies and Cumulative Stage Frequencies for Qasrawi's (1966) Grasshopper Data Collected in 1964

Day	Stage frequencies					Cumulative stage frequencies				
	1	2	3	4	5	1	2	3	4	5
0	0					0				
4	7.0					7.0				
9	8.4					8.4				
13	14					14	0			
18	10	1				11	1	0		
25	7	5	1			13	6	1		
30	1	10	0			11	10	0	0	
33	1	8	1	1		11	10	2	1	
37	3	8	4	2		17	14	6	2	
40	7	12	6	0		25	18	6	0	
44	0	7	6	6		19	19	12	6	0
47	1	1	6	4	1	13	12	11	5	1
54		1	3	2	1	7	7	6	3	1
58		4	4	4	5	17	17	13	9	5
61		0	1	3	2	6	6	6	5	2
65		1	1	5	6	13	13	12	11	6
69		1	1	2	5	9	9	8	7	5
72				0	6	6	6	6	6	6
75				0	6	6	6	6	6	6
81				1	6	7	7	7	7	6
87				1	1	2	2	2	2	1
90					3	3	3	3	3	3
95					3	3	3	3	3	3
100					5	5	5	5	5	5
104					3	3	3	3	3	3
109					4	4	4	4	4	4
115					2	2	2	2	2	2
118					2	2	2	2	2	2
124					2	2	2	2	2	2
130					1	1	1	1	1	1
134					0	0	0	0	0	0

stage 1 and represented with a period for other stages. Because $\hat{w}_j$ and $\hat{a}_j$ cannot be estimated for stage q, periods are used for those values as well. First consider the estimates of A^*, D^*, and B^*, given in Table 13.11.

Recall that $\hat{A}^*$ is the area under the stage frequency curve. Look again at the observed stage 1 and 5 frequency curves plotted in Figures 13.3 and 13.4, respectively. For emphasis, we repeat that a valid measure of area under the curves requires the first and last values to be zero. That is why we added zeroes for the

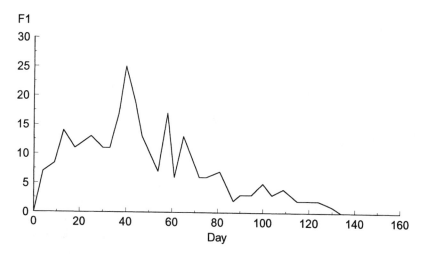

Figure 13.4. Stage–frequency curve for the first stage of Qasawri's (1966) 1964 grass-hopper data.

first and last dates. The estimates of the remaining parameters are given in Table 13.12.

Notice that the information in the last column of Table 13.12 (the number entering each stage) compiled over several years is essential for the key factor analysis discussed earlier. However, other characteristics of the species are also estimated. For example, about 3.5% of the population dies each day. The probability of surviving a stage ranges from 0.70 to 0.75. Stages are estimated to last from 8.2 days for the 3rd stage to 10.1 days for the second day.

The assumption of a constant survival rate can be assessed by regressing the natural logarithm of total frequency against time for those points at or after the peak (*see* Figure 13.5). In our case, the total sample frequencies peaked at 25 on day 40 (*see* Table 13.9). Therefore, the points for days 40 to 130 were included

Table 13.11. Estimates of A*, D*, and B* for Qasrawi's (1966) Grasshopper Data Collected in 1964

j	$\hat{A}*$	$\hat{D}*$	$\hat{B}*$
1	1055.8	51,358.7	48.64
2	774	46,104.5	59.57
3	543.5	37,483	68.97
4	406.5	31,107.5	76.53
5	282	24,162	85.68

Table 13.12. Estimates of Parameters θ, w₁a₁, *and* M₁ *for Qasrawi's (1966)*
Grasshopper Data Collected in 1964

j	$\hat{\theta}$	$\hat{w}$	$\hat{a}$	$\hat{M}$
1	0.965	0.733	8.711	38
2		0.702	9.919	28
3		0.748	8.148	19
4		0.694	10.259	14
5				10

in the regression. Based on the regression, the r^2 is 0.83, and the slope is -0.029. This yields an estimate of $e^{-0.029} = 0.973$ as the constant survival rate, a value that is quite close to the model estimate of 0.965. Thus, the assumption of a constant survival rate seems reasonable.

The model has been fit to the data and estimates of parameters have been obtained. How much variability is there in the estimates? Does the model fit? To address these questions with the help of ECOSTAT, press the *Fit* button. You are asked to identify two data sets. The first will have the estimated stage frequencies for each time period based on the model. The second will have estimated means and standard deviations of some of the estimated parameters.

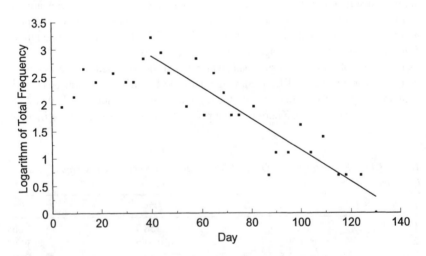

Figure 13.5. Total sample frequencies for the plotted against time for Qasawri's (1966) 1964 grasshopper data. The regression line was fitted to the log frequencies after the peak and is used to assess the validity of the assumption of constant survival.

First, to derive the estimated stage frequencies for each time period, the adjusted estimates of the number entering stage 1 for each time period (b_i^*) are estimated. These may be negative. To obtain nonnegative estimates that sum to the estimated number entering the stage (37.5), it is first assumed that no entries are made after the time for which entries were observed in the sample. For our data, the last nonzero value for stage 1 was observed at 49.5 days, so all the b_i^* values after 49.5 are estimated to be zero. To avoid changing the total, all estimated b_i^* values at these later times are added to the $b_{49.5}^*$. If any estimated b_i^* for the remaining sampling times is less than zero, its value is added to the b_i^* of the preceding time period and then set to zero. For these data, the results are in Table 13.13.

This algorithm gives differing values from those of Manly (1990) but satisfies the same conditions of always having nonzero values that sum to the estimated number entering the stage. It is further assumed here that all new entries occur in the center of the interval even though it would probably be more realistic to spread the entries over the interval (*see* Manly, 1990). Once the entries have been determined, the estimated model parameters are used to project the population forward through time. For example, the number alive on day ($i + 1$) is 0.965 times the number alive on day i. All those entering stage 1 at a given time and surviving 8.9 days enter stage 2. The projected numbers for each sample time are displayed on the screen, in the first output data set named, and in Table 13.14.

To measure the fit of the model, G is computed as in equation (13.41) using the observed values f_{ij} and the estimated values $f_j(t_i)$, as given in Table 13.14. From ECOSTAT, we have $G = 807.08$. To gain insight into the information this value provides on the fit of the model, 100 data sets are simulated. For each data set, the estimated frequencies in the table above are taken as the population values. For each, a Poisson random variate is generated from a distribution with mean that of the estimated frequency. Once this is done for all times i and stages j, G is computed for these simulated values. This process is completed 100 times. The 100 G values from the simulated data sets ranged from 145.33 to 419.61 with an average of 242.1. The observed 834.18 is significantly larger, indicating that either the model is inappropriate or heterogeneity is present. The heterogeneity factor is estimated as $H = 834.18/242.1 = 3.45$. The observed and estimated stage frequencies are shown in Figure 13.6. These suggest that the model may not be appropriate.

To obtain a measure of the precision with which the parameters of the model are estimated, another type of simulation is conducted. Here the observed stage frequencies at each time are taken as the population values. For each, a Poisson random variate with mean equal to that of the observed stage frequency is generated. The analysis is conducted on the simulated data. The process is repeated 100 times. The mean and variance of the parameter estimates from the simulated data are computed and displayed on the screen. They are also stored in the second output data set named. For this example, the results are shown in Table 13.15.

Table 13.13. Estimtaed Entries to Stage 1 for Qasrawi's (1966) 1964 Grasshopper Data

Time	b_i	b_i^*	b_i for simulation
0–1	7.00	7.51	7.51
1–2	2.54	2.78	2.78
2–3	6.72	7.21	6.43
3–4	−0.71	−0.78	0.00
4–5	4.43	5.00	5.00
5–6	0.12	0.13	0.13
6–7	1.12	1.18	1.18
7–8	7.46	8.01	8.01
8–9	9.72	10.25	6.58
9–10	−2.68	−2.87	0.00
10–11	−4.07	−4.29	0.00
11–12	−3.13	−3.54	
12–13	10.93	11.73	
13–14	−9.28	−9.78	
14–15	7.80	8.37	
15–16	−2.27	−2.44	
16–17	−2.09	−2.20	
17–18	0.61	0.64	
18–19	2.16	2.39	
19–20	−3.65	−4.06	
20–21	1.20	1.27	
21–22	0.49	0.53	
22–23	2.49	2.72	
23–24	−1.34	−1.43	
24–25	1.49	1.63	
25–26	−1.23	−1.37	
26–27	0.20	0.21	
27–28	0.39	−0.68	
28–29	−0.61	−0.59	
29–30	−0.87	−0.93	

Notice that all the values agree well with the original estimates based on the data. This differs from Manly (1990) whose simulations indicated that the numbering entering stage j from the simulated data tended to be about 10% less than that estimated from the original observed stage frequencies.

KNM Iterative Method

One of the limitations of the KNM method is that sampling is conducted at frequent intervals throughout the generation. If the first sample has individuals in the second or a higher stage or if the last sample has a sizeable number of indi-

Table 13.14. Model Projections of the Stage Frequencies Based on the Estimated Parameters and Entries in Stage 1 for Qasrawi's (1966) Grasshopper Data

Time	Stage				
	1	2	3	4	5
0	0				
4	6.99				
9	8.39				
13	8.19	5.07			
18	5.01	6.09			
25	4.42	5.34	3.31		
30	3.82	0.00	4.47	2.77	
33	1.22	3.32	4.01	2.49	
37	8.42	2.98	2.54	3.09	
40	13.80	2.67	0.00	3.13	1.94
44	5.41	6.64	2.24	1.98	2.41
47	4.86	5.90	2.08	1.78	2.17
54		3.79	4.65	1.57	3.08
58			7.27	1.41	2.67
61			6.12	0.45	3.62
65			2.56	3.10	3.17
69				4.60	3.06
72				4.13	2.75
75				1.79	4.40
81					5.00
87					4.03
90					3.62
95					3.03
100					2.54
104					2.20
109					1.84
115					1.49
118					1.34
124					1.08
130					0.87
134					0.76

viduals still alive, the simple expediency of adding zeroes before the first sampling date and after the last sampling date does not yield satisfactory results. To address this issue, Manly (1985) proposed an iterative version of the KNM. The iterative approach adjusts the estimates of A^* and D^*, so that only the individuals that enter the population between times t_1 and t_n are included in the estimation process. To do this, the expected contributions to A^* and D^* from individuals who are in the population at time t_n are added in, taking into account the stage duration and

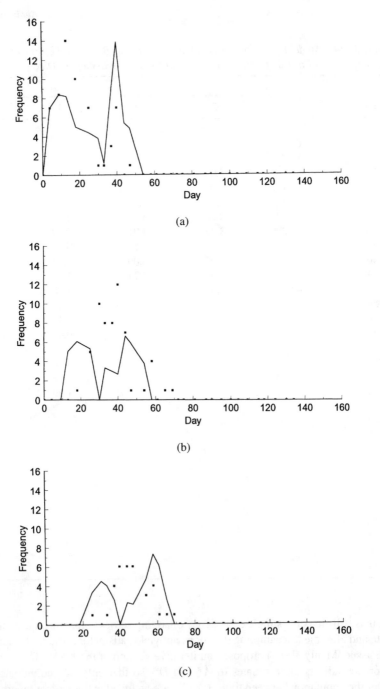

Figure 13.6. Observed (■) and expected (−) frequencies for the KNM model fitted to the grasshoppers in (a) instar 1, (b) instar 2, (c) instar 3, (d) instar 4, and (e) the adult stage for Qasawri's (1966) 1964 grasshopper data.

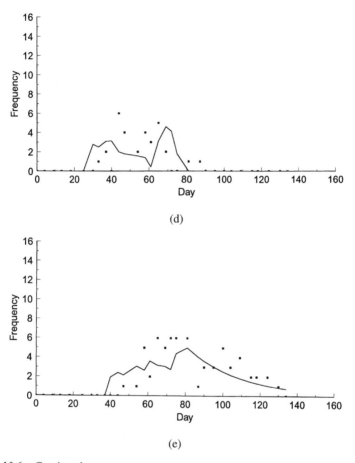

(d)

(e)

Figure 13.6. Continued.

daily survival rates. This adjusts for stopping sampling too early. Again taking into account stage duration and daily survival rates, the expected contributions in A^* and D^* are adjusted for individuals present in the population before time t_1. Because these adjustments depend on unknown parameters, an iterative approach is needed. The approach is as follows:

1. Obtain estimates of the parameters using equations (13.33) to (13.38), ignoring the fact that sampling began after individuals were in the population and/or ended before all were dead.

2. Using parameter estimates where needed, adjust the estimates of A^* and D^*, using

Table 13.15. Estimated Mean and Variance of Parameter Estimates for the KNM Model Applied to Qasrawi's (1966) Grasshopper Data[a]

Stage j	$e^{-\theta}$		$\hat{w}_j$		$\hat{a}_j$		$\hat{M}_j$	
	Mean	SD	Mean	SD	Mean	SD	Mean	SD
1	0.965	0.003	0.739	0.026	8.55	1.15	37.71	3.35
2			0.705	0.032	9.86	1.32	27.86	2.64
3			0.755	0.041	7.93	1.50	19.65	2.02
4			0.691	0.051	10.38	1.65	14.81	1.49
5							10.21	1.06

[a]The estimates are based on 100 simulations using the observed stage frequencies as the population parameters for the Poisson distribution.

$$\hat{A}_j = \hat{A}_j^* + \frac{1}{\theta}\left(F_{nj} - F_{1j} + \sum_{k=1}^{j-1}(f_{nk} - f_{1k})e^{-\theta l_{kj}}\right) \tag{13.42}$$

and

$$\hat{D}_j = \hat{D}_j^* + \frac{1}{\theta^2}(F_{nj}(t_n\theta + 1) - F_{1j}(t_1\theta + 1))$$
$$+ \frac{1}{\theta^2}\sum_{k=1}^{j-1}e^{-\theta l_{ki}}(f_{nk}[(t_n + l_{kj})\theta + 1]$$
$$- f_{1k}[(t_1 - l_{kj})\theta + 1]). \tag{13.43}$$

In the above, l_{kj} is the average time for an individual in stage k to enter stage j ($k < j$). Assuming that the individuals are halfway through stage k, $l_{kj} = a_k/2 + a_{k+1} + a_{k+2} + \ldots + a_{j-1}$.

3. Population parameters are recalculated using the corrected values $\hat{A}_j'$ and $\hat{D}_j'$.

4. If the revised estimates are very different from the previous estimates (0.001 or more in ECOSTAT) from the previous estimates, then the process returns to step (2); otherwise, iteration stops.

Notice that $\hat{A}_j'$ and $\hat{D}_j'$ are obtained each time by adjusting $\hat{A}_j^*$ and $\hat{D}_j^*$, respectively, not by adjusting the last estimates of these quantities.

Example 13.9

Once more we consider Qasrawi's (1966) grasshopper data. However, now suppose that the last sampling date was July 31, 1964. The numbers in the adult stage had peaked, but a number were still alive so that we are no longer com-

Table 13.16.   Estimates of Parameters θ, $w_j a_j$, and M_j for Qasrawi's (1966)
Grasshopper Data Collected Through August 14, 1964

j	$\hat{\theta}$	$\hat{w}$	$\hat{a}$	$\hat{M}$
1	0.971	0.737	10.460	31
2		0.709	11.824	23
3		0.756	9.621	16
4		0.731	10.765	12
5				9

fortable simply adding a zero to the end. The iterative method is now appropriate. To use ECOSTAT, follow the same process as for the noniterative KNM method except select the *Iterative Calculations* option. The results are given in Table 13.16.

Compare these results to those in Example 13.8 that were based on the full data set. Here the estimated daily and stage-specific survival rates are higher. Estimated stage durations are longer, and fewer are estimated to be entering each stage, except the last one. The impact on the estimates tend to decrease as the proportion of the generation viewed through sampling increases. The same measures of fit as in the noniterative approach can also be taken here. Table 13.17 gives the results of the 100 simulations using the observed frequencies as the means of the Poisson variates to obtain estimates of the means and standard deviations of the estimates. From ECOSTAT, we have $G = 745.29$ for the fitted model. The G-values from the simulated data sets ranged from 105.79 to 365.67 and averaged 200.08. The heterogeneity factor is estimated as $745.29/200.08 = 3.72$. The observed and estimated stage frequencies are shown in Figure 13.7.

Table 13.17.   Estimated Mean and Variance of Parameter Estimates for the KNM
Model Applied to Qasrawi's (1966) Grasshopper Data Collected Through July 31,
1964[a]

Stage j	$e^{-\hat{\theta}}$ Mean	SD	$\hat{w}_j$ Mean	SD	$\hat{a}_j$ Mean	SD	$\hat{M}_j$ Mean	SD
1	0.971	0.004	0.738	0.029	10.69	2.11	31.44	5.18
2			0.709	0.034	12.04	1.94	23.19	3.77
3			0.762	0.045	9.45	1.65	16.40	2.50
4			0.736	0.046	10.63	1.49	12.42	1.51
5							9.09	0.74

[a]Estimates are based on 100 simulations using the observed stage frequencies as the population parameters for the Poisson distribution.

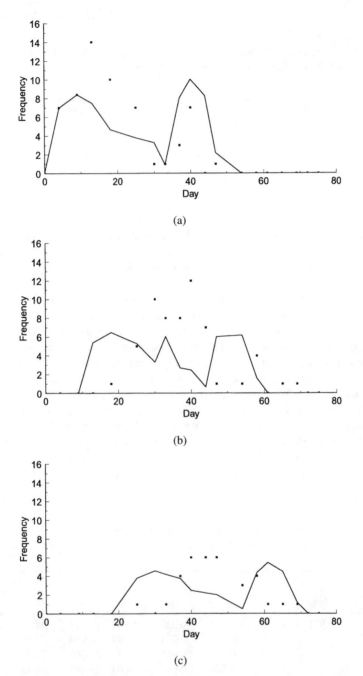

Figure 13.7. Observed (■) and expected (−) frequencies for the KNM model fitted to grasshoppers in (a) instar 1, (b) instar 2, (c) instar 3, (d) instar 4, and (e) the adult stage for Qasawri's (1966) grasshopper data, assuming that sampling terminated on July 31, 1964.

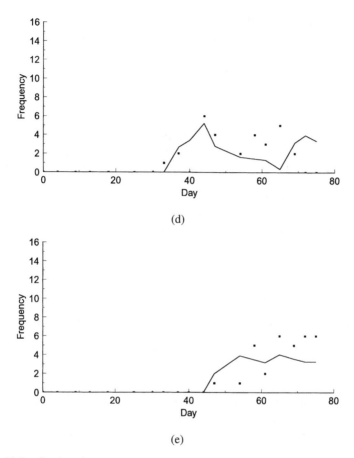

Figure 13.7. Continued.

Again the results are similar to those obtained for the full data set. Not surprisingly, the largest changes occurred in the last stage that was not sampled completely.

It should be noted that the iterative method does not always converge; that is, it is not always possible to iterate until the parameter estimates do not change very much from one iteration to the next. However, the method does tend to work well when sampling terminates too soon (Manly, 1990). Convergence is more likely to become an issue if sampling begins too late so that some individuals are already in stage 2.

The assumptions for the KNM method are restrictive. Other methods relax some of these conditions. The Kempton (1979) method allows for time-dependent survival rates and permits a distribution, such as the gamma, to model stage duration.

The Bellows and Birley (1981) model allows for a different survival rate for each stage as well as a distribution for stage durations. The latter two models require assumptions about the form of the distribution of entry times into stage 1. The Kempton and Bellows and Birley models will be described briefly.

Kempton Method

Read and Ashford (1968) presented the first truly stochastic model for stage frequency data, and this was discussed further by Ashford et al. (1970). Assuming an Erlangian distribution for the times spent in different stages, they used maximum likelihood estimators to fit models to a couple of data sets. Their distributional assumptions led to computational difficulties. Kempton (1979) overcame these problems by modifying the assumptions slightly, and we discuss one of his three models here.

Kempton stated that models of life-stage data must include three components:

1. Survival rates
2. Entry times to stage 1
3. Duration of stages $1, 2, \ldots, q$

To develop Kempton's model assume that the survival per unit time is constant and that the probability of surviving to time t is

$$w(t) = e^{-\theta t}, \qquad t > 0. \tag{13.44}$$

Suppose that the distribution of the time to entry to stage 1 is gamma with parameters r_0 and λ and that the distribution of stage duration for stage j is gamma with parameters r_j and λ, for $j = 1, 2, \ldots, q$. Further assume these gamma distributions are independent. Notice that there is a constant scale parameter λ. This is largely for computational ease because the sum of independent gammas with a constant parameter λ is gamma with that same parameter λ and r equal to the sum of the individual r_i values (*see* Chapter 1). Therefore, the time to entry to stage j will follow a gamma distribution with parameters c_j and λ where

$$c_j = k_0 + k_1 + \ldots + k_{j-1}. \tag{13.45}$$

The average time of entry to stage j will be the mean of the distribution which is $\mu_j = c_j/\lambda$. Then the mean duration of stage j is $a_j = r_j/\lambda$. Therefore,

$$\mu_j = \mu_1 + a_1 + a_2 + \ldots + a_{j-1}; \tag{13.46}$$

that is, the mean entry time to stage j is equal to the mean entry time to stage 1 and the average time spent in the first $(j - 1)$ stages.

In general, the probability that an individual is in stage j at time t is

$$p_j(t) = w(t) \int_0^t (g_j(y) - g_{j+1}(y))dy, \qquad j < q, \qquad (13.47)$$

and

$$p_q(t) = w(t) \int_0^t g_q(y)dy, \qquad (13.48)$$

where t is measured relative to the process starting time of zero, $w(t)$ is the probability of surviving to time t, and $g_j(t)$ is the probability density function (pdf) of the stage j entry time. Here we are assuming that $g_j(t)$ is a gamma pdf with parameters c_j and λ. Notice that the integral from 0 to t of $g_j(y)$ gives the probability that an individual has entered stage j by time t. Similarly, the integral from 0 to t of $g_{j+1}(y)$ gives the probability that an individual has entered stage $(j + 1)$ by time t. The difference gives the probability that an individual has entered but not left stage j by time t. Multiplying this by the probability of surviving to time t gives the desired probability.

Some implicit assumptions are made in the above computations. The first assumption is that the probability of survival depends only on age and not on development rate. Further, it is assumed that the time spent in a stage is independent of time spent in previous or subsequent stages. Therefore, if it is known that an individual passes quickly through the first stage, this provides no additional information on the survival probability for that individual or on the time it will take that individual to pass through subsequent stages. Although it seems more likely that individuals who are slow (or quick) to develop in one stage will be slow (or quick) to develop in other stages, this is not allowed for in the model. Kempton recognized this but noted that seldom would the available data permit this feature to be accounted for when fitting models.

Now the problem becomes one of estimating the gamma scale parameter λ, the mean time of entry to the first stage μ_1, the survival parameter θ, and the durations of stages 1 to $(q - 1)$, $a_1, a_2, \ldots, a_{q-1}$. This assumes that the process starting time t_0 is known so that the sample time $t_1, t_2, \ldots, t_n$ can be taken relative to this. If t_0 is not known, then some realistic point before the first sample should be taken. The choice of starting time is probably not too crucial when estimating stage durations or survival rates (Munholland, 1988).

To fit the model, Kempton assumed that the sample count of the number of individuals in stage j at time t_i is a random variate from a Poisson distribution with mean $M_0 p_j(t_i)$, where M_0 is the total number of individuals at time 0. He then developed the likelihood equations which he used to obtain the MLE's of the model parameters. Manly (1990) noted that under this assumption, the distribution of the stage frequencies conditional on the observed total frequency across all samples is multinomial and used an alternative approach to estimate the model parameters. The software MAXLIK (Manly, 1990) implements the latter approach.

In either case, time could be measured on a calendar or physiological time scale. It is also possible to use one time scale for stage duration and the other for the survival function.

Kempton (1979) also considered the Weibull and lognormal distributions for the survival function and the normal distribution for the distribution of stage duration. These are further discussed by Manly (1990).

Bellows and Birley Model

The Bellows and Birley model (1981) is an extension of a discrete time model first proposed by Birley (1977, 1979). It permits a different survival rate for each stage. Either the distribution of entry times to stage must be known or it must be estimated assuming its parametric form is known. Bellows and Birley provided equations for the development of the individuals in a single cohort entering stage j at the same time, and Manly extended these to cover the development of several cohorts entering a stage at different times.

A foundational equation of the model is

$$f_j(t) = M_1 \sum_{i=0}^{t} P_j(i)\phi_j^{t-1}[1 - H_j(t - i)]. \tag{13.49}$$

To understand this equation, notice that by multiplying the total number of individuals entering stage 1, M_i, by the proportion of those individuals entering stage j at time i, $P_j(i)$, we have the number of individuals in stage j. These will still be in stage j at time t if (1) they survive with probability ϕ_j^{t-i} and (2) do not pass into stage $(j + 1)$ with probability $1 - H_j(t - i)$. Assuming that survival is independent of stage duration, we obtain the results in equation (13.49). Notice here that $H_j(t)$ is the probability that an individual is in stage j for t or less time units. Now, if $J_j(t)$ denotes the probability of a stage duration of t time units, then

$$H_j(t) = h_j(0) + h_j(1) + \ldots + h_j(t). \tag{13.50}$$

Another basic equation for this model computes the proportion of the population entering stage $(j + 1)$ at time t as follows:

$$P_{j+1}(t) = \sum_{t=0}^{t} P_j(i)\phi_j^{t-i}h_j(t - i). \tag{13.51}$$

The ith term is the product of $P_j(i)$ (the proportion entering stage j at time i), ϕ_j^{t-i} (the probability of surviving the time interval from i to t), and $h_j(t - i)$ (the probability of staying in stage j for $(t - i$ time units).

To complete the model, the parametric forms of the distribution of stage duration and of the distribution of entry times to stage 1 must be specified. Following Manly (1990), we assume the Weibull distribution in each case. Therefore,

$$H_j(t) = 1 - e^{-(t/Q_j)^{\alpha_j}} \qquad (13.52)$$

and

$$P_1(t) = e^{-[(t-1)/Q_0]^{\alpha_0}} - e^{-(t/Q_0)^{\alpha_0}}, \qquad (13.53)$$

for $i = 1, 2, \ldots, t$ and $P_i(0) = 0$. Here Q_0 and Q_1 are scale parameters that need to be estimated.

The stage-specific survival rate for stage j is

$$w_j = \frac{\sum\limits_{i=1}^{n} P_{j+1}(i)}{\sum\limits_{i=1}^{n} P_j(i)}. \qquad (13.54)$$

Sampling must continue long enough for $P_{j+1}(n)$ to be estimable. Then estimates of w_j may be obtained by using estimates in the right-hand side of equation (13.54). The number entering stage j is

$$M_j = M_1 w_1 w_2 \ldots w_{j-1}. \qquad (13.55)$$

As with the Kempton model, this model can be fit using MAXLIK (Manly 1990).

Single Cohort Stage-Frequency Data

In the previous section, we considered multicohort data; that is, the individuals were entering the population over an extended period of time. Here we assume that the individuals represent a single cohort that enters the population at a single point in time. It would seem that if one did not have to account for individuals entering at various points in time that the modeling of the data would be easier. This is, in fact, the case. We need to decide whether the assumption that the sample stage frequencies are proportional to the population stage frequencies is valid. If so, we can estimate survival rates and distribution of stage duration as before. If not, we can estimate the distribution of stage duration but not survival rates. We consider only the first case.

Analysis Using Multi-Cohort Methods

Each of the methods used for analyzing multicohort data may be used with the single cohort data with perhaps some slight modifications. The Kiritani–Nakasuji–

Manly method may be used as before provided that the unit time survival rate is constant. With the single cohort, the entry distribution for stage 1 does not need to be estimated for either the Kempton or Bellows and Birley models. Further, the number entering stage 1, M_1, is often known and does not need to be estimated. Of the three, the Bellows and Birley (1981) model has the greatest flexibility and thus may be the most suitable for single cohort data. Notice that for this model, removing the entry time distribution to stage 1 is equivalent to setting $P_1(0) = 1$ and $P_1(i) = 0$ for $i > 0$ in equation (13.49).

Example 13.10

Bellows and Birley (1981) conducted a study of *Callosobruchus chinensis* (Linné), a bruchid pest of stored pulses. The adults of this species lay their eggs on the surface of dried peas or beans. Upon hatching, the larvae burrow into the cotyledon. The four larval stages are passed within a single bean, and pupation occurs in the cavity created by larval feeding. The adults emerge from the bean 1 or 2 days after eclosion. Although more than one larva can develop within a bean, there is some competition for space. In the Bellows and Birley study, adult *C. chinensis* were isolated with cowpeas (held in a large plastic box) for 24 hours at 30°C (86°F) and 70% relative humidity. After 24 hours, the adults were removed, and the peas returned to these experimental conditions. Subsequently, each day cowpeas were taken from the box. The number of hatched eggs on each pea was recorded. Then each pea was dissected to determine the number and stage of the *C. chinensis* present. Enough cowpeas were sampled so that at least 50 living individuals were recorded. During the time of the experiment, an average of around 70 hatched eggs were counted each day. Sampling continued until all the observed survivors were in the emerged adult stage. After investigating the possible effects of larval density, cowpeas with five and fewer hatched eggs were included in the life table development. The results of each day's sampling were presented in percentages and are given in Table 13.18.

Only a single cohort is present because all eggs were laid during a 24-hour period. Sampling ceased before all the emerged adults had died so the iterative version of the KNM method is needed. Although at day 0, we know that no eggs were present on the cowpeas, no mortality was considered in the egg stage. Thus only stages 2 to 7 are analyzed. ECOSTAT was used to estimate the model parameters using the iterative KNM method. The data are in eggs.dat. The output from ECOSTAT is presented in Table 13.19.

One obvious problem exists. The estimates of M_j are more than 100 for stages 2 to 6. Because the table values are percentages, these cannot be valid. To obtain valid estimates, the stage-specific survival rates were applied to the known number (100%) entering instar I. For example, the percentage entering instar III was estimated to be the percentage entering instar II (96.8) multiplied by the stage-specific survival rate for instar II (0.973), giving 94.2. These values are given in

Table 13.18.   Bellows and Birley (1981) Callosobruchus chinensis Data Collected from a Box of Cowpeas

| Day | Eggs | Larvae | | | | Pupae | Adults | |
		I	II	III	IV		In pea	Adults
1	100.00							
2	100.00							
3	100.00							
4	59.00	41.00						
5		96.77						
6		91.78						
7		23.08	71.79					
8		4.84	74.19	17.74				
9		1.05	57.89	40.00				
10		1.85	3.70	89.81	0.93			
11			4.76	41.67	48.81			
12			0.00	3.53	90.59			
13			1.16	0.00	96.30			
14				0.00	100.00			
15				1.16	82.56	15.12		
16					37.68	57.97		
17					10.61	74.24		
18					10.47	81.40		1.16
19					2.08	70.83	18.75	0.00
20						22.62	72.62	1.19
21						8.22	75.34	16.44
22						4.82	38.55	53.01
23							12.68	85.92
24							8.51	91.49
25							1.96	88.24
26								88.41

From Bellows and Dirley, 1981, by permission of *Researches in Population Ecology*

the "Adjusted $\hat{M}_j$" column of Table 13.19. The observed stage frequencies indicate the percentages entering stages 6, 7, and 8 should be about 90; however, the estimates of those entering these stages are below 90. This probably results from the low total frequencies (84.9 on Day 17, as an example) in the middle of the sampling period. It could also be due, in part, to the restrictive assumption of the KNM method that daily survival must be constant.

ECOSTAT can also be used to test the model for fit. G for the fitted model is 32,893.26. The G values from 100 simulated data sets ranged from 1079.38 to 3551.24 with an average of 2231.78. The heterogeneity factor is estimated to be $32,893.26/2231.78 = 14.74$. The observed stage frequencies and model based

Table 13.19.   Estimates of Parameters θ, w_j, a_j, and M_j for Bellows and Birley's (1981)
Callosobruchus chinensis Data

j	$\hat{\theta}$	$\hat{w}$	$\hat{a}$	$\hat{M}$	Adjusted $\hat{M}$
2	0.90	0.975	2.4	110	100.0
3		0.979	2.0	108	97.5
4		0.981	1.9	105	95.5
5		0.952	4.8	103	93.6
6		0.964	3.5	98	89.1
7		0.975	2.4	95	85.9
8				92	83.8

estimates are illustrated in Figure 13.8 and point to regions in which the model
fits poorly (and others where it fits well). Bellows and Birley (1981) fit their
model to the data and found the model fit well.

Nonparametric Estimation

Pontius et al. (1989a,b) developed a non-parametric approach that is applicable
when sampling begins when a cohort starts developing and continues until all
individuals are in the last stage. As before, sample times are $t_1, t_2, \ldots, t_n$. Define
the following terms:

$g_j(x)$:   probability density function of the entry time to stage j

$G_j(x)$:   probability that the entry time to stage j is after x

p_{ij}:   the proportion of individuals in the sample at time t_i that are not yet
             in stage j

μ_j: mean time to reach stage j

n_i: total sample frequency at time i

All individuals are in stage 1 at time t_0, which is prior to t_1, and they are in
stage q at time t_n. Given the pdf of the entry time to stage j, $g_j(x)$, the probability
that entry time to stage j is after x may be computed by integrating $g_j(x)$ from x
to infinity and will be denoted by $G_j(x)$. Then the mean entry time to stage j is

$$\mu_j = \int_0^\infty xg_j(x)dx = \int_0^\infty G_j(x)dx. \tag{13.56}$$

Using the trapezoidal rule to approximate this integral in (13.56), we have

$$\mu_j \approx \frac{1}{2} \sum_{i=0}^{n-1} [G_j(t_i) + G_j(t_{i+1})](t_{i+1} - t_i). \tag{13.57}$$

To obtain an estimate of μ_j, note that the proportion of individuals in the sample at time t_j that are not yet in stage i is an estimate of $G_j(t)$, with $p_{i1} = 0$, for all i. Then using m_j to denote an estimate of μ_j, we have

$$m_j = \frac{1}{2} \sum_{i=0}^{n-1} (p_{ij} + p_{i+1,j})(t_{i+1} - t_i) \tag{13.58}$$

which may be written as

$$m_j = \frac{1}{2} t_i + \frac{1}{2} \sum_{i=1}^{n-1} p_{ij}(t_{i+1} - t_{i-1}). \tag{13.59}$$

Assuming the sample proportions p_{ij} are binomially distributed and independent for different sample times, the variance of m_j may be estimated by

$$s_{m_j}^2 = \frac{1}{4} \sum_{i=1}^{n-1} \frac{p_{ij}(1 - p_{ij})(t_{i+1} - t_{i-1})^2}{n_i} \tag{13.60}$$

where n_i is the total sample frequency at time t_i.

The simplest of two equations given by Pontius et al. for estimating the variance of the entry time to stage j is

$$s_j^2 = \frac{1}{2} \sum_{i=0}^{n-1} (p_{ij} + p_{i+1,j})(t_{i+1}^2 - t_i^2) - \mu_j^2 \tag{13.61}$$

where the J is used to denote the entry time.

For this model, the duration of stage j can be estimated by

$$a_j = m_{j+1} - m_j. \tag{13.62}$$

Again assuming that proportions p_{ij} are binomially distributed, the variance of the estimated duration of stage j is estimated by

$$s_{a_j}^2 = \frac{1}{4} \sum_{i=1}^{n-1} \frac{(p_{i,j+1} - p_{ij})(1 - p_{i,j+1} + p_{ij})(t_{i+1} - t_{i-1})^2}{n_i}. \tag{13.63}$$

If the observed variability exceeds that for the binomial distribution, this estimate of the variance of a_j will under estimate the true variance. Assuming that the times spent in each stage are independent for each individual, the estimate of the variance of the duration of stage j is

$$s_j^2 = s_{J+1}^2 - s_J^2. \tag{13.64}$$

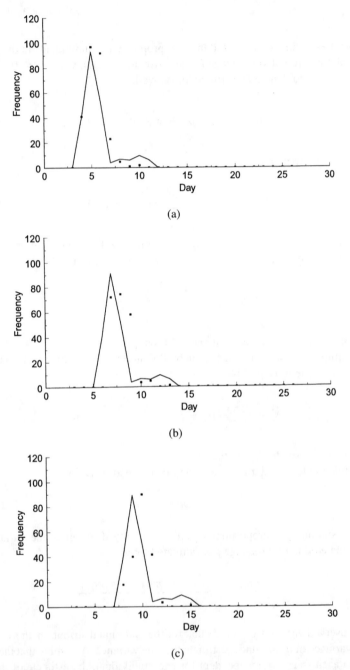

Figure 13.8. Observed (■) and expected (−) frequencies for the KNM model fitted to *Callosobruchus chinensis* in (a) larval stage I, (b) larval stage II, (c) larval stage III, (d) larval stage (IV), (e) pupal stage, (f) adults in peas, and (g) emerged adults based on Bellows and Birley's (1981) data. *(continued)*

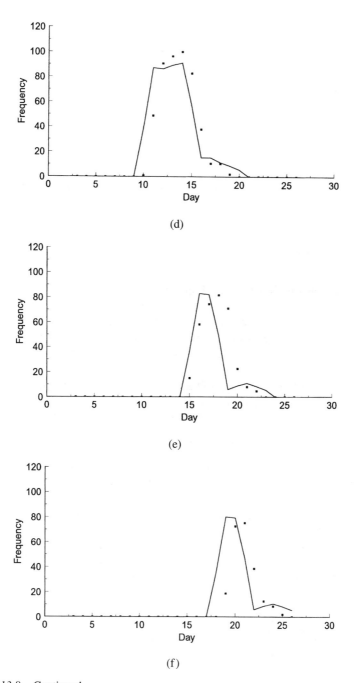

(d)

(e)

(f)

Figure 13.8. Continued.

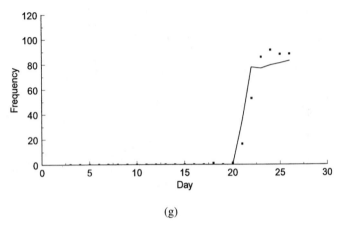

(g)

Figure 13.8. Continued.

Matrix Models for Reproducing Populations

In the earlier sections, we used data collected on either single or multiple cohorts to draw inference about survival rates, duration of life stages, and perhaps entry time into various stages. Here the focus shifts to continuously reproducing populations. Given the structure of the population in one time period, we want to estimate the numbers in each life stage during the next time period or, even better, at some more distant time. We look at three approaches to this problem: the Bernardelli–Leslie–Lewis model, the Lefkovitch model, and the Usher model.

Bernardelli–Leslie–Lewis Model

Although Leslie (1945, 1948) usually receives credit for the matrix approach to modeling population dynamics, Bernardelli (1941) and Lewis (1942) independently proposed it earlier. Hence we follow Manly (1990) in giving credit to all three and refer to what is often called the Leslie method as the Bernardelli–Leslie–Lewis (BLL) method.

For this approach, a population is viewed at discrete points in time 0, 1, 2, ... Individuals are in age groups 0, 1, 2, . . ., k, where age group x is comprised of all individuals aged from x to just less than $(x + 1)$. Only females are counted. The following notation is used for this model:

$n(x, t)$: number of females in age group x at time t.

$p(x)$: probability that a female in group x at time t will survive to be in age group $(x + 1)$ at time $(t + 1)$.

$B(x)$: mean number of female offspring born to females aged x to $(x + 1)$ in a unit of time that survive to the end of that period.

Based on these definitions, the number of females in age group 0 at time (t + 1) is the sum of the offspring from females of all ages produced at time t; that is,

$$n(0, t + 1) = B(0)n(0, t) + B(1)n(1, t) + \ldots + B(k)n(k, t). \quad (13.65)$$

Further, the number of females of age (x + 1) at time (t + 1) is given by

$$n(x + 1, t + 1) = p(x)n(x, t) \quad (13.66)$$

for $x = 0, 1, \ldots, k - 1$. These may be summarized in matrix form as

$$\begin{pmatrix} n(0, t + 1) \\ n(1, t + 1) \\ \vdots \\ n(k, t + 1) \end{pmatrix} = \begin{pmatrix} B(0) & B(1) & \ldots & B(k - 1) & B(k) \\ p(0) & 0 & \ldots & 0 & 0 \\ \vdots & & & & \\ 0 & 0 & \ldots & p(k - 1) & 0 \end{pmatrix} \begin{pmatrix} n(0, t) \\ n(1, t) \\ \vdots \\ n(k, t) \end{pmatrix} \quad (13.67)$$

or

$$N_{t+1} = MN_t. \quad (13.68)$$

Therefore, if this relationship holds through the study period, the number in each age group at time t can be computed using

$$N_t = M^t N_0. \quad (13.69)$$

The matrix **M** is usually called the Leslie matrix. Subject to mild assumptions, it may be shown that a population following this model will reach a stable distribution of the relative numbers of individuals in the various age classes. Further, the population will be growing or declining at a constant rate. The long-term behavior of the population is determined by the dominant eigenvalue of **M** as discussed by Pollard (1973, chapter 4).

Plant (1986) considered the problem of estimating the elements of **M** when modeling a population with a relatively small number of age groups. He developed estimates of survival probabilities and fecundity rates for grouped ages.

Lefkovitch Model

The BLL model requires the population to be grouped by ages. Lefkovitch (1963, 1964a, 1964b, 1965) modified the model to allow grouping by life stages instead. He did this by allowing the number in stage j at time (t + 1) to depend on the numbers in life stages $0, 1, 2, \ldots, j - 1$, and not just the preceding stage. Therefore, denoting the number of individuals in stage j at time t by $f_j(t)$, the Lefkovitch model can be expressed in matrix notation as

$$
\begin{pmatrix} f_1(t + 1) \\ f_2(t + 1) \\ \vdots \\ f_q(t + 1) \end{pmatrix} = \begin{pmatrix} m_{11} & m_{12} & & m_{1q} \\ m_{21} & m_{22} & \cdots\cdots\cdots & m_{2q} \\ \vdots & & & \\ m_{q1} & m_{q2} & & m_{qq} \end{pmatrix} \begin{pmatrix} f_1(t) \\ f_2(t) \\ \vdots \\ f_q(t) \end{pmatrix}
\tag{13.70}
$$

or

$$
F_{t+1} = MF_t.
\tag{13.71}
$$

This leads to

$$
F_t = M^t F_0.
\tag{13.72}
$$

Matrix M is more complex than the Leslie matrix. However, the long-term behavior of the population will be determined by the eigenvalue with the largest absolute value and its corresponding eigenvector. In fact, the proportions of the populations in the various life stages should converge to the proportions in the eigenvector, and the numbers in each stage should increase by a factor equal to the eigenvalue. A complex value for the eigenvalue implies cyclic population changes. However, this may be caused by variability in the m_{ij} values during the sampling period. A negative value of this dominant eigenvalue is biologically meaningless. Manly (1990) discusses estimation of the m_{ij} values.

Usher Model

The primary problem with using Lefkovitch's model is the large number of coefficients in M that must be estimated. Usher (1966, 1969) recognized that this estimation problem could be simplified if the time between samples was small enough so that the possibility an individual passes through more than one stage during that time could be discounted. This led him to suggest the following simplified form of the Lefkovitch model:

$$
\begin{pmatrix} f_1(t + 1) \\ f_2(t + 1) \\ f_3(t + 1) \\ \vdots \\ f_{q-1}(t + 1) \\ f_q(t + 1) \end{pmatrix} = \begin{pmatrix} B_1 & B_2 & B_3 & \cdots & B_{q-1} & B_q \\ b_1 & a_2 & 0 & \cdots & 0 & 0 \\ 0 & b_2 & a_3 & \cdots & 0 & 0 \\ \vdots & & & & & \\ 0 & 0 & 0 & \cdots & a_{q-1} & 0 \\ 0 & 0 & 0 & \cdots & b_{q-1} & a_q \end{pmatrix} \begin{pmatrix} f_1(t) \\ f_2(t) \\ f_3(t) \\ \vdots \\ f_{q-1}(t) \\ f_q(t) \end{pmatrix}.
\tag{13.73}
$$

Here B_j is the contribution of stage j at time t to stage 1 at time $(t + 1)$. If $j = 1$, this contribution consists of those individuals who were in stage 1 at time t who remain in stage 1 at time $(t + 1)$ or the offspring produced by stage 1 individuals from t to $(t + 1)$. For $j > 1$, all contributions to stage 1 are through reproduction. Because we assume that an individual can pass through at most one stage between sampling periods, the numbers in j at time t consist of those who move from stage $(j - 1)$ at time $(t - 1)$ to stage j by time t and those who remain in stage j from time $(t - 1)$ to time t. Therefore, b_j represents the probability that an individual in stage j at time t will move to stage $(j + 1)$ at time $(t + 1)$, and a_j is the probability that an individual in stage j at time t will still be in stage j at time $(t + 1)$. The behavior of the model is determined by the eigenvalue with the largest absolute value and its corresponding eigenvector just as it was with Lefkovitch's model. That is, the numbers of individuals in the various life stages will be proportional to the corresponding values of the eigenvector when the population becomes stable. Further, if the eigenvalue is real, then the population size will be multiplied by the eigenvalue each time unit once these stable proportions have been attained.

As Manly (1990) noted, more work is needed to assess the best way to estimate the transition matrix M. We will estimate the parameters $B_1, B_2, \ldots, B_q, a_2, a_3, \ldots, a_q, b_1, b_2, \ldots, b_{q-1}$ by performing q regressions of the frequencies at time $(t + 1)$ on those at time t. For stage 1, we have

$$f_1(t + 1) = B_1 f_1(t) + B_2 f_2(t) + \ldots + B_q f_q(t), \tag{13.74}$$

and for $j > 1$, we have

$$f_j(t + 1) = b_{j-1} f_{j-1}(t) + a_j f_j(t). \tag{13.75}$$

Notice that an intercept is not included in any of these regressions.

Summary

To understand the population dynamics of an organism, life history information is often collected in the form of a life table. Information on survival rates for various stages are summarized in these tables.

By accumulating life table information for several generations, the factors causing fluctuations in the population size can be evaluated. The life stage that has the greatest influence on population dynamics is the first key factor. The life stage having the next most influence is the second key factor, and so on. Determination of key factors has relied heavily on observations of graphical relationship between the k_j and K values (Varley and Gradwell, 1960) and on regression of the k_j values on K (Podoler and Rogers, 1975). Manly (1977b) developed an alternative key factor analysis that assesses the effect of each stage on the variability in the number entering the final stage (actually the variance of the log of the number estimated to enter the last stage). Each approach provides a different perspective on the data, and we encourage a parallel analysis using each approach as we did with the tawny owl data.

Sometimes, the information needed for life tables is not readily discernible from sampling. Organisms may be entering the population through time. Therefore, on a given sampling date, organisms may be observed in more than one stage. Also, sampling points may be close enough that an organism is observed in the same stage on more than one sampling occasion. Models for these data are generally broken into two classes: multi-cohort models and single cohort models.

The Kiritani–Nakasuji–Manly KNM model is a simple model that can be used both for multi- and single cohorts. It is assumed that sampling begins at the time the first organism enters the population, or shortly before that time, and continues until all the organisms in that generation have died. An iterative method has been developed for use when sampling begins late or ends early. If sampling begins after some organisms are in the second or a later stage, convergence of the iterative method may be a problem. The model is based on the assumptions that the survival rate per unit time is the same in all stages for the entire sampling period and that migration is negligible. Alternative models have been developed that are more flexible than the KNM.

For continuously reproducing populations, population projection models have proven useful. These rely on the development of age or stage-based estimates of fecundity, survival, and transition from one stage to the next.

Our experience is that modeling life stages and the development of population projection matrices is challenging. Care needs to be taken to collect all pertinent data. Fitting the models then requires careful thought about both the statistics and the biology involved.

Exercises

1. Leslie et al. (1955) studied the Orkney vole (*Microtus orcadensis* Millais). The life histories of 42 males and 42 females, living as mated pairs were observed from the age of nine weeks onward. In addition, 18 males and 18 females of known age were kept separately. These were used to replace individuals in the original mated stock as deaths occurred. Because no natural deaths occurred in the stock cages before the 36 voles entered the study, they are included in the data we use to develop a life table. Some accidental deaths occurred and a few females were purposely killed. These are recorded in the a_x column. To adjust for the accidental deaths, Leslie et al. subtracted one-half of the accidental deaths during a time interval from the number

alive at the beginning of the time interval for the purposes of estimating the mortality rate. The data are given in the table below:

Age in weeks (x)	Males observed deaths a_x	Males observed deaths d_x	Females observed deaths a_x	Females observed deaths d_x
9				1
15				
21		1		1
27				3
33				1
39		1		
45		2	1	
51	1	2	1	1
57				
53		1		1
69		3		
75				1
81		4		1
87		7		
93		6		6
99		10	1	4
105		6		5
111		6		5
117		3		9
123		2	6	7
129				2
135		1	1	
141				
147				
153			1	
159			1	
Total	1	59	12	48

A. Construct a life table for adult female Orkney voles based on the data above.

B. Construct a life table for adult male Orkney voles based on the data above.

C. Discuss the results in A and B above.

2. Deevy (1947) reviewed a portion of the study by Murie (1944) of the wolves of Mt. McKinley. Most of the deaths of the Dall mountain sheep (*Ovis d. dalli*) in the Park were thought to be the result of predation by wolves. The two principal methods of defense against wolves are flight to higher ele-

vations where wolves cannot pursue and group action or herding. Numerous skulls showed evidence of a necrotic bone disease, but it was not possible to determine whether death was due solely to the disease or whether the disease ensured death by predation. Murie picked up the skulls of 608 Dall mountain sheep that had died at some time previous to his 1937 arrival as well as 221 skulls during his 4-year stay. The age of the sheep at death was determined by counting the annual rings on the horns. Because skulls of young sheep are fairly perishable, the "lamb" and "yearling" classes are probably under-represented. A few skulls without horns were found. By their osteology, these were judged to belong to sheep nine years old or older and were apportioned among the older age classes. Assuming a stable age distribution and a rate of population increase of 1 (i.e., it is not increasing), Deevy presented the following values for the l_x column based on the initial sample of 608 sheep.

Age in years (x)	l_x	Age in years (x)	l_x	Age in years (x)	l_x
0–0.5	1.000	4–5	0.764	9–10	0.439
0.5–1	0.946	5–6	0.734	10–11	0.252
1–2	0.801	6–7	0.688	11–12	0.096
2–3	0.789	7–8	0.640	12–13	0.006
3–4	0.776	8–9	0.571	13–14	0.003

    A.   Develop the remaining columns of the life table.

    B.   What conclusions can be drawn from the life table?

3.   Edmondson (1945) studied the sessile rotifer *Floscularia conifera*. This species lives attached to water plants, especially *Utricularia,* surrounded by a tube constructed by the rotifer out of pellets of detritus. The tube is added to at the top continuously throughout life. By dusting the *Utricularia* plant with a suspension of powdered carmine, Edmondson was able to identify all the rotifer population members living in a pond. On subsequent visits, the *Floscularia* that had been dusted were conspicuously marked by bands of carmine-stained pellets in the walls of their tubes. Each band was surmounted by new construction of varying widths. Among other things, Edmondson determined that the expected life of solitary individuals was only half that of members of colonies of two or more. Deevey (1947) combined the information for solitary and colonial individuals to obtain the following portion of a life table based on 50 rotifers:

Age in days (x)	l_x	Age in days (x)	l_x	Age in days (x)	l_x
0–1	1.000	4–5	0.720	8–9	0.080
1–2	0.980	5–6	0.420	9–10	0.040
2–3	0.780	6–7	0.280	10–11	0.020
3–4	0.720	7–8	0.220	—	—

A. Complete other columns of a life table for these data.

B. What conclusions can you draw from these data?

4. Farner (1945) conducted a study based on U.S. Fish and Wildlife Service data on 855 banded American robins (*Turdus migratorius* Linnaeus). The birds were banded as young within the breeding rage of the type subspecies between 1920 and 1940, and subsequently recovered dead. Because banded nestlings are likely to be picked up near the banding stations or not at all, mortality in the first year of life can not be estimated with any accuracy. Farner begins his life table on November 1. Deevey (1947) reports the following partial life table for 568 birds banded as nestlings and known to be alive on November 1 of their first year.

Age in years (x)	l_x	Age in days (x)	l_x	Age in days (x)	l_x
0–1	1.000	3–4	0.099	5–6	0.010
1–2	0.497	4–5	0.036	6–7	0.006
2–3	0.229				

A. Complete a life table for these data.

B. What conclusions can you draw from these data?

5. In a study of the brown rat (*Rattus norvegicus*), Leslie et al. (1953) report life-table information developed by Wiesner and Sheard (1935) for albinos of the Wistar strain housed in Edinburgh. The life table is based on 1,456 female brown rats that were alive at the age of 31 days. A portion of the life table is given below:

Age in days (x)	l_x	Age in days (x)	l_x	Age in days (x)	l_x
51	0.99728	451	0.92107	851	0.40763
101	0.99245	501	0.89151	901	0.31895
151	0.98502	551	0.85153	951	0.22544
201	0.98164	601	0.80254	1001	0.15799
251	0.97208	651	0.75005	1051	0.10269
301	0.95955	701	0.67862	1101	0.05020
351	0.94838	751	0.58762	1151	0.20824
401	0.93783	801	0.49508	1201	0.02824

A. Develop other columns of the life table.

B. What conclusions can you draw from these data.

6. Hassell et al. (1987) report on a study of the population dynamics of the viburnum whitefly (*Aleurotrachelus jelinekii* (Frauenf.)). *Viburnum tinus* L. is the principal host in Britain for this whitefly. It is an evergreen shrub. During the study period, more than 90% of the leaves normally survived

into the next season although prolonged and severe frosts can lead to severe defoliation. This happened in the winter of 1962–1963, resulting in only 384 leaves on study bush B in April/May of 1963. After that, the bush grew rapidly. An annual census was taken prior to adult emergence in April/May of each study year. From the start of the study to the 1969–1970 generation, every leaf on bush B was examined. The number, instar, and condition of the larvae wee recorded, giving a total count of second, third, and fourth instar larvae. Beginning in 1970–1971, the bush had grown so that a complete count was no longer possible. From the 1970–1971 generation to the 1975–1976 generation, every third leaf was counted, from 1976–1977 to 1977–1978 every fifteenth leaf was counted, and in the final year every 25th leaf was counted. Partial counts were scaled to give an estimate for the full bush. The remaining life-table data were determined using information from a more detailed census of 30 leaves bearing whiteflies. For example, the proportion of adults emerging on these labelled leaves was applied to the total number of living fourth instar larvae found during the April/May census, providing an estimate of the total number of adults emerging from the bush. Similarly, the number of eggs and the number of first instar larvae were the number of second instar larvae and the survival rate of the eggs and first instar larvae on the thirty leaves. Finally, the potential fecundity was estimated to be 55 eggs/adult based on 10 replicates of a clip-cage experiment. The results are given in the table below.

| Year | Eggs | Larvae | | | | Adults | Leaves |
		I	II	III	IV		
1962–1963						213	384
1963–1964	646	603	202	201	168	56	4,133
1964–1965	430	421	208	208	195	128	9,032
1965–1966	402	393	199	197	144	31	6,214
1966–1967	254	249	137	137	134	79	6,436
1967–1968	652	638	587	587	582	300	7,346
1968–1969	3,665	3,585	1,061	1,054	1,021	624	9,842
1969–1970	4,006	3,918	2,961	2,784	2,640	1,367	13,609
1970–1971	10,852	10,765	8,447	7,079	6,209	3,716	13,841
1971–1972	17,376	17,249	14,754	13,585	12,236	4,815	19,265
1972–1973	22,041	21,556	12,904	10,411	6,957	2,135	18,238
1973–1974	38,786	38,274	10,028	8,438	6,756	2,973	22,114
1974–1975	13,668	13,526	10,565	10,199	9,374	5,624	223,622
1975–1976	17,810	17,753	12,794	11,940	10,864	7,800	23,055
1976–1977	28,455	27,829	21,948	21,043	18,961	10,271	17,610
1977–1978	90,452	88,462	58,238	55,381	49,654	24,205	39,181
1978–1979	107,897	105,523	69,470	60,095	52,141	34,126	23,600

From Hassell, Southwood, and Reader, 1987, by permission of Blackwell Science Ltd.

Notice that the potential number of eggs has not been included in the above table but can be computed based on the number of adults on the bush the preceding season. Because the adult *A. jelinekii* show little migratory activity and the bush was 130 m from the nearest neighboring viburnum bush, the population was viewed to be closed, allowing for the potential number of eggs to be estimated in this manner.

A. Using the counts for the total bush (and being sure to add the potential number of eggs), do a key factor analysis on the data. Be sure to interpret your results.

B. Using the mean counts per leaf and again accounting for the potential number of eggs, do a factor analysis on the data. Be sure to interpret your results.

C. Compare the results of A and B with the number of leaves on a bush. Does it matter whether the total or mean counts are used? Is this always true or does this provide further insight into the population?

7. Manley (1979, 1990) presented life tables for the winter moth *Operophlera brumeta* in Wytham Wood, Berkshire, based on Table F and Figure 7.4 of Varley et al. (1973). The frequencies are per m². There are seven stages. The first is the number of eggs produced from the adults in the previous year, this is taken to be 56.25 eggs per adults. Stage 2 is the number of fully grown larvae. Stages 3, 4, and 5 are the larvae after parasitism by *Cyzensis albicans,* after parasitism by non-specific Diptera and Hymenoptera, and after parasitism by *Plistophora operophterae,* respectively. Stage 6 is the pupae surviving predation. Stage 7 is the adults after parasitism by *Cratichneumon culex.* The data are given in the table below.

	No. of survivors to stage						
Year	1	2	3	4	5	6	7
1950	4,365.0	112.2	87.1	79.4	70.8	14.5	7.41
1951	417.0	117.5	114.8	109.6	102.3	17.4	13.8
1952	758.6	55.0	54.6	47.5	42.4	7.50	7.03
1953	389.0	18.2	17.8	17.0	15.5	7.08	4.90
1954	275.4	158.5	157.0	146.6	127.6	23.8	20.2
1955	1,122.0	77.6	77.1	71.9	65.6	14.7	11.9
1956	645.7	95.5	89.1	87.0	83.2	28.2	14.8
1957	831.8	275.4	263.0	257.0	229.1	37.2	23.4
1958	1,288.0	190.5	162.2	154.9	141.3	21.4	14.8
1959	812.8	57.5	45.7	41.7	39.8	8.71	6.17
1960	346.7	21.4	20.4	17.8	16.6	3.16	1.12
1961	61.7	7.59	7.59	6.61	6.03	3.63	3.02
1962	166.0	13.5	12.9	11.2	10.5	6.03	5.25

(*continued*)

(*continued*)

			No. of survivors to stage				
Year	1	2	3	4	5	6	7
1963	288.4	40.7	40.7	37.2	36.3	14.5	11.0
1964	602.6	131.8	130.3	127.4	124.5	22.6	16.4
1965	891.3	269.2	251.2	245.5	239.9	32.4	24.0
1966	1,249.0	51.3	46.8	46.2	44.2	6.10	2.85
1967	154.9	9.77	9.55	8.91	8.91	2.82	2.82
1968	154.9	10.0	10.0	9.77	9.12	3.02	3.02

From Manly, 1979, by permission of *Researches on Population Ecology*.

Conduct a complete key factor analysis for these data.

8. As part of a study on the comparative demography of Collembola, van Straalen (1982) collected data on the 1979 spring generation of the litter-inhabiting spring-tail *Orchesella cincta* (Linné). The generation could be separated from the one before and after it. The study region was a 1,500-m^2 site in a coniferous forest stand. Sampling was conducted every two weeks. On each sampling occasion, 36 litter-humus cores (with a total area of 0.261 m^2) were collected at random from the site. The Collembola were extracted from the cores using a modified Tullgren apparatus. They were counted and assigned to size classes with class width 0.45 mm. The data are given in the table below:

					Stage			
i	t_i	1	2	3	4	5	6	7
1	22	34	0	0	0	0	0	0
2	24	588	124	47	1	0	0	0
3	26	378	199	55	33	1	2	0
4	28	175	302	193	73	11	2	0
5	30	15	89	149	86	42	7	0
6	32	25	53	85	137	117	42	1
7	34	0	12	24	51	89	65	3
8	36	0	0	2	14	44	25	0
9	38	0	0	0	27	31	34	6
10	40	0	0	0	18	34	52	13
11	42	0	0	0	7	2	5	1
12	46	0	0	0	4	2	3	1
13	50	0	0	0	0	0	1	1

From van Stranlan, 1982, by permission of Blackwell Science Ltd.

First notice that data for $t_i = 44$ and 48 are not included in the table. Use linear interpolation to add these weeks to the data set. Also, assume that no members of this generation were observed in week 20. Then conduct an analysis of the data. Be sure to test any model for fit. Also discuss the results in the context of the study.

9. Southwood and Jepson (1962) reported results of a study of the frit fly (*Oscinella frit* L.) during 1958 and 1959. A 2-acre field near Ascot Berkshire, was planted with oats. In an effort to maximize the populations of *O. frit*, the oats were sowed late each year. Two generations were studied, the tiller generation and the panicle generation. For the 1958 panicle generation, the mean number of eggs, larvae, and puparia per row foot observed during each 4-day sampling period is given in the table below:

Sampling period	Eggs	Larvae	Puparia	Empty puparia
July 4–8	10			
July 9–13	9			
July 14–18	49			
July 19–23	129	2	2	
July 24–28	43	9	2	
July 29–August 2	4	28	13	
August 3–7		37	49	
August 8–12		21	88	4
August 13–17		6	33	7
August 18–22		2	62	10
August 23–27		2	85	20

A. For the June 30–July 3 sampling period, assume that the number of eggs was zero. Analyze these data. Be sure to assess the goodness-of-fit. State any conclusions that you can draw from the analysis.

B. Suppose we are unwilling to assume that the number of eggs for the June 30–July 3 sampling period was zero. Reanalyze the data. How does this impact the estimates? State any conclusions you can draw from the analysis.

10. Southwood and Jepson (1962) reported on a study of the frit fly *Oscinella frit* L. The study region was a 2-acre field near Ascot, Berkshire. Late sowing dates in 1958 and 1959 were chosen so that the populations of frit fly would be maximized. During 1958, two-day sampling periods were used for the tiller generation. The results are presented in the table below. The figures reported are the mean number per square yard, and the dash (—) is used to denote times for which no data were collected.

Sampling period	Eggs	Larvae I	II	III	Puparia
May 13–14	—	—	—	—	—
May 15–16	0.0	—	—	—	—
May 17–18	—	—	—	—	—
May 19–20	7.0	—	—	—	—
May 21–22	3.5	1.8	0.0	0.0	0.0
May 23–24	—	—	—	—	—
May 25–26	—	—	—	—	—
May 27–28	42.5	10.5	1.6	0.0	0.0
May 29–30	21.0	3.5	7.0	0.0	0.0
May 31–June 1	—	—	—	—	—
June 2–3	28.0	14.0	3.5	0.0	0.0
June 4–5	38.5	14.0	19.2	5.3	1.8
June 6–7	—	3.5	10.5	10.5	0.0
June 8–9	14.0	0.0	10.5	10.5	0.0
June 10–11	—	3.5	3.5	3.5	2.1
June 12–13	0.0	0.0	14.5	15.7	13.1
June 14–15	—	—	—	—	—
June 16–17	0.0	0.7	11.2	20.2	9.8
June 18–19	0.0	0.0	10.5	25.2	3.0
June 20–21	—	0.0	0.0	12.6	4.2

Conduct an analysis of these data. Be sure to assess goodness-of-fit. Discuss the results of the analysis, drawing conclusions where possible.

11. Ashford et al. (1970) conducted a life-stage analysis on data collected during 1965. The study was of a natural population of *Neophilaenus lineatus* conducted on a 1,000-m^2 site on East Budleigh Common, Devon. This insect is univoltine. Larvae hatch in the spring from eggs laid the previous year. This species has five nymphal instars, each of which encloses itself in a froth exudate ("spittle") derived from the plant sap on which it feeds. It molts into an adult stage that is more mobile and does not produce spittle. The adults mate, deposit their eggs, and die before winter. The sampling fraction was 0.0015, except as follows: 0.0005 on May 17, 0.0008 on May 24, 0.001 on June 26, and 0.002 on September 16 and October 5. The data collected are given in the table below.

Sampling date	Nymphal instar I	II	III	IV	V	Adult
April 22	1.43					
April 27	28.6					
April 30	22.9	4				
May 3	25.7	4				*(continued)*

(*continued*)

Sampling date	Nymphal instar					Adult
	I	II	III	IV	V	
May 6	58.6	17				
May 10	80.0	31	2			
May 13	35.7	41	11			
May 17	2.86	18	5	2		
May 20	8.57	32	44	7		
May 24	1.43	14	25	9		
May 27		18	30	10	1	
May 31		4	40	41	8	
June 3			23	57	21	
June 8			6	31	37	
June 11			1	22	29	
June 17				4	34	
June 21				2	35	8
June 26				1	21	12
June 30					16	24
July 7						42
July 14						33
July 21						20
August 3						29
August 11						17
August 18						29
August 25						25
September 1						20
September 10						14
September 16						14
September 27						7
October 5						4

From Ashford, Read, and Vickers, 1970, by permission of Blackwell Science Ltd.

A. Conduct an analysis of these data. Be sure to assess goodness-of-fit. Discuss the results of the analysis, drawing conclusions where possible.

B. Contrast your results with those of Ashford et al.

12. As discussed in beginning in Example 13.7, Qasrawi (1966) studied the grasshopper *Chorthippus parallelus* on East Budleigh Common, Devon, during 1964 and 1965. The study region was 3,500 m². The entire area was sampled in 1964, but only five-sevenths of the area was sampled in 1965. Eggs are laid in late summer and do not hatch until the following spring. The insect passes through four instars before reaching the adult stage. It may be killed at any stage by predators or die from other causes. The adult females lay eggs, and both male and female adults die before the winter season. For 1965, the sampling fraction was 0.0028 up to August 23, and

it was 0.0040 from August 26th onward. Data for 1965 are given in the table below:

			Larval instar		
Date	I	II	III	IV	Adults
May 14	2				
May 17	0				
May 20	1				
May 24	4	1			
May 27	2	2			
May 31	1	1			
June 3	2	1			
June 7	10	2			
June 11	5	0	1		
June 17	4	6	1		
June 21	8	4	2		
June 24	4	4	2		
June 28	3	6	4		
July 1	1	2	4	1	
July 5	3	5	5	3	
July 8	2	3	3	0	
July 14	2	2	5	2	
July 16	1	2	5	3	
July 19	0	0	3	3	1
July 22	1	1	3	5	2
July 26	0	1	0	2	1
July 29	1	1	2	2	2
August 3		0	1	1	2
August 5		1	0	2	1
August 9			0	2	3
August 12			1	1	1
August 16				1	3
August 19				0	6
August 23				0	5
August 26				1	2
September 1					4
September 6					2
September 9					4
September 13					1
September 16					1
September 20					0
September 27					3

A. Conduct an analysis of these data. Be sure to assess goodness-of-fit. Discuss the results of the analysis, drawing conclusions where possible.

B.   Contrast your results with those of Ashford et al. (1970).

13.   Crouse et al. (1987) developed an Usher model to help understand the population dynamics of loggerhead sea turtles (*Caretta caretta*). Noting that
Frazer (1983) had derived life-table information based on stages, Crouse et
al. also considered stage classes based on size. The first class is comprised
of eggs and hatchlings. Small and large juveniles make up the second and
third stage classes, respectively. Subadults are in the fourth class. Stage
classes 5, 6, and 7 consist of novice breeders, first-year remigrants, and
mature breeders, respectively. The last three stages are kept separately because of large differences in fecundity despite similarities in survival rates.
Using Frazer's life table information, the following population matrix was
developed.

$$
\begin{pmatrix}
0 & 0 & 0 & 0 & 127 & 4 & 80 \\
0.6747 & 0.7370 & 0 & 0 & 0 & 0 & 0 \\
0 & 0.0486 & 0.6610 & 0 & 0 & 0 & 0 \\
0 & 0 & 0.0147 & 0.6907 & 0 & 0 & 0 \\
0 & 0 & 0 & 0.0518 & 0 & 0 & 0 \\
0 & 0 & 0 & 0 & 0.8091 & 0 & 0 \\
0 & 0 & 0 & 0 & 0 & 0.8091 & 0.8089
\end{pmatrix}
$$

The dominant eigenvalue for this matrix is 0.9450, indicating a long-term
trend of a reduction in the total population size of about 5% per year. Crouse
et al. noted that most conservation efforts were aimed toward protecting the
nests so that eggs could hatch. They were interested in the effects of the
other life stages.

A.   Determine long-term survival if 100% survival is attained for stages 1,
5, and 6 without changing the other survival rates.

B.   Evaluate the effect of increasing stage 3 survival to 77%.

C.   Evaluate the effect of increasing stage 4 and stage 7 survival by 18.5%
and 17%, respectively.

D.   Discuss the implications that the results of parts A through C can have
on loggerhead sea turtle conservation programs.

14

Probit and Survival Analysis

Introduction

In previous chapters, we have estimated survival rates for a given time period. Here, we want to return to the issue of survival. We will be concentrating on two aspects of survival. First, we will look at estimating survival as a function of some dose. The dose may be used in controlling the population such as applying insecticide or herbicide to reduce the insect or weed populations, respectively. From both the economic and environmental perspectives, we want to keep the number of applications to a minimum, leading us to concentrate on distributions and sampling (*see* Chapters 1 to 6). However, once it has been determined that control is needed, the question arises, "How much is enough?" The statistical approach to interpreting bioassay results has historically been probit analysis. More recently, logistic regression has become popular. Both probit analysis and logistic regression are considered here. We should also note that the same methods of estimating the probability of survival as a function of some dose are appropriate even though the dose may not have been purposefully applied. It could be the result of contamination in the environment. The question then becomes, "How much can the individuals tolerate?" Although the question differs the statistical approach to the problem is the same. Finney (1973) is the classical work in this area. Morgan (1992) provides a comprehensive treatment of quantal response data.

Determining length of life and measuring survival probabilities are important elements in understanding the biology of a species. The open population models in Chapter 10 provided estimates of survival probabilities in open populations. In Chapter 13, survival rates were discussed in the context of life tables and modeling stage frequency distributions. In this chapter, two other aspects of survival are reviewed: (1) estimation of the probability of nest success, and (2) estimation of survival from radio telemetry data. Birds' nests are susceptible to predation and weather effects, making the time between nest initiation and the fledgling leaving the nest critical to the population dynamics. Therefore, effort has been expended on estimating nest success, the probability that an egg will hatch and the young bird will subsequently leave the nest. The initial intuitive approaches adopted by scientists are flawed statistically, leading researchers to provide improved methods. Radio telemetry is being used increasingly to study animal behavior. In addition to documenting daily and seasonal movement patterns, these studies permit survival to be estimated. Although the applications appear quite different, the statistical methods are similar, and we look at estimating both nest success and survival from radio telemetry data.

Probit Analysis

Suppose that an individual is exposed to some agent that could cause death. Let T be the critical level of the agent; that is, T is the minimum dosage level that will result in death. If the dose is less than T, the individual will survive. If the dose is T or greater, the individual will die. Because of variation in fitness, genetic makeup, and other factors, the critical value is not the same for each individual. Now assume we have a population of individuals, and the critical dosage level is determined for each. If the population is large, we could draw a histogram of these critical values. Commonly, the histogram has the shape of a normal, logistic, or extreme value distribution. The goal of probit analysis may be (1) to estimate the parameters of the distribution of critical values, or (2) to estimate certain quantile values of the distribution.

Unfortunately, it is not possible to observe the critical value for an individual. If the individual is given a dose and survives, then we know that the critical value is above the dose given. That individual's critical value will most likely be different after the dose. The individual may be weakened, causing the critical dose to be lower. Alternately, the dose may help the individual to gain resistance, and the critical value rises. In either event, the critical dose has changed; therefore, the same individual should not be used for further testing. Suppose instead the individual dies. Then we know the dose was at or above the critical level for that individual. The critical dose could have been less, but again that information is no longer possible to obtain. Therefore, we are not able to observe the random variable T. We are only able to observe the response of an individual to a particular dose. Let the random variable Y be 1 if the individual dies and 0 if he survives. Then Y is a Bernoulli random variable with parameter π equal to the probability that a randomly selected individual has a critical value less than the specified dose x; that is,

$$P(Y = 1) = P(T \leq x) = \pi(x) \qquad (14.1)$$

(*see* Figure 14.1). Traditionally, the goal has been to determine the level of dosage

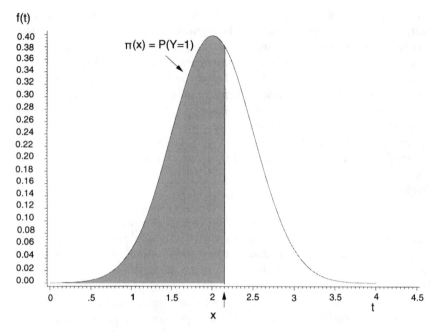

Figure 14.1. Proportion of a population with critical values (*t*) below *x* assuming a probit model.

that would result in 50% mortality, the median of the distribution of *T*. This is often denoted by LD_{50}. However, other quantile values or the parameters of the critical-dose distribution may be of interest. To summarize, although the critical values are assumed to have a continuous distribution, they cannot be observed. Instead, the discrete random variable *Y* is observed. We may want to estimate the parameters of the distribution of *T* or certain quantiles of that distribution, especially the median.

Before discussing methods of estimating the distribution of *T* or its quantiles, we will describe a common experimental design used in these studies. For example, in toxicology experiments, test organisms are divided into groups, often but not always, of equal size. Each group is subjected to a known level *x* of a toxin. (In other cases, it could be a level *x* of dose or of stimulant.) The dose differs from group to group, but is assumed to be constant within each group. For the *i*th group, the administered dose x_i, the number treated within a group n_i, and those group members surviving y_i are recorded. Our goal is to model the proportion π_x surviving at dose *x* as a function of *x*. This is often done for *x* on the logarithmic scale.

For these experiments, modern methods of analysis date from Bliss (1935). For the probit model, it is assumed that the distribution of *T* is normal; that is,

$$\pi_x = \Phi(\alpha + \beta x) \tag{14.2}$$

where $\Phi(\cdot)$ is the standard normal distribution function and α and β are unknown parameters to be estimated. Notice that based on this model, estimates of π will lie between 0 and 1. This is important because as a probability π must be in this range. Linear regression models could, and often do, produce estimates of π less than zero or greater than one, making them unsuitable for this purpose. Also, note that if $\beta > 0$, the survival probability is an increasing function of the applied dose; it is a decreasing function if $\beta < 0$.

From equation (14.2), we have

$$\Phi^{-1}(\pi_x) = \alpha + \beta x. \tag{14.3}$$

The dose x is known. Parameters α and β are to be estimated. π_x is the probability that a randomly selected member from the population has a critical value at or below x. Because π_x is the unknown to be modeled, $\Phi^{-1}(y_i/n_i)$ is an intuitive response variable. However, at low doses, all group members may survive, and they may all die at high doses. Use of $\Phi^{-1}(y_i/n_i)$ in these cases leads to infinite estimates for α and/or β. To avoid this problem, various modified empirical transformations are used, such as $\Phi^{-1}\{(y_i + 0.5)/(n_i + 1)\}$, but the choice of modification is largely arbitrary.

The logistic density has the form

$$f(x) = \frac{e^x}{(1 + e^x)^2}, \qquad -\infty < x < \infty. \tag{14.4}$$

The cumulative distribution function is

$$\begin{aligned} F(x) &= P(X \le x) \\ &= 1 - \frac{1}{1 + e^x}, \qquad -\infty < x < \infty. \end{aligned} \tag{14.5}$$

The pth quantile of the distribution occurs at

$$x_p = ln\left(\frac{p}{1 - p}\right). \tag{14.6}$$

Often p in equation (14.6) represents the probability of an event (e.g., the death of an animal) occurring. Therefore, the ratio of p to $(1 - p)$ represents the odds of an event occurring; that is, a ratio of 3 means the event is three times as likely to occur as it is not to occur. Thus x_p is called the log odds.

The logit model is

$$ln\left(\frac{\pi_x}{1 - \pi_x}\right) = \alpha + \beta x \qquad (14.7)$$

where x is the dose, α and β are unknown parameters, and π_x is the proportion of the population with critical value less than or equal to x. Notice that the log odds of the dose being at or above the critical value is linearly related to dose, with dose generally measured on the logarithmic scale.

Historically, the normal distribution has served as the model of critical dosages; however, increasingly the logistic distribution is being used. This is due to several factors. First, estimation is easier for the logit than for the probit model. Second, the logistic is easily interpreted in terms of the log odds. Last, the two distributions differ little in the range from 0.1 to 0.9, making it difficult to choose between the two on the basis of goodness-of-fit tests.

The models presented here are the traditional and simplest. However, with the increased computing power available today, covariates may be added to the linear relationship of dose. Also, often some mortality will occur without the presence of any treatment. If the expected proportion of deaths due to natural mortality is C, this can be accounted for in the model as

$$\pi = Pr(Y = 1) = C + (1 - C)F(\alpha + \beta x) \qquad (14.8)$$

where F is the normal, logistic, or some other distribution function being used in the modeling process.

Estimation of the parameters is generally based on maximum likelihood methods. SAS® (1990b) may be used to estimate these parameters using PROC PROBIT or PROC LOGISTIC.

Example 14.1

Bliss (1935) studied the mortality of adult flour beetles (*Tribolium confusu*) after a 5-hour exposure to gaseous carbon disulfide (CS_2). The data are given in Table 14.1. Notice that the logarithm of dose is taken using base 10 logarithms. We begin by plotting the data and looking for the S-shaped curve. We have done this for both the original dose scale (Figure 14.2) as well as the log dose scale (Figure 14.3). In each case, the S-shape is visible, but it is more pronounced for the log dose scale.

We will analyze the data using SAS® (1990b). The probit model can be fit with the following set of statements:

```
PROC LOGISTIC DATA = BEETLE;
MODEL KILLED/INSECTS = LOGDOSE/LINK = NORMIT;
OUTPUT OUT = PREDNOR  P = PREDNOR  L = L  U = U;
```

Table 14.1. Mortality of Bliss's (1935) Flour Beetles at Various Levels of CS_2

Dose (CS_2 mg/L)	Log of dose	No. of beetles tested	No. of beetles killed
49.06	1.691	49	6
52.99	1.724	60	13
56.91	1.755	62	18
60.84	1.784	56	28
64.76	1.811	63	52
68.69	1.837	59	53
72.61	1.861	62	61
76.54	1.884	60	60

The option LINK = NORMIT in the model statement indicates that the probit and not the logit model is to be fit. The OUTPUT statement produces a new data set (having a name specified by OUT =) that has the original data as well as the predicted values, the lower 95% confidence limits, and the upper 95% confidence limits as indicated through $P =$, $L =$ and $U =$, respectively. In each case, the word after the equals sign is the specified name for the new variable.

The output indicates that the estimated intercept is -34.94 and that the coefficient for LOGDOSE is estimated to be 19.73. These are the estimates for α and β, respectively in equation (14.2). Therefore, the estimated model is

$$\hat{\Phi}^{-1}(\pi(x)) = -34.94 + 19.73x$$

or

$$\hat{\pi}(x) = \Phi(-34.94 + 19.73x)$$

where x is the logarithm of the dose. Suppose that the log dose is 1.78 (or dose is $10^{1.78} = 60.26$. Then

$$\hat{\pi}(x) = \Phi(-34.94 + 19.73(1.78)) = \Phi(0.18) = 0.57$$

That is, we estimate that 57% of the adult flour beetle population would die after a 5-hour exposure to 60.26 mg/L of CS_2. The value of the distribution function at 0.18 can be found either using tables of the standard normal distribution function or using ECOSTAT (*see* Chapter 1). PROC PROBIT could have also been used for this analysis. The observed, predicted and 95% confidence limits based on the probit model are shown in Figure 14.4.

Now we will fit the logit model. The SAS® statements need only a couple of adjustments. First, the logit model is the default in PROC LOGISTIC so the option LINK = is no longer needed. Second, a new data set name, and perhaps

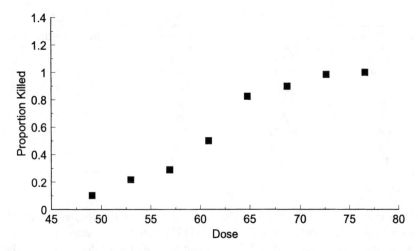

Figure 14.2. Proportion of Bliss's (1935) flour beetles killed for various levels of CS_2 dose.

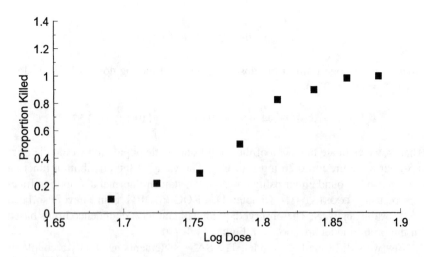

Figure 14.3. Proportion of Bliss's (1935) flour beetles killed for various levels of CS_2 log dose.

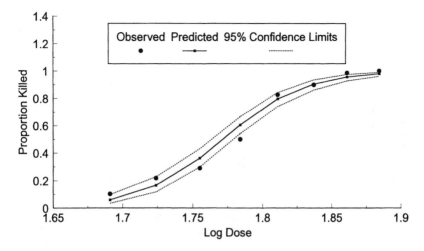

Figure 14.4. Observed and predicted mortality rates with 95% confidence limits based on the probit model.

different variable names, could be specified for the output data set. Noting that the intercept is α and the coefficient on the log dose is β in equation (14.7), we obtain the following estimated model:

$$\ln\left(\frac{\hat{\pi}}{1 - \hat{\pi}}\right) = \hat{\alpha} + \hat{\beta}x = -60.72 + 34.27x.$$

Solving for the estimate of π, we have

$$\hat{\pi}(x) = \frac{e^{-60.72 + 34.27x}}{1 + e^{-60.72 + 34.27x}}$$

The observed and predicted probabilities of mortality as well as 95% confidence limits on the predicted probability of mortality are shown in Figure 14.5. If the log dose is 1.78 (or the dose is $10^{1.78} = 60.26$, the estimated probability of mortality at that dose is

$$\hat{\pi}(x) = \frac{e^{-60.72 + 34.27(1.78)}}{1 + e^{-60.72 + 34.27(1.78)}} = 0.57.$$

Notice, that for this level of dosage, the probit and the logit models provide the same estimates. This emphasizes that the two are very similar within the range

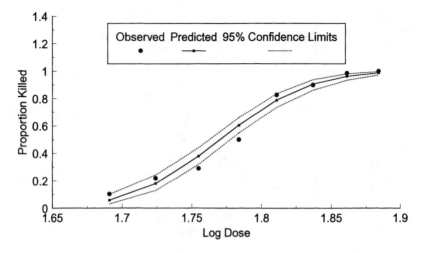

Figure 14.5. Observed and predicted mortality rates with 95% confidence limits based on the logit model.

of 0.1 to 0.9. Distinguishing between the two requires large sample sizes so that the tail areas can be estimated with reasonable precision.

Nest Survival Analysis

A nest survival study usually begins with a search for nests. New nests may be identified throughout the study. Often the nests are not found before eggs have been laid and incubation begun. Therefore, if a nest fails, the age is unknown. If the nest succeeds, the age can be determined afterward. Originally, ornithologists reported the proportion of the observed nests that were successful; that is, the number of succeeding nests was viewed as a binomial random variable.

Mayfield Method

Mayfield (1961, 1975) was the first to note that this approach overestimated the probability of nest success. He pointed out that nest mortality is a function of time. If a nest is found immediately after eggs have been laid, then the probability of observing failure is higher than if the nest is found shortly before the eggs hatch. Further, for nests containing large young when found, fledging success is nearly 100%. If all nests are found immediately before the eggs hatch or when large young are present, the difficulty of using the observed proportions is obvious. However, if nests are found at various times, then the error may not be noticed. Mayfield developed a model based on the number of days that nests are exposed (or at risk) instead of the number of nests. The estimate of the probability of failure is

$$1 - \hat{p} = \frac{\text{number of nests lost}}{\text{number of nest days}} \tag{14.9}$$

where the number of nest days is the number of days a nest is at risk, totaled over the number of nests.

To develop the model, Mayfield noted that nest success can be viewed in five stages: (1) survival during nest construction, (2) survival during egg-laying, (3) survival during incubation, (4) hatching that is assumed to occur when the first young bird breaks free of the shell, and (5) survival of young to fledging. The last three stages are modeled. Hatching is used to separate periods of incubation and young in the nest. Survival in these two periods is modeled separately because their survival rates tend to differ.

Consider first the time of incubation. Let p be the probability that a nest survives a day. It is assumed that the probability of nest survival is constant and that nest survival on any given day is independent of survival on other days. Then if a nest is found and visited t days later, the probability it survived that interval is p^t. The probability that it survived k days and was then destroyed is $p^k(1 - p)$. If visits are daily, then the probability a nest will be destroyed between visits is $(1 - p)$, and the probability of its surviving is p. If visits are made every t days and the nest is destroyed between visits, Mayfield assumed that the nest was at risk for one-half of the interval; that is, it survived for one-half the interval minus 1 day and then was destroyed. The probability of this event is $p^{t/2 - 1}(1 - p)$. The probability a nest survives the interval is p^t.

Suppose a sample of nests is found at various stages. Let n_t be the number of nests that was due to hatch t days later or were visited t days later. Of these, h_t hatched or remained viable, and f_t failed by the subsequent visit. Assuming that all failures occurred at the midpoint of the interval, Johnson (1979) showed that the maximum likelihood estimator of the proportion failing is

$$1 - \hat{p} = \frac{f_1 + \sum f_t}{h_1 + \sum th_t + f_1 + 0.5 \sum tf_t}, \tag{14.10}$$

which is equivalent to Mayfield's estimator.

Johnson (1979) determined the variance of the estimated survival rate using large sample properties of the maximum likelihood estimator and found it to be

$$s_{\hat{p}}^2 = \frac{(\text{exposure} - \text{losses}) \times \text{losses}}{(\text{exposure})^3} \tag{14.11}$$

where

$$\text{Exposure} = h_1 + \sum th_t + f_1 + 0.5 \sum tf_t \tag{14.12}$$

and

$$\text{Losses} = f_1 + \sum f_r. \tag{14.13}$$

Standard normal-based confidence intervals for the survival rate were recommended by Johnson.

Bart and Robson (1982) suggested a square-root transformation to obtain better confidence intervals. Let x be the average interval length,

$$x = \frac{\sum Ln_L}{\sum n_L} \tag{14.14}$$

where n_L is the total number of nests (survivals and failures) observed in interval L, and L is the interval length in days. Then the transformed daily survival rate is

$$\hat{p}_t = \sqrt{1 - \hat{p}^x}. \tag{14.15}$$

The estimated standard error of $\hat{p}_t$ is

$$s_{\hat{p}_t} = \frac{x\hat{p}^{x-1}}{2\sqrt{1 - \hat{p}^x} \sqrt{\sum_L \frac{n_L L^2 \hat{p}^{L-2}}{1 - \hat{p}^L}}}. \tag{14.16}$$

Concern has been expressed about some features of the Mayfield model. Green (1977) criticized the model assumption that the population is homogeneous, arguing that some nests were more prone to destruction than others. However, Johnson (1979) found the Mayfield model to be robust to this heterogeneity.

Miller and Johnson (1978) determined that if the time between observations was more than a day, then bias was introduced by assuming that all nests failed at the midpoint between observations. They recommended using the 40% point as the day of failure estimate if visits are infrequent. Johnson (1979) suggested that this correction was needed when the intervals between visits exceeded an average of about 15 days; otherwise, destroyed nests are assumed to have too much exposure. Johnson also obtained estimates for a more general model that did not require an assumption about the time of nest failure. However, he concluded that unless a large number of nests were included in the sample, that the Mayfield model, with perhaps the 40% adjustment for infrequent visits, worked well and was computationally much simpler.

Example 14.2

Mayfield (1961) reported on a nest study for Kirtland's Warbler (*Dendroica kirtlandii*). Data were collected on 154 nests. During incubation, these nests had observed total exposure of 878 nest-days. Because the average incubation is 10.3 nest days, Mayfield noted that nearly twice this exposure would have been observed if the nests had all been observed from initiation. Further only 113 nests had known outcomes, but all nests were included in the study. In 878 nest-days exposure during incubation, 35 nests were lost, with 19 destroyed and 16 deserted. Therefore, the estimated mortality rate is

$$\frac{35}{878} = 0.040.$$

The daily survival rate is then $1 - 0.040 = 0.960$. From equation (14.11), we have an estimate of the variance in the estimated survival rate is

$$s_{\hat{p}}^2 = \frac{(878 - 35)(35)}{878^3} = 0.0000436$$

or an estimated standard error of 0.0066.

Pollock Method

Pollock (1984) and Pollock and Cornelius (1988) proposed a discrete survival model that allows estimation of the survival distribution from the time of nest origination. We will use the following notation:

Parameters:

J: number of days required for incubation and fledging

T_i: age at death of ith nest; $T_i = 1, 2, \ldots, J$

P_i: probability that a nest fails at age i units of time

δ_i: probability of an intact nest is first encountered at time i

Statistics:

n_{iH}:   number of nests found of age i that later succeed

$n_H = \Sigma n_{iH}$:   total number of successful nests encountered

n_{iF}:   number of nests of unknown age that are observed for i units of time and then fail

$n_F = \Sigma n_{iF}$:   total number of unsuccessful nests encountered

For this model, the number of days J required for incubation and fledging is assumed to be known. Further, let T_i denote the age at death of the ith nest. Note that T_i can assume integer values $1, 2, \ldots, J$. Define

$$P_i = P(T = 1), \qquad i = 1, 2, 3, \ldots, J \tag{14.17}$$

where P_i is the probability a nest fails at age i units of time. The probability of a nest succeeding is

$$P_{J+1} = 1 - P_1 - P_2 - \ldots - P_J. \tag{14.18}$$

Therefore, J distinct survival parameters are in the model. Also notice that the units of time could be days, 3 days, weeks, or whatever unit of time is most appropriate.

The set of $\delta_1, \delta_2, \ldots, \delta_J$ is composed of J nuisance parameters representing the probability of an intact nest being first encountered at age i, $i = 1, 2, \ldots, J$. A critical assumption to the model is that the encounter probabilities are independent of the survival probabilities. That is, nests that are discovered early in the cycle. are neither more or less likely to survive future time units than nests discovered late in the cycle.

On the basis of these assumptions, Pollock (1984) developed maximum likelihood estimators based on three conditional multinomial distributions. By imposing the constraint that $\Sigma\delta = 1$, the estimators of the encounter probabilities take the simple form:

$$\hat{\delta}_i = \frac{n_{iH}}{n_H}, \qquad i = 1, 2, \ldots, J. \tag{14.19}$$

Notice that $\hat{\delta}_J = 1 - \hat{\delta}_1 - \hat{\delta}_2 - \ldots - \hat{\delta}_{J-1}$ due to the constraint. Estimators for the survival probabilities are obtained by solving a system of J equations in J unknowns (see Pollock, 1984).

The model is based on several assumptions in addition to assuming that encounter and survival probabilities are independent. One is that the observed nests represent a random sample with respect to survival. If the nests included in the sample are the ones that are easier to find and thus more susceptible to predation, then this assumption may be violated.

Further, it is assumed that visiting a nest has no impact on its probability of survival. Bart and Robson (1982) noted that visiting a nest may temporarily reduce the probability of its survival as a visitor could lead predators to the nest. They suggested a modification in the Mayfield model to account for this temporary depression in survival probability.

Bart and Robson (1982) also state that nests should be visited on a regular schedule. If a researcher believes the nest is about to fail, there is a natural tendency to visit the nest more frequently. However, this practice is to be avoided (to avoid disturbing the nest).

No seasonal component is included in the model. All nests are treated as a cohort from time of initiation. Stratified sampling could be used to divide the nests into, for example, early, middle and late season if the sample size is large enough.

Cornelius and Pollock (1987) and Pollock and Cornelius (1988) analyzed data sets on great-tailed grackles and mourning doves using the Pollock method. They noted that their results are comparable to those determined under the Mayfield method. Intuitively, Pollock's method of modeling survival from the time of nest initiation should be better than Mayfield's method of estimating nest survival based on days of observed exposure. However, at this time, no comprehensive comparison has been made. Mayfield's method is simple and appears to be robust to departures from the assumptions. Therefore, we will adopt that method. Further study in this area, particularly through simulation, would be useful.

Analysis of Radiotelemetry Data

Radio-tagging animals is becoming an increasingly popular method of following animals in their natural habitat. After catching an animal by trap, dart gun, or some other method, it is fitted with a small radio transmitter and released. The transmitters may be attached to collars put around the neck of larger mammals such as deer or bear. Implantation of the transmitters may be needed for other animals such as fish. Each transmitter has a unique signal so that information can be kept on each individual. After release, the animal's signal may be monitored until the animal dies or its signal is lost. If the signal is lost or if the study terminates before the animal dies, the data are *censored,* in that we know that the animal was alive at the time of censoring, but we do not know how much longer the animal lived. Because only marked animals are followed, it is usually possible to estimate only survival rates, not population sizes or recruitment rates. Here we discuss discrete methods of analyzing the data as well as the Kaplan-Meier distribution free method and the staggered entry model.

Trent and Rongstad Method

Trent and Rongstad (1974) proposed the same approach for the analysis of radiotelemetry data that Mayfield (1961, 1975) proposed for nest survival data. These investigators counted the number of days of exposure to death for all animals and the number of deaths. From this the daily probability of death, and hence the daily probability of survival, is estimated. For this model, survival is assumed to be constant for each animal day, and the probability of survival is constant from day to day. The model can be generalized to allow survival to vary by dividing the study period into time periods and using data from each time period to estimate the survival probability for that period.

Heisey and Fuller (1985) generalized the Trent and Rongstad (1974) method to allow for different causes of death. They developed the microcomputer program MICROMORT for the model. White (1983) proposed analyzing radio telemetry data in a manner similar to bird banding data. This allows for survival rates to vary with period. He developed the program SURVIV for the analysis of data based on his approach.

Survival Analysis Model

Pollock (1984) and Pollock et al. (1989) applied a traditional survival analysis to radio telemetry data. Five basic model assumptions are made:

1. A random sample of size n animals is drawn from the population.
2. The survival time for one animal is independent of that for other animals.
3. Survival is not affected by tagging.
4. Time of death is known exactly.
5. The censoring mechanism is random.

The first three assumptions were also made in the capture–recapture models (Chapters 9 and 10) and the band-return models (Chapter 10). Usually, radio-tagged animals are monitored on a daily basis. Thus, from a practical standpoint, time of death is known exactly. Also implicit in this assumption is that the study region has a fixed area, so that if an animal has a functional radio it is found (with probability one). The last assumption may be violated. For example, a predator could destroy the censoring mechanism with the animal.

Three possible causes of censoring are (1) transmitter failure, (2) emigration of the radio-tagged animal, and (3) survival of the radio-tagged animal beyond the end of the study. Emigration may sometimes be ignored based on biological considerations.

Two sets of conceptual random variables are present in the development of the survival analysis model:

1. A set of survival times from tagging to death, $T_1, T_2, \ldots, T_n$, is observable if there is no censoring. Further, these are assumed to constitute a random sample from some probability distribution.
2. A set of censoring times, $C_1, C_2, \ldots, C_n$, is observable if there are no deaths. These too are assumed to constitute a random sample from some probability distribution.

Then based on the above assumptions, maximum likelihood estimates could be obtained assuming some probability distribution such as exponential, gamma, or Weibull. Alternatively, the Kaplan–Meier (1958) product limit estimate does not require any distributional assumptions. For this model, the estimated survival rate for the period $\hat{S}_K$ is

$$\hat{S}_K = \prod_{i=1}^{D} \left(1 - \frac{1}{N_i} \right)$$

where N_i is the number of individuals still at risk after death i has occurred and D is the total number of deaths in the time period of the study. The variance of $\hat{S}_K$ is estimated as

$$s_{\hat{S}_K}^2 = \frac{\hat{S}_K^2 (1 - \hat{S}_K)}{N_D}$$

where N_D is the number of individuals still at risk immediately after the last death $(i = D)$ has occurred.

If important covariates such as age or weight at tagging are available, they may be incorporated into the Cox proportional hazards model (Cox, 1972). Pollock et al. (1989) applied this model to a study in which young and adult female black ducks were radio-tagged in New Jersey in winter. They found that the condition index of adult birds was significantly related to survival. Birds in poorer condition at the start of winter were less likely to survive. However, they found no relationship in the condition index and survival for young birds, a phenomenon they were unable to explain in biological terms.

Example 14.3

Trent and Rongstad (1964) studied radio-collared rabbits with the results shown in Table 14.2. The Trent–Rongstad estimate of daily survival for the first period is

Table 14.2. Trent and Rongstad's (1964) Data on Radio-Collared Rabbits

Time period	Days in period	Rabbit days	Deaths
March–April	61	617	3
May–June	61	820	3
July–August	62	1,047	1
September–October	61	1,660	6
November–December	51	1,447	9
January–February	59	657	5

$$\frac{617 - 3}{617} = 0.9951.$$

The estimated variance of the estimate assuming deaths were recorded daily is

$$s_{\hat{p}}^2 = \frac{(617 - 61)(61)}{617^3} = 0.000144$$

for an estimated standard error of 0.012. Notice that insufficient information is present for us to use the Kaplan–Meier product limit estimator. This should be kept in mind when collecting data.

Summary

In agriculture, pesticides are often applied to achieve control of a pest population. Determining the amount that will be effective, but not excessive, is important. Probit analysis and logistic regression are common tools for analyzing data from studies designed for this purpose. The same analysis is useful when determining the effect of contaminants in the environment. The probit and logit models are commonly applied in this setting. Newer developments allow these models to be extended (*see* Morgan, 1992).

Early methods for estimating nest success are similar to those for estimating survival from radio telemetry model. The Mayfield method for estimating nest success is appealing because it is easily applied. The same may be said for the Trent and Rongstad approach to analyzing radio telemetry data. In each case, Pollock (1984) has suggested a more general statistical model. The nest success model requires the solution of the solution of J equations in J unknowns, where J is the number of days from nest initiation to fledging. For radiotelemetry data, ideas from survival analysis, such as the Kaplan–Meier product limit estimate, are suggested. The Cox proportional hazards model permits covariates to be included in the model. The staggered entry model accounts for animals being tagged over time, a common occurrence in practice.

Exercises

1. Kooijman (1981) studied the mortality rates for *Daphnia magna* in water containing various levels of cadmium chloride. After 18 days, the accumulated numbers of deaths at various levels of cadmium chloride are given in the table below.

Dose (µg/L)	No. of Daphnia treated	No. of Daphnia dead after 18 Days
0.0	50	0
3.2	50	1
5.6	49	5
10.0	50	10
18.0	53	38
32.0	50	50
56.0	59	59

Reprinted from *Water Research* **15,** Kooljman, Statistical aspects of the determination of mortality rates in bioassays, pp 749–759, 1983, with kind permission from Elsevier Science Ltd. The Boulevard, Langford Lane, Kidlington OX5 16B, UK.

A. Plot the data on the dose and log dose scales.

B. Analyze the data using a probit model.

C. Analyze the data using a logit model.

D. Discuss your conclusions in the context of the study.

2. Mayfield (1961) reported on nest survival of Kirtland's Warbler (*Dendroica kirtlandii*) during the nestling period. A total of 144 nests were observed during the nestling period. During the 735 nest-days of exposure, 22 nests were lost, 19 were destroyed and 3 deserted.

   A. Estimate the daily mortality and survival rates for Kirtland's Warbler's during the nestling period.

   B. The time from hatching until the young leave the nest is about nine days. Estimate the probability that a Kirtland's Warbler will survive to leave the nest.

   C. Assuming survival during incubation (*see* Example 14.2) is independent of that during the fledging stage, estimate the probability a Kirtland's Warbler survives from egg laying to the time the young leaves the nest.

   D. Set a 95% confidence interval on the daily survival rate.

3. Look at the data in Example 14.3 again.

   A. Estimate the survival rate and its standard error for the remaining five time periods.

   B. Discuss the results in the context of the study. What future research direction might you pursue.

15

Chaos

Introduction

In Chapter 3, we considered models of immigration and movement within a given ecosystem. In that and other chapters, the focus has been on precise estimation of population density. This is an extremely important aspect of applied ecology. Consequently, many population models have been developed to give rise to a probability distribution that will fit the distribution of counts obtained in the sampling process. For example, the negative binomial consistently fits insect data. Unfortunately, there are at least 16 models giving rise to the negative binomial, many of which could be given a biological interpretation (Boswell and Patil, 1970). Therefore, the fact that a particular probability distribution fits does not necessarily provide insight into the underlying biological forces.

A further, and perhaps more pressing, concern is that the models giving rise to probability distributions tend to provide a picture of what might have occurred at a particular instant in time. Yet the agroecosystem is not static. Insects continue to enter through birth or immigration, move within the field, and leave through death or emigration. These models are not able to reflect these dynamics.

Fluctuations in population size have been modeled using both difference and differential equations. Stochastic versions of the traditional population equations have been developed. Increasingly, the impact of chaos theory on modeling population dynamics has become evident. This chapter gives an overview of these ideas.

Population Models

Many species have well-defined, non-overlapping generations that lead to population growth in discrete steps. For small organisms, the time between steps may be short, and continuous approximations to these discrete steps may be reasonable. However, a year is a common time between steps. In these cases, reflecting the discrete nature of the generations seems appropriate. Models then relate the population size at time $(t + 1)$, N_{t+1}, to the population size at time t, N_t. This leads to the study of difference equations (*see* May 1974, 1976; Murray 1993).

Difference equations, or discrete models have the form

$$N_{t+1} = f(N_t) \tag{15.1}$$

where $f(N_t)$ is in general a nonlinear function of N_t. Suppose the population one step later is proportional to the current population; that is,

$$N_{t+1} = rN_t. \tag{15.2}$$

Given an initial population of size N_0, the population at time N_t can be found by repeatedly applying equation (15.2), giving

$$N_t = r^t N_0. \tag{15.3}$$

For this model, the population grows or declines geometrically depending on whether $r > 1$ or $r < 1$, respectively. Although this model is not very realistic for most populations or for long time spans, it has been effectively used to model early stages of population growth for some bacteria (Murray, 1993).

Other discrete models have been proposed to reflect a population's response to crowding and self-regulation. Suppose K is the population's *carrying capacity;* that is, K is the largest population that the environment can sustain. A variety of functions $f(N_t)$ have been used in biological situations (a list may be found in May and Oster, 1976). One of these models, sometimes called the Verhulst process, has the form

$$N_{t+1} = rN_t\left(1 - \frac{N_t}{K}\right), \qquad r > 0, \qquad K > 0. \tag{15.4}$$

By defining $X_t = N_t/K$, we can rewrite equation (15.4) as

$$X_{t+1} = rX_t(1 - X_t), \qquad r > 0. \tag{15.5}$$

An obvious weakness of this model is that if $N_t > K$, $N_{t+1} < 0$. We shall restrict our attention only to values $0 < N_t < K$, which implies $0 < X_t < 1$.

Although more realistic models of population growth that incorporate a carrying capacity have been developed, the one in equation (15.5) is of particular interest to us as it gives us insight into the meaning of chaos. First, suppose $0 < r < 1$. In this case, the population eventually becomes extinct because $X_t \to 0$

as $t \rightarrow \infty$. A steady state occurs when $X_t^* = 0$, but this is not very interesting. It simply says if the population size is 0, it will continue to be 0. This is then a steady state for all values of r. At $r = 1$, X_t changes its behavioral character. Such critical values are called *bifurcation* values.

In the range $1 < r < 3$, a positive steady state occurs at $X_t^* = (r - 1)/r$, as well as 0. $r = 3$ is the next bifurcation value. For $r > 3$, $X_t = (r - 1)/r$ becomes unstable, and two new stable periodic solutions are

$$X_{2t} = \frac{(r + 1) \pm \sqrt{(r + 1)(r - 3)}}{2r}. \tag{15.6}$$

The solutions given in equation (15.6) are *stable periodic solutions of period 2*. This means that if the population size is initially one of the X_t^* values in equation (15.6), the population size returns to that value every two periods; that is, $X_0 = X_2 = \ldots = X_{2t}$. Denote the next bifurcation value as r_4. The subscript 4 is used because at that point, the stable periodic solutions of period 2 become unstable, and four stable periodic solutions of period 4 can be determined. At the following bifurcation value r_8, the period 4 stable solutions become unstable, but eight new stable periodic solutions of period 8 may be found. This continues so that the sequence of stable solutions have periods 2, 2^2, 2^3, 2^4, ... To summarize, as r increases through successive bifurcations, each even p-periodic solution branches into a $2p$-periodic solution.

As r continues to increase, it reaches a critical value r_c at which instability occurs for all periodic solutions of period 2^p. For $r > r_c$, locally attracting cycles with periods k, $2k$, $4k$, . . . occur, but now k is odd. The critical value r_c for the model in equation (15.3) is approximately 3.828. Sarkovskii (1964) proved that if a solution of odd (≥ 3) period exists for a value r_c, aperiodic or chaotic solutions exist for $r > r_c$. These chaotic solutions oscillate in an apparently random manner. Although we have discussed the behavior for this particular model, models of the form in equation (15.1), which incorporate a maximum carrying capacity K, exhibit bifurcations to higher periodic solutions eventually leading to chaos. We will return to chaos, its meaning, and its impact on biology shortly.

Model extensions to account for intraspecific and interspecific competition have been proposed, and work is continuing in this area.

Example 15.1

Suppose the population is 10% of its maximum and $r = 1.5$. The population grows rapidly and then remains more or less static or steady. With the population at 0.1 and $r = 2.75$, the population rises rapidly and then starts to vacillate slightly up and down. This is what might be expected from populations that have reached a balance between birth rate and limiting factors, such as predation and disease. If X_0 remains at 0.1 and r increases toward 4, the peaks and troughs of the population vacillations deepen. Notice that if $r = 2$ and $X_0 = 0.5$, $X_t = 0.5$ for all t, verifying that $(r - 1)/r$ is a stable solution for this value of r.

Sometimes, generations overlap or the time between generations is short. The discrete models described above are no longer appropriate. Models that can reflect the constantly changing population dynamics are needed. In 1798, Robert T. Malthus, an English clergyman, wrote an essay on the *Principle Population* (Cambel, 1993, p. 82). In this essay, he reasoned that the human population grows geometrically, but the necessary resources to sustain man grows only arithmetically. He predicted that humans would soon use up all the natural resources in the world and that there would be widespread hunger, famine, and disease. We now know that the increase in technology and humans ability to adapt have prevented this happening on the scale and in the time frame Malthus predicted. We must admit that in certain areas of the world, there have been elements of this prediction that have come true, at least partially. Based on Malthus's theory, the instantaneous change in population growth is

$$\frac{dN}{dt} = rN \tag{15.7}$$

where N is the population size, t is time, and r is the intrinsic growth rate. The solution to equation (15.7) is known as the Malthusian equation for exponential increases in population size:

$$N_t = N_0 e^{rt}. \tag{15.8}$$

In the presence of limited resources, the population would eventually approach the carrying capacity of the environment and density-dependent effects would become important. The simplest model accounting for carrying capacity is the Verhulst-Pearl logistic equation

$$\frac{dN}{dt} = N(\lambda - sN). \tag{15.9}$$

A criticism of the differential and difference equations approach to modeling population growth is their deterministic nature. Therefore, stochastic versions of the deterministic population growth models have also been developed using birth and death processes. For example, let λ be the birth rate and μ be the death rate. Then $\lambda - \mu$ is the intrinsic rate of natural increase. The stochastic form of this simple birth and death process is

$$\frac{dp_N(t)}{dt} = -N(\lambda + \mu)p_N(t) + \lambda(N - 1)p_{N-1}(t) + \mu(N + 1)p_{N+1}(t) \tag{15.10}$$

where $p_N(t)$ is the probability that the population size will be N at time t. Stochastic versions of logistic growth have also been developed. Further extensions of these models attempt to more nearly reflect reality.

Murray (1993) has a thorough discussion of both difference and differential equations and their biological applications. Stochastic versions of these can be found in Karlin and Taylor (1975, 1981). Our experience is that these models tend to fit only very restrictive cases. Will we be able to develop accurate population models? What, if anything, does chaos have to do with our ability to construct such models? We now turn our attention to answering these questions.

Chaos

The complexity of ecological systems is staggering and to define and explain these systems is almost beyond human ability. The smallest ecosystem has large numbers of interacting organisms and conditions. Ecologists have catalogued the animals and plants in many types of ecosystems. The dominant animals or plants as well as the prevailing climate have been identified. Food chains and groups of food chains have been listed. This process has sometimes led us to think that when we get all the ecosystems catalogued and when we can define how all the food chains work, we will be able to build models for these systems. This Newtonian deterministic approach to nature has been wide spread in science, primarily because of the tremendous success of physics in explaining and modeling many natural phenomena. Differential equations have been used to model events such as tidal and star movements, but they have not been very useful for modeling populations of organisms.

One of the ways designed to help explain ecological systems is *chaos theory*. Chaos has no set definition. As it applies to ecology, and biology in general, it is a series of nonlinear events. Any complex system has some aspects of nonlinearity, and all ecosystems are complex. Even the even-aged monocultures occurring in agriculture are complex and will be chaotic during at least some period of their existence. This has grim implications for our ability to model these systems for any period of time. Like predicting the weather, any ecological model will only be accurate for short periods. The reason for this is that in complex systems, small changes in one event may cause large changes over time. This is the so-called *butterfly effect*.

The butterfly effect was first used by Edward Lorenz, the father of modern chaos theory (Lorenz et al., 1963a,b, 1964). The butterfly effect signifies that small changes in starting conditions can dramatically impact outcomes of simple or complex events. Lorenz encountered this effect in his study of long-range weather forecasting and concluded that such efforts are largely futile. After World War II, it was thought that it was only a matter of time until weather could be not only accurately predicted, but also controlled (Gleick, 1987). All that was needed was enough sensors that could take accurate data. However, Lorenz realized that if sensors could be placed in every cubic foot of the earth's atmosphere from the ground to the stratosphere, small events, such as the beating of a butterfly's wings at a point in between the sensors, would go undetected. Although the effect would be initially small, through time it would result in dramatic changes in the weather. Because of this, Lorenz concluded that accurate weather prediction beyond 2 or 3 days was futile.

Lorenz came to recognize the butterfly effect while developing a weather model using a primitive computer. He had to disrupt the program. When he started it again, he entered the starting value accurately to three decimal points. When he returned, the output had changed dramatically. At first, the differences were extremely small, but they continued to increase until the two outcomes were very different. Thus, a butterfly beating its wings in Peking, China, can affect the weather in New York (Gleick, 1987). Thus the butterfly effect is more formally defined as a sensitivity to initial conditions (Cambel, 1993). Because many ecological processes are weather-driven, our inability to predict the weather does not bode well for our efforts to model population dynamics.

Ecology is filled with examples where sensitivities to initial conditions affect the outcome. It might be said that the whole of ecological events is a collection of rare events that were determined by some initial condition. A small change in the initial condition would have led to an entirely different set of observed events. Think of a bird sitting on a branch. What is the probability that this particular bird would be on this branch at precisely this moment? How many small happenings could have occurred to prevent it? What is the likelihood of your being in a position to be watching this bird?

This points out another feature that ecologists have noted. Events vary in their sensitivity. This sensitivity of initial and subsequent events is the reason that variety tests for crops are conducted over several years at a number of locations before recommendations are made on their usage. Testing in several locations for several years does not always ensure that the variety will do well on the date and in the location that a crop is planted. A conspicuous example is Comanche wheat that was released in Oklahoma. The variety had been tested for five years in many parts of the state. In each year, it had produced outstanding yields and quality. The year that it was released was a much wetter year than any year it had been tested. The results were disastrous. Comanche wheat was very susceptible to the initial condition of wet, cool soils.

Anyone who has followed variety testing at any experiment station has seen examples of location, time of planting and variety interaction. Home gardeners experience this sensitivity when they have multiple plantings of corn or tomatoes. Home gardeners often plant corn at weekly intervals to produce a longer period for fresh corn eating. In many years, these weekly plantings do not show consistent growth differences. If the same varieties of tomatoes are planted in a garden, the results vary from planting to planting. The same variety will produce different quantities and qualities of tomatoes in different years and at different planting times.

Even though small changes in initial conditions can result in large differences in outcomes, the outcomes in nature tend to be smooth. Regular cycles are observable. It would seem ecologists are like two-dimensional creatures in a three-dimensional world. There must be many things we can not see, or if we see them, we do not know what they mean. What we need is a breakthrough, a new way of approaching ecosystem studies, so that the regularity of nature can be more readily quantified and understood. As with other discoveries, it is very likely that the answers will be simple and obvious in retrospect. We have only to seek and to find them.

Summary

Difference and differential equations have provided a foundation for modeling population dynamics within agroecosystems. Deterministic models have dominated, but stochastic versions of many deterministic models are available. These have been instructive in helping understand population dynamics, but they have not been widely useful in providing good models of complex systems. Chaos theory provides insight into why modeling population dynamics over extended time periods has been largely unsuccessful. Lorenz recognized the futility in long-range weather prediction because extremely small changes in initial conditions resulted in large differences in outcomes over time. Ecosystems are largely weather-driven and thus are subject to sensitivity in initial conditions. Yet, weather is only one of many elements in the ecosystem that is sensitive to slight changes in conditions. In spite of this, nature exhibits regular patterns. Some new approach is needed to describing population dynamics in ecosystems.

References

Akaike, H. 1973. Information theory and an extension of the maximum likelihood principle. In B.N. Petran and F. Csáki (eds.): *International Symposium on Information Criterion*, 2nd Ed. Akadémiai Kiadi, Budapest, Hungary, pp. 267–281.

Albrecht, P. 1980. On the correct use of the chi-square goodness-of-fit test. *Scandinavian Actuarial Journal* 149–160.

Alho, J.M. 1990. Logistic regression in capture–recapture models. *Biometrics* **46:** 623–635.

Allen, J.C. 1976. A modified sine wave method for calculating degree days. *Environmental Entomology* **5:** 388–396.

Allen, J.C., D. Gonzalez, and D.V. Gokhale. 1972. Sequential sampling plans for the bollworm, *Heliothis zea. Environmental Entomology* **1:** 771–780.

Anderson, D.R., and R.S. Pospahala. 1970. Correction of bias in belt transects of immotile objects. *Journal of Wildlife Management* **34:** 141–146.

Anderson, D.R., and R.T. Sterling. 1974. Population dynamics of molting pintail drakes banded in south-central Saskatchewan. *Journal of Wildlife Management* **38:** 266–274.

Anderson, D.R., J.L. Laake, B.R. Crain, and K.P. Burnham. 1979. Guidelines for line transect sampling of biological populations. *Journal of Wildlife Management* **43:** 70–78.

Anderson, D.R., K.P. Burnham, and B.R. Crain. 1979. Line transect estimation of population size: The exponential case with grouped data. *Communications in Statistics— Theory and Methods* **A8:** 487–507.

Anderson, D.R., K.P. Burnham, and B.R. Crain. 1980. Some comments on Anderson's and Pospahala's correction of bias in line transect sampling. *Biometrical Journal* **22:** 513–524.

Anderson, J. 1962. Roe-deer census and population analysis by means of modified marking release technique. In E.D. Le Cren and M.W. Holdgate (eds): *The Exploitation of Natural Animal Populations* Blackwell Scientific, Oxford, pp. 72–82.

Anscombe, F.J. 1949. The statistical analysis of insect counts based on the negative binomial distribution. *Biometrics* **5:** 165–173.

Anscombe, F.J. 1950. Sampling theory of the negative binomial and logarithmic series distributions. *Biometrika* **37**: 358–382.

Armitage, P. 1947. Some sequential tests of Student's hypotheses. *Journal of the Royal Statistical Society Suppl.* **9**: 250–263.

Armitage, P. 1950. Sequential analysis with more than two alternative hypotheses, and its relation to discriminant function analysis. *Journal of the Royal Statistical Society B* **12**: 137–144.

Armitage, P. 1957. Restricted sequential procedures. *Biometrika* **44**: 9–26.

Arnason, A.N., and C.J. Schwartz. 1986. POPAN-3. *Extended Analysis and Testing Features for POPAN-2.* Charles Babbage Research Center, St. P'erre, Manitoba.

Arnold, C.Y. 1959. The determination and significance of the base temperature in a linear heat unit system. *Proceedings of the American Society of Horticulture Science* **74**: 430–435.

Ashford, J.R., K.L.Q. Read, and G.G. Vickers. 1970. A system of stochastic models applicable to animal population dynamics. *Journal of Animal Ecology* **39**: 29–50.

Atchley, W.R., C.T. Gaskins, and D. Anderson. 1976. Statistical properties of ratios. I. Empirical results. *Systematic Zoology* **25**: 137–148.

Ayre, G.L. 1962. Problems in using the Lincoln index for estimating the size of ant colonies (Hymenoptera: Formicidae), *Journal of the New York Entomological Society* **70**: 159–166.

Bailey, N.T.J. 1951. On estimating the size of mobile populations from capture–recapture data. *Biometrika* **38**: 293–306.

Bailey, N.T.J. 1952. Improvements in the interpretation of recapture data. *Journal of Animal Ecology* **21**: 120–127.

Bailey, G.N.A. 1968. Trap-shyness in a woodland population of Bank voles (*Clethrionomys glareolus*). *Journal of Zoology London* **156**: 517–521.

Bain, L.J., and M. Engelhardt. 1992. *Introduction to Probability and Mathematical Statistics,* 2nd Ed. Duxbury Press, New York.

Banerjee, B. 1976. Variance to mean ratio and the spatial distribution of animals. *Experientia.* **32**: 993–994.

Bart, J., and D.S. Robson. 1982. Estimating survivorship when the subjects are visited periodically. *Ecology* **63**: 1078–1090.

Bartlett, M.S. 1964. The spectral analysis of two-dimensional point processes. *Biometrika* **51**: 299–311.

Baskerville, C.L., and P. Emin. 1969. Rapid estimation of heat accumulation from maximum and minimum temperatures. *Ecology* **50**: 514–517.

Beall, G. 1939. Methods of estimating the population of insects in a field. *Biometrika* **30**: 422–439.

Beall, G. 1942. The transformation of data from entomological field experiments so that the analysis of variance becomes applicable. *Biometrika* **32**: 243–262.

Bellows, T.S., and M.H. Birley. 1981. Estimating developmental and mortality rates and stage recruitment from insect stage-frequency data. *Researches on Population Ecology* **23:** 232–244.

Bernardelli, H. 1941. Population waves. *Journal of the Burma Research Society* **31:** 1–18.

Besag, J.E. 1974. Spatial interaction and the statistical analysis of lattice systems. *Journal of the Royal Statistical Society B* **36:** 192–225.

Besag, J.E., and J.T. Gleaves. 1973. On the detection of spatial pattern in plant communities. *Bulletin of the International Statistical Institute* **45 Book 1:** 153–158.

Beverton, R.J.H., and S.J. Holt. 1957. *On the Dynamics of Exploited Fish Populations.* Her Majesty's Stationery Office, London.

Beyer, W.H. (ed.). 1968. *Handbook of Tables for Probability and Statistics,* 2nd ed. The Chemical Rubber Co., Cleveland, OH.

Bibby, C.J., and S.T. Buckland. 1987. Bias of bird census results due to detectability varying with habitat. *Acta Ecologica* **8:** 103–112.

Billard, L. 1977. Optimum partial sequential tests for two-sided tests of the binomial parameter. *Journal of the American Statistical Association* **72:** 197–201.

Billard, L., and M.K. Vagholkar. 1969. A sequential procedure for testing a null hypothesis against a two-sided alternative hypothesis. *Journal of the Royal Statistical Society, Series B* **31:** 285–294.

Birley, M. 1977. The estimation of insect density and instar survivorship functions from census data. *Journal of Animal Ecology* **46:** 497–510.

Birley, M. 1979. The estimation and simulation of variable development period, with applications to the mosquito *Aedes aegypti* (L.). *Researches on Population Ecology* **21:** 68–80.

Bliss, C.I. 1934. The method of probits. *Science* **79:** 38–39.

Bliss, C.I. 1935. The calculation of the dosage–mortality curve. *Annals of Applied Biology* **22:** 134–167.

Bliss, C.I., and R.A Fisher. 1953. Fitting the negative binomial distribution to biological data and note on the efficient fitting of the negative binomial. *Biometrics* **9:** 176–200.

Bliss, C.I. and A.R.G. Owens. 1958. Negative binomial distributions with a common k. *Biometrika* **45:** 37–58.

Bliss, C.I., and A.R.G Owens. 1958. Negative binomial distributions with a common k. *Biometrika* **45:** 37–58.

Boswell, M.T., and G.P. Patil. 1970. Chance mechanisms generating the negative binomial distribution. In G.P. Patil (ed.): *Random Counts in Scientific Work.* Vol 1. Pennsylvania State University, University Park, pp. 3–22.

Braner, M., and N.G. Hairston. 1989. From cohort data to life table parameters via stochastic modelling. In I.L. McDonald, B.F.J. Manly, J.A. Lockwood, and J.A. Logan (eds.): *Estimation and Analysis of Insect Populations.* pp 81–92. Springer-Verlag Lecture Notes in Statistics Vol. 5. Springer-Verlag, Berlin.

Brock, V.E. 1954. A preliminary report on a method of estimating reef fish populations. *Journal of Wildlife Management* **18:** 297–308.

Brownie, C., and D.S. Robson. 1983. Estimation of time-specific survival rates from tag-resighting samples: A generalization of the Jolly–Seber model. *Biometrics* **39:** 437–453.

Brownie, C., D.R. Anderson, K.P. Burnham, and D.S. Robson 1985. *Statistical Inference from Band Recovery Data—A Handbook*. 2nd Ed. Resource Publ. No. 156. U.S. Department of the Interior, Fish and Wildlife Service, Washington, D.C., 305 pp.

Buckland, S.T. 1980. A modified analysis of the Jolly–Seber capture–recapture survival model. *Biometrics* **36:** 419–435.

Buckland, S.T. 1982. A mark-recapture survival analysis. *Journal of Animal Ecology* **51:** 833–847.

Buckland, S.T. 1985. Perpendicular distance models for line transect sampling. *Biometric* **41:** 177–195.

Buckland, S.T., D.R. Anderson, K.P. Burnham, and J.L. Laake. 1993. *Distance Sampling: Estimating Abundance of Biological Populations*. Chapman & Hall, London.

Burnham, K.P. 1972. Estimation of Population Size in Multiple Capture–Recapture Studies When Capture Probabilities Vary Among Animals. Ph.D. dissertation. Oregon State University, Corvallis, OR.

Burnham, K.P., and D.R. Anderson. 1992. Data-based selection of an appropriate biological model: The key to modern data analysis. In D.R. McCullough and R.H. Barrett (eds.): *Wildlife 2001: Populations*. Elsevier Science Publishers, London, pp. 16–30.

Burnham, K.P., and W.S. Overton. 1978. Estimation of the size of a closed population when capture probabilities vary among animals. *Biometrika* **65:** 625–633.

Burnham, K.P., and W.S. Overton. 1979. Robust estimation of population size when capture probabilities vary among animals. *Ecology* **60:** 927–936

Burnham, K.P., D.R. Anderson, and J.L. Laake. 1980. Estimation of density from line transect sampling of biological populations. *Wildlife Monograph* **72:** 1–202.

Burnham, K.P., D.R. Anderson, G.C. White, C. Brownie, and K.H. Pollock. 1987. *Design and Analysis Methods for Fish Survival Experiments Based on Release–Recapture*. Monograph No. 5. American Fisheries Society.

Byth, K., and B.D. Ripley. 1980. On sampling spatial patterns by distance methods. *Biometrics* **36:** 279–284.

Cambel, A.B. 1993. *Applied Chaos Theory*. Academic Press. San Diego.

Campbell, A., B.D. Fraser, N. Gilbert, A.P. Gutierrez, and M. Mackaner. 1974. Temperature requirements of some aphids and their parasites. *Journal of Applied Ecology* **11:** 431–438.

Carothers, A.D. 1973. Capture–recapture methods applied to a population with known parameters. *Journal of Animal Ecology* **42:** 125–146.

Carothers, A.D. 1979. Quantifying unequal catchability and its effect on survival estimates in an actual population. *Journal of Animal Ecology* **48:** 863–869.

Caughley, G. 1966. Mortality patterns in mammals. *Ecology* **47:** 906–918.

Caughley, G. 1977a. *Analysis of Vertebrate Populations*. John Wiley & Sons, London.

Caughley, G. 1977b. Sampling in aerial survey. *Journal of Wildlife Management* **41:** 605–615.

Chakravarti, I.M., R.G. Laha, and J. Roy. 1967. *Handbook of Methods of Applied Statistics,* Vol. 1. John Wiley & Sons, New York.

Chapman, D.G. 1948. A mathematical study of confidence limits of salmon populations calculated from sample tag ratios. *International Pacific Salmon Fisheries Community Bulletin* **2:** 69–85.

Chapman, D.G. 1951. Some properties of the hypergeometric distribution with applications to zoological censuses. *University of California Publications in Statistics* **1:** 131–160.

Chapman, D.G. 1952. Inverse multiple and sequential sample censuses. *Biometrics* **8:** 286–306.

Chapman, D.G. 1954. The estimation of biological populations. *Annals of Mathematical Statistics* **25:** 1–15.

Chapman, D.G. 1955. Population estimation based on change of composition caused by selective removal. *Biometrika* **42:** 279–290.

Chapman, D.G. 1965. The estimation of mortality and recruitment from a single-tagging experiment. *Biometrics* **21:** 529–542.

Chow, Y.S., and H. Robbins. 1965. On the asymptotic theory of fixed-width confidence intervals for the mean. *Annals of Mathematical Statistics* **36:** 457–462.

Chung, J.H., and D.B. DeLury. 1950. *Confidence Limits for the Hypergeometric Distribution.* University of Toronto Press, Toronto.

Cliff, A.D., and J.K. Ord. 1973. *Spatial Autocorrelation.* Pion Limited, London.

Cliff, A.D., and J.K. Ord. 1981. *Spatial Processes: Models and Applications.* Pion Limited, London.

Cochran, W.G. 1977. *Sampling Techniques.* 3rd Ed. John Wiley & Sons, New York.

Cohen, A.C., Jr. 1960. An extension of a truncated Poisson distribution. *Biometrics* **16:** 446–450.

Conner, M.C., and R.F. Labisky. 1985. Evaluation of radioisotope tagging for estimating abundance of raccoon populations. *Journal of Wildlife Management* **49:** 326–332.

Connor, M.C., R.A. Lancia, and K.H. Pollock. 1986. Precision of the change-in-ratio technique for deer population management. *Journal of Wildlife Management* **50:** 125–129.

Conover, W.J. 1972. A Kolmogorov goodness-of-fit test for discontinuous distributions. *Journal of the American Statistical Association* **67:** 591–596.

Conroy, M.J., and B.K. Williams. 1984. A general methodology for maximum likelihood inference from band-recovery data. *Biometrics* **40:** 739–748.

Cormack, R.M. 1964. Estimates of survival from the sighting of marked animals. *Biometrika* **51:** 429–438.

Cormack, R.M. 1966. A test for equal catchability. *Biometrics* **22:** 330–342.

Cormack, R.M. 1968. The statistics of capture-recapture methods. *Oceanography and Marine Biology, An Annual Review* **6:** 455–506.

Cormack, R.M. 1979. Spatial aspects of competition between individuals. In R.M. Cormack and J.K. Ord (eds.): *Spatial and Temporal Analysis in Ecology.* International Cooperative Publishing House, Fairland, MD, pp. 151–212.

Cormack, R.M. 1989. Loglinear models for capture–recapture. *Biometrics* **45:** 395–413.

Cornelius, W.L., and K.H. Pollock. 1987. *A Review of Nest Survival Models.* Technical Report 1908, North Carolina State University, Department of Statistics, Raleigh, NC.

Corneliussen, A.H., and D.W. Ladd. 1970. On sequential tests of the binomial distribution. *Technometrics* **12:** 635–646.

Cox, D.R. 1972. Regression models and life tables (with discussion). *Journal of the Royal Statistical Society, Series B* **34:** 187–220.

Crain, B.R., K.P. Burnham, D.R. Anderson, and J.L. Laake. 1978. *A Fourier Series Estimator of Population Density for Line Transect Sampling.* Utah State University Press, Logan, UT.

Crain, B.R., K.P. Burnham, D.R. Anderson, and J.L. Laake. 1979. Nonparametric estimation of population density for line transect sampling using Fourier series. *Biometrical Journal* **21:** 731–748.

Cressie, N.A.C. 1993. *Statistics for Spatial Data.* John Wiley & Sons, New York, 900 pp.

Cressie, N., and D.M. Hawkins. 1980. Robust estimation of the variogram. I. *Journal of the International Association for Mathematical Geology* **12:** 115–125.

Cressie, N., and T.R.C. Read. 1984. Multinomial goodness-of-fit tests. *Journal of the Royal Statistical Society* **46:** 440–464.

Cressie, N., and T.R.C. Read. 1989. Pearson χ^2 and the loglikelihood ratio statistic G^2: A comparative review. *International Statistical Review* **57:** 19–43.

Crick, F.H.C., and P.A. Lawrence. 1975. Compartments and polychones in insect development. *Science* **189:** 340–347.

Crouse, D.T., L.B. Crowder, and H. Caswell. 1987. A stage-based population model for loggerhead sea turtles and implications for conservation. *Ecology* **68:** 1412–1423.

Dacy, M.F. 1965. The geometry of central place theory. *Geografiska Annaler* **47B:** 505–515.

Darroch, J.N. 1958. The multiple-recapture census. I. Estimation of a closed population. *Biometrika* **45:** 343–359.

Darroch, J.N. 1959. The multiple-recapture census. II. Estimation when there is immigration or death. *Biometrika* **46:** 336–351.

David, F.N. and P.G. Moore. 1954. Notes on contagious distributions in plant populations. *Annals of Botany London N.S.* **18:** 47–53.

Deevey, E.S. Jr. 1947. Life tables for natural populations of animals. *Quarterly Review of Biology* **22:** 2283–2314.

DeLury, D.B. 1947. On the estimation of biological populations. *Biometrics* **3:** 145–167.

Derron, M. 1962. Mathematische problem der automobilversichervng. *MVSV* **62:** 103–123.

Dice, L.R. 1938. Some census methods for mammals. *Journal of Wildlife Management* **2:** 119–130.

Diggle, P.J. 1975. Robust density estimation using distance methods. *Biometrika* **62:** 39–48.

Diggle, P.J. 1977. The detection of random heterogeneity in plant populations. *Biometrics* **33:** 390–394.

Diggle, P.J. 1979. On parameter estimation and goodness-of-fit testing for spatial point patterns. *Biometrics* **35:** 87–101.

Diggle, P.J. 1983. *Statistical Analysis of Spatial Point Patterns.* Academic Press, London, 148 pp.

Diggle, P.J., J.E. and Besag, and J.T. Gleaves. 1976. Statistical analysis of spatial point patterns by means of distance methods. *Biometrics* **32:** 659–667.

Douglas, J.B. 1955. Fitting the neyman type A (two parameter) contagious distribution. *Biometrics* **11:** 149–158.

Douglas, J.B. 1975. Clustering and aggregation. *Sankhyā B* **37:** 398–417.

Drummer, T.D. 1990. Estimation of proportions and ratios from line transect data. *Communications in Statistics—Theory and Methods* **A19:** 3069–3091.

Drummer, T.D., and L.L. McDonald. 1987. Size bias in line transect sampling. *Biometrics* **43:** 13–21.

Eberhardt, L.L. 1968. A preliminary appraisal of line transect. *Journal of Wildlife Management* **42:** 1–31.

Eberhardt, L.L. 1969. Population estimates from recapture frequencies. *Journal of Wildlife Management* **33:** 28–39.

Edmondson, W.T. 1945. Ecological studies of sessile Rotatoria. Part II. Dynamics of populations and social structures. *Ecological Monographs* **15:** 141–172.

Edwards, R.Y. 1954. Comparison of an aerial and ground census of moose. *Journal of Wildlife Management* **18:** 403–404.

Edwards, W.R., and L.L. Eberhardt. 1967. Estimating cottontail abundance from live trapping data. *Journal of Wildlife Management* **31:** 87–96.

Evans, W. 1975. Methods of Estimating Densities of White-Tailed Deer. Ph.D. dissertation. Texas A&M University, College Station, TX.

Evans, M.A., D.G. Bonett, and L.L. McDonald. 1994. A general theory for modeling capture–recapture data from a closed population. *Biometrics* **50:** 396–405.

Fairfield Smith, H. 1938. An empirical law describing heterogeneity in the yields of agricultural crops. *Journal of Agricultural Science (Cambridge)* **28:** 1–23.

Farner, D.S. 1945. Age groups and longevity in the American robin. *Wilson Bulletin* **57:** 56–74.

Feller, W. 1945. On the normal approximation to the binomial distribution. *Annals of Mathematical Statistics* **16:** 319–329.

Feller, W. 1968. *An Introduction to Probability Theory and Its Applications.* Vol. 1. 3rd Ed. John Wiley & Sons, New York.

Fienberg, S.E. 1972. The multiple recapture census for closed populations in incomplete 2^k contingency tables. *Biometrika* **59:** 591–603.

Finney, D.J. 1973. *Statistical Method in Biological Assay* 2nd Ed. Hafner, New York.

Fischler, K.J. 1965. The use of catch-effort, catch-sampling, and tagging data to estimate a population of blue crabs. *Transactions of the American Fisheries Society* **94:** 287–310.

Fisher, R.A. 1941. The negative binomial distribution. *Annals of Eugenics, London* **11:** 182–187.

Fisher, R.A., H.G. Thornton, and W.A. Mackenzie. 1922. The accuracy of the plating method of estimating the density of bacterial populations, with particular reference to the use of Thornton's agar medium with soil samples. *Annals of Applied Biology* **9:** 325–329.

Fisz, M. 1963. *Probability Theory and Mathematical Statistics.* John Wiley & Sons, New York.

Flyger, V.F. 1959. A comparison of methods for estimating squirrel populations. *Journal of Wildlife Management* **23:** 220–223.

Forrester, G.E. 1994. Influences of predatory fish on the drift dispersal and local density of stream insects. *Ecology* **75:** 1208–1218.

Fowler, G.W. 1985. The use of Wald's sequential probability ratio test to develop composite three-decision sampling plans. *Canadian Journal of Forest Research* **15:** 326–330.

Fowler, G.W., and A.M. Lynch. 1987. Sampling plans in insect pest management based on Wald's sequential probability ratio test. *Environmental Entomology* **16:** 345–354.

Frazer, N.B. 1983. Demography and Life History Evolution of the Atlantic Loggerhead Sea Turtle, *Caretta caretta.* Ph.D. dissertation. University of Georgia, Athens, GA.

Freeman, M.F. and J.W. Tukey. 1950. Transformations related to the angular and the square root. *Annals of Mathematical Statistics* **27:** 607–611.

Funderburk, J.E., L.G. Higley, and L.P. Pedigo. 1984. Seedcorn maggot (Diptera: Anthomyiidae) phenology in central Iowa and examination of a thermal unit system to predict development under field conditions. *Environmental Entomology* **13:** 105–109.

Gaddum, J.H. 1933. Reports on biological standards. III. Methods of biological assay depending on a quantal response. Special Report Series No. 183. Medical Research Council, London.

Gates, C.E. 1969. Simulation study of estimation for the line transect sampling method. *Biometrics* **25:** 317–328.

Gates, C.E. 1979. Line transect and related issues. In R.M. Cormack, G.P. Patil, and D.S. Robson (eds.): *Sampling Biological Populations.* International Co-operative Publishing House, Fairland, MD, pp. 71–154.

Gates, C.E. 1989. Discrete, a computer program for fitting discrete frequency distributions. In L. McDonald, B. Manly, J. Lockwood and J. Logan, eds. *Lecture Notes in Statistics: Estimation and Analysis of Insect Populations.* Springer-Verlag, Berlin, pp. 458–466.

Gates, C.E., and F.G. Ethridge. 1972. A generalized set of discrete frequency distributions with FORTRAN program. *International Association for Mathematical Geology* **4:** 1–24.

Gates, C.E., and P.W. Smith. 1980. An implementation of the Burnham-Anderson distribution free method of estimating wildlife densities from line transect data. *Biometrics* **36:** 155–160.

Gates, C.E., W.H. Marshall, and D.P. Olson. 1968. Line transect method of estimating grouse population densities. *Biometrics* **24:** 135–145.

Gates, C.E., W. Evans, D.R. Gober, et al. 1985. Line transect estimation of animal densities from large data sets. In S.L. Beasom, and S.F. Roberson (eds.): *Game Harvest Management.* Caesar Kleberg Wildlife Research Institute, Texas A&I University, Kingsville, Texas. pp. 37–50.

Gates, C.E., F.G. Ethridge, and J.D. Geaghan. 1987. *Fitting Discrete Distributions. User's Documentation for the* FORTRAN *Computer Program* DISCRETE. Texas A&M University, College Station, TX.

Geary, R.C. 1954. The contiguity ratio and statistical mapping. *The Incorporated Statistician* **5:** 115–145.

Geis, A.D. 1955. Trap response of the cottontail rabbit and its effect on censusing. *Journal of Wildlife Management* **19:** 466–472.

Gerking, S.D. 1967. Statistics of the fish population of Gordy Lake, Indiana. In *The Biological Basis of Freshwater Fish Production.* Blackwell Scientific, London.

Ghosh, M. and N. Mukhopadhyay. 1979. Sequential point estimation of the mean when the distribution is unspecified. *Communications in Statistics-Theory and Methods* **A8:** 637–652.

Gibbs, J.W. 1902. *Elementary Principles in Statistical Mechanics.* Reprinted 1960. Dover, New York.

Gleick, J. 1987. *Chaos Making a New Science.* Penguin, New York.

Gotway, C.A. 1991. Fitting semivariogram models by weighted least squares. *Computers and Geosciences* **17:** 171–172.

Green, R.F. 1977. Do more birds produce fewer young? A comment on Mayfield's measure of nest success. *Wilson Bulletin* **89:** 173–175.

Green, R.G. and C.A. Evans. 1940. Studies on a population cycle of snowshoe hares on the Lake Alexander area. I. Gross annual census, 1932–1939. *Journal of Wildlife Management* **4:** 220–238.

Gulland, J.A. 1963. On the analysis of double-tagging experiments. North Atlantic Fish Marking Symposium. I.C.N.A.F. Special Publication, **4:** 228–229.

Gurland, J. 1959. Som Applications of the negative binomial and other contagious distributions. *American Journal of Public Health* **49:** 1388–1399.

Hartsack, A.W. 1973. *USDA–ARS Annual Report, Part I, Mothz-2.* Texas A&M University, College Station, TX.

Hartsack, A.W., J.A. Witz, J.P. Hollingsworth, R.L. Ridgway, and J.D. Lopez. 1976. MOTHZV-2: A computer simulation of *Heliothis zea* and *Heliothis virescens* population dynamics. *USDA/ARS Usser's Manual* ARS-S-127.

Hassell, M.P., T.R.E. Southwood, and P.M. Reader. 1987. The dynamics of the viburnum whitefly (*Aleurotrachelus jelinekii*): A case study of population regulation. *Journal of Animal Ecology* **56:** 283–300.

Hayes, R.J., and S.T. Buckland. 1983. Radial-distance models for the line-transect method. *Biometrics* **39:** 29–42.

Hayne, D.W. 1949a. An examination of the strip census method for estimating animal populations. *Journal of Wildlife Management* **13**: 145–157.

Hayne, D.W. 1949b. Two methods for estimating population from trapping records. *Journal of Mammalogy* **30**: 399–411.

Hefferman, P.M. 1996. Improved sequential probability ratio tests for negative binomial populations. *Biometrics* **52**: 152–157.

Heimbuch, D.G., and J.M. Hoenig 1989. Change-in-ratio estimators for habitat usage and relative population size. *Biometrics* **45**: 439–451.

Heisey, D.M., and T.K. Fuller. 1985. Evaluation of survival and cause specific mortality rates using telemetry data. *Journal of Wildlife Management* **49**: 668–674.

Hemingway, P. 1971. Field trials of the line transect method of sampling large populations of herbivores. In E. Duffey and A.S. Watts (eds.): *The Scientific Management of Animal and Plant Communities for Conservation.* Blackwell Scientific, Oxford, pp. 405–411.

Higley, L.G., and R.K.D. Peterson. 1994. Initializing sampling programs. In L.P. Pedigo and G.D. Buntin (eds.): *Handbook of Sampling Methods for Arthropods in Agriculture.* pp. 119–136, CRC Press, Boca Raton, FL.

Higley, L.G., L.P. Pedigo, and K.R. Ostlie. 1986. DEGDAY: A program for calculating degree-days, and assumptions behind the degree-day approach. *Environmental Entomology* **15**: 999–1016.

Hill, B.G., R.W. McNew, J.H. Young, and W.E. Ruth. 1975. The effects of sampling-unit size in some southwestern Oklahoma cotton insects. *Environmental Entomology* **4**: 491–494.

Hines, W.G.S., and R.J.O. Hines. 1979. The Eberhardt index and the detection of non-randomness of spatial point distributions. *Biometrika* **66**: 73–79.

Hinz, P., and J. Gurland. 1970. A test of fit for the negative binomial and other contagious distributions. *Journal of the American Statistical Association* **65**: 887–903.

Hoeffding, W. 1965. Asymptotically optimal tests for the multinomial distribution. *Annals of Mathematical Statistics* **36**: 369–401.

Hogg, R.V., and A.T. Craig. 1995. *Introduction to Mathematical Statistics.* 5th Ed. Prentice-Hall, Upper Saddle River, NJ.

Holbrook, S.J., and R.J. Schmitt. 1989. Resource overlap, prey dynamics, and the strength of competition. *Ecology* **70**: 1943–1953.

Holgate, P. 1965. Tests of randomness based on distance methods. *Biometrika* **52**: 345–353.

Hope, A.C.A. 1968. A simplified Monte Carlo significance test procedure. *Journal of the Royal Statistical Society B* **30**: 582–598.

Hopkins, B. 1954. A new method of determining the type of distribution of plant individuals. *Annals of Botany* **18**: 213–226.

Howe, R.W. 1967. Temperature effects on embryonic development in insects. *Annual Review of Entomology* **12**: 15–42.

Hubbard, D.J., and O.B. Allen. 1991. Robustness of the SPRT for a negative binomial to misspecification of the dispersion parameter. *Biometrics* **47**: 419–427.

Huchinson, T.P. 1979. The validity of the chi-squared test when expected frequencies are small: A list of recent research references. *Communications in Statistics Theory and Methods* **A8:** 327–335.

Huffman, M.D. 1983. An efficient approximate solution to the Kiefer–Weiss problem. *Annals of Statistics* **11:** 306–316.

Huggins, R.M. 1989. On the statistical analysis of capture experiments. *Biometrika* **76:** 133–140.

Huggins, R.M. 1991. Some practical aspects of a conditional likelihood approach to capture experiments. *Biometrics* **47:** 725–732.

Isaaks, E.H., and R.M. Srivastava. 1989. *An Introduction to Applied Geostatistics.* Oxford University Press, New York.

Ito, Y. 1972. On the methods for determining density-dependence by means of regression. *Oecologia* **10:** 347–372.

Iwao, S. 1968. A new regression method for analyzing the aggregation pattern of animal populations. *Research Population Ecology* **10:** 1–20.

Iwao, S. 1970a. Probability of spatial distribution in animal population ecology. In G.P. Patil (ed.): *Random Counts in Scientific Work,* Vol. 2. Pennsylvania State University Press, Philadelphia, pp. 117–149.

Iwao, S. 1970b. Analysis of contagiousness in the action of mortality factors on the western tent caterpillar population by using the m*–m relationship. *Researches on Population Ecology* **12:** 100–110.

Iwao, S. 1975. A new method of sequential sampling to classify populations relative to a critical density. *Researches on Population Ecology* **16:** 281–288.

Johnson, D.H. 1979. Estimating nest success: The Mayfield method and an alternative. *Auk* **96:** 651–661.

Johnson, N.L., S. Kotz, and A.W. Kemp. 1992. *Discrete Distributions.* 2nd Ed. John Wiley & Sons, New York.

Jolly, G.M. 1965. Explicit estimates from capture–recapture data with both death and immigration-stochastic model. *Biometrika* **52:** 225–247.

Jolly, G.M. 1982. Mark-recapture models with parameters constant in time. *Biometrics* **38:** 301–321.

Journel, A.G., and C.J. Huijbregts. 1978. *Mining Geostatistics.* Academic Press, London.

Kac, M. 1983. Marginalia. *American Scientist* **71:** 405–406.

Kaplan, E.L., and P. Meier. 1958. Nonparametric estimation from incomplete observations. *Journal of the American Statistical Association* **53:** 457–481.

Karadinos, M.G. 1976. Optimum sample size and comments on some published formulae. *Bulletin of the Entomology Society of America* **22:** 417–421.

Karlin, S., and H.M. Taylor. 1975. *A first Course in Stochastic Processes.* 2nd Ed. Academic Press, San Diego.

Karlin, S., and H.M. Taylor. 1981. *A Second Course in Stochastic Processes.* Academic Press, San Diego.

Kathirgamatamby, N. 1953. Note on the Poisson index of dispersion. *Biometrika* **40:** 225–228.

Kelker, G.H. 1940. Estimating deer population by a differential hunting loss in the sexes. *Proceedings of the Utah Academy of Science, Arts and Letters* **17:** 6–69.

Kelker, G.H. 1944. Sex-ratio equations and formulas for determining wildlife populations. *Proceedings of the Utah Academy of Science, Arts and Letters* **19–20:** 189–198.

Kelker, G.H. 1945. Measurement and Interpretation of Forces that Determine Populations of Managed Deer. Ph.D. dissertation. University of Michigan, Ann Arbor, MI.

Kempton, R.A. 1979. Statistical analysis of frequency data obtained from sampling an insect population grouped by stages. In J.K. Ord, G.P. Patil, and C. Taillie (eds.): *Statistical Distributions in Scientific Work* International Cooperative Publishing House: maryland. pp. 401–418.

Kennedy, W.J., Jr., and J.E. Gentle. 1980. *Statistical Computing.* Marcel Dekker, New York.

Kiritani, K., and F. Nakasuji. 1967. Estimation of the stage-specific survival rate in the insect population with overlapping stages. *Researches on Population Ecology* **9:** 143–152.

Kirk, J.R.V., and M.T. Aliniazee. 1981. Determining low temperature threshold for pupal development of the estern cherry fruit fly for use in phenology models. *Environmental Entomology* **10:** 968–971.

Kish, L. 1965. *Survey Sampling.* John Wiley & Sons, New York.

Kleczkowshi, A. 1949. The transformation of local lesion counts for statistical analysis. *Annals of Applied Biology* **36:** 139–152.

Kogan, M., and D.C. Herzog. 1980. *Sampling Methods in Soybean Entomology.* Springer-Verlag: New York.

Kooijman, S.A.L.M. 1983. Statistical aspects of the determination of mortality rates in bioassays. *Water Research* **17:** 749–759.

Krebs, C.J. 1989. *Ecological Methodology.* Harper & Row, Hagerstown, MD.

Kronmal, R., and M. Tarter. 1968. The estimation of probability densities and cumulatives by Fourier series methods. *Journal of the American Statistical Association* **63:** 925–952.

Kuno, E. 1969. A new method of sequential sampling to obtain the population estimates with a fixed level of precision. *Research in Population Ecology* **11:** 127–136.

Kuno, E. 1971. Sampling error as a misleading artifact in key factor analysis. *Researches on Population Ecology* **13:** 28–45.

Lamb, R.J. 1992. Development rate of *Acyrthosiphon pisum* (Homoptera: Aphididae) at low temperatures: Implications for estimating rate parameters for insects. *Environmental Entomology* **21:** 10–19.

Lamb, R.J., G.H. Gerber, and G.F. Atkinson. 1984. Comparison of development rate curves applied to egg hatching data of *Entomoscelis americana* (Coleoptera: Chrysomelidae). *Environmental Entomology* **13:** 868–872.

Laplace, P.S. 1786. Sur les naissances, les mariages et les morts. In *Histoire de l'académie royale des sciences.* Année. 1783, Paris.

Larntz, K. 1978. Small-sample comparisons of exact levels for chi-squared goodness-of-fit statistics. *Journal of the American Statistical Association* **73:** 253–263.

Lebreton, J.D., K.P. Burnham, J. Clobert, and D.R. Anderson. 1992. Modeling survival and testing biological hypotheses using marked animals: A unified approach with case studies. *Ecological Monographs* **62:** 67–118.

Lee, S.-M., and A. Chao. 1994. Estimating population size via sample coverage for closed capture-recapture models. *Biometrics* **50:** 88–97.

Lefkovitch, L.P. 1963. Census studies on unrestricted populations of *Laxioderma serricorne* (F.)(Coleoptera: Anobiidae). *Journal of Animal Ecology* **32:** 221–231.

Lefkovitch, L.P. 1964a. The growth of restricted populations of *Lasioderma serricorne* (F.)(Coleoptera: Anobiidae). *Bulletin of Entomological Research* **55:** 87–96.

Lefkovitch, L.P. 1964b. Estimating the Malthusian parameter from census data. *Nature* **204:** 810.

Lefkovitch, L.P. 1965. The study of population growth in organisms grouped by stages. *Biometrics* **21:** 1–18.

Lehmann, E.L. 1959. *Testing Statistical Hypotheses.* John Wiley & Sons, New York.

Leopold, A. 1933. *Game Management.* Charles Schribner's Sons: New York.

Leslie, P.H. 1945. On the use of matrices in certain population mathematics. *Biometrika* **33:** 182–212.

Leslie, P.H. 1948. Some further notes on the use of matrices in population mathematics. *Biometrika* **35:** 213–245.

Leslie, P.H. 1952. The estimation of population parameters from data obtained by means of the capture-recapture method. II. The estimation of total numbers. *Biometrika* **39:** 363–388.

Leslie, P.H. 1958. Statistical appendix. *Journal of Animal Ecology* **27:** 84–86.

Leslie, P.H., and D. Chitty. 1951. The estimation of population parameters from data obtained by means of the capture-recapture method. I. The maximum likelihood equations for estimating the death rate. *Biometrika* **38:** 269–292.

Leslie, P.H., and D.H.S. Davis. 1939. An attempt to determine the absolute number of rats on a given area. *Journal of Animal Ecology* **8:** 94–113.

Leslie, P.H., D. Chitty, and H. Chitty. 1953. The estimation of population parameters from data obtained by means of the capture–recapture method. III. An example of the practical applications of the methods. *Biometrika* **40:** 137–169.

Leslie, P.H., J.S. Tener, M. Vizoso, and H. Chitty. 1955. The longevity and fertility of the Orkney vole, *Microtus oreadensis,* as observed in the laboratory. *Proceedings of the Zoology Society London* **125:** 115–125.

Lewis, E.G. 1942. On the generation and growth of a population. *Sankhya* **6:** 93–96.

Lin, S.K. 1985. Characterization of Lightning as a Disturbance to the Forest Ecosystem in East Texas. M.S. Thesis. Texas A&M University, College Station, TX.

Lincoln, F.C. 1930. Calculating waterfowl abundance on the basis of banding returns. *Circular of the U.S. Department of Agriculture* No. 118: 1–4.

Little, T.M., and F.J. Hills. 1978. *Agricultural Experimentation Design and Analysis.* John Wiley & Sons, New York.

Lloyd, M. 1967. Mean crowding. *Journal of Animal Ecology* **36:** 1–30.

Loery, G., and J.D. Nichols. 1985. Dynamics of a black-capped chickadee population, 1958–1983. *Ecology* **66:** 1038–1044.

Loery, G., K.H. Pollock, J.D. Nichols, and J.E. Hines. 1987. Age-specificity of avian survival rates: An analysis of capture-recapture data for a black-capped chickadee population, 1958–1983. *Ecology* **68:** 1038–1044.

Lorden, G. 1976. 2-SPRT and the modified Kiefer–Weiss problem of minimizing the expected sample size. *Annals of Statistics* **4:** 281–291.

Lorden, G. 1980. Structure of sequential tests minimizing an expected sample size. *Zeitschrift für Wahrscheinlichkeits theorie und Verwandte Gebete* **51:** 291–302.

Lorenz, E.N. 1963a. Deterministic nonperiodic flow. *Journal of Atmospheric Science* **20:** 131–141.

Lorenz, E.N. 1963b. The mechanics of vacillarion. *Journal of Atmospheric Science* **20:** 448–461.

Lorenz, E.N. 1964. The problem of deducing the climate from the governing equations. *Tellus* **16:** 1–11.

Louda, S.M., and M.A. Potvin. 1995. Effect of inflorescence-feeding insects on the demography and lifetime fitness of a native plant. *Ecology* **76:** 229–245.

Louda, S.M., M.A. Potvin, and S.K. Collings. 1995. Seed predation and seedling competition in the recruitment and population dynamics of Platte thistle in Sandhills Prairie. *American Midland Naturalist* **124:** 105–113.

Lowry, R.K., and D.A. Ratkowsky. 1983. A note on models of poikilotherm development. *Journal of Theoretical Biology* **105:** 453–459.

Manly, B.F.J. 1971. A simulation study of Jolly's method for analyzing capture-recapture data. *Biometrics* **27:** 415–424.

Manly, B.F.J. 1976. Extensions to Kiritani and Nakasuji's method for the analysis of stage frequency data. *Researches on Population Ecology* **17:** 191–199.

Manly, B.F.J. 1977a. A further note on Kiritani and Nakasuji's model for stage–frequency data including comments on Tukey's jackknife technique for estimating variances. *Researches on Population Ecology* **18:** 177–186.

Manly, B.F.J. 1977b. The determination of key factors from life table data. *Oecologia* **31:** 111–117.

Manly, B.F.J. 1977c. A simulation experiment on the application of the jackknife with Jolly's method for the analysis of capture-recapture data. *Acta Theriologica* **22:** 215–223.

Manly, B.F.J. 1979. A note on key factor analysis. *Researches on Population Ecology* **21:** 30–39.

Manly, B.F.J. 1984. Obtaining confidence limits on parameters of the Jolly–Seber model for capture-recapture data. *Biometrics* **40**: 749–758.

Manly, B.F.J. 1985. Further improvements to a method for analysing stage-frequency data. *Researches on Population Ecology* **27**: 325–332.

Manly, B.F.J. 1990. *Stage-Structured Populations: Sampling, Analysis and Simulation.* Chapman & Hall, London.

Marriott, F.H.C. 1979. Monte Carlo tests: How many simulations? *Applied Statistics* **28**: 75–77.

Matheron, G. 1962. *Traité de geostatisque appliquée.* Vol. I. *Mémoires du bureau de Recherches geologiques et minières No. 14.* Editions Technip, Paris.

May, M.L. 1979. Insect thermoregulation. *Annual Review of Entomology* **24**: 313–349.

May, R.M. 1974. Biological populations with nonoverlapping generations: Stable points, stable cycles, and chao. *Science* **186**: 645–647.

May, R.M. 1975. Biological populations obeying difference equations: Stable points, stable cycles and chaos. *Journal of Theoretical Biology* **51**: 511–524.

May, R.M. 1976. Simple mathematical models with very complicated dynamics. *Nature* **261**: 459–467.

May, R.M., and G.F. Oster. 1976. Bifurcations and dynamic complexity in simple ecological models. *American Naturalist* **110**: 573–599.

Mayfield, H.F. 1961. Nesting success calculated from exposure. *Wilson Bulletin* **73**: 255–261.

Mayfield, H.F. 1975. Suggestions for calculating nest success. *Wilson Bulletin* **87**: 456–466.

Mayfield, H.F. 1981. Problems in estimating population size through counts of singing males. In C.J. Ralph and J.M. Scott (eds.): *Estimating Numbers of Terrestrial Birds. Studies in Avian Biology No. 6.* Cooper Ornithological Society, pp. 220–224.

McBratney, A.B., and R. Webster. 1981. Detection of ridge and furrow pattern by spectral analysis of crop yield. *International Statistical Review* **49**: 45–52.

McDonald, L.L. and B.F.J. Manly. 1989. Calibration of biased sampling procedures. In L. McDonald, B. Manly, J. Lockwood, and J. Logan (eds.): *Estimation and Analysis of Sect Populations.* pp. 467–483. Springer-Verlag Lecture Notes in Statistics Vol. 5. Springer-Verlag, Berlin.

McGuire, J.U., T.A. Brindley, and T.A. Bancroft. 1957. The distribution of corn borer larvae *Pyrausta nubilalis* (HBN.), in field corn. *Biometrics* **13**: 65–78.

Menkens, G.E., Jr., and S.H. Anderson. 1988. Estimation of small-mammal population size. *Ecology* **69**: 1952–1959.

Mercer, W.B., and A.D. Hall. 1911. The experimental error of field trials. *Journal of Agricultural Science (Cambridge)* **4**: 107–132.

Miller, H.W., and D.H. Johnson. 1978. Interpreting the results of nesting studies. *Journal of Wildlife Management* **42**: 471–476.

Milne, A. 1959. The centric systematic area-sample treated as a random sample. *Biometrics* **15**: 270–297.

Moivre, A. de. 1733. Approximatio ad Summam Ferminorum Binomii $(a + b)^n$ in Seriem expansi. *Supplementum II to Miscellanae Analytica* 1–7.

Moran, P.A.P. 1950. Notes on continuous stochastic phenomena. *Biometrika* **37**: 17–23.

Moran, P.A.P. 1951. A mathematical theory of animal trapping. *Biometrika* **38**: 307–311.

Morgan, B.J.T. 1992. *Analysis of Quantal Response Data.* Chapman & Hall, London.

Morgan, M.E.P., P. MacLeod, E.O. Anderson, and C.I. Bliss. 1951. A sequential procedure for grading milk by microscopic counts. *Connecticut (Storrs) Agricultural Experimental Station Bulletin* **276**: 1–35.

Morista, M. 1959. Measuring of the dispersion and analysis of distribution patterns. *Memoires of the Faculty of Science, Kyushu University, Series E. Biology* **2**: 215–235.

Morris, R.F. 1954. A sequential sampling technique for spruce budworm egg surveys. *Canadian Journal of Zoology* **32**: 302–313.

Morris, R.F. 1957. The interpretation of mortality data in studies of population dynamics. *The Canadian Entomologist* **89**: 49–69.

Morris, R.F. 1959. Single factor analysis in population dynamics. *Ecology* **40**: 580–588.

Mountford, M.D. 1961. On E.C. Pielou's index of non-randomness. *Journal of Ecology* **49**: 271–275.

Mukhopadhyay, N. 1978. *Sequential Point Estimation of the Mean When the Distribution is Unspecified.* Technical Report No. 312, University of Minnesota.

Munholland, P.L. 1988. Statistical Aspects of Field Studies on Insect Populations. Ph.D. Thesis. University of Waterloo, Ontario, Canada.

Murie, A. 1944. *The Wolves of Mount McKinley.* Fauna of the national Parks of the U.S., Fauna Series No. 5. U.S. Department of the Interior, National Park Service, Washington, D.C., 238 pp.

Murray, J.D. 1993. *Mathematical Biology.* 2nd ed. Springer-Verlag, New York.

Nadás, A. 1969. An extension of a theorem of Chow and Robbins on sequential confidence intervals for the mean. *Annals of Mathematical Statistics* **40**: 667–671.

Nass, C.A.G. 1959. The χ^2 test for small expectations in contingency tables, with special reference to accidents and absenteeism. *Biometrika* **46**: 365–385.

Newman, J.E. 1971. Measuring corn maturity with heat units. *Crop Soils* **23**: 11–14.

Neyman, J. 1939. On a new class of "contagious" distributions, applicable in entomology and bacteriology. *Annals of Mathematical Statistics* **10**: 35–57.

Nicot, P.C., D.I. Rouse, and B.S. Yandell. 1984. Comparison of statistical methods for studying spatial patterns of soilborne plant pathogens in the field. *The American Phytopathological Society* **74**: 1399–1402.

Nowierski, R.M., A.P. Gutierrez, and J.S. Yaninek. 1983. Estimation of thermal threshold and age-specific life table parameters for the walnut aphid (Homoptera: Aphididae) under field conditions. *Environmental Entomology* **12**: 680–686.

Numata, M. 1961. Forest vegetation in the vicinity of Choshi. Coastal flora and vegetation at Choshi, Shiba Prefecture. IV. *Bulletin of the Choshi Mariane Laboratory of Chiba University* **3**: 28–48. (in Japanese).

Nyrop, J.P. and G.A. Simmons. 1984. Errors incurred when using Iwao's sequential decision rule in insect sampling. *Environmental Entomology* **13**: 1459–1465.

Oakland, G.B. 1950. An application of sequential analysis to whitefish sampling. *Biometrics* **6:** 59–67.

Onsager, J.A. 1976. The rationale of sequential sampling, with emphasis on its use in pest management. Technical Bulletin 1526. U.S. Department of Agriculture, Washington, D.C.

Otis, D.L. 1980. An extension of the change in ratio method. *Biometrics* **36:** 141–147.

Otis, D.L., K.P. Burnham, G.C. White, and D.R. Anderson. 1978. Statistical inference from capture data on closed animal populations. *Wildlife Monographs.* Vol. **62.** pp. 1–135.

Pathria, R.K. 1972. *Statistical Mechanics.* Pergamon Press/Braunschweig.

Patil, G.P. 1960. On the evaluation of the negative binomial distribution with examples. *Technometrics* **2:** 501–505.

Patil, G.P., and W.M. Stiteler. 1974. Concepts of aggregation and their qualification: A critical review with some new results and applications. *Researches on Population Ecology* **15:** 238–254.

Patil, G.P., C. Taillie, and J.K. Ord, eds. 1979. *Statistical Distributions in Scientific Work.* Vol. 4. International Co-operative Publishing House, Fairland, MD.

Patil, G.P., M.T. Boswell, S.W. Joshi, and M.V. Ratnaparkhi. 1984. *Dictionary and Classified Bibliography of Statistical Distributions in Scientific Work.* Vol. 1. International Co-operative Publishing House. Fairland, MD.

Paulik, G.J., and D.S. Robson. 1969. Statistical calculations for change-in-ratio estimators of population parameters. *Journal of Wildlife Management* **33:** 1–27.

Payendeh, B. 1970. Comparison of methods for assessing spatial distribution of trees. *Forest Science* **16:** 312–317.

Payton, M.E. 1991. *An Examination of Sequential Procedures for the Testing of Three Hypotheses.* Ph.D. dissertation, Oklahoma State University, Stillwater, OK.

Pearson, K. 1900. On the criterion that a given system of deviations from the probable in the case of a correlated system of variables is such that it can be reasonably supposed to have arisen from random sampling. *Philosophical Magazine Series* **50:** 157–175.

Pedigo, L.P., and M.R. Zeiss. 1996. *Analyses in Insect Ecology and Management.* Iowa State University Press: Ames.

Perry, J.N., and R. Mead. 1979. On the power of the index of dispersion test to detect spatial pattern. *Biometrics* **35:** 613–622.

Pettitt, A.N., and M.A. Stephens. 1977. The Kolmogorov-Smirnov goodness-of-fit statistic with discrete and grouped data. *Technometrics* **19:** 205–210.

Pielou, E.C. 1959. The use of point-to-plant distances in the pattern of plant populations. *Journal of Ecology* **47:** 607–613.

Pielou, E.C. 1977. *Mathematical Ecology.* John Wiley & Sons, New York.

Pieters, E.P., and W. Sterling. 1974. A sequential sampling plan for the cotton fleahopper, *Pseudatomoscelis seriatus. Environmental Entomology* **3:** 102–106.

Plant, R.E. 1986. A method for computing the elements of the Leslie matrix. *Biometrics* **42:** 933–939.

Podoler, H., and D. Rogers. 1975. A new method for the identification of key factors from life table data. *Journal of Animal Ecology* **44:** 85–115.

Pollard, J.H. 1973. *Mathematical Models for the Growth of Human Populations.* Cambridge University Press: Cambridge.

Pollock, K.H. 1974. The Assumption of Equal Catchability of Animals in Tag–Recapture Experiments. Ph.D. dissertation. Cornell University, Ithaca, NY, 82 pp.

Pollock, K.H. 1975. A K-sample tag-recapture model allowing for unequal survival and catchability. *Biometrika* **62:** 577–583.

Pollock, K.H. 1981. Capture–recapture models allowing for age-dependent survival and capture rates. *Biometrics* **37:** 521–529.

Pollock, K.H. 1984. Estimation of survival distributions in ecology. *Internation Biometrics Conference* **12:** 187–195.

Pollock, K.H. 1991. Modeling capture, recapture, and removal statistics for estimation of demographic parameters for fish and wildlife populations: Past, present, and future. *Journal of the American Statistical Association* **86:** 225–238.

Pollock, K.H., and W.L. Cornelius. 1988. A distribution-free nest survival model. *Biometrics* **44:** 397–404.

Pollock, K.H., and M.C. Otto. 1983. Robust estimation of population size in closed animal populations from capture–recapture experiments. *Biometrics* **39:** 1035–1049.

Pollock, K.H., J.E. Hines, and J.D. Nichols. 1985a. Goodness-of-fit tests for open capture-recapture models. *Biometrics* **41:** 399–410.

Pollock, K.H., R.A. Lancia, M.C. Conner, and B.L. Wood. 1985b. A new change-of-ratio procedure robust to unequal catchability of types of animal. *Biometrics* **41:** 653–662.

Pollock, K.H., J.D. Nichols, C. Brownie, and J.E. Hines. 1990. Statistical inference for capture–recapture experiments. *Wildlife Monographs* **107.** pp. 1–97.

Pollock, K.H., S.R. Winterstein, C.M. Bunck, and P.D. Curtis. 1989. Survival analysis in telemetry studies: The staggered entry design. *Journal of Wildlife Management* **53:** 7–15.

Poston, F.L., R.B. Hammond, and L.P. Pedigo. 1977. Growth and development of the painted lady on soybeans (Lepidoptera: Nymphaliade). *Journal of the Kansas Entomological Society* **50:** 31–36.

Press, W.H., B.P. Flannery, S.A. Teukolsky, and W.T. Vetterling. 1986. *Numerical Recipes: The Art of Scientific Computing.* Cambridge University Press, Cambridge.

Preston, F.W. 1948. The commonness and rarity of species. *Ecology* **29:** 254–283.

Pruess, K.P. 1983. Day-degree methods for pest management. *Environmental Entomology* **12:** 613–619.

Pucek, Z. 1969. Trap response and estimation of numbers of shrews in removal catches. *Acta Theriologica* **14:** 403–426.

Qasrawi, H. 1966. A study of the energy flow in a natural population of the grasshopper *Chorthippus parallebelus* Zett. (Orthoptera Acridae). Ph.D. thesis. University of Exeter, UK.

Quang, P.X. 1991. A nonparametric approach to size-biased line transect sampling. *Biometrics* **47:** 269–279.

Quinn, T.J., II. 1979. The effects of school structure on line transect estimators of abundance. In G.P. Patil and M.L. Rosenzweig (eds.): *Contemporary Quantitative Ecology and Related Econmetrics.* International Co-operative Publishing House, Fairland, MD, pp. 473–491.

Rao, C.R. 1947. The problem of classification and distance between two populations. *Nature* **159**: 30.

Rao, C.R. 1970. *Advanced Statistical Methods in Biometric Research.* Hafner, New York.

Read, K.L.Q., and J.R. Ashford. 1968. A system of models for the life cycle of a biological organism. *Biometrika* **55**: 211–221.

Read, T.R.C., and N.A.C. Cressie. 1988. *Goodness-of-Fit Statistics for Discrete Multivariate Data.* Springer-Verlag, New York.

Rexstad, E., and K. Burnham. 1991. *Users' Guide for Interactive Program CAPTURE: Abundance Estimation of Closed Animal Populations.* Colorado Cooperative Fish and Wildlife Research Unit. Colorado State University, Fort Collins, CO, 29 pp.

Richards, D.G. 1981. Environmental acoustics and censuses of singing birds. In C.J. Ralph and J.M. Scott (eds.): *Estimating Numbers of Terrestrial Birds. Studies in Avian Biology No. 6.* Cooper Ornithological Society, pp. 297–300.

Ricker, W.E. 1958. Handbook of computations for biological statistics of fish populations. *Bulletin of the Fisheries Board of Canada* **119**: 300 pp.

Ricker, W.E. 1975. Computations and interpretation of biological statistics of fish populations. *Bulletin of the Fisheries Research Board of Canada No. 191.* 382 pp.

Ring, D.R. 1978. Biology of the Pecan Weevil, Emphasizing the Period from Oviposition to Larval Emergence. M.S. thesis. Texas A&M University, College Station, TX.

Ripley, B.D. 1976. The second-order analysis of stationary point processes. *Journal of Applied Probability* **13**: 255–266.

Ripley, B.D. 1977. Modelling spatial patterns (with discussion). *Journal of the Royal Statistical Society B* **39**: 172–212.

Ripley, B.D. 1981. *Spatial Statistics.* John Wiley & Sons, New York, 252 pp.

Robinnette, W.L., C.M. Loveless, and D.A. Jones. 1974. Field tests of strip census methods. *Journal of Wildlife Management* **38**: 81–96.

Robson, D.S. 1969. Mark-recapture methods of population estimation. In *New Developments in Survey Sampling,* N.L. Johnson and H. Smith, eds. John Wiley & Sons, New York, pp. 120–140.

Robson, D.S., and H.A. Regier. 1968. Estimation of population number and mortality rates. In W.E. Ricker (ed.): *Methods for Assessment of Fish Production in Fresh Waters.* IBP Handbook No. 3. Blackwell Scientific, Oxford, pp. 124–158.

Robson, D.S., and W.D. Youngs. 1971. Statistical analysis of reported tag-recaptures in the harvest from an exploited population. BU-369-M. Biometrics Unit, Cornell University, Ithaca, NY, 15 pp.

Roff, D.A. 1973. On the accuracy of some mark-recapture estimators. *Oecologia (Berline)* **12**: 15–34.

Rojás, B.A. 1964. La binomial negativa y la estimación de intensidad de plagas en el suelo. *Fitotecnia Latinamerica* **1:** 27–36.

Rossi, R.E., D.J. Mulla, A.G. Journel, and E.H. Granz. 1992. Geostatistical tools for modeling and interpreting ecological spatial dependence. *Ecological Monographs* **62:** 277–314.

Royama, T. 1977. Population persistence and density-dependence. *Ecological Monographs* **47:** 1–35.

Rudd, W.G. 1980. Sequential estimation of soybean arthropod population densities. In: M. Kogan and D.C. Herzog (eds.): *Sampling Methods in Soybean Entomology.* Springer-Verlag, New York, pp. 94–104.

Rupp, R.S. 1966. Generalised equation for the ratio method of estimating population abundance. *Journal of Wildlife Management* **30:** 523–526.

Sanborn, S.M., J.A. Wyman, and R.S. Chapman. 1982. Threshold temperature and heat unit summations for seedcorn maggot development under controlled conditions. *Annals of the Entomological Society of America* **75:** 103–106.

Sarkovskii, A.N. 1964. Coexistence of cycles of a continuous map of a line into itself. (In Russian.) *Ukrainskii Matematicheskii Zhurnal* **16:** 61–71.

SAS® Institute Inc. 1990a. *SAS User's Guide: Procedures, Version 6.* SAS® Institute, Cary, NC.

SAS® Institute Inc. 1990b. *SAS User's Guide: Statistics, Vol. 1, Version 5.* SAS Institute, Cary, NC.

Satterthwaite, F.E. 1946. An approximate distribution of estimates of variance components. *Biometrics Bulletin* **2:** 110–114.

Schnabel, Z.E. 1938. The estimation of the total fish population of a lake. *American Mathematical Monthly* **45:** 348–352.

Schneider, S.M. 1989. Problems associated with life cycle studies of a soil-inhabiting organism. In L. McDonald, B. Manly, J. Lockwood, and J. Logan (eds.): *Estimation and Analysis of Insect Populations.* pp. 156–166. Springer-Verlag Lecture Notes in Statistics Vol. 5. Springer-Verlag, Berlin.

Schwarz, C.J., K.P. Burnham, and A.N. Arnason. 1988. Post-release stratification of band-recovery models. *Biometrics* **44:** 765–785.

Seber, G.A.F. 1965. A note on the multiple-recapture census. *Biometrika* **52:** 249–259.

Seber, G.A.F. 1970a. Estimating time-specific survival and reporting rates for adult birds from band returns. *Biometrika* **57:** 313–318.

Seber, G.A.F. 1970b. A note on the multiple recapture census. *Biometrika* **52:** 249–259.

Seber, G.A.F. 1970c. The effects of trap response on tag-recapture estimates. *Biometrika* **26:** 13–22.

Seber, G.A.F. 1982. *The Estimate of Animal Abundance and Related Parameters.* 2nd Ed. Macmillan, New York.

Seber, G.A.F. 1986. A review of estimating animal abundance. *Biometrics* **42:** 267–292.

Seber, G.A.F., and J.F. Whale. 1970. The removal method for two and three samples. *Biometrics* **26:** 393–400.

Seebeck, K. 1989. A Computer Program to Develop and Evaluate a Wald's Sequential Probability Ratio Test for the Parameters of Three Discrete Distributions. M.S. report, Oklahoma State University, Stillwater, OK.

Sharpe, J.H., and D.W. DeMichele. 1977. Reaction kinetics of poikilotherm development. *Journal of Theoretical Biology* **64:** 649–660.

Sinclair, A.R.E. 1972. Long term monitoring of mammal populations in the Sereni: Census of non-migratory ungulates, 1971. *East African Wildlife Journal* **10:** 387–398.

Skellam, J.G. 1952. Studies in statistical ecology: I. Spatial pattern. *Biometrika* **39:** 346–362.

Slade, N.A. 1977. Statistical detection of density dependence from a series of sequential censuses. *Ecology* **58:** 1094–1102.

Smith, C.A.B. 1947. Some examples of discrimination. *Annals of Eugenics, London* **13:** 372.

Smith, R.H. 1973. The analysis of intra-generation change in animal populations. *Journal of Animal Ecology* **42:** 611–622.

Smith, T.B. 1987. Bill size polymorphism and intraspecific niche utilization in an African finch. *Nature* **329:** 717–719.

Snedecor, G.W., and W.G. Cochran. 1989. *Statistical Methods.* 8th Ed. Iowa State University Press, Ames, IA.

Sobel, M., and A. Wald. 1949. A sequential decision procedure for choosing one of three hypotheses concerning the unknown mean of a normal distribution. *Annals of Mathematical Statistics* **20:** 502–522.

Sokal, R.R., and N.L. Oden. 1978a. Spatial autocorrelation in biology. 1. Methodology. *Biological Journal of the Linnean Society* **10:** 199–228.

Sokal, R.R., and N.L. Oden. 1978b. Spatial autocorrelation in biology 2. Some biological implications and four applications of evolutionary and ecological interest. *Biological Journal of the Linnaean Society* **10:** 229–249.

Somerville, P.N. 1957. Optimum sampling in binomial populations. *Journal of the American Statistical Association* **52:** 494–502.

Southern, H.N. 1970. The natural control of a population of tawny owls (*Strix aluco*). *Journal of Zoology, London* **162:** 197–285.

Southwood, T.R.E. 1978. *Ecological Methods.* 2nd Ed. Chapman & Hall, London.

Southwood, T.R.E., and W.F. Jepson. 1962. Studies of the populations of *Oscinella frit* L. (Dipt: Chloropidae) in the oat crop. *Journal of Animal Ecology* **31:** 481–495.

Stiteler, W.M., and G.P. Patil. 1971. Variance to mean ratio and Morista's index as measures of spatial pattern in ecological populations. In G.P. Patil, E.C. Pielou, and W.E. Waters (eds.): Statistical Ecology. Vol. 1. Pennsylvania State University Press, University Park, pp. 423–459.

Steel, R.G.D., and J.H. Torrie. 1980. *Principles and Procedures of Statistics, a Biometical Approach.* McGraw-Hill, New York.

Stein, C. 1945. A two-sample test for a linear hypothesis whose power is independent of the variance. *Annals of Mathematical Statistics* **16:** 243–258.

Stephens, M.A. 1974. EDF statistics for goodness-of-fit and some comparisons. *Journal of the American Statistical Association* **69:** 730–737.

Stokes, S.L. 1984. The Jolly–Sever method applied to age-stratified populations. *Journal of Wildlife Management* **48:** 1053–1059.

Strabala, M.A. 1984. Monte Carlo Simulations of the Analysis of Variance For Discrete Data. M.S. thesis. Oklahoma State University, Stillwater, OK.

Strauss, D.J. 1975. A model for clustering. *Biometrika* **62:** 467–475.

Stuart, A., and J.K. Ord. 1987. *Kendall's Advanced Theory of Statistics.* Vol. 1. Griffin, London.

Student. 1907. On the error of counting with a haemacytometer. *Biometrika* **3:** 351–360.

Sukhatme, P.V., and B.V. Sukhatme. 1970. *Sampling Theory of Surveys with Applications,* 2nd Ed. Iowa State University Press, Ames, IA.

Sylvia, T.D. 1995. *Riparian Habitats of the Central Platte as a Corridor for Dispersal of Small Mammals in Nebraska.* M.S. thesis. University of Nebraska, 69 pp.

Sylwester, D. 1974. A Monte Carlo study of multidimensional contingency table analysis. *Biometrics* **30:** 386.

Tanaka, R. 1956. On differential response to live traps of marked and unmarked small mammals. *Annotal Zoology Japan* **29:** 44–51.

Tanaka, R. 1963. On the problem of trap-response types of small mammal populations. *Research on Population Ecology* **5:** 139–146.

Taylor, L.R. 1961. Aggregation, variance and the mean. *Nature* **189:** 732–735.

Taylor, L.R. 1965. A natural law for the spatial disposition of insects. In *Proceedings of the Twelfth International Congress on Entomology* 396–397.

Taylor, L.R. 1970. Aggregation and the transformation of counts of *Aphis fabae Scop.* on beans. *Ann. Appl. Biol.* **65:** 181–189.

Taylor, L.R. 1971. Aggregation as a species characteristic. In G.P. Patil, Pielou, E.C. and W.E. Waters (eds.): *Statistical Ecology.* Vol. 1. Pennsylvania State University Press, Philadelphia, pp. 357–377.

Taylor, L.R. 1984. Assessing and interpreting the spatial distributions of insects populations. *Annual Review of Entomology* **29:** 321–357.

Taylor, L.R., I.P. Woiwod, and J.N. Perry. 1978. The density-dependence of spatial behavior and the rarity of randomness. *Journal of Animal Ecology* **47:** 383–406.

Thomas, M. 1949. A generalization of Poisson's binomial limit for use in ecology. *Biometrika* **36:** 18–25.

Thompson, S.K. 1992. *Sampling.* John Wiley & Sons, New York.

Tolstov, G.P. 1962. *Fourier Series.* Prentice-Hall: Englewood Cliffs, NJ, 336 pp.

Tostowaryk, W., and J.M. McLeod. 1972. Sequential sampling for egg clusters of the Swaine jackpine sawfly. *Neodiprion swainei* (Hymenoptera: Diprionidae). *Canadian Entomologist* **104:** 1343–1347.

Trent, T.T., and O.J. Rongstad. 1974. Home range and survival of cottontail rabbits in southwestern Wisconsin. *Journal of Wildlife Management* **38:** 459–472.

Tukey, J.W. 1977. *Exploratory Data Analysis.* Addison-Wesley, Reading, MA.

Udevitz, M.S. 1989. Change-in-Ratio Estimators for Estimating the Size of Closed Populations. Ph.D. dissertation. North Carolina State University, Raleigh, NC.

Usher, M.B. 1966. A matrix approach to the management of renewable resources, with special reference to selection forests. *Journal of Applied Ecology* **3**: 355–367.

Usher, M.B. 1969. A matrix model for forest management. *Biometrics* **25**: 309–315.

van Straalen, N.M. 1982. Demographic analysis of arthropod populations using a continuous stage-variable. *Journal of Animal Ecology* **51**: 769–783.

Varley, G.C., and G.R. Gradwell. 1960. Key factors in population studies. *Journal of Animal Ecology* **29**: 399–401.

Varley, G.C., and G.R. Gradwell. 1970. Recent advances in insect population dynamics. *Annual Review of Entomology* **15**: 1–24.

Varley, G.C., G.R. Gradwell, and M.P. Hassell. 1973. *Insect Population Ecology.* Blackwell Scientific, Oxford.

Wagner, T.L., H. Wu, P.J.H. Sharpe, R.M. Schoolfield and R.N. Coulson. 1984. Modeling insect development rates: A literature review and application of a biophysical model. *Annals of the Entomology Society of America* **77**: 208–225.

Wagner, T.L., R.L. Olson, and J.L. Willers. 1991. Modeling arthropod development time. *Journal of Agricultural Entomology* **8**: 251–270.

Wald, A. 1943. Tests of statistical hypothesis concerning several parameters when the number of observations is large. *Transactions of the American Mathematical Society* **54**: 426–482.

Wald, A. 1945. Sequential tests of statistical hypotheses. *Annals of Mathematical Statistics* **16**: 117–186.

Wald, A. 1947. *Sequential Analysis.* John Wiley & Sons, New York.

Wald, A., and J. Wolfowitz. 1948. Optimum character of the sequential probability ratio test. *Annals of Mathematical Statistics* **19**: 326–339.

Wall, S.B. Vander. 1994. Seed fate pathways of antelope bitterbrush: Dispersal by seed-caching yellow pine chipmunks. *Ecology* **75**: 1911–1926.

Wang, J.Y. 1960. A critique of the heat unit approach to plant response studies. *Ecology* **41**: 785–790.

Waters, W.E. 1955. Sequential sampling in forest insect surveys. *Forest Science* **1**: 68–79.

Waters, W.E. 1959. A quantitative measure of aggregation in insects. *Journal of Econ. Entomology* **52**: 1180–1184.

Weiss, L. 1953. Testing one simple hypothesis against another. *Annals of Mathematical Statistics* **24**: 273–281.

Welch, B.L. 1939. Note on discriminant functions. *Biometrika* **31**: 218.

Wetherill, G.B ., and K.D. Glazebrook. 1986. *Sequential Methods in Statistics,* 3rd ed. Chapman & Hall, New York.

White, G.C. 1983. Numerical estimation of survival rates from band-recovery and biotelemetry data. *Journal of Wildlife Management* **47:** 716–728.

White, G.C., D.R. Anderson, K.P. Burnham, and D.L. Otis. 1982. *Capture–Recapture and Removal Methods for Sampling Closed Populations.* Los Alamos National Laboratory, LA 8787-NERP, Los Alamos, NM, 235 pp.

Whittle, P. 1954. On stationary processes in the plane. *Biometrika* **41:** 434–449.

Wilkinson, G.N., S.R. Eckert, T.W. Hancock, and O. Mayo. 1983. Nearest neighbor (NN) analysis with field experiments. *Journal of the Royal Statistical Society B* **45:** 151–178.

Willson, L.J. 1981. *Estimation and Testing Procedures for the Parameters of the Negative Binomial Distribution.* Ph.D. Thesis. Oklahoma State University, Stillwater, Oklahoma.

Willson, L.J., and J.H. Young. 1983. Sequential estimation of insect population densities with a fixed coefficient of variation. *Environmental Entomology* **12:** 669–672.

Willson, L.J., J.L. Folks, and J.H. Young. 1984. Multistage estimation compared with fixed sample size estimation of the negative binomial distribution. *Biometrics* **40:** 109–117.

Willson, L.J., J.L. Folks, and J.H. Young. 1985. Complete sufficiency and maximum likelihood estimation for the two-parameter negative binomial distribution. *Metrika* **33:** 349–362.

Willson, L.J., J.H. Young, and J.L. Folks. 1987. A biological application of Bose–Einstein statistics. *Communications in Statistics—Theory and Methods* **A16:** 445–459.

Wilson, L.F. 1959. Branch "Tip" sampling for determining abundance of spruce budworm egg masses. *Journal of Economic Entomology* **52:** 618–621.

Wittes, J.T. 1972. On the bias and estimated variance of Chapman's two-sample capture–recapture population estimate. *Biometrics* **28:** 592–597.

Wood, G.W. 1963. The capture-recapture technique as a means of estimating populations of climbing cutworms. *Canadian Journal of Zoology* **41:** 47–50.

Worner, S.P. 1992. Performance of phenological models under variable temperature regimes: Consequences of the Kaufmann or rate summation effect. *Environmental Entomology* **21:** 689–699.

Wu, B. 1985. A Monte Carlo Study of Five Goodness-of-Fit Tests. M.S. thesis. Oklahoma State University, Stillwater, OK.

Young, L.J. 1994. Computation of some exact properties of Wald's SPRT when sampling from a class of discrete distributions. *Biometrical Journal* **36:** 627–635.

Young, J.H., and L.J. Willson. 1987. The use of Bose-Einstein statistics in population dynamics models of arthropods. *Ecological Modeling* **37:** 456–467.

Young, L.J., and J.H. Young. 1989. A model of arthropod movement within agroecosystems. In L. McDonald, B. Manly, J. Lockwood and J. Logan (eds.): *Lecture Notes in Statistics: Estimation and Analysis of Insect Populations.* Springer-Verlag, Berlin, pp. 378–386.

Young, L.J., and J.H. Young. 1994. Statistics with agricultural pests and environmental impacts. In G.P. Patil and C.R. Rao (ed.): *Handbook of Statistics,* Vol. 12. pp. 735–770. Elsevier Science B.V. Amsterdam.

Zippin, C. 1956. An evaluation of the removal method of estimating animal populations. *Biometrics* **12:** 163–169.

Zippin, C. 1958. The removal method of populations estimation. *Journal of Wildlife Management* **22:** 82–90.

Index